Structural Steel Design

THIRD EDITION

Jack C. McCormac

CLEMSON UNIVERSITY

HARPER & ROW, PUBLISHERS, New York
Cambridge, Hagerstown, Philadelphia, San Francisco,
London, Mexico City, São Paulo, Sydney

1817

Sponsoring Editor: Carl McNair
Production Manager: Marion A. Palen
Compositor: Science Typographers
Printer and Binder: The Maple Press Company
Art Studio: J & R Technical Services Inc.

Structural Steel Design, Third Edition

Library of Congress Cataloging in Publication Data

McCormac, Jack C.
    Structural steel design.

    Includes bibliographical references and index.
        1. Building, Iron and Steel.   2. Steel, Structural.
I. Title.
TA684.M25   1981   624.1'821   80-23328
ISBN 0-06-044344-8

Contents

Preface, ix

Chapter 1 Introduction to Structural Steel Design, 1
1-1. Advantages of Steel as a Structural Material, 1
1-2. Disadvantages of Steel as a Structural Material, 2
1-3. Early Uses of Iron and Steel, 3
1-4. The Structural Designer, 6
1-5. Objectives of the Structural Designer, 6
1-6 Factor of Safety, 7
1-7. Failures of Engineering Structures, 9
1-8. Specifications, 10
1-9. Variations in Designs, 12
1-10. Calculation Accuracy, 12
1-11. Design Aids, 13

Chapter 2 Properties of Structural Steel, 14
2-1. Stress-Strain Relationships in Structural Steel, 14
2-2. Elastic and Plastic Design Methods Defined, 17
2-3. Load and Resistance Factor Design, 18
2-4. Steel Sections, 19
2-5. Modern Structural Steels, 22
2-6. Furnishing of Structural Steel, 28
2-7. Design of Steel Members, 30
2-8. SI Units, 32

Chapter 3 Tension Members, 35
3-1. Introduction, 35
3-2. Allowable Tensile Stresses and Loads, 38
3-3. Net Areas, 40
3-4. Effect of Staggered Holes, 42
3-5. Effective Net Areas, 47
3-6. Design of Tension Members, 49
3-7. Splices for Tension Members, 55
3-8. Design of Rods and Bars, 57
3-9. Eye-Bar Tension Members, 61
3-10. Design for Fatigue Loads, 62
Problems, 64

Chapter 4 Introduction to Compression Members, 70

4-1. General, 70
4-2. Residual Stresses, 72
4-3. Sections Used for Columns, 73
4-4. Development of Column Formulas, 77
4-5. Derivation of the Euler Formula, 78
4-6. Revisions to Euler Formula, 84
Problems, 86

Chapter 5 Design of Compression Members, 89

5-1. Practical Design Formulas, 89
5-2. Straight-Line Formulas, 89
5-3. Parabolic Formulas, 92
5-4. Gordon-Rankine Formula, 94
5-5. The Secant Formula, 96
5-6. The AISC and AASHTO Formulas, 97
5-7. Effective Column Lengths, 100
5-8. Maximum Slenderness Ratios, 103
5-9. Design of Columns with AISC Formulas, 103
5-10. Lacing and Tie Plates, 108
5-11. Stiffened and Unstiffened Elements, 112
5-12. Base Plates for Concentrically Loaded Columns, 114
Problems, 118

Chapter 6 Design of Beams, 124

6-1. Types of Beams, 124
6-2. The Flexure Formula, 124
6-3. Selection of Beams, 125
6-4. Compact Sections, 129
6-5. Holes in Beams, 132
6-6. Lateral Support of Beams, 135
6-7. Design of Laterally Unsupported Beams, 137
6-8. Design of Continuous Members, 144
Problems, 148

Chapter 7 Design of Beams (Continued), 154

7-1. Shear, 154
7-2. Web Crippling, 158
7-3. Vertical Buckling of Webs, 160
7-4. Maximum Deflections of Beams, 160
7-5. Unsymmetrical Bending, 163
7-6. Design of Purlins, 169
7-7. The Shear Center, 171
7-8. Torsion, 176
7-9. Lintels, 181
7-10. Beam-Bearing Plates, 183
Problems, 185

Chapter 8 Bending and Axial Stress, 191

8-1. Occurrence, 191
8-2. Calculation of Stresses, 192
8-3. Specifications for Combined Stresses, 194
8-4. Design for Axial Compression and Bending, 196
8-5. AISC Requirements, 197
8-6. Design of Beam-Columns (AISC), 203

8-7.   Combined Axial Tension and Bending, 209
8-8.   Further Discussion of Effective Lengths of Columns, 210
Problems, 218

Chapter 9 Bolted Connections, 224

9-1.   Introduction, 224
9-2.   Types of Bolts, 224
9-3.   History of High-Strength Bolts, 226
9-4.   Advantages of High-Strength Bolts, 227
9-5.   Installation of High-Strength Bolts, 227
9-6.   Load Transfer and Types of Joints, 231
9-7.   Failure of Bolted Joints, 234
9-8.   Specifications for High-Strength Bolts, 235
9-9.   Spacing and Edge Distances of Bolts, 237
9-10. Bearing-Type Connections—Loads Passing Through Center of Gravity of Connections, 239
9-11. Friction-Type Connections—Loads Passing Through Center of Gravity of Connections, 243
9-12. Bolts Subjected to Eccentric Shear, 245
9-13. Tension Loads on Bolted Joints, 251
9-14. Prying Action, 253
9-15. Bolts Subjected to Combined Shear and Tension, 257
Problems, 261

Chapter 10 Historical Notes on Riveted Connections, 271

10-1. General, 271
10-2. Types of Rivets, 273
10-3. Strength of Riveted Connections—Rivets in Shear, 274
10-4. Rivets Subjected to Eccentric Shear, 277
10-5. Rivets Subjected to Combined Shear and Tension, 277
10-6. Rivets in Tension, 281
Problems, 282

Chapter 11 Welded Connections, 290

11-1. General, 290
11-2. Advantages of Welding, 291
11-3. Types of Welding, 292
11-4. Welding Inspection, 295
11-5. Classification of Welds, 297
11-6. Welding Symbols, 299
11-7. Groove Welds, 301
11-8. Fillet Welds, 303
11-9. Strength of Welds, 305
11-10. AISC Requirements, 305
11-11. Design of Simple Fillet Welds, 310
11-12. Design of Fillet Welds for Truss Members, 314
11-13. Shear and Torsion, 318
11-14. Shear and Bending, 321
Problems, 323

Chapter 12 Building Connections, 331

12-1. Selection of Type of Fastener, 331
12-2. Types of Beam Connections, 332
12-3. Standard Riveted or Bolted Beam Connections, 337
12-4. Semirigid and Rigid Riveted or Bolted Connections, 341

12-5. Types of Welded Beam Connections, 345
12-6. Welded Web Angles, 346
12-7. Design of Welded Seated Beam Connections, 349
12-8. Welded Stiffened Beam Seat Connections, 353
12-9. Welded Moment-Resistant Connections, 356
Problems, 361

Chapter 13 Design of Steel Buildings, 365

13-1. Introduction, 365
13-2. Types of Steel Frames Used for Buildings, 365
13-3. Common Types of Floor Construction, 369
13-4. Concrete Slabs on Open-Web Steel Joists, 370
13-5. One-Way and Two-Way Reinforced Concrete Slabs, 372
13-6. Composite Floors, 374
13-7. Concrete-Pan Floors, 376
13-8. Structural Clay Tile, Gypsum Tile, and Concrete-Block Floors, 377
13-9. Steel-Decking Floors, 378
13-10. Flat Slabs, 378
13-11. Precast Concrete Floors, 379
13-12. Types of Roof Construction, 381
13-13. Exterior Walls and Interior Partitions, 382
13-14. Design Live Loads (Vertical), 383
13-15. Wind and Earthquake Loadings, 385
13-16. Fireproofing of Structural Steel, 387

Chapter 14 Introduction to Steel Bridges, 389

14-1. General, 389
14-2. Through, Deck, and Half-Through Bridges, 390
14-3. Erection Methods for Bridges, 391
14-4. The Beam Bridge, 393
14-5. The Plate Girder Bridge, 393
14-6. Truss Bridges for Medium Spans, 394
14-7. Subdivided Truss Bridges, 395
14-8. Continuous Bridge Trusses, 396
14-9. Cantilever Bridges, 397
14-10. Steel Arches, 399
14-11. Suspension Bridges, 402
14-12. Miscellaneous Bridge Types, 403
14-13. Summary of Bridge Types, 406
14-14. Comparison of High-Level Bridges and Low-Level Movable Bridges, 406
14-15. Movable Bridges, 408
14-16. Selection of Bridge Type, 410
14-17. Spacing of Piers and Abutments, 410
14-18. Live Loads for Highway Bridges, 411
14-19. Live Loads for Railway Bridges, 414

Chapter 15 Composite Design, 416

15-1. Composite Construction, 416
15-2. Advantages of Composite Construction, 417
15-3. Discussion of Shoring, 418
15-4. Effective Flange Widths, 419
15-5. Stress Calculations for Composite Sections, 420
15-6. Shear Transfer, 426
15-7. Proportioning Composite Sections, 434

15-8. Design of Encased Sections, 440
15-9. Design of Composite Sections Using AASHTO Specifications, 444
15-10. Miscellaneous, 448
Problems, 450

Chapter 16 Built-up Beams and Plate Girders, 453
16-1. Cover-Plated Beams, 453
16-2. Introduction to Plate Girders, 455
16-3. Comments on Specifications, 458
16-4. Proportions of Plate Girders—AASHTO and AREA, 458
16-5. Cutting off Cover Plates—AASHTO, AREA, and AISC, 465
16-6. Cover-Plate and Flange Connectors—AASHTO, AREA, and AISC, 468
16-7. Stiffeners—AASHTO, 471
16-8. Longitudinal Stiffeners—AASHTO, 474
16-9. Web Splices—AASHTO, AREA, and AISC, 475
16-10. Proportions of Webs—AISC, 480
16-11. Proportions of Flange—AISC, 483
16-12. Design of Stiffeners—AISC, 484
16-13. Example Plate Girder Design—AISC, 485
Problems, 494

Chapter 17 Design of Roof Trusses, 496
17-1. Introduction, 496
17-2. Types of Roof Trusses, 497
17-3. Selection of Type of Roof Truss, 500
17-4. Spacing and Support of Roof Trusses, 501
17-5. Estimated Weight of Roof Trusses, 502
17-6. Discussion of Roof Truss Analysis, 503
17-7. Example Analysis of a Roof Truss, 505
17-8. Design of a Roof Truss, 509
17-9. Trusses for Industrial Buildings, 512
17-10. Roof Truss Bracing, 513
Problems, 515

Chapter 18 Design of Bridges, 517
18-1. Introduction, 517
18-2. Bridge Floor System, 519
18-3. Live-Load Truss Forces, 525
18-4. Truss Dead Loads, 529
18-5. Selection of Truss Members, 531
18-6. Lateral Bracing, 534
18-7. Bridge-Truss Deflections, 536
18-8. End Bearings for Bridges, 537
Problems, 541

Chapter 19 Design of Rigid Frames, 543
19-1. Introduction, 543
19-2. Supports for Rigid Frames, 543
19-3. Rigid-Frame Knees, 546
19-4. Approximate Analysis of Rigid Frames, 548
19-5. "Exact" Analysis of Rigid Frames, 551
19-6. Preliminary Design, 556
19-7. Final Design and Details, 565
19-8. Plastic Design of Rigid Frames, 567
Problem, 567

Chapter 20 Multistory Buildings, 568
20-1. Introduction, 568
20-2. Column Splices, 570
20-3. Discussion of Lateral Forces, 572
20-4. Types of Lateral Bracing, 575
20-5. Analysis of Buildings with Diagonal Wind Bracing for Lateral Forces, 579
20-6. Moment-Resisting Joints, 580
20-7. Analysis of Buildings with Moment-Resisting Joints for Lateral Loads, 581
20-8. Analysis of Buildings for Gravity Loads, 587
20-9. Design of Members, 590

Chapter 21 Plastic Analysis, 591
21-1. Introduction, 591
21-2. Theory of Plastic Analysis, 593
21-3. The Platsic Hinge, 593
21-4. The Plastic Modulus, 595
21-5. Factors of Safety and Load Factors, 597
21-6. The Collapse Mechanism, 598
21-7. Plastic Analysis by the Equilibrium Method, 599
21-8. The Virtual-Work Method, 607
21-9. Location of Plastic Hinge for Uniform Loadings, 610
Problems, 612

Chapter 22 Plastic Analysis and Design, 618
22-1. Introduction to Plastic Design, 618
22-2. AISC Requirements for Plastic Design, 623
22-3. Continuous Beams, 626
22-4. Plastic Analysis of Frames, 632
Problems, 638

Chapter 23 Miscellaneous Topics, 645
23-1. Introduction, 645
23-2. Moment Resisting Column Bases, 645
23-3. Ponding, 650
23-4. Load and Resistance Factor Design, 652
23-5. Conclusion of Text, 657

Index, 659

Preface

The purpose of this book is unchanged: to present the basic information necessary to design simple steel structures in a way that creates interest in the subject. The primary objective of this edition is to update the text to reflect the latest specification revisions of the American Institute of Steel Construction in 1978 and the American Association of State Highway and Transportation Officials in 1977.

The specification changes have particularly affected the design of connections and tension members. In addition to changes in these areas the author has included or expanded the topics of ponding, torsion, load and resistance factor design, and moment resisting base plates. The homework problems have been changed appreciably and their number increased. SI units have been introduced to show readers that they will have little difficulty learning and using them.

The author continues to be grateful to those people who have taken the time to write to him concerning the text. He is especially grateful to the following people for their very helpful reviews of this edition of the book: Frank D. DeFalco, Worcester Polytechnic Institute, Robert Coffin Garson, University of Pittsburgh, and Louis C. Tartaglione, University of Lowell.

Jack C. McCormac

Chapter 1
Introduction
to Structural
Steel Design

1-1. ADVANTAGES OF STEEL AS A STRUCTURAL MATERIAL

A person traveling in the United States might quite understandably decide that steel was the perfect structural material. He would see an endless number of steel bridges, buildings, towers, and other structures— comprising, in fact, a list too lengthy to enumerate. After seeing these numerous steel structures he might be quite surprised to learn that steel was not economically made in the United States until the middle of the nineteenth century and the first wide-flange beams were not rolled until 1908.

His assumption of the perfection of this metal, perhaps the most versatile of structural materials, would appear to be even more reasonable when he considered its great strength, light weight, ease of fabrication, and many other desirable properties. These and other advantages of structural steel are discussed in detail in the following paragraphs.

High Strength

The high strength of steel per unit of weight means dead loads will be small. This fact is of great importance for long-span bridges, tall buildings, and for structures having poor foundation conditions.

Uniformity

The properties of steel do not change appreciably with time as do those of a reinforced-concrete structure.

Elasticity

Steel behaves closer to design assumptions than most materials because it follows Hooke's law up to fairly high stresses. The moments of inertia of a steel structure can be definitely calculated while the values obtained for a reinforced concrete structure are rather indefinite.

1

Permanence

Steel frames that are properly maintained will last indefinitely. Research on some of the newer steels indicates that under certain conditions no painting maintenance whatsoever will be required.

Ductility

The property of a material by which it can withstand extensive deformation without failure under high tensile stresses is said to be its *ductility*. When a mild steel member is being tested in tension, a considerable reduction in cross section and a large amount of elongation will occur at the point of failure before the actual fracture occurs. A material that does not have this property is probably hard and brittle and might break if subjected to a sudden shock.

In structural members under normal loads, high stress concentrations develop at various points. The ductile nature of the usual structural steels enables them to yield locally at those points, thus preventing premature failures. A further advantage of ductile structures is that when overloaded their large deflections give visible evidence of impending failure (sometimes jokingly referred to as "running time").

Additions to Existing Structures

Steel structures are quite well suited to having additions made to them. New bays or even entire new wings can be added to existing steel frame buildings, and steel bridges may often be widened.

Miscellaneous

Several other important advantages of structural steel are: (a) ability to be fastened together by several simple connection devices including welds, bolts, and rivets, (b) adaptation to prefabrication, (c) speed of erection, (d) ability to be rolled into a wide variety of sizes and shapes as described in Section 2-4 of this book, (e) toughness and fatigue strength, (f) possible reuse after a structure is disassembled, and (g) scrap value even though not reusable in its existing form.

1-2.   DISADVANTAGES OF STEEL AS A STRUCTURAL MATERIAL

In general steel has the following disadvantages.

Maintenance Costs

Most steels are susceptible to corrosion when freely exposed to air and water and must therefore be periodically painted. The use of weathering steels, in suitable design applications, tends to eliminate this cost.

Fireproofing Costs

Although structural members are incombustible their strength is tremendously reduced at temperatures commonly reached in fires when the other materials in a building burn. Many disastrous fires have occurred in empty buildings where the only fuel for the fires were the buildings themselves. Furthermore steel is an excellent heat conductor such that nonfireproofed steel members may transmit enough heat from a burning section or compartment of a building to ignite materials with which they are in contact in adjoining sections of the building. As a result of these facts the steel frame of a building must be fireproofed if the building is to have an appreciable fire rating.

Susceptibility to Buckling

The longer and slenderer compression members, the greater the danger of buckling. As previously indicated, steel has a high strength per unit weight and when used for steel columns is sometimes not very economical because considerable material has to be used merely to stiffen the columns against buckling.

Fatigue

Another undesirable property of steel is that its strength may be reduced if it is subjected to a large number of stress reversals or even to a large number of variations of stress of the same character (i.e., tension or compression). The present practice is to reduce the estimated strengths of such members if it is anticipated that they will have more than a prescribed number of cycles of stress variation.

1-3. EARLY USES OF IRON AND STEEL

Although the first metal used by human beings was probably some type of copper alloy such as bronze (made with copper and tin and perhaps some other additives) the most important metal developments throughout history have occurred in the manufacture and use of iron and its famous alloy named steel. Today, iron and steel comprise almost 95% of all the tonnage of metal produced in the world.[1]

Despite diligent efforts for many decades archaeologists have been unable to discover when iron was first used. They did find an iron dagger and an iron bracelet in the Great Pyramid in Egypt which they claim had been there undisturbed for at least 5000 years. The use of iron has had a great influence on the course of civilization since the earliest times and may very well continue to do so in the centuries ahead. Since the beginning of the iron age in about 1000 B.C. the progress of civilization in peace

---

[1]American Iron and Steel Institute, *The Making of Steel* (Washington, D.C., not dated), p. 6.

and war has been heavily dependent on what people have been able to make with iron. On many occasions its use has decidedly affected the outcome of military engagements. For instance in 490 B.C. in Greece at the Battle of Marathon the greatly outnumbered Athenians killed 6400 Persians and only lost 192 of their own men. Each of the victors wore 57 pounds of iron armor in the battle. (This was the battle where the runner Pheidippides ran the approximately 25 miles to Athens and died while shouting news of the victory.) This battle supposedly saved Greek civilization for many years.

According to the classic theory concerning the first production of iron in the world there was once a great forest fire on Mount Ida in Ancient Troy (now Turkey) near the Aegean Sea. The land surface supposedly had a rich content of iron and the heat of the fire is said to have produced a rather crude form of iron which could be hammered into various shapes. Many historians believe however that human beings first learned to use iron that fell to the earth in the form of meteorites. Frequently the iron in meteorites is combined with nickle with the result that a harder metal is produced. Perhaps early human beings were able to hammer and chip this material into the shape of crude tools and weapons.

Steel is defined as a combination of iron with a small amount of carbon usually less than 1%. It also contains small percentages of some other elements. Although some steel has been made for at least 2000 or 3000 years there was really no economical production method available until the middle of the nineteenth century.

The first steel was surely obtained when the other elements necessary for producing it were accidentally present when iron was heated. As the years went by steel was probably made by heating iron in contact with charcoal. The surface of the iron absorbed some carbon from the charcoal which was then hammered into the hot iron. Repeating this process several times resulted in a case-hardened exterior of steel. In this way the famous swords of Toledo and Damascus were produced.

The first large volume process for producing steel was named after Sir Henry Bessemer of England. He received an English patent for his process in 1855 but his efforts to obtain a U.S. patent for the process in 1856 were unsuccessful as it was proved that William Kelly of Eddyville, Kentucky, had made steel by the same process seven years before Bessemer applied for his English patent. Although Kelly was given the patent the name Bessemer was used for the process.[2]

Kelly and Bessemer learned that a blast of air through molten iron burned out most of the impurities in the metal. Unfortunately at the same time the blow eliminated some desirable elements such as carbon and manganese. It was later learned that these needed elements could be restored by adding spiegeleisen which is an alloy of iron, carbon, and

[2]American Iron and Steel Institute, *Steel '76* (Washington, D.C., 1976), pp. 5–11.

manganese. It was further learned that the addition of limestone in the converter resulted in the removal of the phosphorus and most of the sulfur.

The Bessemer converter was commonly used in the United States until after the turn of the century but since that time it has been replaced with better methods such as the open-hearth process and the basic oxygen process.

Erection of tower for The Holy Name Church, Edensburg, Pa. (Courtesy of The Lincoln Electric Company).

As a result of the Bessemer process structural carbon steel could be produced in quantity by 1870 and by 1890 steel had become the principal structural metal used in the United States.

The first use of metal for a sizable structure occurred in England in Shropshire (about 140 miles northwest of London) in 1779 when cast iron was used for the construction of the 100-ft Coalbrookdale Arch Bridge over the River Severn. It is said that this bridge (which still stands) was a turning point in engineering history because it changed the course of the Industrial Revolution by introducing iron as a structural material. This iron was supposedly four times as strong as stone and thirty times as strong as wood.[3]

A number of other cast iron bridges were constructed in the following decades; but soon after 1840 the more malleable wrought iron began to replace cast iron. Then the development of the Bessemer process and subsequent advances such as the open-hearth process permitted the manufacture of steel at competitive prices which encouraged the beginning of the almost unbelievable developments of the last 100 years with structural steel.

1-4. THE STRUCTURAL DESIGNER

The structural designer can take great pride in his part in the development of our country. Cities, farmlands, and industrial areas of the United States are filled with the amazing structures designed by members of his profession. But even this remarkable array of structures will be merely child's play compared to the structural endeavors of the next few generations. These structures of the future should provide endless opportunities in the structural field for young engineers.

The structural designer arranges and proportions structures and their parts so that they will satisfactorily support the loads to which they may feasibly be subjected. It might be said that he is involved with the following: the general layout of structures; studies of the possible structural forms that can be used; consideration of loading conditions; analysis of stresses, deflections, etc.; design of parts; and the preparation of design drawings. More precisely, the word *design* pertains to the proportioning of the various parts of a structure after the forces have been calculated, and it is this process which will be emphasized throughout the text using structural steel as the material.

1-5. OBJECTIVES OF THE STRUCTURAL DESIGNER

The structural designer must learn to arrange and proportion the parts of his structures so that they can be practically erected and will have

[3] M. H. Sawyer, "World's First Iron Bridge," *Civil Engineering* (New York: ASCE, December 1979), pp. 46–49.

sufficient strength and reasonable economy. These items are discussed briefly below.

Safety

Not only must the frame of a structure safely support the loads to which it is subjected, but it must support them in such a manner that deflections and vibrations are not so great as to frighten the occupants or cause unsightly cracks.

Cost

The designer needs to keep in mind the items causing lower cost without sacrifice of strength. These items which are discussed in more detail throughout the text include the use of standard-size members, simple connections and details, and the use of members and materials that will not require an unreasonable amount of maintenance through the years.

Practicality

Another objective is the design of structures that can be fabricated and erected without great problems arising. Designers need to understand fabrication methods and should try to fit their work to the fabrication facilities available.

Designers should learn everything possible about the detailing, the fabrication, and the field erection of steel. The more the designer knows about the problems, the tolerances, and the clearances in shop and field the more probable it is that reasonable, practical, and economical designs will be produced. This knowledge should include information concerning the transportation of the materials to the job site (such as the largest pieces that can practically be transported by rail or truck), labor conditions, and the equipment available for erection. Perhaps the designer should ask himself the question, "Could I get this thing together if I were sent out to do it?"

Finally he needs to proportion the parts of the structure so that they will not unduly interfere with the mechanical features of the structure (pipes, ducts, etc.) or the architectural effects.

1-6. FACTOR OF SAFETY

The factor of safety of a structural member is defined as the ratio of strength of the member to its maximum anticipated stress. The strength of a member used in determining the factor of safety may be thought of as being the ultimate strength of the member, but often some lesser value is used. For instance, failure may be assumed to occur when the members become excessively deformed. If this is the case the safety factor might be determined by dividing the yield-point stress by the maximum anticipated stress. For ductile materials the safety factor is usually based on yield-point stresses while for brittle materials it is probably based on ultimate strengths.

The student may feel that it is quite foolish to build a structure with a strength of several times that which is theoretically required. As the years go by, however, he will learn that safety factors are subject to so many uncertainties that he may spend sleepless nights wondering if those he has used are sufficient (and he may join other designers in calling them "factors of ignorance" rather than factors of safety). Some of the uncertainties affecting safety factors are

1. Material strengths may initially vary appreciably from their assumed values and they will vary more with time due to creep, corrosion, and fatigue.

2. The methods of analysis are often subject to appreciable errors.

3. The so-called beggaries of nature or acts of God (hurricanes, earthquakes, etc.) cause conditions difficult to predict.

4. The stresses produced during fabrication and erection are often severe. Workmen in shop and field seem to treat steel shapes with reckless abandon. They drop them. They ram them. They force the members into position to line up the bolt holes. In fact, the stresses during fabrication and erection may exceed those which occur after the structure is completed. The floors for the rooms of apartment houses and office buildings are probably designed for live loads varying from 40 to 80 psf (pounds per square foot). During the erection of such buildings the contractor may have 10 ft of bricks or concrete blocks or other construction materials or equipment piled up on some of the floors, causing loads of several hundred pounds per square foot. This discussion is not intended to criticize the practice (not that it is a good one), but rather to make the student aware of the things that happen during construction. (It is probable that the majority of steel structures are overloaded somewhere during construction but hardly any of them fail.)

5. There are technological changes which affect the magnitude of live loads. The constantly increasing traffic loads applied to bridges through the years is an illustration.

6. Although the dead loads of a structure can usually be estimated quite closely, the estimate of the live loads is more inaccurate. This is particularly true in estimating the worst possible combination of live loads occurring at any one time. For instance, in estimating the load supported by a column in the bottom level of a 30-story building—would 100% live load be assumed to exist on every one of the 30 floors at the same time or is some lesser percentage more realistic?

7. Other uncertainties are the presence of residual stresses and stress concentrations, variations in dimensions of member cross sections, etc.

The magnitude of the safety factors used will be affected by the preceding uncertainties and also by the answers to the following questions.

1. Is the structure to be permanent or temporary?
2. Is it a public or private structure?
3. What is the penalty for failure? Will there be loss of life or extensive property damage or great inconvenience while the structure is out of use?
4. Is a particular member a main one or a secondary one? (It may be reasonable to use high safety factors for the design of main members and low values for secondary members.)

1-7. FAILURES OF ENGINEERING STRUCTURES

Many people who are superstitious do not discuss flat tires or make their wills because they are afraid that by doing so they are tempting fate. These same people would probably not care to discuss the subject of engineering failures. Despite the prevalence of this superstition the author feels that an awareness of the items which have most frequently caused failures in the past is invaluable to experienced and inexperienced designers alike. Perhaps a study of past failures is more important than a study of past successes. Benjamin Franklin supposedly made the observation that "a wise man learns more from failures than from success."[4]

The designer of little experience particularly needs to know where she should give the most attention and where she may need outside advice. The vast majority of designers, experienced and inexperienced, select members of sufficient size and strength. The collapse of structures is usually due to insufficient attention to the details of connections, deflections, erection problems, and foundation settlements. Rarely if ever do steel structures fail due to faults in the material; but rather to its improper use.

A frequent fault displayed by designers is that after carefully designing the members of a structure they carelessly select connections which may or may not be of sufficient size. They may even turn the job of selecting the connections over to draftsmen who may not have sufficient backgrounds to understand the difficulties that can arise in connection design. Perhaps the most common mistake made in connection design is to neglect some of the forces acting on the connection, such as twisting moments. In a truss for which the members have been designed for axial forces only, the connections may be eccentrically loaded, resulting in moments that cause increasing stresses. These secondary stresses are occasionally so large that they need to be considered in design.

Another source of failure occurs where beams supported on walls have insufficient bearing or anchorage. Imagine a beam of this type supporting a flat roof on a rainy night when the roof drains are not

[4] Maurice J. Rhude, "Research Needed in Wood Structures," *Proc. ASCE* **93**, no. ST2 (April, 1967), p. 79.

functioning properly. As the water begins to form puddles on the roof it tends to cause the beam to sag in the middle, causing a pocket to catch more rain which will cause more beam sag, etc. As the beam deflects it pushes out against the walls, causing possible collapse of walls or slippage of beam ends off the wall. Picture a 60-ft steel beam supported on a wall with only an inch or two of bearing when the temperature drops 50 or 60 degrees overnight. A collapse due to a combination of beam contraction, outward deflection of walls, and vertical deflection due to precipitation loads is not difficult to visualize; furthermore, actual cases in engineering literature are not difficult to find.

Foundation settlements cause a large number of structural failures, probably more than any other factor. Most foundation settlements do not result in collapse but they very probably cause unsightly cracks and depreciation of the structure. If all parts of the foundation of a structure settle equally, the stresses in the structure theoretically will not change. The designer, usually not able to prevent settlement, has the goal of designing foundations in such a manner that equal settlements occur. Equal settlements may be an impossible goal and consideration should be given to the stresses that would be produced if settlement variations occurred. The student's background in structural analysis will tell him that uneven settlements in statically indeterminate structures may cause extreme stress variations. Where foundation conditions are poor it is desirable, if feasible, to use statically determinate structures whose stresses are not appreciably changed by support settlements. (The student will actually learn in subsequent discussions that the ultimate strength of steel structures is affected only slightly by uneven support settlements.)

Some other sources of structural failures occur because inadequate attention is given to deflections, fatigue of members, bracing against swaying, vibrations, and the possibility of buckling of compression members or the compression flanges of beams. The usual structure when completed is sufficiently braced with floors, walls, connections, and special bracing, but there are times during construction when many of these items are not present. As previously indicated, the worst conditions may well occur during erection and special temporary bracing may be required.

1-8. SPECIFICATIONS

For most structures the designer is controlled by specifications. Even if not so controlled she will probably refer to them as a guide. No matter how many structures she has designed it is impossible for her to have encountered every situation, and by referring to specifications she is making use of the best available material on the subject. Engineering specifications are developed by various engineering organizations and present the best opinion of these organizations as to what represents good engineering practice.

Examples of specifications that are referred to in detail in later chapters are

American Institute of Steel Construction (AISC)
American Welding Society (AWS)
American Association of State Highway and Transportation Officials
    (AASHTO)
American Railway Engineering Association (AREA)
American Society for Testing and Materials (ASTM)

Municipal and state governments concerned with the safety of the public have established building codes by which they control the construction of various structures under their jurisdiction. These codes, which are actually ordinances, specify design loads, allowable stresses, construction types, material quality, and other factors. They vary considerably from city to city, a fact which causes some confusion among architects and engineers.

Several organizations publish recommended practices for regional or national use. Their specifications are not legally enforceable unless they are embodied in the local building code or made a part of a particular contract. Among these organizations are the AISC, AASHTO, and others listed earlier in this section. Nearly all municipal and state building codes have adopted the AISC Specification and nearly all state highway departments have adopted the AASHTO Specifications. These codes generally include some additional provisions pertaining to their local needs.

Many people feel that specifications prevent the engineer from thinking for himself—and there may be some basis for the criticism. They say that the ancient engineers who built the great pyramids, the Parthenon, and the great Roman bridges were controlled by few specifications, which is certainly true. On the other hand, it should be said that only a few score of these great projects were built over many centuries and they were apparently built without regard to cost of material, labor, or human life. They were probably built by intuition and by certain rules of thumb developed by observing the minimum size or strength of members which would just fail under given conditions. Their probably numerous failures are not recorded in history; only their successes endured.

Today, however, there are hundreds of projects being constructed at any one time in the United States which rival in importance and magnitude the famous structures of the past. It appears that if all engineers in our country were allowed to design projects such as these without restrictions there would be many disastrous failures. *The important thing to remember about specifications, therefore, is that they are not written for the purpose of restricting engineers but for the purpose of protecting the public.*

No matter how many specifications are written it is impossible for them to cover every possible situation. As a result, no matter which code or specification is or is not being used, the ultimate responsibility for the design of a safe structure lies with the structural designer.

1-9. VARIATIONS IN DESIGNS

Designs of the same structures using the same specifications can be surprisingly different. The differences are primarily caused by the uncertainties mentioned in the preceding paragraphs which call for the application of the engineer's judgment. Two engineers designing a bridge might make entirely different estimates of future traffic, wind loads, snow loads, ice loads, etc. In selecting steel beams for a certain bending moment it will be discovered that there may possibly be several beams of entirely different dimensions and shapes which will support approximately the same moments.

Furthermore, different engineers will probably give different weight to what happens if the structure fails. For instance, in designing a line of power poles would the engineer design the poles to withstand twice the greatest wind force ever recorded anywhere on earth, or will he use some lesser loading condition, assuming that under very unusual storm conditions some poles may come down?

1-10. CALCULATION ACCURACY

A most important point which many students with their superb pocket calculators have difficulty in understanding is that structural design is not

Hackensack River Bridge, New Jersey Turnpike (Courtesy of American Bridge Division, U.S. Steel Corporation).

an exact science for which answers can confidently be calculated to eight places. The reasons for this fact have already been discussed; the methods of analysis are based on partly true assumptions, the strengths of materials used vary appreciably, and maximum loadings can only be approximated. With respect to this last sentence, how many of the users of this book could estimate within 10% the maximum load in pounds per square foot that will ever occur on the building floor which they are now occupying? Calculations to more than two or three significant figures are obviously of little value and may actually be harmful in that they mislead the student by giving him a fictitious sense of precision.

1-11.   DESIGN AIDS

The use of various manuals, tables, charts, monographs, and computer programs provides a great deal of assistance in design. These devices give information about forms and details which are commonly used by the engineering profession and will thus help to provide economy both in the design office and in field construction.

It is probably undesirable, however, for the beginning engineer to make extensive use of most tables and charts until he has had some experience. One of the famous rules of engineering (and a very good one) is that the engineer should not use a table or chart which he does not understand sufficiently to reconstruct. He needs a sound knowledge of the basic principles of structural design in order to see the limitations of these devices. (The author would like to hedge a little on this subject by saying that he does not mean to imply that the young engineer going to work in a design office should refuse to use their design aids with which he is unfamiliar until he has worked them out on company time. They are in business to make money and need to have their personnel work as efficiently as possible.)

Chapter 2
Properties
of Structural
Steel

2-1. STRESS-STRAIN RELATIONSHIPS IN STRUCTURAL STEEL

To understand the behavior of steel structures it is absolutely essential for the designer to be familiar with the properties of steel. Stress-strain diagrams present a valuable part of the information necessary to understand how steel will behave in a given situation. Satisfactory steel design methods cannot be developed unless complete information is available concerning the stress-strain relationships of the material being used.

If a piece of mild structural steel is subjected to a tensile force it will begin to elongate. If the tensile force is increased at a constant rate the amount of elongation will increase constantly within certain limits. In other words, elongation will double when the stress goes from 6,000 to 12,000 psi (pounds per square inch). When the tensile stress reaches a value roughly equal to one-half of the ultimate strength of the steel the elongation will begin to increase at a greater rate without a corresponding increase in the stress.

The largest stress for which Hooke's law applies or the highest point on the straight line portion of the stress-strain diagram is the *proportional limit*. The largest stress which a material can withstand without being permanently deformed is called the *elastic limit*. This value is seldom actually measured and for most engineering materials including structural steel is synonomous with the proportional limit. For this reason the term *proportional elastic limit* is sometimes used.

The stress at which there is a decided increase in the elongation or strain without a corresponding increase in stress is said to be the *yield point*. It is the first point on the stress-strain diagram where a tangent to the curve is horizontal. The yield point is probably the most important property of steel to the designer as the elastic design procedures are based on this value (with the exception of compression members where buckling may be a factor). The allowable stresses used in these methods are usually taken as some percentage of the yield point. Beyond the yield point there is

Newport Bridge between Jamestown and Newport, R.I. (Courtesy of Bethlehem Steel Company).

a range in which a considerable increase in strain occurs without increase in stress. The strain that occurs before the yield point is referred to as the *elastic strain*; the strain that occurs after the yield point, with no increase in stress, is referred to as the *plastic strain*. These latter strains usually vary from 10 to 15 times the elastic strains.

Yielding of steel without stress increase may be thought to be a severe disadvantage when in actuality it is a very useful characteristic. It has often performed the wonderful service of preventing failure due to omissions or mistakes on the designer's part. Should the stress at one point in a ductile steel structure reach the yield point, that part of the structure will yield locally without stress increase, thus preventing premature failure. This ductility allows the stresses in a steel structure to be readjusted. Another way of describing this phenomenon is to say that very high stresses caused by fabrication, erection, or loading will tend to equalize themselves. It might also be said that a steel structure has a reserve of plastic strain that enables it to resist overloads and sudden shocks. If it did not have this ability, it might suddenly fracture, like glass or other vitreous substances.

Following the plastic strain there is a range where additional stress is necessary to produce additional strain and this is called *strain-hardening*. This portion of the diagram is not too important to today's designer. A familiar stress-strain diagram for mild structural steel is shown in Fig. 2-1. Only the initial part of the curve is shown here because of the great

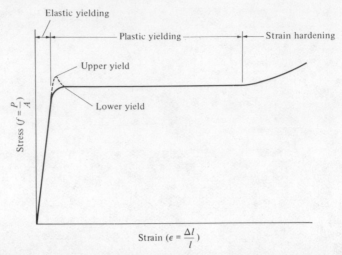

Figure 2-1 Typical stress–strain diagram for a mild structural steel.

deformation which occurs before failure. At failure in the mild steels the total strains are from 150 to 200 times the elastic strains. The curve will actually continue up to its maximum stress value and then "tail off" before failure. A sharp reduction in the cross section of the member takes place (called "necking") followed by failure.

The stress-strain curve of Fig. 2-1 is typical of the usual ductile structural steel and is assumed to be the same for members in tension or compression. (The compression members must be short because long compression members subjected to compression loads tend to bend laterally and their properties are greatly affected by the bending moments so produced.) The shape of the diagram varies with the speed of loading, the type of steel, and the temperature. One such variation is shown in the figure by the dotted line which is marked *upper yield*. This shape stress-strain curve is the result when a mild steel has the load applied rapidly while the *lower yield* is the case for slow loading.

A very important property of a structure which has not been stressed beyond its yield point is that it will return to its original length when the loads are removed. Should it be stressed beyond this point it will return only part of the way to its original position. This knowledge leads to the possibility of testing an existing structure by loading and unloading and measuring deflections. If after the loads are removed the structure will not resume its original dimensions, it has been stressed beyond its yield point.

Steel is an alloy consisting almost entirely of iron (usually over 98%). It also contains small quantities of carbon, silicon, manganese, sulfur, phosphorus, and other elements. Carbon is the material that has the greatest effect on the properties of steel. The hardness and strength increase as the carbon percentage is increased but unfortunately the resulting steel is more brittle and its weldability is adversely affected. A

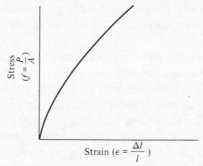

Figure 2-2   Typical stress–strain diagram for a brittle steel.

smaller amount of carbon will make the steel softer and more ductile but also weaker. The addition of such elements as chromium, silicon, and nickel produces steels with considerably higher strengths. These steels, however, are appreciably more expensive and are often not so easy to fabricate.

A typical stress-strain diagram for a brittle steel is shown in Fig. 2-2. Such a material shows little or no permanent deformation at fracture. Unfortunately, low ductility or brittleness is a property usually associated with high strengths in steels (although not entirely confined to high-strength steels). As it is desirable to have both high strength and ductility, the designer may have to decide between the two extremes or compromise between them. A brittle steel may fail suddenly without warning when overstressed, and during erection could possibly fail due to the shock of erection procedures.

2-2. ELASTIC AND PLASTIC DESIGN METHODS DEFINED

The majority of steel structures designed in the past and of those being designed today are handled by the *elastic-design* methods. The designer estimates the "working loads," or loads that the structure may feasibly have to support, and proportions the members on the basis of certain allowable stresses. These allowable stresses are usually some fraction of the specified minimum yield point of the steel. Although the term "elastic design" is very commonly used to describe this method the terms *allowable-stress design* or *working-stress design* are definitely more appropriate. Many of the provisions of the specifications for this method are actually based on plastic or ultimate-strength behavior and not on elastic behavior.

The ductility of steel has been shown to give it a reserve strength and the realization of this fact is the theory behind *plastic design*. In this method the working loads are estimated and multiplied by certain load or overcapacity or safety factors and the members designed on the basis of collapse strengths. Other names for this method are *limit design* and *collapse design*. Although only a few thousand structures have been designed around the world by the plastic-design methods, the profession is definitely moving in that direction. This trend is particularly reflected in the AISC Specification.

The design profession has long been aware that the major portion of the stress-strain curve lies beyond the steel's elastic limit. Furthermore, tests through the years have made it clear that steels can resist stresses appreciably larger than their yield points, and that in cases of overload, statically indeterminate structures have the happy facility of spreading the load out due to the steel's ductility. On the basis of this information many plastic-design proposals have been made in recent decades. It is un- doubtedly true that for certain types of structures, plastic design results in a more economical use of steel than does elastic design. Chapters 21 and 22 are devoted entirely to this subject.

2-3. LOAD AND RESISTANCE FACTOR DESIGN

At the present time the AISC Specification consists of Part 1, which deals with allowable stress design and Part 2, which deals with plastic design. It is anticipated that within a very short time the AISC will add a Part 3 to their specification which will permit Load and Resistance Factor Design (or it may be that Part 2 will be replaced with this method). This method, abbreviated LRFD, is a design procedure that combines the calculation of ultimate or limit states of strength and serviceability with a probability based approach to safety.

The basic criterion of LRFD is that the theoretical or nominal capacity of each member multiplied by an undercapacity or resistance factor less than 1.0 (to account for variations in material properties and member dimensions) must at least equal an analysis factor greater than 1.0 (to account for uncertainties in structural analysis) multiplied by each load effect times its load overcapacity factor. This criterion can be expressed as follows.

(resistance factor)(nominal capacity of member)

$\geq$(analysis factor)(Σ of load effects times their respective load factors)

In Canada a probability-based LRFD method has already been adopted for both hot-rolled and cold-formed steel structures while much progress has been made in Europe toward the goal of formulating LRFD procedures for various national codes. Efforts are being made to prepare similar procedures for timber and reinforced concrete structures in both Europe and America.[1] As a result great increases in the use of LRFD will seemingly be made during the next decade. The subject of LRFD is continued in Chapter 23.

[1]M. K. Ravindra and T. V. Galambos, "Load and Resistance Factor Design for Steel," *ASCE Journal of Structural Division* **104**, no. ST9 (September, 1978), pp. 1337–1353.

Steel erection for the Chase Manhattan Bank Building, New York City (Courtesy of Bethlehem Steel Company).

2-4. STEEL SECTIONS

The first structural shapes made in the United States were angle irons rolled in 1819. Steel I beams were first rolled in the United States in 1884 and the first skeleton frame structure using steel (the Home Insurance Company Building of Chicago) was erected that same year.[2]

During these early years the various mills rolled their own individual shapes and published catalogs which provided the dimensions, weight, and other properties of these shapes. In 1896 the Association of American Steel Manufacturers (now the American Iron and Steel Institute, AISI) made the first efforts to standardize shapes. Today nearly all structural shapes are standardized though their exact dimensions may vary just a little from mill to mill.[2]

Structural steel can be economically rolled into a wide variety of shapes and sizes without appreciably changing its physical properties. Usually the most desirable members are those which have large section moduli in proportion to their areas. The I, T, and [shapes, so commonly used, fall into this class.

[2]W. McGuire, *Steel Structures* (Englewood Cliffs, N.J.: Prentice-Hall, 1968), pp. 19–21.

Steel sections are usually designated by the shapes of their cross sections. As examples, there are angles, tees, zees, and plates. It is necessary, however, to make a definite distinction between American standard beams (called S beams) and wide-flange beams (called W beams) as they are both I-shaped. The inner surface of the flange of a W section is either parallel to the outer surface or nearly so with a maximum slope of 1 to 20 on the inner surface, depending on the manufacturer.

The S beams, which were the first beam sections rolled in America, have a slope on their inside flange surfaces of 1 to 6. It might be noted that the constant, or nearly constant, thickness of W beam flanges as compared to the tapered S beam flanges may facilitate connections. Wide-flange beams comprise nearly 50% of the tonnage of structural steel shapes rolled today. The W and S sections are shown in Fig. 2-3 together with several other familiar steel sections. The uses of these various shapes will be discussed in detail in the chapters to follow.

Constant reference is made throughout this book to the *Manual of Steel Construction* published by the AISC. This manual, which provides detailed information for structural steel shapes, is referred to hereafter as the AISC Manual or the Steel Handbook. This manual, which provides detailed information for structural steel shapes, has been said to be the most used handbook ever produced by a single industry and over a million copies are now in circulation. Reference is made herein to the 8th edition of the handbook which is based on the November 1978 AISC Specification. Structural shapes are abbreviated by a certain system described in the handbook for use in drawings, specifications, and designs. This system is standardized so that all steel mills can use the same identification for purposes of ordering, billing, etc. In addition so much work is handled today with computers and other automated equipment that it is necessary to have a letter and number system which can be printed out with a standard keyboard (as opposed to the old system where certain symbols were used for angles, channels, etc.). Examples of this abbreviation system

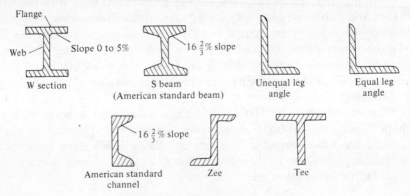

Figure 2-3 Rolled-steel shapes.

are as follows:

1. A W27×114 is a W section approximately 27 in. deep weighing 114 lb/ft.
2. An S12×35 is an S section 12 in. deep weighing 35 lb/ft.
3. An HP12×74 is a bearing pile section which is approximately 12 in. deep weighing 74 lb/ft. Bearing piles are made with the regular W rolls but with thicker webs to provide better resistance to the impact of pile driving.
4. An M8×6.5 is a miscellaneous section approximately 8 in. deep weighing 6.5 lb/ft. It is one of a group of doubly-symmetrical H-shaped members which cannot by dimensions be classified as a W, S, or HP section.
5. A C10×30 is a channel 10 in. deep weighing 30 lb/ft.
6. An MC18×58 is a miscellaneous channel which cannot be classified as a C shape by dimensions.
7. An L6×6×$\frac{1}{2}$ is an equal leg angle, each leg being 6 in. long and $\frac{1}{2}$ in. thick.
8. A WT18×140 is a tee obtained by splitting a W36×280. This type of section is known as a structural tee.

The student should refer to the AISC Manual for information concerning other rolled shapes, such as the distinction between bars and plates, designation of open web joists, etc. Additional sections will be mentioned herein as it becomes necessary.

In Part 1 of the AISC Manual the dimensions and properties of W, S, C, and other shapes are given. The dimensions of the members are given in

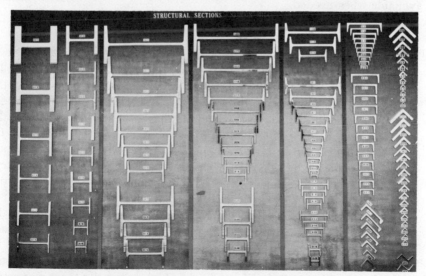

Structural shapes rolled at Bethlehem Steel Company's Bethlehem, Pa., plant (Courtesy of Bethlehem Steel Company).

decimals (for the use of designers) and in fractions to the nearest sixteenth of an inch (for the use of craftsmen and steel detailers or draftsmen). Also provided for the use of designers are such items as moments of inertia, section moduli, radii of gyration, and other cross-sectional properties discussed later in this text.

There are variations present in any manufacturing process and the steel industry is certainly no exception. As a result the cross-sectional dimensions of steel members may vary somewhat from the values specified in the AISC Manual. Maximum tolerances are specified in the manual for the rolling of steel shapes. As a result stress calculations can be made on the basis of the properties given in the manual regardless of the manufacturer.

Through the years there have been changes in the sizes of steel sections. For instance, there may be insufficient demand to continue rolling a certain shape; an existing shape may be dropped because a similar sized but more efficient shape has been developed; and so forth. Occasionally the engineer may need to know the properties of one of the discontinued shapes which are no longer listed in her edition of the Steel Handbook or in other tables normally available to her. As an example, it may be desired to add another floor to an existing building which was constructed with shapes no longer rolled. In 1953 the AISC published a book entitled *Iron and Steel Beams* 1873 *to* 1952 which gives a complete listing of iron and steel beams and their properties rolled in the United States during that period. Since that book was published there have been many additional shape changes. As a result the wise structural engineer will carefully preserve old editions of the manual so as to have them available when needed.

2-5. MODERN STRUCTURAL STEELS

As previously described steel is an alloy of iron and the nonmetallic element carbon. Its properties can be greatly changed by varying the quantities of carbon present and by adding other elements such as silicon, nickle, manganese, and copper. A steel that has a significant amount of the latter elements is referred to as an alloy steel. Although these elements do have a great effect on the properties of steel the actual quantities of carbon or other alloying elements are quite small. For instance the carbon content of steel is almost always less than 0.5% by weight and is normally from 0.2% to 0.3%.

The chemistry of steel is extremely important in its effect on the properties of the steel such as weldability, corrosion resistance, resistance to brittle fracture, and so on. The carbon present in steel increases its hardness and strength but at the same time reduces its ductility as do the impurities phosphorus and sulfur. The ASTM specifies the exact maximum percentages of carbon, manganese, silicon, etc., which are permissible for a number of structural steels. Although the physical and mechanical properties of steel sections are primarily determined by their chemical composition they are also influenced to a certain degree by the rolling process and by their stress history and heat treatment.

Most of the steel used in engineering structures in the United States is a structural carbon steel designated as A36 by the ASTM, but many other steels are available and demand for them is rising rapidly.

In recent decades the engineering and architecture professions have been continually requesting stronger steels, steels with more corrosion resistance, steels with better welding properties, and various other requirements. Research by the steel industry during this period has supplied several groups of new steels which satisfy many of the demands and today there are quite a few structural steels designated by the ASTM and included in the AISC Specification.

One half of a 170-ft clear span roof truss for the Athletic and Convention Center, Lehigh University, Bethlehem, Pa. (Courtesy of Bethlehem Steel Company).

These steels can be grouped into three major classifications: the carbon steels, the high-strength low-alloy steels and the quenched and tempered alloy steels. There are indeed other groups of high-strength steels such as the ultra high-strength steels (having yield stresses in the 160,000 to 300,000 psi range) but these are not included in this section because they are not today generally covered by the ASTM and thus are not included in the AISC Specification.

Carbon Steels

These steels have as their principal strengthening agents carefully controlled quantities of carbon and manganese. In detail the carbon steels are those which have the following maximum amounts of elements: 1.7% carbon, 1.65% manganese, 0.60% silicon, and 0.60% copper. These steels are divided into four categories depending on carbon percentages as follows.

1. Low carbon steel $< 0.15\%$.
2. Mild carbon steel 0.15–0.29%. (The structural carbon steels fall into this category.)
3. Medium carbon steel 0.30–0.59%.
4. High carbon steel 0.60–1.70%.

The common everyday American steel, A36, which has a yield stress of 36,000 psi, is suitable for riveted, bolted, or welded bridges and buildings. As it is the most important steel today and may very well be for a good many years to come it is used for the majority of the design problems in this text. Other carbon steels include A500, and A501 (both applying to tubing), and A529 (pertaining to some thinner plates and shapes). Table 2-1 presents a brief summary of the various carbon and higher strength steels considered in the AISC Specification. The A242 and A588 steels are particularly corrosion resistant and are so entitled in Table 2-1.

High-Strength Low-Alloy Steels

There are a large number of high-strength low-alloy steels and they are included under several ASTM numbers. In addition to carbon and manganese these steels obtain their higher strengths and other properties by the addition of one or more alloying agents such as columbium, vanadium, chromium, silicon, copper, nickle, and others. Included are steels with yield stresses as low as 40,000 psi and as high as 70,000 psi. These steels (including A242, A440, A441, A572, and A588) generally have much greater atmospheric corrosion resistance than the carbon steels.

A steel of particular interest in this group was developed in 1933 by the U.S. Steel Corporation to provide resistance for the very severe corrosive conditions of railroad coal cars. This steel oxidizes and forms a very tightly adherent film on its surface which prevents further oxidation and thus eliminates the need for painting. After this process takes place in from about 18 months to 3 years depending on the type of exposure (rural, industrial, etc.) the steel reaches a deep reddish to brown to black color.

You can see the many uses which can be made of such a steel particularly for structures with exposed members that are difficult to paint —bridges, electrical transmission towers, and others. This steel is not considered to be satisfactory for use in locations where it is frequently subject to salt water sprays or fogs, or continually submerged in water (fresh or salt) or the ground or in locations where there are severe corrosive industrial fumes.

Quenched and Tempered Alloy Steels

These steels have alloying agents in excess of those used in the carbon steels and they are heat-treated by quenching and tempering to obtain strong and tough steels with yield strengths in the 80,000 to 110,000 psi range. Quenching consists of rapid cooling of the steel with water or oil from temperatures of at least 1650°F down to about 300 or 400°F. In tempering the steel is reheated to at least 1150°F and allowed to cool.

The quenched and tempered steels do not exhibit well-defined yield points as do the carbon steels and the high-strength low-alloy steels. As a result their yield strengths are usually defined as the stress at 0.2% offset strain. (In other words a line is drawn parallel to the straight line portion of the stress strain diagram from a strain of 0.002 until it intersects the stress strain curve. The stress at that point is the assumed yield point for 0.2% offset strain.) The quenched and tempered steels in Table 2-1 are listed under the ASTM designation A514 which has yield stresses of 90,000 or 100,000 psi depending on thickness.

It is said that there are today more than 200 steels on the market which provide yield stresses in excess of 36,000 psi. The steel industry is now experimenting with steels with yield stresses varying from 200,000 to 300,000 psi and this may only be the beginning. Many people in the steel industry feel that steels with 500,000 psi yield strengths will be made available within a few years. The theoretical binding force between iron atoms has been estimated to be in excess of 4,000,000 psi.[3]

Although the prices of steels increase with increasing yield points the percentage of price increase does not keep up with the percentage of yield-point increase. The result is that the use of the stronger steels will quite frequently be economical for tension members, beams, and columns. Perhaps the greatest economy can be realized with tension members (particularly those without rivet or bolt holes). They may provide a great

[3] L. S. Beedle et al. *Structural Steel Design* (New York: Ronald Press, 1964), p. 44.

Table 2-1 Properties of Structural Steels Considered in AISC Specifications

ASTM Designation	Type of steel	Shapes available	Recommended uses	Yield stress psi	Atmospheric corrosion resistance
A36	Carbon	Shapes and bars	Riveted, bolted or welded bridges and buildings, and other structural uses	36,000 up through 8 in. (32,000 above 8 in.)	
A500	"	Cold-formed welded and seamless tubing	"	33,000 for Grade A 42,000 for Grade B	
A501	"	Hot-formed welded and seamless tubing PLs and bars up through $\frac{1}{2}$ in. and some light shapes	"	36,000	
			Riveted, bolted and welded buildings, and related construction	42,000	
A440	High-strength	" " "	Riveted and bolted bridges and buildings, and other structural uses; not recommended for welding	50,000 up through $\frac{3}{4}$ in. 46,000 above $\frac{3}{4}$ in. through $1\frac{1}{2}$ in. 42,000 above $1\frac{1}{2}$ in. through 4 in.	Approximately equal to two times that of structural carbon steel
A441	High-strength low-alloy	" " "	May be used for welded, riveted, and bolted structures but primarily intended for welded bridges and buildings	50,000 up through $\frac{3}{4}$ in. 46,000 above $\frac{3}{4}$ in. through $1\frac{1}{2}$ in. 42,000 above $1\frac{1}{2}$ in. through 4 in. 40,000 above 4 in. through 8 in.	"

Table 2-1 (Continued)

ASTM	Type	Forms	Recommended uses	Minimum yield point, psi	Corrosion resistance
A572	"	" " "	Grades 42, 45, and 50 especially intended for riveted, bolted or welded bridges, buildings, and other structures. Grades 55, 60, and 65 especially intended for riveted or bolted bridges, and for other structures which are riveted, bolted or welded	42,000 / 45,000 / 50,000 / 55,000 / 60,000 / 65,000 (varying thickness, chemical compositions, etc. (see ASTM))	
A242	Corrosion-resistant High-strength low-alloy	Shapes and bars	Welded, riveted or bolted construction; welding technique very important	50,000 up through $\frac{3}{4}$ in. 46,000 above $\frac{3}{4}$ in. through $1\frac{1}{2}$ in. 42,000 above $1\frac{1}{2}$ in. through 4 in.	Approximately equal to four times that of structural carbon steel with copper
A588	"	" " "	May be used for welded, riveted or bolted construction but primarily intended for welded bridges and buildings; welding technique very important	50,000 up through 4 in. 46,000 above 4 in. through 5 in. 42,000 above 5 in. through 8 in.	Approximately equal to four times that of structural carbon steel with copper
A514	Quenched and tempered alloy	PLs up through 4 in, and a limited number of shapes	Primarily intended for use in welded bridges and other structures; welding technique very important	100,000 up through $2\frac{1}{2}$ in. 90,000 above $2\frac{1}{2}$ in. through 4 in.	

NOTE: At this point the student should carefully examine pages 1-5 and 1-6 of the AISC Manual for more information regarding the different steels available. Particular attention should be given to the determination of yield stresses for the different grades of steels (for plates and bars see page 1-6, and for the rolled sections see page 1-6). Notice carefully the division of shapes into groups and the resulting effects on yield stresses.

deal of saving for beams if deflections are not important or if deflections can be controlled (as by cambering, use of composite construction, etc). In addition considerable economy can frequently be achieved with high-strength steels for short and medium length stocky columns. Another application which can provide considerable saving is in hybrid construction. In this type of construction two or more steels of different strengths are used, the weaker steels being used where stresses are smaller and the stronger steels where stresses are higher.

Among the other factors that might lead to the use of high-strength steels are the following.

1. Superior corrosion resistance.
2. Possible savings in shipping, erection, and foundation costs caused by weight saving.
3. Use of shallower beams permitting smaller floor depths.
4. Possible savings in fireproofing because smaller members can be used.

The first thought of most engineers in choosing a type of steel is the direct cost of the members. Such a comparison can be made quite easily, but the economy of which strength grade to use cannot be obtained unless consideration is given to weights, sizes, deflections, maintenance, and fabrication. To make an accurate general comparison of the steels is probably impossible—rather it is necessary to have a specific job to consider.

2-6. FURNISHING OF STRUCTURAL STEEL

The furnishing of structural steel consists of the rolling of the steel shapes, the fabrication of the shapes for the particular job (including cutting to the proper dimensions and punching of holes necessary for field connections), and their erection. Occasionally a company will perform all three of these functions, but the average company performs only one or two of them. For instance, many companies fabricate structural steel and erect it while others may only be steel fabricators or steel erectors. There are approximately 400 companies in the United States which make up the fabricating industry for structural steel. Most of them do fabrication and erection.

Structural steel is said to be handled by mill or stock orders. The mill orders are the most economical as the fabricator orders the shapes to certain lengths directly from the rolling mill. The process of ordering directly, although requiring more time, saves handling costs and reduces the costs of storage because the material is not kept in stock for a time with the resulting interest and storage charges. Some fabricators carry certain sizes in stock. These sizes can be obtained immediately but will cost more because of the extra storage, interest, and handling costs previously mentioned.

Steel erection for Transamerica Pyramid, San Francisco, Calif. (Courtesy of Kaiser Steel Corporation).

The design of structural steel is usually made by an engineer in collaboration with an architectural firm. The designer makes design drawings that show member sizes, controlling dimensions, and any unusual connections. The company that is to fabricate the steel makes the detailed drawings which give all the information necessary to fabricate the members correctly. These details include all of the dimensions for each member, the locations and sizes of holes, the position and sizes of connections, and the like.

The erection of steel buildings is more a matter of assembly than nearly any other line of construction work. Each of the members is marked in the shop with letters and numbers to distinguish them from the other members to be used. The erection is performed in accordance with a set of erection plans. These plans are not detailed drawings but are simple line diagrams showing the position of the various members in the building. Each of the lines shown has a number and letter on it which corresponds to the same designation on the fabricated steel. This lettering system enables the workmen in the field easily to find the position of each member. (People performing the steel erection are often called *ironworkers*, which is a name held over from the days before structural steel.)

2-7. DESIGN OF STEEL MEMBERS

The design of a steel member involves much more than a calculation of the properties required to support the loads and the selection of the lightest section providing these properties. Although at first glance this procedure would seem to give the most economical designs, many other factors need to be considered. Among these are

1. The designer needs to select steel sections of sizes which are usually rolled. Steel beams and bars and plates of unusual sizes will be difficult to obtain during boom periods and will be expensive during any period. A little study on the designer's part will enable him to avoid these expensive shapes. Steel fabricators are constantly supplied with information from the steel companies and the steel warehousemen as to the sizes of sections available. This information includes the section sizes as well as the standard lengths available. (Most structural shapes can be obtained in lengths from 60 to 75 ft depending on the producer, while it is possible under certain conditions to obtain some shapes up to 120 ft in length.)

2. A blind assumption that the lightest section is the cheapest one may be in considerable error. A building frame designed by the "lightest-section" procedure will consist of a large number of different shapes and sizes of members. Trying to connect these many-sized members and fit them in the building will be quite complicated and the pound price of the steel will in all probability be rather high. A more reasonable approach would be to smooth out the sizes by selecting many members of the same sizes although some of them may be slightly overdesigned.

3. The beams usually selected for the floors in buildings will be the deeper sections because these sections for the same weights have the largest moments of inertia and the greatest resisting moments. As building heights increase, however, it may be economical to modify this practice. As an illustration of this fact the erection of a 20-story building, for which each floor has a minimum clearance, is considered. It is assumed that the depths of the floor beams may be reduced by 6 in. without an unreasonable increase in beam weights. The beams will cost more but the building height will be reduced by 20×6 in. $= 120$ in. or 10 ft, with resulting savings in walls, elevator shafts, column heights, plumbing, wiring, and footings.

4. For larger sections, particularly the built-up ones, the designer needs to have information pertaining to transportation problems. The desired information includes the greatest lengths and depths that can be

shipped by truck or rail, clearance available under bridges and power lines leading to the project, and allowable loads on bridges. It may be possible to fabricate a steel roof truss in one piece, but is it possible to transport it to the job site and erect it in one piece?

5. Sections should be selected which are reasonably easy to erect and which have no conditions present that will make them difficult to main-

Steel erection for Transamerica Pyramid, San Francisco, Calif. (Courtesy of Kaiser Steel Corporation).

tain. As an example, it is necessary to have access to all exposed surfaces of steel bridge members so that they may be periodically painted.

6. Buildings are often filled with an amazing conglomeration of pipes, ducts, conduits, and other items. Every effort should be made to select steel members that will fit in with the requirements made by these items.

7. The members of a steel structure are often exposed to the public, particularly in the case of steel bridges and auditoriums. Appearance may often be the major factor in selecting the type of structure to be used as where a bridge is desired which will fit in and actually contribute to the appearance of an area. Exposed members may be surprisingly graceful when a simple arrangement and perhaps curved members are used, but other arrangements may create a terrible eyesore. The student has certainly seen illustrations of each case. It is very interesting to know that beautiful structures in steel are usually quite reasonable in cost.

2-8.   SI UNITS

The International Bureau of Weights and Measures has as its goal the establishment of a rational and coherent worldwide system of units. In 1960 they named this system the "International System of Units," with the abbreviation SI in all languages. SI units, which are currently being adopted by quite a few countries including most English-speaking nations (Britain, Australia, Canada, South Africa, and New Zealand), differ from the metric system now being used in most European countries and in many other parts of the world.

The SI system has the very important advantage that only one unit is given for each physical quantity, such as the meter (m) for length, the kilogram (kg) for mass, the second (s) for time, the newton (N) for force, and so on. From these basic units other units are derived as follows.

area | square meter (m^2)
acceleration | meter per second squared (m/s^2)
force | kilogram meter per second squared $(kg \cdot m/s^2) = $ newton (N)
stress | newton per square meter $(N/m^2) = $ pascal (Pa)

The multiples of these units are in the decimal system and are expressed in powers of ten that are multiples of three, as shown in Table 2-2, where length units are illustrated. In other words, multiple and submultiple steps of one thousand are recommended. It will be noted that the centimeter is not shown because its value (which would be 10^{-2} in the table) is inconsistent with the theory of having prefixes that are ternary powers of 10.

In many countries of the world the comma is used to indicate a decimal; thus to avoid confusion in the SI system, spaces rather than commas are used. For a number having four or more digits, the digits are

Table 2-2 SI Decimal Multiples

SI Symbol	Name	Multiplier	Example (meters)
G	giga	1 000 000 000	Gm = 1 000 000 000
M	mega	1 000 000	Mm = 1 000 000
k	kilo	1 000	km = 1 000
m	milli	0.001	mm = 0.001
μ	micro	0.000 001	μm = 0.000 001
n	nano	0.000 000 001	nm = 0.000 000 001

separated into groups of threes, counting both right and left from the decimal. For example, 3,245,621 is written as 3 245 621 and 2,015.3216 is written as 2 015.321 6.

When units are to be multipled together, a dot is used to separate them. Bending moment is a term frequently used in structural design. In SI units it is expressed in newton meters. It is to be written as N·m (not mN, which represents millinewton). There is a great premium in the SI system on symbology. As a result, the designer must be exceptionally careful to use the correct symbols. For example, it is correct to write 100 meters, but in abbreviation 100 m must be used and not 100 ms, which would be 100 milliseconds.

Table 2-3 gives conversion values for some customary measurements to the SI system. Perhaps if you were to memorize a few approximate values from this table, you would begin to get a feel for the relative values between the two systems. For instance, 25 mm is approximately equal to 1 in., and 300 mm is approximately equal to 1 ft.

Table 2-3 Common Values for Some Common Units

Customary units	SI units
1 in.	25.400 mm = 0.025 400 m
1 in.^{2}	645.16 mm^2 = 6.451 600 m$^2 \times 10^{-4}$
1 ft	304.800 mm = 0.304 800 m
1 lb	4.448 222 N
1 kip	4 448.222 N = 4.448 222 kN
1 psi	6.894 757 kN/m^2 = 0.006 895 MN/m^2 = 0.006 895 N/mm^2
1 psf	47.880 N/m^2 = 0.047 880 kN/m^2
1 ksi	6.894 757 MN/m^2 = 6.894 757 MPa
1 in.-lb	0.112 985 N·m
1 ft-lb	1.355 818 N·m
1 in.-k	112.985 N·m
1 ft-k	1 355.82 N·m = 1.355 82 kN·m

The unit of stress in the SI system is the newton per square meter (N/m^2), which is also called a pascal (Pa). To have a feel for the magnitude of this number, a megapascal (MPa) is equal to 0.145 kip per square inch. Thus, 138 MPa is equal to 20 ksi.

In the same fashion, the terms in.-k and ft-k have been constantly used. In the SI system they are usually expressed in $kN \cdot m$. One $kN \cdot m$ is equal to 0.738 ft-k. From Table 2-3 it can be seen that 1 pound is equal to 4.448 222 newtons. Thus, 1 newton is equal to about one-fourth of a pound, which is roughly equal to the weight of one apple.

In a few places throughout the text the author has shown the SI units in parentheses. He particularly wants the reader to understand that from the standpoint of the designer there is no problem in learning and using SI units. The few units ($mm, m, MPa = N/mm^2$, and $kN \cdot m$) can be learned in a few minutes and can be applied with no difficulties. The steel and construction industries, however, are in a different situation as they face a difficult and expensive problem in converting their sizes, shapes, tables, etc., into the new units. The present tables of the AISC Manual are not given in SI units although some conversion tables are provided.

Chapter 3
Tension Members

3-1. INTRODUCTION

Tension members are found in bridge and roof trusses, towers, bracing systems, and in situations where they are used as tie rods. The selection of a section to be used as a tension member is one of the simplest problems encountered in design. As there is no danger of buckling, the calculations simply involve the division of the load by the allowable tensile stress to give the effective cross-sectional area required ($A_{reqd.} = T/F_t$) and the selection of a steel shape which furnishes the required area.

The type of member used may depend more upon the type of end connection than on any other factor. One of the simplest forms of tension members appears to be the circular rod, but there is some difficulty in connecting it to many structures. The rod has been used frequently in the past but has only occasional uses today in bracing systems, light trusses, and in timber construction. One important reason why rods are not popular with designers is that they have been used improperly so often in the past that they have a bad name; but if they are designed and installed correctly they are satisfactory for many situations.

The average size rod has very little stiffness and may quite easily sag under its own weight injuring the appearance of the structure. The threaded rods formerly used in bridges often worked loose and rattled. Another disadvantage of rods is the difficulty of fabricating them with the exact lengths required and the consequent difficulties of installation.

When rods are used in wind bracing it is a good practice to produce initial tension in them, as this will tighten up the structure and reduce rattling and swaying. To obtain initial tension the members may be detailed shorter than their required lengths, a method that gives the steel fabricator very little trouble. A common rule of thumb used is to detail the rods about $\frac{1}{16}$ in. short for each 20 ft of length. (Approximate stress $f = \epsilon E = [\frac{1}{16}/(12)(20)] \ (29 \times 10^6) = 7550$ psi.) Another very satisfactory method involves tightening the rods with some sort of sleeve nut or turnbuckle. Part 4 of the AISC Manual provides detailed information for these devices.

The preceding discussion on rods should illustrate why rolled shapes such as angles have supplanted rods for most applications. In the early days of steel structures, tension members consisted of rods, bars, and perhaps cables. Today, although the use of cables is increasing for suspended-roof structures, tension members usually consist of single angles, double angles, tees, channels, W sections, or sections built up from plates or rolled shapes. These members look better than the old ones, are stiffer and easier to connect. Another type of tension section often used is the welded tension plate or flat bar which is very satisfactory for use in transmission towers, signs, foot bridges, and similar structures.

The tension members of steel roof trusses may consist of single angles as small as $2\frac{1}{2} \times 2 \times \frac{1}{4}$ ($64 \times 51 \times 6.4$ mm) for minor members. A more satisfactory member is made from two angles placed back to back with sufficient space between them to permit the insertion of plates for connection purposes. Where steel sections are used back-to-back in this manner, they should be connected every 4 or 5 ft to prevent rattling, particularly in bridge trusses. Single angles and double angles are probably the most common types of tension members in use. Structural tees make very satisfactory chord members for welded trusses because web members can conveniently be connected to them.

For bridges and large roof trusses tension members may consist of channels, W or S shapes, or even sections built up from some combination of angles, channels, and plates. Single channels are frequently used as they have little eccentricity and are conveniently connected. Although, for the same weight, W sections are stiffer than S sections, they may have a connection disadvantage in their varying depths. For instance, the W12 × 79, W12×72 and W12×65 all have slightly different depths (12.38 in., 12.25 in., and 12.12 in. respectively), while the S sections of a certain nominal size all have the same depths. For instance, the S12×50, the S12×40.8, and the S12×35 all have 12.00-in. depths.

Although single structural shapes are a little more economical than built-up sections, the latter are occasionally used when the designer is unable to obtain sufficient area or rigidity from single shapes. Where built-up sections are used it is important to remember that field connections will have to be made and paint applied; therefore, sufficient space must be available to accomplish these things.

Members consisting of more than one section need to be tied together. Tie plates (also called tie bars) located at various intervals or perforated cover plates serve to hold the various pieces in their correct positions. These plates serve to correct any unequal distribution of loads between the various parts. They also keep the slenderness ratios (to be discussed) of the individual parts within limitations and they may permit easier handling of the built-up members. Long individual members such as angles may be

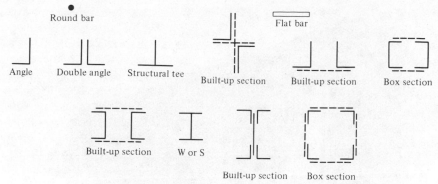

Figure 3-1   Types of tension members.

inconvenient to handle due to flexibility, but when four angles are laced together into one member as shown in Fig. 3-1, the member has considerable stiffness. None of the tie plates may be considered to increase the effective areas of the sections. As they do not theoretically carry portions of the stress in the main sections, their sizes are usually governed by specifications and perhaps by some judgment on the designer's part. Perforated cover plates (see Fig. 5-7) are an exception to this rule as part of their areas can be considered as being effective in resisting axial load.

Erection of Federal Reserve bank building, Minneapolis, Minn. (Courtesy of Inryco, Inc.).

Federal Reserve bank building, Minneapolis, Minn. (Courtesy of Inryco, Inc.)

A few of the various types of tension members in general use are illustrated in Fig. 3-1. In this figure the dotted lines represent the tie plates or bars used to connect the shapes.

Steel cables are made with special steel alloy wire ropes which are cold-drawn to the desired diameter. The resulting wires with strengths of about 200,000 to 250,000 psi can be economically used for suspension bridges, cable supported roofs, ski lifts, and other similar applications.

Normally to select a cable tension member the designer uses a manufacturer's catalogue. From the catalogue she determines the yield point of the steel and selects the cable size required for the load using an appropriate safety factor. From the catalogue she can also select clevises or other devices to use for connectors at the cable ends.

3-2.   ALLOWABLE TENSILE STRESSES AND LOADS

A tension member without holes or threads has a tensile strength equal to its yield stress times its gross cross-sectional area, A_g. Its allowable total tensile load is equal to its yield stress divided by an acceptable safety factor or, that is, its allowable tensile stress, F_t, times its gross area.

$$T = \frac{F_y}{S.F.} A_g = F_t A_g$$

When tension members have holes punched in them for rivets or bolts the reduced area after the holes are taken out is called the net area, A_n. The allowable tensile strength at the section through the holes can then be expressed as follows.

$$T = F_t A_n$$

Although the maximum strength of a member is often controlled by yielding at its net section sometimes the gross section away from the holes may yield before the ultimate tensile stress, F_u, is reached. This means that actual failure may occur when the yield stress is reached on the gross section. Therefore the AISC Specification provides one allowable stress $0.60F_y$ on the gross section and another allowable stress of $0.50F_u$ on the effective net area. Values for F_u, which is usually referred to in specifications as the specified minimum tensile strength, are given for various steels in Table 2 of Appendix A of the AISC Specification. For A36 steel, 58 ksi is the minimum specified F_u value.

As a result of the preceding, the AISC Specification gives allowable stresses of $0.60F_y$ on the gross cross-sectional areas of members at sections where there are no holes. For sections where there are holes for bolts or rivets, the allowable tensile stress is $0.50F_u$, applicable to the effective net areas. The actual effective net area, A_e, which can be considered to resist tension at the section through the holes may be somewhat smaller than the actual net area, A_n, because of stress concentrations and other factors which are discussed in Section 3-5 of this chapter. Thus the allowable capacity of a tensile member with bolt or rivet holes permitted by the AISC Specification is equal to the smaller of the following two values.

$$T = 0.60F_y A_g \quad \text{or} \quad T = 0.50F_u A_e$$

The preceding values are not applicable to threaded steel rods nor to members with pin holes (as in eyebars). For members with pin holes the allowable tensile stress permitted by the AISC is equal to $0.45F_y$ and is applied to the calculated net area, A_n, of the section. Tension members made with threaded steel rods and their allowable stresses are discussed in Section 3-8 of this chapter.

Sections 3-3 through 3-5 of this chapter are devoted to the calculation of net areas and effective net areas as defined by the AISC. Sections 3-6 through 3-9 are concerned with the design of various types of tension members.

The bridge specifications allow somewhat smaller allowable tensile stresses than does the AISC. For instance the 1977 AASHTO Specifications permit an allowable tensile stress of $0.55F_y$ for members without holes and the smaller of $0.55F_y$ or $0.46F_u$ applied to the calculated net areas for members with bolt or rivet holes. The 1969 AREA Specifications permit 20,000 psi on net sections for A36 steel members and $0.55F_y$ for higher strength steel members.

It is not likely that stress fluctuations will be a problem in the average building frame because the changes in load in such structures usually occur only occasionally and produce relatively minor stress variations. Full design wind or earthquake loads occur so infrequently that they are not considered in fatigue design. Should there, however, be frequent variations or even reversals in stress the matter of fatigue must be considered. This subject is presented in Section 3-10 of this chapter.

3-3.   NET AREAS

The presence of a hole obviously increases the unit stress in a tension member even if the hole is occupied by a rivet or bolt. (When high-strength bolts are used there may be some disagreement with this statement under certain conditions.) There is less area of steel to which the load can be distributed and there will be some concentration of stress along the edges of the hole.

Tension is assumed to be uniformly distributed over the net section of a tension member, although photoelastic studies show there is a decided increase in stress intensity around the edges of holes, sometimes equaling several times the stresses if the holes were not present. For ductile materials, however, a uniform stress distribution assumption is reasonable when the material is loaded beyond its yield point. Should the fibers around the holes be stressed to their yield point they will yield without further stress increase, with the result that there is a redistribution or balancing of stresses. At ultimate load it is reasonable to assume a uniform stress distribution. The subject of plastic theory will be discussed at length in later chapters. The importance of ductility on the strength of riveted or bolted tension members has been clearly demonstrated in tests. Tension members (with rivet or bolt holes) made from ductile steels have proved to be as much as one-fifth to one-sixth stronger than similar members made from brittle steels with the same ultimate strengths.

It is possible by a laborious mathematical process to consider stress concentrations occurring around holes and at other sudden changes in the dimensions of the cross section. The resulting stresses are quite approximate and it is doubtful if they are appreciably more valuable than those obtained by dividing the total load by the cross-sectional area after any holes are subtracted.

This discussion is applicable only for tension members subjected to relatively static loading. Should tension members be designed for structures subjected to fatigue-type loadings considerable effort should be made to minimize the items causing stress concentrations such as points of sudden change of cross section, sharp corners, etc. In addition as mentioned in Section 3-2 allowable stresses might have to be reduced.

The term "net cross-sectional area" or simply "net area" refers to the gross cross-sectional area of a member minus any holes, notches, or other indentations. In considering the area of such items as these it is important to realize that it is usually necessary to subtract an area a little larger than the actual hole. For instance, in fabricating structural steel which is to be connected with rivets or bolts the holes are usually punched $\frac{1}{16}$ in. larger than the diameter of the rivet or bolt. Furthermore, the punching of the hole is assumed to damage or even destroy $\frac{1}{16}$ in. (1.6 mm) more of the surrounding metal; therefore, the area of the holes subtracted is $\frac{1}{8}$ in. (3 mm) larger than the diameter of the rivet or bolt. The area of the holes subtracted is rectangular and equals the diameter of the hole times the thickness of the metal.

For steel much thicker than the rivet or bolt diameters it is difficult to punch out the holes to the full sizes required without excessive deformation of the surrounding material. These holes may be subpunched (with diameters $\frac{3}{16}$ in. undersized) and then reamed out to full size after the pieces are assembled. Very little material is damaged by this quite expensive process as the holes are even and smooth and it is considered unnecessary to subtract the $\frac{1}{16}$ in. for damage to the sides. Sometimes when very thick pieces are being connected the holes may be drilled to the diameter of the bolts or rivets plus $\frac{1}{32}$ in. This is a very expensive process and should be avoided if possible.

It may be necessary to have an even greater latitude in meeting dimensional tolerances during erection and for high-strength bolts larger than $\frac{5}{8}$ in. in diameter, holes larger than the standard ones may be used without reducing the performance of the connections. These oversized holes can be short-slotted or long-slotted as described in Chapter 10.

Example 3-1 illustrates the calculations necessary for determining the net area of a plate-type of tension member.

Example 3-1

Determine the net area of the $\frac{3}{8} \times 8$-in. plate shown in Fig. 3-2. The plate is connected at its end with two lines of $\frac{3}{4}$-in. bolts.

SOLUTION

$$\text{net area} = A_n = \left(\tfrac{3}{8}\right)(8) - (2)\left(\tfrac{3}{4} + \tfrac{1}{8}\right)\left(\tfrac{3}{8}\right) = 2.34 \text{ in.}^2(1\ 510 \text{ mm}^2)$$

The connections of tension members should be arranged so that no eccentricity is present. (An exception to this rule is permitted by the AISC for certain welded connections as described in Chapter 11.) If this arrangement is possible the stress is assumed to be spread uniformly across the net section of a member. Should the connections have eccentricities, moments

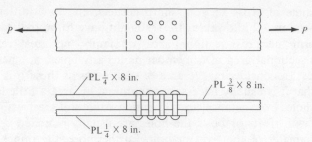

Figure 3-2

will be produced which will cause additional stresses in the vicinity of the connection. Unfortunately it is often quite difficult to arrange connections without eccentricity. Although specifications cover some situations, the designer may have to give consideration to eccentricities in some cases with estimates of her own.

The lines of action of truss members meeting at a joint are assumed to coincide. Should they not coincide, eccentricity is present and secondary stresses are the result. The centers of gravity of truss members are assumed to coincide with the lines of action of their respective forces. No problem is present in a symmetrical member as its center of gravity is at its center line, but for unsymmetrical members the problem is a little more difficult. For these members the center line is not the center of gravity, but the usual practice is to arrange the members at a joint so their gage lines coincide. If a member has more than one gage line, the one closest to the actual center of gravity of the member is used in detailing. Figure 3-3 shows a truss joint in which the gage lines of all the members pass through the same point.

3-4. EFFECT OF STAGGERED HOLES

Should there be more than one row of bolt or rivet holes in a member it is usually desirable to stagger them in order to provide as large a net area as possible at any one section to resist the load. In the preceding paragraphs

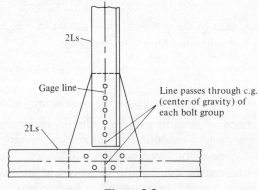

Figure 3-3

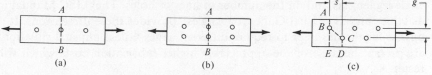

Figure 3-4

tensile members have been assumed to fail transversely as along line AB in either Fig. 3-4(a) or 3-4(b). Figure 3-4(c) shows a member in which a failure other than a transverse one is possible. The holes are staggered, and a failure along section $ABCD$ is possible unless the holes are a large distance apart.

To determine the critical net area in Fig. 3-4(c), it might seem logical to compute the area of a section transverse to the member (as AE) less the area of one hole and then the area along section $ABCD$ less two holes. The smallest value obtained along these sections would be the critical value, but this method is at fault. Along the diagonal line from B to C there is a combination of direct stress and shear and a somewhat smaller area should be used. The strength of the member along section $ABCD$ is obviously somewhere between the strength obtained by using a net area computed by subtracting one hole from the transverse cross-sectional area and the value obtained by subtracting two holes from section $ABCD$.

Tests on joints show that little if anything is gained by using complicated and theoretical formulas to consider the staggered-hole situation, and the problem is usually handled with an empirical equation. The AISC and other specifications use a very simple method for computing the net width of a tension member along a zigzag section.[1] The method is to take the gross width of the member regardless of the line along which failure might occur, subtract the diameter of the holes along the zigzag section being considered, and add for each inclined line the quantity given by the expression $s^2/4g$.

In this expression s is the longitudinal spacing (or pitch) of any two holes and g is the transverse spacing (or gage) of the same holes. The values of s and g are shown in Fig. 3-4(c). There may be several paths any one of which may be critical at a particular joint. Each possibility should be considered and the one giving the least value should be used. The smallest net width obtained is multiplied by the plate thickness to give the net area, A_n. Example 3-2 illustrates the method of computing the critical net area of a section which has three lines of bolts. (For angles, the gage for holes in opposite legs is considered to be the sum of the gages from the back of the angle minus the thickness of the angle.)

Holes for bolts or rivets are normally punched in steel angles at certain standard locations. These locations or gages are dependent on the

[1]V. H. Cochrane, "Rules for Riveted Hole Deductions in Tension Members," *Engineering News-Record*, (New York, November 16, 1922), pp. 847–848.

angle leg lengths and on the number of lines of holes. The AISC Manual in its table entitled "Usual Gages for Angles" provides these dimensions. It is unwise for the designer to require different gages unless unusual situations are present because of the appreciably higher fabrication costs which will result.

Example 3-2

Determine the critical net area of the $\frac{1}{2}$-in. thick plate shown in Fig. 3-5 using the AISC Specification (Section 1.14.2). The holes are punched for $\frac{3}{4}$-in. bolts.

SOLUTION

The critical section could possibly be $ABCD$, $ABCEF$, or $ABEF$. Hole diameters to be subtracted are $\frac{3}{4} + \frac{1}{8} = \frac{7}{8}$ in. The net widths for each case are as follows:

$$ABCD = 11 - (2)\left(\frac{7}{8}\right) = 9.25 \text{ in.}$$

$$ABCEF = 11 - (3)\left(\frac{7}{8}\right) + \frac{(3)^2}{(4)(3)} = 9.125 \text{ in.} \qquad \text{(controls)}$$

$$ABEF = 11 - (2)\left(\frac{7}{8}\right) + \frac{(3)^2}{(4)(6)} = 9.625 \text{ in.}$$

$$A_n = (9.125)\left(\tfrac{1}{2}\right) = 4.56 \text{ in.}^2 \qquad\qquad\qquad Ans.$$

The problem of determining the minimum pitch of staggered bolts such that no more than a certain number of holes need be subtracted to determine the net section is handled in Example 3-3. The Steel Handbook has a chart entitled "Net Section of Tension Members," which can be used to determine the values of $s^2/4g$. This chart can also be used to handle the type of problem solved in Example 3-3.

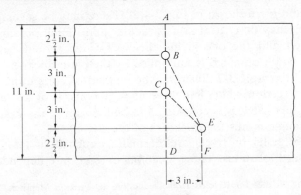

Figure 3-5

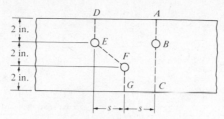

Figure 3-6

Example 3-3

For the two lines of bolt holes shown in Fig. 3-6 determine the pitch which will give a net area *DEFG* equal to the one along *ABC*. The problem may also be stated as follows: determine the pitch that will give a net area equal to the gross area less one bolt hole. The holes are punched for $\frac{3}{4}$-in. bolts.

SOLUTION

$$ABC = 6 - (1)\left(\frac{7}{8}\right) = 5.125 \text{ in.}$$

$$DEFG = 6 - (2)\left(\frac{7}{8}\right) + \frac{s^2}{(4)(2)} = 4.25 + \frac{s^2}{8}$$

$$ABC = DEFG$$

$$5.125 = 4.25 + \frac{s^2}{8}$$

$$s = 2.65 \text{ in.} \qquad \qquad Ans.$$

The $s^2/4g$ rule is merely an approximation or simplification of the complex stress variations which occur in members with staggered arrangements of bolts or rivets. Steel specifications can only provide minimum standards and designers will have to logically apply such information to complicated situations which the specifications could not cover in their attempts at brevity and simplicity. The next few paragraphs present a discussion and numerical examples of the $s^2/4g$ rule applied to situations not specifically addressed in the AISC Specification.

The AISC Specification does not include a method to be used for determining the net widths of sections other than plates and angles. For channels, W sections, S sections, and others the web and flange thicknesses are not the same. As a result it is necessary to work with net areas rather than net widths. If the holes are placed in straight lines across such a member the net area can simply be obtained by subtracting the cross-sectional areas of the holes from the gross area of the member. If the holes are staggered it is necessary to multiply the $s^2/4g$ values by the applicable thickness to change it to an area. Such a procedure is illustrated for a W section in Example 3-4 where bolts pass through the web only.

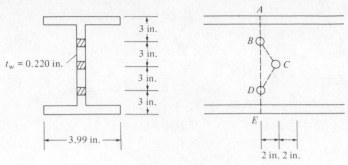

Figure 3-7

Example 3-4

Determine the net area of the W12×16 (A=4.71 in.²) shown in Fig. 3-7 assuming the holes are for 1-in. bolts. Use AISC Specification.

SOLUTION
Net areas

$$ABDE = 4.71 - (2)\left(1\tfrac{1}{8}\right)(0.220) = 4.21 \text{ in.}^2$$

$$ABCDE = 4.71 - (3)\left(1\tfrac{1}{8}\right)(0.220) + (2)\frac{(2)^2}{(4)(3)}(0.220) = 4.11 \text{ in.}^2$$

If the zigzag line goes from a web hole to a flange hole the thickness changes at the junction of the flange and web. In Example 3-5 the author has computed the net area of a channel that has bolt holes staggered in its flanges and web. The channel is assumed to be flattened out into a single plate as shown in parts (b) and (c) of Fig. 3-8. The net area along route $ABCDEF$ is determined by taking the area of the channel minus the area of the holes along the route in the flanges and web plus the $s^2/4g$ values

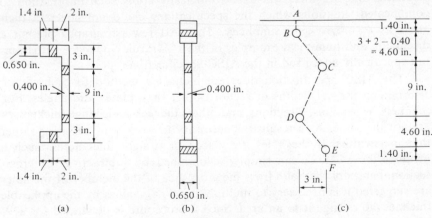

Figure 3-8

for each zigzag line times the appropriate thickness. For line CD, $s^2/4g$ has been multiplied by the thickness of the web. *For lines BC and DE (which run from holes in the web to holes in the flange) an approximate procedure has been used in which the $s^2/4g$ values have been multiplied by the average of the web and flange thicknesses.*

Example 3-5

Determine the net area along route $ABCDEF$ for the C15×33.9 ($A = 9.96$ in.^{2}) shown in Fig. 3-8. Holes are for $\frac{3}{4}$-in. bolts.

SOLUTION

approximate net A along $ABCDEF = 9.96 - (2)\left(\frac{7}{8}\right)(0.650) - (2)\left(\frac{7}{8}\right)(0.400)$

$$+ \frac{(3)^2}{(4)(9)}(0.400) + (2)\frac{(3)^2}{(4)(4.60)}$$

$$\times \left(\frac{0.650 + 0.400}{2}\right)$$

$$= 8.736 \text{ in.}^2$$

3-5. EFFECTIVE NET AREAS

When a member other than a flat plate or bar is loaded in axial tension until failure occurs across its net section its actual tensile failure stress will probably be less than the coupon tensile strength of the steel. *This will normally be the case unless all the various elements which make up the section are connected so stress is transferred uniformly across the section.* The reason for the reduced strength of the member is the concentration of shear stress, called *shear lag,* in the vicinity of the connection. In such a situation the flow of tensile stress between the full member cross section and the smaller connected cross section is not 100% effective. As a result the AISC (1.14.2.2) specifies that the effective net area, A_e, of such a member is to be determined by multiplying its net area, A_n, by a reduction coefficient, C_t, as follows.

$$A_e = C_t A_n$$

Unless all of the segments of the profile of sections other than flat plates or bars are connected so as to provide a uniformly distributed stress transfer the values of C_t shown in Table 3-1 are to be used in computation (unless larger values can be justified on the basis of tests or recognized theory). It will be noted from the values in the table that as the number of fasteners in a line is increased the shear lag decreases.

For short bolted or riveted tension members consisting of gusset or splice plates or other connection fittings where the elements effectively fall

Table 3-1   Effective Net Area, A_n (adapted from AISC-1.14.2.2 and 1.14.2.3)

Types of members	Minimum number of fasteners per line	Special requirements	Effective net area A_e
(a) Full length tension members having all cross-sectional elements connected to transmit the tensile force	1	None	A_n
(b) Short tension member fittings, such as splice plates, gusset plates, or beam-to-column fittings	1	None	A_n but not exceeding $0.85A_g$
(c) W, M, or S rolled shapes	3	$\dfrac{\text{flange width}}{\text{section depth}} \geqslant \frac{2}{3}$ Connection is to flange or flanges	$0.90A_n$
(d) Structural tees cut from sections meeting requirements of (c) above	3	Connection is to flange or flanges	$0.90A_n$
(e) W, M, or S shapes not meeting the conditions of (c), and other shapes, including built-up sections, having unconnected segments not in the plane of the loading	3	None	$0.85A_n$
(f) All shapes in (c), (d), or (e)	2	None	$0.75A_n$

SOURCE: C. G. Salmon and J. E. Johnson, *Steel Structures Design and Behavior* (New York: Harper & Row, 1980), p.74.

in a single plane the AISC Specification states that the effective net area is to be taken as the actual net area, A_n, but may not be more than 85% of the gross area of the section. Several decades of research have shown that such joints seldom have an efficiency greater than 85% even if the holes represent a very small percentage of the gross area.

Example 3-6 illustrates the calculations necessary for determining the effective net area of a W section which is connected only to its flanges. In addition the allowable total tensile load that the member can support is determined.

Example 3-6

Determine the allowable tensile load a W10×45 with two lines of $\frac{3}{4}$ in. bolts in each flange can support using A36 steel and the AISC Specification. There are assumed to be at least three bolts in each line and the bolts are not staggered with respect to each other.

<u>Using</u> W10×45 ($A_g = 13.3$ in.^{2}, $d = 10.10$ in., $b_f = 8.02$ in., $t_f = 0.620$ in.)

(a) $T = 0.60F_y A_g = (0.60)(36)(13.3) = 287.3$ k←
(b) $A_n = 13.3 - (4)(\frac{7}{8})(0.620) = 11.13$ in.^{2}
  $C_t = 0.90$ since $b_f > \frac{2}{3}d$
  $A_e = C_t A_n = (0.90)(11.13) = 10.02$ in.^{2}
  $T = 0.50F_u A_e = (0.50)(58)(10.02) = 290.6$ k

Allowable $T = 287.3$ k (1 278 kN)

The AASHTO specification requirements for computing the effective areas of bolted or riveted tension members are somewhat different from the AISC. For single angle, double angle, or structural tee members which are connected back to back on the same side of a gusset plate only the net area of the connected legs plus one-half of the areas of the unconnected legs may be considered as being effective in resisting tension. Section 1.7.8 of the 1977 AASHTO Specifications gives rules for computing effective tensile areas for other connection arrangements.

3-6. DESIGN OF TENSION MEMBERS

The preceding paragraphs have shown that the determination of the required net area for tension members is not difficult; however, it will be seen that the selection of the exact proportions of the members is not as simple. Although the designer has considerable freedom in the selection, the resulting members should have the following properties; (a) compactness, (b) dimensions that fit into the structure with reasonable relation to the dimensions of the other members, and (c) connections to as many parts of the sections as possible to minimize shear lag.

The choice of member type is often affected by the type of connections used for the structure. Some steel sections are not very convenient to bolt or rivet together with the required gusset plates, while the same sections may be welded together with little difficulty. Tension members consisting of angles, channels, W or S sections will probably be used when the connections are made with bolts or rivets, while plates, channels, and structural tees might be used for welded structures.

Various types of sections are selected for tension members in the examples to follow and in each case where bolts or rivets are used some allowance is made for holes. Should the connections be made *entirely* by welding, no holes have to be added to the net areas to give the required gross areas. *The student should realize, however, that very often welded members may have holes punched in them for temporary bolting during field erection before the permanent field welds are made. These holes need to be considered in design.*

The slenderness ratio of a member is the ratio of its unsupported length to its least radius of gyration. Steel specifications give upper permissible values of slenderness ratios for both tension and compression members. The purpose of such limitations for tension members is to ensure the use of sections with sufficient stiffness to prevent undesirable lateral deflections or vibrations. Although tension members are not subject to buckling under normal loads stress reversal may occur during shipping and erection and perhaps due to wind or other loads. The specifications require that slenderness ratios be kept below certain maximum values in order that some minimum compressive strengths be provided in the members. For tension members other than rods the AISC *recommends* maximum permissible slenderness ratios of 240 for main members and 300 for bracing and other secondary members. The AASHTO values are 200 and 240, respectively, and are *mandatory*. The *mandatory* maximum for all tension members is 200 in the AREA.

Example 3-7 illustrates the design of a tension member with a W section while Example 3-8 illustrates the selection of a single angle tension member. In both cases the AISC Specification is used from which the allowable tensile load T is the least of (a) $0.60F_y A_g$ or (b) $0.50F_u A_e$.

(a) To satisfy the first of these expressions the minimum gross area must be at least equal to the following.

$$\min A_g = \frac{T}{0.60F_y} \tag{1}$$

(b) To satisfy the second expression the minimum value of A_e must be at least

$$\min A_e = \frac{T}{0.50F_u}$$

And since $A_e = C_t A_n$ the minimum value of A_n is

$$\min A_n = \frac{\min A_e}{C_t} = \frac{T}{0.50F_u C_t}$$

Then the minimum A_g for the second expression must at least equal the minimum value of A_n plus the estimated hole areas.

$$\min A_g = \frac{T}{0.50F_u C_t} + \text{estimated hole areas} \tag{2}$$

The designer can substitute into Equations 1 and 2 taking the larger value of A_g so obtained for her initial size estimate. It is also well to notice that for main members the maximum preferable slenderness ratio l/r by the AISC is 240. From this value it is possible to compute the least permissible value of r for a particular design or that is the value of r for which the slenderness ratio is 240. It is undesirable to consider a section whose least r is less than this value because its slenderness ratio would exceed the preferable value.

$$\min r = \frac{l}{240}$$

Example 3-7

Select a W12 section to resist a 390-k tensile load using A36 steel and the AISC Specification. The member is to be 30 ft long and is to be connected through its flanges only with at least three $\frac{7}{8}$-in. bolts in each line. Assume that there can be as many as four bolt holes at any one cross section (two in each flange).

SOLUTION

(a) $\min A_g = \dfrac{T}{0.60F_y} = \dfrac{390}{(0.60)(36)} = 18.06$ in.^{2} ←

(b) $C_t = 0.90$ from Table 3-1

Assume flanges are about $\frac{5}{8}$ in. thick after some study of the W12 sections in the AISC Manual.

$$\min A_g = \frac{T}{0.50F_u C_t} + \text{estimated hole areas}$$

$$= \frac{390}{(0.50)(58)(0.90)} + (4)(1.0)\left(\frac{5}{8}\right) = 17.44 \text{ in.}^2$$

(c) $\min r = \dfrac{l}{240} = \dfrac{(12 \times 30)}{240} = 1.50$

Try W12×65 $\left(A = 19.1 \text{ in.}^2, t_f = 0.605 \text{ in.}, r_y = 3.02 \text{ in.} \right)$

$T = (0.60)(36)(19.1) = 412.6$ k > 390 k   OK

$A_e = C_t A_n = 0.90[19.1 - (4)(1.0)(0.605)] = 15.01$ in.^{2}

$T = (0.50)(58)(15.01) = 435.3$ k > 390 k   OK

$\dfrac{l}{r} = \dfrac{(12 \times 30)}{3.02} = 119 < 240$ OK

Use W12×65

Example 3-8

Design a 9-ft single-angle tension member to support a total tensile load of 65.8 k. The member is to be connected to one leg only with $\frac{7}{8}$-in. bolts (at least three in a line). Assume that only one bolt is to be located at any one cross section. Use the AISC Specification and A36 steel.

Discussion

There are many different angles listed in the AISC Manual which will support the load for the conditions described. As a result it may seem rather difficult to arrive at the absolutely lightest satisfactory section. It is possible, however, to set up a table with which the various possible sections may be methodically considered for various angle thicknesses. In the table presented with this solution the author has listed various possible thicknesses of angles which will provide sufficient areas. He has then found the lightest equal leg angle for each thickness and then repeated the process for angles with unequal leg lengths. Finally a glance at the completed table reveals the lightest angle or the one with the smallest cross-sectional area. A similar process could be followed for designing tension members from other steel sections.

SOLUTION

This problem is one which can be conveniently handled with the use of a table and such a procedure is followed here for a few angle sizes. The limiting values are calculated as follows.

(a) $\min A_g = \dfrac{T}{0.60F_y} = \dfrac{65.8}{(0.6)(36)} = 3.05$ in.^{2}

(b) From Table 3-1 $C_t = 0.85$

$$\min A_n = \dfrac{T}{0.50F_u C_t} = \dfrac{65.8}{(0.50)(58)(0.85)} = 2.67 \text{ in.}^2$$

(c) $\min r = \dfrac{l}{240} = \dfrac{(12 \times 9)}{240} = 0.45$

The lightest section is $L5 \times 3\frac{1}{2} \times \frac{3}{8}$ from the table on page 53 which has a least $r = 0.762 > 0.45$ minimum.

Use $\underline{\underline{L5 \times 3\frac{1}{2} \times \frac{3}{8}}}$

Example 3-9 illustrates the design of a single-angle tension member connected by one leg using the AASHTO Specifications.

Angle t (in.)	Area of one 1-in. bolt hole (in.^{2})	Gross area reqd. = larger of $T/0.60F_y$ or $T/0.50F_uC_t$ + hole area	Lightest angles available (equal and unequal legs) and their areas (in.^{2})
$\frac{5}{16}$	0.312	3.05	$6\times6\times\frac{5}{16}(3.65)$*
$\frac{3}{8}$	0.375	3.05	$5\times5\times\frac{3}{8}(3.61)$ $5\times3\frac{1}{2}\times\frac{3}{8}(3.05)\leftarrow$
$\frac{7}{16}$	0.438	3.11	$4\times4\times\frac{7}{16}(3.31)$* $5\times3\times\frac{7}{16}(3.31)$*
$\frac{1}{2}$	0.500	3.17	$3\frac{1}{2}\times3\frac{1}{2}\times\frac{1}{2}(3.25)$* $4\times3\times\frac{1}{2}(3.25)$

*These angles may not be readily available from material suppliers.

Example 3-9

Select a single-angle tension member to resist a total force of 51,000 lb using the AASHTO Specifications and an allowable stress of 20,000 psi. One leg of the angle is to be connected with $\frac{7}{8}$-in. bolts and it is assumed to be possible for two holes to occur at one section. The length of the member is 12 ft.

SOLUTION

$$\text{net area required} = \frac{51,000}{20,000} = 2.55 \text{ in.}^2$$

Try $L7\times4\times\frac{3}{8}$

$$\text{gross area} = 3.98 \text{ in.}^2$$
$$\text{less hole area} = -(2)(1)(\tfrac{3}{8}) = -0.75 \text{ in.}^2$$
$$\text{less 50\% area of uncon. leg} =$$
$$-(0.50)(4-\tfrac{3}{8})(\tfrac{3}{8}) = -0.68 \text{ in.}^2$$
$$\text{actual net area} = \overline{\quad 2.55 \text{ in.}^2 = \text{reqd. area.}}$$

Example 3-10 presents the review of a tension member built up from two channels. Included in the problem is the design of the necessary tie plates.

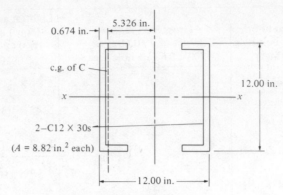

Figure 3-9

Example 3-10

Two C12×30s 12 in. back-to-back (see Fig. 3-9) have been selected to support a 280,000-lb tensile load. The member is 30 ft long, consists of A36 steel and has one $\frac{7}{8}$-in. bolt through each channel flange. Using the AISC Specification and assuming $C_t = 0.85$ determine if the member is satisfactory and design the necessary tie plates. Assume centers of bolt holes are $1\frac{3}{4}$ in. from back of channels.

SOLUTION

Using C12×30s $\left(A = 8.82 \text{ in.}^2 \text{ each, } t_f = 0.501 \text{ in.} \right)$

(a) $T = 0.60 \, F_y A_g = (0.60)(36)(2 \times 8.82) = 381.0 \text{ k} > 280 \text{ k}$ OK

(b) $C_t = 0.85$

$A_n = (2)(8.82) - (4)(1.00)(0.501) = 15.64 \text{ in.}^2$

$T = 0.50 F_u A_e = (0.50)(58)(0.85)(15.64) = 385.5 \text{ k} > 280 \text{ k}$ OK

Slenderness ratio:

$$I_x = (2)(162) = 324 \text{ in.}^4$$

$$I_y = (2)(5.14) + (2)(8.82)(5.33)^2 = 511 \text{ in.}^4$$

$$r_x = \sqrt{\frac{324}{17.64}} = 4.29 \text{ in.}$$

$$\frac{l}{r} = \frac{(12 \times 30)}{4.29} = 83.9 < 240 \qquad\qquad \text{OK}$$

Design of tie plates (AISC Sec. 1.18.3.2):

distance between lines of bolts $= 12.00 - (2)\left(1\frac{3}{4}\right) = 8.50 \text{ in.}$

minimum length of tie plates $= \left(\frac{2}{3}\right)(8.50) = 5.67 \text{ in.}$ $\left(\underline{\text{say 6 in.}} \right)$

minimum thickness of tie plates $= \left(\frac{1}{50}\right)(8.50) = 0.17 \text{ in.}$ $\left(\text{say } \frac{3}{16} \text{ in.} \right)$

minimum width of tie plates $=8.50+(2)\left(1\frac{1}{2}\right)=11.5$ in.
(Section 1.16.5)

Maximum permissible spacing of tie plates:

$$\text{least } r \text{ of one } C = 0.763$$

$$\text{max permissible} \frac{l}{r} = 240$$

$$\frac{12 \times l}{0.763} = 240$$

$$l = 15.3 \text{ ft}$$

Use tie PLs $\frac{3}{16} \times 6 \times 11\frac{1}{2}$, maximum spacing 15 ft

3-7.   SPLICES FOR TENSION MEMBERS

A tension member splice has the purpose of replacing the member at the point where it is cut. At first glance the problem may seem to be an unusually simple one but there are several difficulties which can arise, primarily caused by eccentricities of loads.

Tension members are often more than large enough to resist the calculated maximum force. This fact causes the following question to arise: "Should the splice be designed for the maximum calculated force or for the actual strength of the member?" The AREA says that the splice must be made for the full strength of the member—that is, the full net effective area of the member must be replaced. The AASHTO says a splice should be designed for the calculated maximum force or 75% of the strength whichever is larger; while the AISC requires the design to be made for the calculated maximum force or 50% of the member strength, whichever is larger.

Eccentricity of the loads is a common problem with splices as illustrated by the single-angle splice shown in Fig. 3-10. In this type of splice

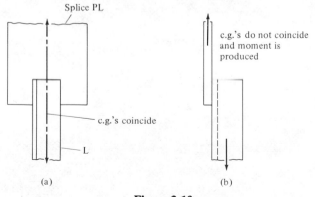

Figure 3-10

it may be possible to get the centers of gravities lined up in one direction as shown in part (a) of the figure but not in the other direction as shown in part (b) of the same figure.

Perhaps the usual practice in splice design is to neglect the effect of eccentricities, but designers need to be on the lookout for extreme cases so that they can be allowed for. Actually the single-angle splice shown in the preceding figure with one plate is one of the worst types and should be avoided. In this regard it is desirable to splice all parts of tension members. For instance, if an angle is being spliced, each leg should be spliced. As described in Section 3-5, the AASHTO permits the use of only one-half of the area of an unconnected angle leg in resisting stress. A specification such as this one was written to make a rough allowance for the effect of eccentricity.

A few possible types of tension member splices are shown in Fig. 3-11. The rather poor method of splicing a single angle with a single plate is shown in part (a) of the figure. Part (b) shows a much better method (perhaps the best one) for splicing a single angle. In this method the splice is made with another angle. The splice angle has shorter legs and must therefore have a greater thickness than the angle being spliced to provide the same net area.

Sometimes a lug angle is used as shown in part (c) of the figure. This angle will reduce the eccentric bending stresses in the member and stiffen up the whole connection slightly. Splices may be butt splices or shingle splices. These types of splices are shown in parts (d) and (e) respectively of Fig. 3-11. In the butt splice a W shape or built-up member is cut entirely at one section and splice material must be added to make the splice. Probably the shingle splice for built-up sections is a little better in that it causes a

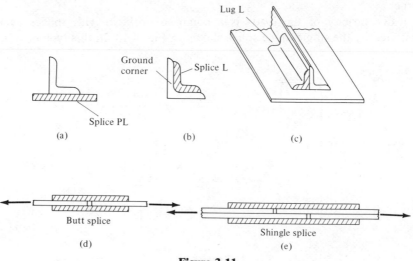

Figure 3-11

reduction of stress concentrations by staggering the splices and also keeps the entire member from being cut at one point.

Sometimes it is necessary to use what is called an "indirect" splice. An indirect splice is one in which there is an intermediate plate between the part being spliced and the splice itself. Most specifications require an increase in the number or amount of connectors (rivets, bolts, or welds) when indirect splices are used. An increase of from one-fourth to one-third the number of connectors is usually required for each intermediate plate.

Many persons do not think that gusset plates should be used as splice plates or as part of them. The truth of the matter probably is that they should not be designed as gusset plates and then have the additional task of serving as splice plates without a proportionate increase in section. In other words, gusset plates that are used for such situations need to be designed for both tasks.

Rarely is it necessary to splice steel rods, as they are usually not of sufficient length to require splicing. Should splicing be necessary, however, turnbuckles may be used. These devices are also of advantage in putting initial tension in the rods and in getting them to fit properly as previously described in this chapter.

3-8. DESIGN OF RODS AND BARS

When rods and bars are used as tension members they may be simply welded at their ends, or they may be threaded and held in place with nuts. The AISC allowable stress for threaded rods is equal to $0.33F_u$ and is to be applied to the gross area of the rod A_D computed with the major thread diameter or that is the diameter to the outer extremity of the thread. The area required for a particular tensile load T can be computed from the following expression.

$$A_D = \frac{T}{0.33F_u}$$

In the tables entitled "Threaded Fasteners Screw Threads" the AISC Manual lists properties of standard threaded rods. For a tensile load of 20 k (it cannot be less than 6 k according to Section 1.15.1 of the AISC Specification) the area and rod size required can be determined as follows.

$$A_D = \frac{20}{(0.33)(58)} = 1.045 \text{ in.}^2$$

Use $1\frac{1}{4}$-in. diameter rod with seven threads per inch $\left(A_D = 1.227 \text{ in.}^2\right)$

Sometimes upset rods are used where the ends are made larger than the regular rod and the threads are placed in the enlarged section so that the area at the root of the thread is larger than that of the regular bar. As a footnote to their Table 1.5.2.1 the AISC Specification states that the tensile capacity of the threaded part of an upset rod ($0.33F_u A_D$) based upon the cross-sectional area at the upset end's major thread diameter must be larger than the value obtained by multiplying the nominal rod area by $0.60F_y$. Upsetting permits the designer to use the entire area of the cross section; however, the use of upset bars is probably not economical and should be avoided unless a large order is being made.

A common example of the use of tension rods occurs in steel-frame industrial buildings which have purlins running between their roof trusses to support the roof surface. These types of buildings will also frequently have girts running between the columns along the vertical walls. Sag rods may be required to provide support for the purlins parallel to the roof surface and vertical support for the girts along the walls. For roofs with steeper slopes than 1 vertically to 4 horizontally, sag rods are often considered necessary to provide lateral support for the purlins, particularly where the purlins consist of steel channels. Steel channels are commonly used as purlins but they have very little resistance to lateral bending. The section modulus needed to resist bending parallel to the roof surface is small but nevertheless an extremely large channel is required to provide such a modulus. The use of sag rods for providing lateral support to purlins made from channels is usually economical because of the bending weakness of channels about their y axes. For light roofs (as where trusses support corrugated steel roofs) sag rods will probably be needed at the one-third points if the trusses are more than 20 ft on centers. Sag rods at the midpoints are sufficient if the trusses are less than 20 ft on centers. For heavier roofs such as those made of slate, cement tile, or clay tile, sag rods will probably be needed at closer intervals. The one-third points will probably be necessary if the trusses are spaced at greater intervals than 14 ft and the mid points will be satisfactory if truss spacings are less than

New Albany Bridge crossing the Ohio River between Louisville, Ky., and New Albany, Ind. (Courtesy of The Lincoln Electric Company.)

14 ft. Some designers assume that the load components parallel to the roof surface can be taken by the roof, particularly if it consists of corrugated steel sheets, and that tie rods are unnecessary. This assumption, however, is open to some doubt and definitely should not be followed if the roof is very steep.

The designer has to use her own judgment in limiting the slenderness values for rods as they will usually be several times the limiting values mentioned for other types of tension members. A common practice of many designers is to use rod diameters no less than 1/500th of their lengths to obtain some rigidity even though stress calculations may permit much smaller sizes.

It is usually desirable to limit the minimum size of sag rods to $\frac{5}{8}$ in. because smaller rods than these are often injured during construction. The threads on smaller rods are quite easily injured by overtightening which seems to be a frequent habit of workmen. Sag rods are designed for the purlins of a roof truss in Example 3-11. The rods are assumed to support the simple beam reactions for the components of the gravity loads (roofing, purlins, snow and ice) parallel to the roof surface. Wind forces are assumed to act perpendicular to the roof surfaces and theoretically will not affect the sag rod forces. The maximum force in a sag rod will occur in the top sag rod because it must support the sum of the forces in the lower sag rods. It is theoretically possible to use smaller rods for the lower sag rods but this reduction in size will probably be impractical.

Example 3-11

Design the sag rods for the purlins of the truss shown in Fig. 3-12. Purlins are to be supported at their one-third points between the trusses which are spaced 21 ft on centers. Use A36 steel and the AISC Specification and assume a minimum size rod of $\frac{5}{8}$ in. is permitted. A clay tile roof weighing 16 psf (0.77 kN/m^2) of roof surface is used and supports a snow load of 20 psf (0.96 kN/m^2) of horizontal projection of roof surface. Details of the purlins and the sag rods and their connections are shown in Figs. 3-12 and 3-13. In these figures the dotted lines represent a common practice of using ties and struts in the end panels in the plane of the roof to give greater resistance to loads located on one side of the roof (a loading situation which might occur when snow is blown off one side of the roof during a severe windstorm).

SOLUTION
Gravity loads in psf of roof surface are as follows.

$$\text{purlins} = \frac{7 \times 11.5}{37.9} = 2.1 \text{ psf}$$

$$\text{snow} = 20\frac{3}{\sqrt{10}} = 19.0 \text{ psf}$$

$$\text{tile roofing} = \frac{16.0 \text{ psf}}{37.1 \text{ psf}}$$

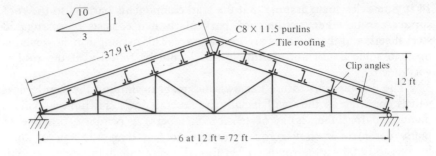

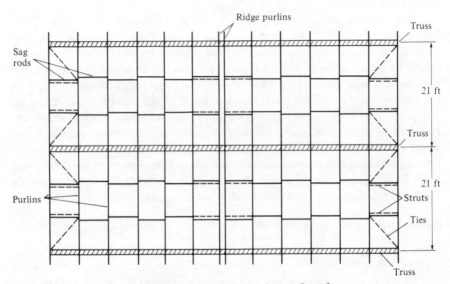

Figure 3-12   Plan of two bays of roof.

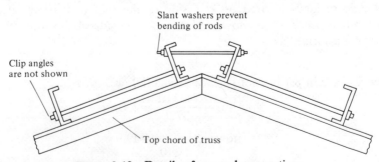

Figure 3-13   Details of sag rod connections.

Component of loads parallel to roof surface $=(1/\sqrt{10}\,)\times37.1=11.7$ psf.
Load on sag rod $=(37.9)(7)(11.7)=3110$ lb (use 6000 lb minimum).

$$A_D=\frac{6.00}{(0.33)(58)}=0.313 \text{ in.}^2$$

Use $\frac{3}{4}$-in. sag rod, 10 threads per inch $\left(A_D=0.442 \text{ in.}^2\right)$

Force in tie rod between ridge purlins (force in inclined sag rods assumed to be a component of this force):

$$T=\frac{\sqrt{10}}{3}\times3110=3270 \text{ lb (use 6000 lb minimum)}$$

$$A_D=\frac{6000}{(0.33)(58)}=0.313 \text{ in.}^2$$

Use $\frac{3}{4}$-in. rod

3-9. EYE-BAR TENSION MEMBERS

Until the early years of the twentieth century nearly all bridges in the United States were pin-connected, but today they are used infrequently because of the advantages of riveted, welded, and bolted connections. One trouble with pin-connected trusses is the wearing of the pins in the holes causing looseness of the joints. Pin-connected eye bars are used occasionally today as tension members for long-span bridge trusses and as hangers for some types of bridges. These types of bridges are subjected to very large dead loads; and the eye bars are, therefore, prevented from rattling under traffic loads. The design of pin-connected members is substantially controlled by empirical specifications which are based on considerations for stress concentrations occurring near the pin holes.

Eye bars are formed by making a rectangular bar or plate with an enlarged head and boring a hole in this head to enable it to be pin-connected. Details of the bar as to the diameter of the head, etc., are probably not worked out by the designer and selection is made from the shapes available from the steel companies because of the high cost of other sizes. The truth of the matter is that eye bars are used infrequently and the cost of any size will be rather large and they will probably have to be specially ordered.

The large trusses for which eye bars are used require large members. Each member will probably be made up of several thicknesses of eye bars each being from 1 to 2 in. thick. From a safety standpoint it is desirable for failure to occur in the members rather than at the joints and the specifications usually require the net area of the bar at the pin to be larger than the net area of the body of the bar. An illustration is the AREA Specification that requires a 40% increase in area at the pin.

3-10.   DESIGN FOR FATIGUE LOADS

It is not likely that fatigue stresses will be a problem in the average building frame because the changes in load in such structures usually occur only occasionally and produce relatively minor stress variations. Should there, however, be frequent variations or even reversals of stress the matter of fatigue must be considered. Fatigue can be a problem in buildings when crane runway girders or heavy vibrating or moving machinery or equipment are supported.

If steel members are subjected to loads that are applied and then removed many thousands of times cracks may occur and they may spread so much as to cause fatigue failures. These cracks tend to occur in places where there are stress concentrations such as where there are holes or rough or damaged edges or poor welds. Furthermore they are more likely to occur in tension members.

The AISC Specification in its Appendix B provides a simple design method for considering repeated loads. With this procedure the number of stress cycles, the expected range of stress (that is the difference between the maximum and minimum estimated stresses) and the type and location of the member are considered. Based on this information a maximum allowable stress range is given.

In detail the maximum calculated stress in a member by the AISC Specification may not exceed the basic allowable stress in that type of member nor may its maximum range of stress exceed the permissible stress range provided in Appendix B of the Specification.

If it is anticipated that there will be less than 20,000 cycles of loading no consideration needs to be given to fatigue. If the number of loading cycles exceeds 20,000 a permissible stress range is determined as follows.

1. The loading condition is determined from Table B1 of Appendix B of the AISC Specification. For instance if it is anticipated that there will be no less than 100,000 cycles of loading (that is roughly 10 applications a day for 25 years) and no more than 500,000 cycles, loading condition #2 is to be used as determined from the table.
2. The type and location of the member or detail is selected from Figure B1 of the Appendix. If a tension member consists of two angles fillet welded to a plate it should be classified as illustrative example 17 in the figure. (The term fillet weld will be defined in Chapter 11. In such a weld one member is lapped over another and they are welded together.)
3. From Table B2 the stress category A, B, C, D, E, or F is selected. For a fillet welded tension connection classified as illustrative example 17 the stress category is E.
4. Finally from Table B3 of the Appendix the allowable stress range for stress category E and loading condition #2 is $F_{sr} = 12.5$ ksi.

Example 3-12 presents the design of a two angle tension member subjected to fluctuating loads using Appendix B of the AISC Specification. Stress fluctuations and reversals are an everyday problem in the design of bridge structures. The 1977 AASHTO Specifications in their Article 1.7.2 provide allowable stress ranges determined in a manner very similar to those of the AISC Specification.

Example 3-12

An 18-ft member is to consist of a pair of equal leg angles with fillet welded end connections. The dead load tensile force is 30 k while it is estimated that the live load force may be applied 250,000 times and may vary from a compression of 12 k to a tension of 65 k. Select the angles using A36 steel and the AISC Specification.

SOLUTION

Referring to Appendix B of the AISC Specification and selecting the following values

From Table B1—loading condition 2
From Fig. B1—illustrative example 17
From Table B2—stress category E
Therefore, from Table B3—allowable stress range $F_{sr} = 12.5$ ksi

Range of loads

Tension = $+30 + 65 = +95$ k
Compression = $+30 - 12 = +18$ k
Therefore, remains in tension

Estimate section size

$$F_t = 0.60 F_y$$

$$A \text{ reqd.} = \frac{95}{(0.5)(58)} = 3.28 \text{ in.}^2$$

Try 2Ls4×4×$\frac{5}{16}$ ($A = 4.80$ in.^{2})

$$\max f_t = \frac{95}{4.80} = 19.79 \text{ ksi}$$

$$\min f_t = \frac{18}{4.80} = 3.75 \text{ ksi}$$

actual stress range = $19.79 - 3.75 = 16.04$ ksi > 12.5 ksi NG

Try 2Ls4×4×$\frac{1}{2}$ ($A = 7.50$ in.^{2}, $r = 1.22$)

$$\max f_t = \frac{95}{7.50} = 12.67 \text{ ksi}$$

$$\min f_t = \frac{18}{7.50} = 2.40 \text{ ksi}$$

actual stress range $= 12.67 - 2.40 = 10.27$ ksi < 12.5 ksi $\qquad$ OK

$$\frac{l}{r} = \frac{(12)(18)}{1.22} = 177 < 240 \qquad \text{OK}$$

Use 2Ls4$\times$4$\times\frac{1}{2}$

PROBLEMS

3-1   to 3-4. Compute the net area of each of the given members.

3-1. An L6$\times$4$\times\frac{3}{4}$ with one line of $\frac{3}{4}$-in. bolts in each leg. (*Ans.* 5.63 in.^{2})

3-2. A pair of 8$\times$6$\times\frac{3}{4}$Ls with two rows of $\frac{7}{8}$-in. bolts in the long legs and one row in the short legs.

3-3. A W27$\times$84 with two holes in each flange and two in the web all for 1-in. bolts. (*Ans.* 20.88 in.^{2})

3-4. The $\frac{3}{4}\times$12PL shown in the accompanying illustration. The holes are for $\frac{3}{4}$-in. bolts.

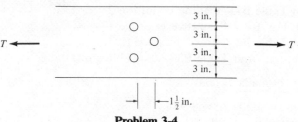

Problem 3-4

3-5. Find the allowable tensile load that can be applied to the connection plate shown in the accompanying illustration if the allowable tensile stress is 20 ksi. The plate is $\frac{7}{8}\times$16 and the holes are punched for 1-in. bolts. Not AISC Specification. (*Ans.* 213 k)

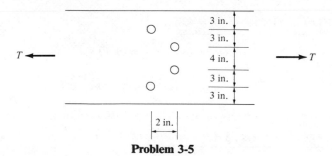

Problem 3-5

3-6. For the plate and bolts shown in the accompanying illustration find the minimum pitch s for which only two bolts need be subtracted at any one section in calculating the net area. Bolts are $\frac{7}{8}$-in.

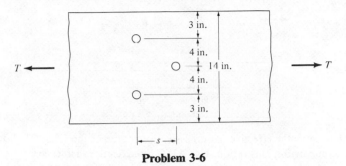

Problem 3-6

3-7. Find the minimum pitch s in the plate of Prob. 3-6 so that only two and one-half bolts need be subtracted at any one section. (*Ans.* 2.00 in.)

3-8. An L8×6×1 is used as a tension member with one gage line of $\frac{7}{8}$-in. bolts in each leg at the usual gage location. What is the minimum amount of stagger necessary so that only one bolt need be subtracted from the gross area of the angle?

3-9. An L8×4×$\frac{3}{4}$ is shown in the accompanying illustration. Two rows of $\frac{7}{8}$-in. bolts are used in the long leg and one in the short leg. Determine the minimum stagger (or pitch s in the figure) necessary so that only two holes need be subtracted in determining the net area. (*Ans.* 2.71 in.)

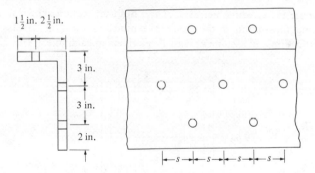

Problem 3-9

3-10. A 6×4×$\frac{1}{2}$ angle has one line of holes in each leg for $\frac{3}{4}$-in. bolts. Determine the minimum pitch so that only one and one-half holes need to be deducted to obtain the net area. (Use the usual gage for angles as given in the AISC Manual.)

3-11. Determine the effective net area of the C12×30 shown in the accompanying illustration. Use the AISC Specification and assume the holes are for 1-in. bolts. (*Ans.* 6.28 in.^{2})

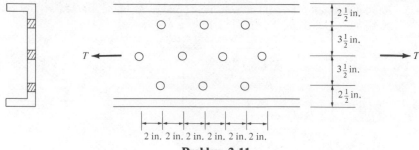

2 in. 2 in. 2 in. 2 in. 2 in. 2 in.

Problem 3-11

3-12. Determine the effective net cross-sectional area of the C12×25 shown in the accompanying illustration. Use AISC Specification. Holes are for $\frac{3}{4}$-in. bolts.

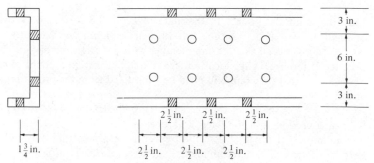

$2\frac{1}{2}$ in. $2\frac{1}{2}$ in. $2\frac{1}{2}$ in.

$1\frac{3}{4}$ in. $2\frac{1}{2}$ in. $2\frac{1}{2}$ in. $2\frac{1}{2}$ in.

Problem 3-12

3-13. Compute the effective net area of the built-up section shown in the accompanying illustration if the holes are punched for $\frac{7}{8}$-in. rivets. Use AISC Specification and assume at least three rivets in each line. (*Ans.* 22.02 in.^{2})

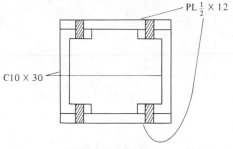

PL $\frac{1}{2}$ × 12

C10 × 30

Problem 3-13

3-14. A C10×15.3 is connected through its web with three gage lines of $\frac{7}{8}$-in. bolts. The gage lines are $2\frac{1}{2}$ in. on centers and the bolts are spaced 3 in. on centers along the gage line. If the center row of bolts is staggered with respect to the outer row determine the effective net cross-sectional area of the channel using the AISC Specification. Assume there are at least three bolts in each line.

3-15. Repeat Prob. 3-5 if A36 steel and the AISC Specification are used. (*Ans.* 302.4 k)

3-16. Using A36 steel and the AISC Specification determine the allowable total tensile load that a W10×60 can support if it has two lines of $\frac{7}{8}$-in. bolts (four in each line) in each flange.

3-17. A single-angle tension member ($7 \times 4 \times \frac{3}{8}$) has two gage lines in its long leg and one in the short leg for $\frac{3}{4}$-in. bolts arranged as shown in the accompanying illustration. Determine the allowable tensile load that the member can support if a steel with $F_y = 50$ ksi is used. A242 steel. AISC Specification. (*Ans.* 181.6 k)

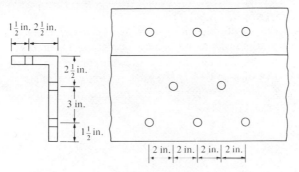

Problem 3-17

3-18. Determine the maximum permissible tensile force that can be supported by the pair of $5 \times 5 \times \frac{7}{8}$ Ls shown in the accompanying illustration if A441 steel and the AISC Specification are used. Standard gages are to be used as determined from the Steel Manual. Holes are for $\frac{3}{4}$-in. bolts.

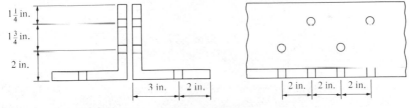

Problem 3-18

3-19. Select the lightest W12 section available to support a tensile load of 400 k using A36 steel and the AISC Specification. The member is to be 25 ft long and is assumed to have two holes for 1-in. bolts in each flange. Assume C_t is 0.90. (*Ans.* W12×65)

3-20. Repeat Prob. 3-19 using A441 steel.

3-21. A W12 shape is to be selected to support a tensile load of 280 k. The member is to be 28 ft long, have an allowable tensile stress of 20 ksi and is assumed to have two holes for $\frac{7}{8}$-in. bolts in each flange. (Maximum l/r is 200.) Not AISC. (*Ans.* W12×58)

3-22. Repeat Prob. 3-21 using A36 steel and the AISC Specification. Assume $C_t = 0.75$.

3-23. Select the lightest American Standard channel that will safely support a tensile load of 175 k. The member is 15 ft long and is assumed to have one hole for a 1-in. bolt through each flange. A36. AISC. $C_t = 0.85$. (*Ans.* C12×30)

3-24. A welded tension member is to consist of two channels placed 12 in. back-to-back with the flanges turned in. Select the channels for a maximum tensile load of 450,000 lbs using A36 steel and the AISC Specification. The member is to be 30 ft long.

3-25. The AISC Specification and A36 steel are to be used in selecting a single angle member to resist a tensile load of 42 k. The member is to be 18 ft long and is assumed to be connected with one row of $\frac{7}{8}$-in. bolts. Assume $C_t = 0.75$. (*Ans.* 1L5×5×$\frac{5}{16}$)

3-26. Repeat Prob. 3-25 selecting a pair of angles and using A36 steel. Assume angle legs are touching and assume one hole for $\frac{7}{8}$-in. bolt in each angle.

3-27. Design member $L_2 L_3$ of the truss shown in the accompanying illustration. It is to consist of a pair of angles with a $\frac{3}{8}$-in. gusset plate at each joint. Use A36 steel and AISC Specification. Assume two $\frac{3}{4}$-in. bolts in each angle. $C_t = 1.0$ (*Ans.* 2L6×3$\frac{1}{2}$×$\frac{3}{8}$)

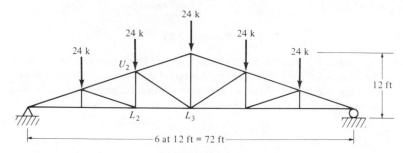

Problem 3-27

3-28. Select a single-angle tension member 12 ft long to support a load of 70 k if the allowable tension is 20 ksi and one leg of the angle is to be connected with a line of $\frac{3}{4}$-in. bolts. Use the AASHTO provision that only one-half of the unconnected leg may be considered as being effective in resisting load. Maximum permissible l/r is 200.

3-29. Select a single angle member to resist a tensile load of 120 k. The member is 15 ft long and is assumed to be connected with two rows of $\frac{7}{8}$-in. bolts. AISC Specification $F_y = 65$ ksi. $C_t = 1.0$. (*Ans.* 1L5×5×$\frac{3}{8}$)

3-30. Repeat Prob. 3-24 assuming that one $\frac{7}{8}$-in. bolt passes through each flange. Design tie plates. AISC Specification. $C_t = 1.0$.

3-31. A tension member is to consist of four equal leg angles arranged as shown in the accompanying illustration to support a tensile load of 560 k. The member is to be 30 ft long and is assumed to have two holes for $\frac{7}{8}$-in. bolts in each angle. A36 steel. AISC Specification. Design the member including the necessary tie plates. $C_t = 1.0$. (*Ans.* 1L5×5×$\frac{3}{4}$)

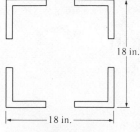

Problem 3-31

3-32. Select a threaded round rod as a hanger to resist a total tensile load of 12,000 lb. Use AISC Specification. A36 steel.

3-33. The horizontal thrust at the base of the three-hinge arch shown in the accompanying illustration is to be resisted by a tie rod of A36 steel. What size threaded round tie rod should be used if the AISC Specification is used? (*Ans.* $2\frac{1}{4}$ in. rod)

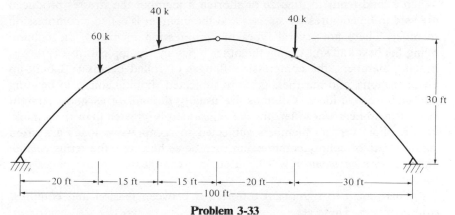

Problem 3-33

3-34. The roof trusses for a particular industrial building are spaced 21 ft on centers, have a roof covering weighing an estimated 6 psf of roof surface, and have purlins spaced as shown in the accompanying illustration and weighing an estimated 3 psf of roof surface. Design sag rods using A36 steel and the AISC Specification, assuming a snow load of 25 psf of horizontal surface. The sag rods are to be used at the one-third points.

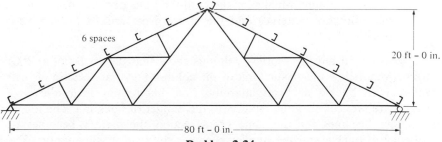

Problem 3-34

Chapter 4
Introduction
to Compression
Members

4-1. GENERAL

When a load tends to squeeze or shorten a member the stresses produced are said to be compressive stresses and the member is called a compression member. There are several types of compression members, the column being the best known. Among the other types are the top chords of trusses, bracing members, the compression flanges of rolled beams and built-up beam sections, and members that are subjected simultaneously to bending and compressive loads. Columns are usually thought of as being straight vertical members whose lengths are considerably greater than their thicknesses. Short vertical members subjected to compressive loads are often called struts or simply compression members; however, the terms *column* and *compression member* will be used interchangeably in the pages that follow.

There are two significant differences between tension and compression members. These are

1. Whereas tensile loads tend to hold a member straight compressive loads tend to bend them out of the plane of the loads (a serious situation).
2. The presence of rivet or bolt holes in tension members reduces the area available for resisting loads; but in compression members the rivets or bolts are assumed to fill the holes (although there may be some very slight initial slippage until the bolts or rivets bear against the adjoining material) and the entire gross area is available for resisting load.

Tests on all but the shortest columns show that they will fail at P/A stresses well below the elastic limit of the column material because of their tendency to buckle or bend laterally. For this reason their allowable stresses are reduced in some relation to the danger of buckling. The longer a column becomes for the same cross section the greater becomes its tendency to buckle and the smaller becomes the load it will support. The tendency of a member to buckle is usually measured by the *slenderness ratio*, which has previously been defined as the ratio of the length of the member to its least radius of gyration. The tendency to buckle is also affected by such factors as the types of end connections, eccentricity of load application, imperfection of column material, initial crookedness of columns, residual stresses from manufacture, etc.

The loads supported by a building column are applied by the column section above and by the connections of other members directly to the column. The ideal situation is for the loads to be applied uniformly across the column with the center of gravity of the loads coinciding with the center of gravity of the column. Furthermore, it is desirable for the column to have no flaws, to consist of a homogeneous material, and to be perfectly straight; but these situations are obviously impossible to achieve.

Loads that are exactly centered over a column are referred to as axial or concentric loads. The dead loads may or may not be concentrically placed over an interior building column and the live loads may never be centered. For an outside column the loading situation is probably even more eccentric as the center of gravity of the loads will usually fall well on the inner side of the column. In other words, it is doubtful that a perfect axially loaded column will ever be encountered in practice.

The other desirable situations are also impossible to achieve because of the following: imperfections of cross-sectional dimensions, residual

Guggenheim Laboratories, Princeton University, Princeton, N.J. (Courtesy of Bethlehem Steel Company.)

stresses, holes punched for rivets or bolts, erection stresses, and transverse loads. The result of all of these imperfections cannot be expressed in a formula and they truthfully require the use of high safety factors.

Slight imperfections in tension members and beams can be safely disregarded as they are of little consequence. On the other hand, slight defects in columns may be of major significance. A column that is slightly bent at the time it is put in place may have serious bending moments. Obviously a column is a more critical member in a structure than is a beam or tension member because minor imperfections in materials and dimensions mean a great deal. This fact can be illustrated by a bridge truss that has some of its members damaged by a truck. The bending of tension members probably will not be serious as the tensile loads will tend to straighten those members; but the bending of any compression members is a serious matter, as compressive loads will tend to magnify the bending in those members.

The preceding discussion should clearly show that column imperfections cause them to bend and the designer must consider stresses due to those moments as well as due to axial loads. Chapters 4 and 5 are limited to a discussion of axially loaded columns while members subjected to a combination of axial loads and bending loads are discussed in Chapter 8.

4-2. RESIDUAL STRESSES

In recent years research at Lehigh University has shown that residual stresses and their distribution are perhaps the most important factors affecting the strength of axially loaded steel columns. These stresses are of particular importance for columns with l/r values varying from approximately 40 to 120, a range that includes a very large percentage of practical columns. A major cause of residual stress is the uneven cooling of shapes after hot-rolling. For instance in a W shape the outer tips of the flanges and the middle of the web cool quickly while the areas at the intersection of the flange and web cool more slowly.

The quicker cooling parts of the sections when solidified resist further shortening while those parts that are still hot tend to shorten further as they cool. The net result is that the areas which cooled more quickly have residual compressive stresses while the slower cooling areas have residual tensile stresses. The magnitude of these stresses varies from about 10 to 15 ksi (69 to 103 MPa) although some values greater than 20 ksi (138 MPa) have been found.

When rolled-steel column sections are tested, their proportional limits are reached at P/A values of only a little more than half of their yield stresses and the stress-strain relationship is nonlinear from there up to the yield stress (see Fig. 4-1). Because of the early localized yielding occurring at some points of the column cross sections, buckling strengths are appre-

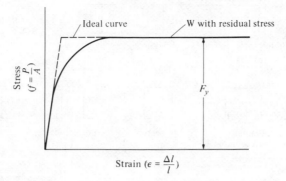

Figure 4-1   Effect of residual stresses on column stress–strain diagram.

ciably reduced. Reductions are greatest for columns with slenderness ratios varying from approximately 70 to 90 and may possibly be as high as 25%.[1]

Residual stresses may also be caused during fabrication when cambering is performed by cold bending and due to cooling after welding. Welding can produce severe residual stresses in columns which actually can approach the yield point in the vicinity of the weld. Another important fact is that columns may actually be appreciably bent by the welding process decidedly affecting their load carrying ability. Figure 4-1 shows the effect of residual stresses due to cooling and fabrication on the stress-strain diagram for a hot-rolled W shape.

4-3. SECTIONS USED FOR COLUMNS

Theoretically an endless number of shapes can be selected to safely resist a compressive load in a given structure. From a practical viewpoint, however, the number of possible solutions is severely limited by such considerations as sections available, connection problems, and type of structure in which the section is to be used. The paragraphs that follow are intended to give a brief resume of the sections which have proved to be satisfactory for certain conditions. These sections are shown in Fig. 4-2 and the letters in parentheses in the paragraphs to follow refer to the parts of this figure.

The sections used for compression members are usually similar to those used for tension members with certain exceptions. The exceptions are caused by the fact that the strengths of compression members vary in some inverse relation to the l/r ratios and stiff members are required. Individual rods, bars, and plates are usually too slender to make satisfactory compression members unless they are very short and lightly loaded.

Single-angle members (a) are satisfactory for bracing and compression members in light trusses. Equal-leg angles may be more economical than unequal-leg angles because their least r values are greater for the same area

[1]L. S. Beedle and L. Tall, "Basic Column Strength," *Proc. ASCE* **86** (July, 1960), pp. 139–173.

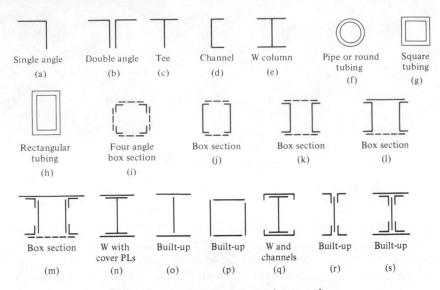

Single angle   Double angle   Tee   Channel   W column   Pipe or round tubing   Square tubing
(a)   (b)   (c)   (d)   (e)   (f)   (g)

Rectangular tubing   Four angle box section   Box section   Box section   Box section
(h)   (i)   (j)   (k)   (l)

Box section   W with cover PLs   Built-up   Built-up   W and channels   Built-up   Built-up
(m)   (n)   (o)   (p)   (q)   (r)   (s)

Figure 4-2   Types of compression members.

of steel. The top chord members of riveted or bolted roof trusses might consist of a pair of angles back-to-back (b). There will often be a space between them for the insertion of a gusset plate at the joints necessary for connections to other members. An examination of this section will show that it is probably desirable to use unequal-leg angles with the long legs back to back to give a better balance between the r values about the x and y axes.

If roof trusses are welded, gusset plates may be unnecessary and structural tees (c) might be used for the top chord members because the web members can be welded directly to the stems of the tees. Single channels (d) are not satisfactory for the average compression member because of their almost negligible r values about their web axes. They can be used if some method of providing extra lateral support in the weak direction is available. The W shapes (e) are probably the most common shapes used for building columns and for the compression members of highway bridges. Their r values although far from being equal about the two axes are much more nearly balanced than for channels.

Several famous bridges constructed during the nineteenth century (such as the Firth of Forth Bridge in Scotland and Ead's Bridge in St. Louis) made extensive use of tube-shaped members. Their use, however, declined due to connection problems and manufacturing costs but with the development of economical welded tubing their use is again increasing.

For small and medium loads, pipe sections or round tubing (f) are quite satisfactory. They are often used as columns in long series of windows, as short columns in warehouses, as columns for the roofs of covered walkways, in the basements and garages of residences, etc. Pipe

Bents fabricated from tubing sections in New Jersey (Courtesy of Bethlehem Steel Company.)

columns have the advantage of being equally rigid in all directions and are usually very economical unless moments are large. The Steel Handbook furnishes the sizes of these sections and classifies them as being standard, extra strong, and double extra strong.

Square and rectangular tubing (g) and (h) have not been used to a great extent for columns until recently. In fact, for many years only a few steel mills manufactured steel tubing for structural uses. Perhaps the major reason why tubing was not used to a great extent was the difficulty of making connections with rivets or bolts. This problem has been fairly well eliminated, however, by the advent of modern welding. The use of tubing for structural purposes by architects and engineers in the years to come will probably be greatly increased for several reasons. These include

1. The most efficient compression member is one that has a constant radius of gyration about its centroid, a property available in round tubing. Square tubing is the next most efficient compression member.
2. Their smooth surfaces permit easier painting.
3. They have excellent torsional resistance.
4. The surfaces of tubing are quite attractive.
5. When exposed the wind resistance of round tubing is only about two-thirds of that of flat surfaces of the same width.

A slight disadvantage that comes into play in certain cases is that the ends of tubing may have to be sealed to protect their inaccessible inside surfaces from corrosion. Although making very attractive exposed members for beams, tubing is at a definite weight disadvantage as compared to the regular rolled beam shapes. Their maximum section modulus is considerably less than for a rolled shape of the same weight per foot.

Where compression members are designed for very large structures it may be necessary to use built-up sections. Built-up sections are needed where the members are long and support very heavy loads and/or when there are connection advantages. Generally speaking, a single shape such as a W section is more economical than a built-up section having the same cross-sectional area. With heavy column loads high strength steels can frequently be used with very economical results if their increased strength permits the use of W sections rather than built-up members.

When built-up sections are used they must be connected on their open sides with some type of lacing (also called lattice bars) to hold the parts together in their proper positions and to assist them in acting together as a unit. The ends of these members are connected with tie plates (also called batten plates or stay plates).

The dotted lines in Fig. 4-2 represent lacing or discontinuous parts and the solid lines represent parts that are continuous for the full length of the members. Four angles are sometimes arranged as shown in (i) to produce large r values. This type of member may often be seen in towers and in crane booms. A pair of channels (j) is sometimes used as a building column or as a web member in a large truss. It will be noted that there is a certain spacing for each pair of channels at which their r values about the x and y axes are equal. Sometimes the channels may be turned out as shown in (k).

A section well suited for the top chords of bridge trusses is a pair of channels with a cover plate on top (l) and with lacing on the bottom. The gusset plates at joints are conveniently connected to the insides of the channels and may also be used as splices. When the largest channels available will not produce a top chord member of sufficient strength a built-up section of the type shown in (m) may be used.

When the rolled shapes do not have sufficient strength to resist the column loads in a building or the loads in a very large bridge truss, their areas may be increased by adding plates to the flange (n). In recent years it has been found that for welded construction a built-up column of the type shown in part (o) is a more satisfactory shape than a W with welded cover plates (n). It seems that in bending (as where a beam frames into the flange of a column) that it is difficult to efficiently transfer tensile force through the cover plate to the column without pulling the plate away from the column. For very heavy column loads a welded box section of the type shown in (p) has proved to be quite satisfactory. Some other built-up sections are shown in parts (q) through (s). The built-up sections shown in parts (n) through (q) have an advantage over those shown in parts (i) through (m) in that they do not require the expense of the lattice work necessary in the former. Lateral shearing forces are negligible for the single column shapes and for the nonlatticed built-up sections, *but they are definitely not negligible for the built-up latticed columns.*

4-4. DEVELOPMENT OF COLUMN FORMULAS

The use of columns goes back before the dawn of history but it was not until 1729 that a paper was published on the subject by Pieter van Musschenbroek, a Dutch mathematician.[2] He presented an empirical column formula for estimating the strength of rectangular columns. A few years later in 1757 Leonhard Euler, a Swiss mathematician, wrote a paper of great value concerning the buckling of columns. He was probably the first person to realize the significance of buckling. The derivation of his formula, the most famous of all column expressions, which is presented in Section 4-5, marked the real beginning of theoretical and experimental investigation of columns.

Engineering literature is filled with formulas developed for ideal column conditions but these conditions are not encountered in actual practice. Consequently practical column design is based primarily on formulas that have been developed to fit with reasonable accuracy test-result curves. The reasoning behind this procedure is simply the fact that the independent derivation of column expressions does not yield formulas that give results comparing closely with test–result curves for all ranges of l/r.

The testing of columns of various ranges of l/r results in a scattered range of values such as those shown by the broad band of dots in Fig. 4-3. The dots will not fall on a smooth curve even if all of the testing is done in the same laboratory because of the difficulty of exactly centering the loads, lack of perfect uniformity of the materials, varying dimensions of the sections, end restraint variations, etc. The usual practice is to attempt to

[2] L. S. Beedle et al., *Structural Steel Design* (New York: Ronald Press, 1964), p. 269.

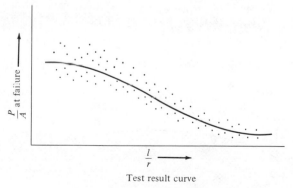

Test result curve

Figure 4-3 Text result curve.

develop formulas which give results represented by an approximate average of the test results with an adequate safety factor applied. The student should also realize that laboratory conditions are not field conditions and column tests probably give the limiting values of column strengths.

4-5.   DERIVATION OF THE EULER FORMULA

The Euler formula is derived in this section for a straight, concentrically loaded, homogeneous, long slender column with rounded ends. It is assumed that this perfect column has been laterally deflected by some means as shown in Fig. 4-4 and that if the concentric load P was removed the column would straighten out completely.

The x and y axes are located as shown in the figure. As the bending moment at any point in the column is $-Py$ the equation of the elastic curve can be written as follows.

$$EI\frac{d^2y}{dx^2} = -Py$$

For convenience in integration both sides of the equation are multiplied by $2\,dy$ and the integration is performed.

$$EI\,2\frac{dy}{dx}d\frac{dy}{dx} = -2Py\,dy$$

$$EI\left(\frac{dy}{dx}\right)^2 = -Py^2 + C_1$$

When $y = \delta$, $dy/dx = 0$, and the value of C_1 will equal $P\delta^2$ and

$$EI\left(\frac{dy}{dx}\right)^2 = -Py^2 + P\delta^2$$

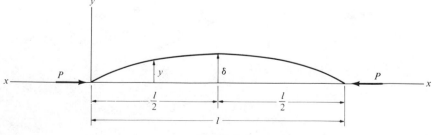

Figure 4-4

The preceding expression is arranged more conveniently as follows.

$$\left(\frac{dy}{dx}\right)^2 = \frac{P}{EI}(\delta^2 - y^2)$$

$$\frac{dy}{dx} = \sqrt{\frac{P}{EI}} \sqrt{\delta^2 - y^2}$$

$$\frac{dy}{\sqrt{\delta^2 - y^2}} = \sqrt{\frac{P}{EI}} \, dx$$

Integrating this expression, the result is

$$\arcsin\frac{y}{\delta} = \sqrt{\frac{P}{EI}} \, x + C_2$$

When $x=0$ and $y=0$, $C_2=0$. The column is bent into the shape of a sine curve expressed by the equation

$$\arcsin\frac{y}{\delta} = \sqrt{\frac{P}{EI}} \, x$$

When $x=l/2$, $y=\delta$, resulting in

$$\frac{\pi}{2} = \frac{l}{2}\sqrt{\frac{P}{EI}}$$

In this expression P is the *critical buckling load* or the maximum load which the column can support before it becomes unstable. Solving for P

$$P = \frac{\pi^2 EI}{l^2}$$

This expression is the Euler formula but is usually written in a little different form involving the slenderness ratio. Since $r=\sqrt{I/A}$ and $r^2 = I/A$ and $I = r^2A$, the Euler formula may be written as

$$\frac{P}{A} = \frac{\pi^2 E}{(l/r)^2}$$

The student should carefully note that the buckling load determined from the Euler equation is independent of the strength of the steel used. This equation is the only purely theoretical column formula of significance which has been developed to date, but is of value only when the end support conditions are carefully considered. The results obtained by application of the formula to specific examples compare very well with test results for centrally loaded, long slender columns with rounded ends. The engineer, however, does not encounter perfect columns of this type. The columns with which he works do not have rounded ends and are not free to rotate because their ends are bolted, riveted, or welded to other members. These practical columns have different amounts of restraint against rotation varying from slight restraint to almost fixed conditions. For the actual cases encountered in practice where the ends are not free to rotate, different length values can be used in the formula and more realistic values will be obtained

To successfully use the Euler equation for practical columns the value of l should be the distance between points of zero moment. This distance is referred to as the *effective length* of the column. For a pinned-end column (whose ends can rotate but cannot translate) the points of zero moment are located at the ends a distance l apart. For columns with different end conditions the effective lengths may be entirely different.

Part (a) of Fig. 4-5 shows the type of bending considered in deriving the Euler equation earlier in this section. [Actually Euler's 1757 publication presented only the so-called flagpole problem shown in Fig. 4-5(c) in which one end is built-in or fixed and the other end is free.] In part (b) of the same figure is shown a column whose ends are fixed. A fixed-end column has points of inflection (P.I.s) at its one-fourth points and the effective length or the distance between the points of inflection equals $l/2$. Obviously the smaller the effective length the greater the estimated strength obtained by substitution in the Euler expression. Although the effective length of the usual practical column falls in between these two values (l and $l/2$), it is possible for it to be greater than l. Should a column be completely free at one end and fixed at the other end its elastic curve will take the shape of the curve of a pinned-end column of twice its length and the effective length will be $2l$ as shown in part (c) of Fig. 4-5.

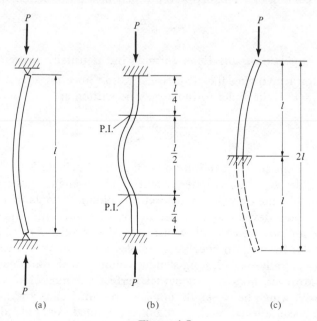

(a)   (b)   (c)

Figure 4-5

It might be interesting to note that for the column shown in part (b) of the figure the end moments actually restrain the column so as to decrease the deflection due to buckling. The result is that the effective length is less than l and the end moments may be thought of as actually strengthening the column. In columns of the type shown in part (c) the end moments actually weaken the column in that they increase the deflection due to buckling. Further discussions of effective lengths are made in Chapters 5 and 8.

Research work in this field by Thomas H. Johnson showed that when practical end conditions were used entirely different values should be used for the Euler formula.[3] He found that when column ends were practically pin connected the value of the Euler formula should be

$$\frac{P}{A} = \frac{16E}{(l/r)^2}$$

For flat or nearly fixed ends he recommended the following:

$$\frac{P}{A} = \frac{25E}{(l/r)^2}$$

It is interesting to note that these values can be obtained by using effective lengths of $0.785l$ and $0.628l$ respectively in the original formula. These revised values of the Euler formula seem to give reasonable values for long slender columns with l/r *values* varying from approximately 120 to 200. The general practice, however, is to use very few members with the high slenderness ratios for which they are applicable. In fact some specifications prohibit the use of main members with l/r values in this range although they may permit such values for secondary members. One of the present AISC column expressions which will be discussed in Chapter 5 is the Euler formula divided by a factor of safety of 1.92 or $23/12$. This formula can be written as

$$\frac{P}{A} = \frac{12\pi^2 E}{23(l/r)^2}$$

A simple application of the Euler formula is given by Example 4-1. It will be remembered that this expression was derived for stresses for which Hooke's law applies and, therefore, is not applicable to stresses above the proportional limit. This situation is encountered in Example 4-1.

Example 4-1

The 4×6 timber column with the dressed dimensions shown in Fig. 4-6 has a modulus of elasticity of 2,000,000 psi and a proportional limit of 5000 psi. If the member is assumed to be pin connected, what are the critical

---

[3] "On the Strength of Columns," *Trans. ASCE* **15**, (July, 1886), p. 522.

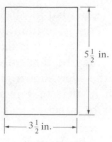

Figure 4-6

axial loads it can support for unsupported lengths of 15 ft and 5 ft according to the Euler formula?

SOLUTION

Properties of column:

$$A = (3.5)(5.5) = 19.25 \text{ in.}^2$$

$$I_x = \left(\tfrac{1}{12}\right)(3.5)(5.5)^3 = 48.53 \text{ in.}^4$$

$$I_y = \left(\tfrac{1}{12}\right)(5.5)(3.5)^3 = 19.65 \text{ in.}^4$$

$$\text{least } r = \sqrt{\frac{I}{A}} = \sqrt{\frac{19.65}{19.25}} = 1.01 \text{ in.}$$

Fifteen-foot column:

$$\frac{l}{r} = \frac{(12)(15)}{1.01} = 178.22$$

$$\frac{P}{A} = \frac{\pi^2 E}{(l/r)^2} = \frac{(\pi^2)(2,000,000)}{(178.22)^2} = 621 \text{ psi} < 5000 \text{ psi} \qquad \text{OK}$$

$$P_{\text{critical}} = (621)(19.25) = 11,954 \text{ lb}$$

Five-foot column:

$$\frac{l}{r} = \frac{12 \times 5}{1.01} = 59.41$$

$$\frac{P}{A} = \frac{(\pi^2)(2,000,000)}{(59.41)^2} = 5593 \text{ psi} > 5000 \text{ psi proportional limit}$$

Euler formula not applicable.

Example 4-2 illustrates the design of a square timber column using the Euler formula. The procedure followed is to assume a column size, compute its allowable axial load, and then to try another size and repeat the procedure until a satisfactory size is obtained. A similar process can be used to select column sections made from steel or other materials. Considerably more information is presented in Chapter 5 concerning other column equations and similar trial and error solutions.

Example 4-2

Select a square timber column 16 ft long to support an axial compression load of 40 k. $E = 1.76 \times 10^6$ psi. Proportional limit = 5000 psi. A safety factor of 3 is to be used.

SOLUTION

Try 6×6 column $\left(\text{dressed dimensions} = 5\frac{1}{2} \times 5\frac{1}{2} \text{ in.}\right)$

$$A = (5.5)(5.5) = 30.25 \text{ in.}^2$$

$$I = \left(\tfrac{1}{12}\right)(5.5)(5.5)^3 = 76.26 \text{ in.}^4$$

$$r = \sqrt{\frac{76.26}{30.25}} = 1.59 \text{ in.}$$

$$\frac{l}{r} = \frac{(12)(16)}{1.59} = 120.75$$

$$\frac{P}{A} = \frac{(\pi^2)(1.76 \times 10^6)}{(120.75)^2} = 1191 \text{ psi} < 5000 \text{ psi} \qquad \text{OK}$$

$$\text{allow } \frac{P}{A} = \frac{1191}{3} = 397 \text{ psi}$$

$$\text{allow } P = (397)(30.25) = 12{,}009 \text{ lb} < 40{,}000 \text{ lb} \qquad \text{NG}$$

Try 8×8 column $\left(\text{dressed dimensions} = 7\frac{1}{2} \times 7\frac{1}{2} \text{ in.}\right)$

$$A = (7.5)(7.5) = 56.25 \text{ in.}^2$$

$$I = \left(\tfrac{1}{12}\right)(7.5)(7.5)^3 = 263.67 \text{ in.}^4$$

$$r = \sqrt{\frac{263.67}{56.25}} = 2.165 \text{ in.}$$

$$\frac{l}{r} = \frac{(12)(16)}{2.165} = 88.68$$

$$\frac{P}{A} = \frac{(\pi^2)(1.76 \times 10^6)}{(88.68)^2} = 2209 \text{ psi} < 5000 \text{ psi} \qquad \text{OK}$$

$$\text{allow } \frac{P}{A} = \frac{2209}{3} = 736.3 \text{ psi}$$

$$\text{allow } P = (736.3)(56.25) = 41{,}417 \text{ lb} > 40{,}000 \text{ lb} \qquad \text{OK}$$

Use 8×8 column

4-6. REVISIONS TO EULER FORMULA

Engineers did not make great use of the Euler equation for a good many years after it was introduced. They felt the values obtained varied too much from test results. The derivation was based on a definite set of conditions (straight columns, concentric loads, etc.). When very careful tests are made under conditions approaching the assumed ones the results compare quite favorably as long as P/A does not exceed the proportional limit of the material. It should be noted here that tests of short columns show that they often fail at P/A stresses well above the proportional limit of the material.

Above the proportional limit the test values and the Euler values vary greatly. The reason for these variations can be understood when it is remembered that the method used to derive the Euler formula was satisfactory only for cases where stresses were proportional to strains. In other words, the modulus of elasticity is not constant outside of the elastic range. His formula with constant E is valid only for the elastic range and must be revised before it is applicable in the inelastic range. As time went by Euler's formula became fairly well accepted in the elastic range but in the inelastic range it was necessary to use certain empirical formulas (a practice still followed). Buckling at stresses above the proportional limit is referred to as *inelastic buckling*, while buckling at stresses below the proportional limit is referred to as *elastic buckling*.

In 1889 Friedrich Engesser proposed the so-called tangent-modulus theory[4] in which the moduli of elasticity for stresses above the proportional limit are determined from the slope of the stress strain curve. Engesser assumed that the column remained straight until failure and that the tangent modulus was constant for the entire cross section of a column. To plot a curve of the Euler equation above the proportional limit with the Engesser proposal it is necessary to assume certain values of P/A, find the tangent modulus for each from a stress-strain curve, and then use the Euler equation to determine the corresponding l/r values for plotting against P/A. The application of this method results in curves that come much closer to test-result curves than does the use of the regular Euler expression.

Unfortunately the application of the tangent-modulus method is rather complex. A particular difficulty with the method is that the tangent modulus has to be determined with great care because in a small distance from the proportional limit to the elastic limit the modulus changes tremendously (from 29×10^6 psi to 0). The importance of determining the tangent modulus with extreme care is seen when it is realized that a great range of practical columns fall into this portion of the curve. The tangent-modulus method is not frequently used because of the difficulties mentioned in the preceding sentences.

[4] *Zeitschrift fur Architekur und Ingenieurwesen*, 1889.

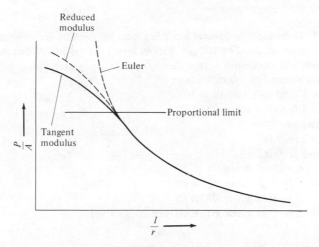

Figure 4-7

During the last 70 years appreciation of the tangent-modulus theory by the engineering profession has made a full circle from considerable respect to little respect and back to considerable respect. After its introduction many engineers claimed that the theory was poor because a very important fact was not considered in its development. The fact supposedly neglected was that the strains on one side of the column were decreasing as were the stresses on that side and that these changes were made in the range of the elastic modulus.

Based on this criticism Engesser revised his old theory and introduced the reduced-modulus or double-modulus theory in 1895. A reduced modulus somewhat greater than the tangent modulus is used and the estimated loads that a column can support are larger than those given by the original Engesser theory. For a good many years the double-modulus theory was accepted as being the correct theory of column action in the inelastic range, but in recent years considerable doubts have been voiced about the double-modulus theory. Actual test results fall in between the values given by the two theories and in fact they tend to be closer to the tangent-modulus values than to the reduced-modulus values. Furthermore, the tangent-modulus values are on the safe side while the double-modulus values are on the unsafe side. F. R. Shanley[5] presented a paper in 1947 which discussed the shortcomings of the double-modulus theory and showed that the original tangent modulus theory was the better of the two. Curves are presented in Fig. 4-7 showing the comparison of the results obtained by using these two formulas and also the Euler formula.

[5] F. R. Shanley, "The Column Paradox," *J. Aeron. Sci.* (May, 1947), p. 26.

PROBLEMS

4-1 to 4-5. Determine the critical buckling load for each of the columns using the Euler equation. $E = 29 \times 10^6$ psi. Proportional limit $= 34{,}000$ psi. Assume simple ends and maximum permissible $l/r = 200$.

4-1. A solid square bar 1.0 in. $\times$ 1.0 in.
    (a) $l = 3$ ft–0 in. (*Ans.* 18.44 k)
    (b) $l = 4$ ft 6 in. (*Ans.* 8.20 k)

4-2. A 1-in. round bar.
    (a) $l = 3$ ft–0 in.
    (b) $l = 5$ ft–0 in.

4-3. The pipe section shown.
    (a) $l = 36$ ft–0 in. (*Ans.* 127.8 k)
    (b) $l = 24$ ft–0 in. (*Ans.* 287.6 k)
    (c) $l = 12$ ft–0 in. (*Ans.* equation not applicable)

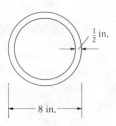

$\frac{1}{2}$ in.

8 in.

Problem 4-3

4-4. A W12×50. $l = 20$ ft–0 in.
4-5. The two C12×30s shown for $l = 36$–ft 0 in. (*Ans.* 497.9 k)

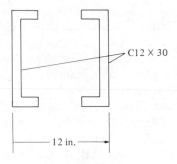

C12 × 30

12 in.

Problem 4-5

4-6. Using SI units determine the critical buckling load of a 50 mm×50 mm solid square bar 2-m long using the Euler equation if $E = 200\,000$ MPa. Assume simple ends and a maximum permissible l/r of 200.

4-7. A 14-ft column with pinned ends is made up of the angles shown in the accompanying illustration. The steel has a modulus of elasticity of 29×10^6 psi and a proportional limit of 34,000 psi. Using Euler's equation (a) determine the critical load of the column (*Ans.* 676.3 k) and (b) determine the critical load of one of the angles acting by itself (*Ans.* 52.05 k).

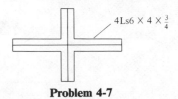

$4 \text{Ls}6 \times 4 \times \frac{3}{4}$

Problem 4-7

4-8. A W8×40 is to serve as an 18 ft pin-connected column. What is the maximum allowable load it can support? Use the Euler expression, a proportional limit of 33,000 psi, $E = 29 \times 10^6$ psi and a safety factor of 2.5.

4-9. The pin-connected wooden column shown in the accompanying illustration is assumed to have an $E = 2 \times 10^6$ psi and a proportional limit of 3000 psi. What is the allowable maximum axial load which it can support if a safety factor of 3 is used according to the Euler expression? Consider lengths of (a) 40 ft, (b) 30 ft, and (c) 20 ft. (*Ans.* (a) 25.22 k, (b) 44.83 k, (c) Equation not applicable)

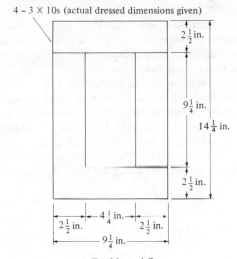

$4 - 3 \times 10\text{s}$ (actual dressed dimensions given)

$2\frac{1}{2}$ in.

$9\frac{1}{4}$ in.

$14\frac{1}{4}$ in.

$2\frac{1}{2}$ in.

$2\frac{1}{2}$ in. $4\frac{1}{4}$ in. $2\frac{1}{2}$ in.

$9\frac{1}{4}$ in.

Problem 4-9

4-10. Using the Euler expression and SI units determine the required cross section for a square timber column 6 m long to support a load of 240 kN. The modulus of elasticity is 12000 MPa, the proportional limit is 34 MPa and a safety factor of 2 is to be used.

4-11. Determine the maximum length for which a W10×22 can be used to support a load of 20 k with a safety factor of 2.5. $E = 29 \times 10^6$ psi. (*Ans.* $l = 21.36$ ft)

4-12. An 8 ft pin-connected column has the cross section shown in the accompanying illustration. Using the Euler equation determine the allowable load which the column can support if the factor of safety is 2. $E = 29 \times 10^6$ psi. Proportional limit = 34,000 psi.

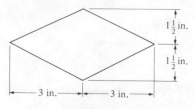

Problem 4-12

4-13. A C10×20 serves as a pin-connected column. Using the Johnson-revised Euler equation for pinned-end columns determine the critical load which it can support for a length of 10 ft. $E = 29 \times 10^6$ psi and the proportional limit is 34,000 psi. (*Ans.* 403.6 kN)

4-14. A S12×31.8 which serves as an 8 ft column has a proportional limit of 34,000 psi and an E of 29×10^6 psi. Using a factor of safety of 2 what is the maximum allowable load which it can support according to the Euler expression, if (a) the column is assumed to be pinned at both ends thus having an effective length equal to l and (b) the column is assumed to have an effective length equal to $2l$?

4-15. Using the Euler expression, determine the required cross section for a square wooden column 20 ft long to support a load of 30 k. The modulus of elasticity is 1.76×10^6 psi, the proportional limit is 4000 psi and a safety factor of 2.5 is to be used. (*Ans.* 8×8 column)

4-16. Using the Euler equation with a safety factor of 3 select a W8 section to support a 60-k load for a 20 ft pin-connected column. $E = 29 \times 10^6$ psi. Proportional limit = 34,000 psi. Maximum permissible $l/r = 240$.

4-17. A W10×49 is to be used as a 34 ft pin-ended column but is braced at middepth in the weak direction. Using a safety factor of 3, $E = 29 \times 10^6$ psi and a proportional limit of 34,000 psi, determine the allowable total axial load which the column can support using the Euler equation. (*Ans.* 156.2 k)

Chapter 5
Design
of Compression
Members

5-1. PRACTICAL DESIGN FORMULAS

Through the years hundreds of different column formulas have been developed with which their originators attempted to approximate test-result curves. Several general types of formulas have been developed such as the straight-line, the parabolic, the Rankine, and others. Some persons say there is no more variation in the results given by the various formulas than occurs in test results. These facts have probably led to the use of simpler formulas, since it is doubtful if the more complicated ones result in appreciably better designs. Several of the better-known column formulas are presented in this chapter.

5-2. STRAIGHT-LINE FORMULAS

The straight-line formulas are the simplest type of column expressions which can be devised for a certain range of l/r. Although there is no theoretical reasoning behind these expressions they were in common use for many years particularly in bridge design. The straight-line formula was first proposed by W. H. Burr in 1882. In 1886 T. H. Johnson proposed a type of straight-line column formula in the *Transactions of the ASCE*.[1]

Straight-line formulas can be devised which will give values approaching test results for l/r values from about 50 to 120. These same formulas, however, give allowable stresses which are much too high when the slenderness ratio is below 50 and much too low when above 120. When column load tests are made there is little decrease in strength as the slenderness ratio increases from zero until it exceeds a value of approximately 50 or 60. Some of the common straight-line formulas took this fact into account. One of these was the 1920 AREA formula (no longer applicable) which follows.

$$\frac{P}{A} = 15,000 - 50\frac{l}{r}$$

[1] "On the Strength of Columns," *Trans. ASCE* **15** (July, 1886), p. 517.

The AREA said the maximum value permissible was 12,500 psi regardless of the value obtained from the formula. In other words for l/r values from 0 to 50 the allowable compression stress was 12,500 psi and for higher l/r values the allowable stress was to be determined from the formula. At that time the AREA did not permit the use of main members with l/r values greater than 100 nor bracing with l/r values greater than 120.

Most of the straight-line formulas were developed during the early days of column tests when test results were quite inconsistent due to variations in testing conditions. During the past few decades these conditions have been greatly improved and column materials made more uniform. These factors have made better test results possible and the straight-line formulas are generally thought to be too inaccurate to represent the strength of steel columns. Consequently these formulas have been almost entirely replaced with parabolic expressions. Straight-line design formulas, however, are used today for some other structural materials.

Example 5-1 illustrates the determination of the allowable compression load for a column using the 1920 AREA formula. In this example the slenderness ratio is determined using the length over the least radius of gyration after which the ratio is substituted into the column formula to determine the allowable stress.

To determine the allowable compression stress to be used for a particular column it is theoretically necessary to compute both l_x/r_x and l_y/r_y. The reader will notice, however, that for most of the steel sections used for columns r_y will be much less than r_x. As a result only l_y/r_y is calculated for most columns and used in the applicable column formulas.

For some columns, particularly the long ones, bracing is supplied perpendicular to the weak axis thus reducing the slenderness or the length free to buckle in that direction. This may be accomplished by framing braces or beams into the sides of a column. For instance horizontal members called *girts* running parallel to the exterior walls of a building frame may be framed into the sides of columns. The result is stronger columns and ones for which the designer needs to calculate both l_x/r_x and l_y/r_y. The larger ratio obtained for a particular column indicates the weaker direction and will be used for calculating the allowable stress in that member.

Steel columns may also be built into substantial masonry walls in such a manner that they are substantially supported in the weaker direction. The designer, however, should be quite careful in assuming complete lateral support parallel to the wall, because a poorly built wall will not provide 100% lateral support. Example 5-10 presented later in this chapter presents an example of a column with different unbraced lengths.

Example 5-1

Using the 1920 AREA formula determine the allowable load which a W14×145 can support for an 18-ft length.

SOLUTION

Using a W14×145 $\left(A = 42.7 \text{ in.}^2, r_x = 6.33 \text{ in.}, r_y = 3.98 \text{ in.} \right)$

$$\max \frac{l}{r} = \frac{(12)(18)}{3.98} = 54.27$$

$$\text{allow } \frac{P}{A} = 15,000 - (50)(54.27) = 12,286 \text{ psi}$$

$$\text{allow } P = (12,286)(42.7) = \underline{\underline{524.6 \text{ k}}}$$

The design applications of the column equations presented in this chapter involve a trial and error process. For a particular design the allowable stress is not known until a column size is selected and vice versa. Once a trial section is selected the least r for that section is known and can be substituted into the column equation to determine the allowable stress.

The designer will usually assume an allowable stress, divide that stress into the column load to give an estimated column area, select a column, and check to see if the section selected is over or understressed, and if appreciably so, try another size. The student, however, may feel that she does not have a sufficient background or knowledge to make a reasonable stress assumption, but if she will read the information contained in the next paragraph she will immediately be able to make excellent estimates.

The slenderness ratios for average columns of 10- to 15-ft lengths will generally fall in a range of about 40 to 60. If for a particular column a slenderness ratio somewhere in this range is assumed and substituted into the column equation it will usually provide a satisfactory allowable stress estimate.

In Example 5-2, which follows, a 12-ft column is selected using the 1920 AREA formula. A slenderness ratio of 50 is assumed, substituted into the formula, and the resulting stress divided into the column load to estimate the area required.

To estimate the slenderness ratio for a particular column the designer may pick a value a little higher than 40 to 60 if the column is appreciably longer than the 10- to 15-ft range and vice versa. A very heavy load, say larger than 500 k (2 224 kN) will yield a rather large column for which the least r will be rather large and the designer may estimate a little smaller value of l/r. For lightly loaded bracing members he will estimate high slenderness ratios perhaps over 100.

Example 5-2

Using the 1920 AREA formula, select a W12 section to support a 315-k load if its unsupported length is to be 15 ft.

SOLUTION

Assuming a slenderness ratio of 50 (based on the information presented immediately preceding this problem) the allowable stress will equal 15,000

$-(50)(50) = 12,500$ psi.

$$A \text{ reqd.} = \frac{315}{12.5} = 25.2 \text{ in.}^2$$

Try W12×87 ($A = 25.6$ in.^{2}, $r = 3.07$)

$$\frac{l}{r} = \frac{(12)(15)}{3.07} = 58.63$$

allow $\dfrac{P}{A} = 15,000 - (50)(58.63) = 12,068$ psi

allow $P = (12.068)(25.6) = 308.9$ k < 315 k   therefore try next larger W12

Try W12×96 ($A = 28.2$ in.^{2}, $r = 3.09$)

$$\frac{l}{r} = \frac{(12)(15)}{3.09} = 58.25$$

allow $\dfrac{P}{A} = 15,000 - (50)(58.25) = 12,087$ psi

allow $P = (12.087)(28.2) = 341$ k > 315 k   OK

Use W12×96

5-3. PARABOLIC FORMULAS

Another group of fairly simple column expressions are the parabolic formulas. These empirical formulas, first proposed by J. B. Johnson,[2] can be made to represent test results fairly well for slenderness ratios varying from 0 to approximately 130 or 140. Above this range they give values that are much too conservative. The formulas (no longer applicable) which are given at the end of this paragraph are taken from the 1969 AASHTO Specifications and are typical of parabolic column formulas. These particular expressions are given for concentrically loaded columns consisting of A36 steel and with l/r values not exceeding 130. It will be noted that most present-day U.S. specifications use F_a as the allowable axial compression stress instead of P/A.

$$F_a = \frac{P}{A} = 16,000 - 0.30\left(\frac{l}{r}\right)^2 \quad \text{(riveted ends)}$$

$$F_a = \frac{P}{A} = 16,000 - 0.38\left(\frac{l}{r}\right)^2 \quad \text{(pinned ends)}$$

[2] J. B. Johnson, F. E. Turneaure, and C. W. Byran, *Modern Framed Structures* (New York: Wiley, 1910), pp. 164–167.

Example 5-3

Select a W14 section to resist a compressive load of 280 k using the 1969 AASHTO formula for pinned ends. The member is to be 20 ft long and consist of A36 steel.

SOLUTION

Assume an l/r of 80 as column is rather long thus giving an estimated allowable stress equal to $16,000 - (0.38)(80)^2 = 13,568$ psi.

$$A \text{ reqd.} = \frac{280}{13.568} = 20.64 \text{ in.}^2$$

<u>Try W14×74 ($A = 21.8 \text{ in.}^2, r = 2.48$)</u>

$$\frac{l}{r} = \frac{(12)(20)}{2.48} = 96.77 < 130 \qquad\qquad\qquad \text{OK}$$

$$F_a = 16,000 - (0.38)(96.77)^2 = 12,442 \text{ psi}$$

$$\text{allow } P = (12.442)(21.8) = 271.2 \text{ k} < 280 \text{ k} \qquad\qquad \text{NG}$$

<u>By examination use W14×82 ($A = 24.1 \text{ in.}^2, r = 2.48$)</u>

Example 5-4

Using the 1969 AASHTO formula for riveted ends determine the allowable compressive load that the member shown in Fig. 5-1 can support if it is 19 ft long and consists of A36 steel.

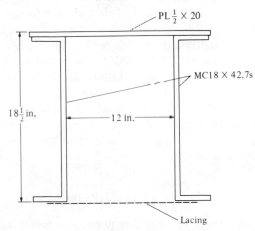

Figure 5-1

SOLUTION

$$A = (2)(12.6) + (20)\left(\tfrac{1}{2}\right) = 35.2 \text{ in.}^2$$

$$\bar{y} \text{ from top} = \frac{(25.2)(9.5) + (10)(0.25)}{35.2} = 6.87 \text{ in.}$$

$$I_x = (2)(554) + (25.2)(2.63)^2 + \left(\tfrac{1}{12}\right)(20)\left(\tfrac{1}{2}\right)^3 + (10)(6.62)^2$$

$$I_x = 1721 \text{ in.}^4$$

$$I_y = (2)(14.4) + (25.2)(6.88)^2 + \left(\tfrac{1}{12}\right)\left(\tfrac{1}{2}\right)(20)^3$$

$$I_y = 1553 \text{ in.}^4$$

$$\text{least } r = \sqrt{\frac{1553}{35.2}} = 6.64 \text{ in.}$$

$$\frac{l}{r} = \frac{12 \times 19}{6.64} = 34.3 < 130 \qquad\qquad\qquad \text{OK}$$

$$F_a = 16,000 - (0.30)(34.3)^2 = 15,650 \text{ psi}$$

allowable $P = (15,650)(35.2) = 551,000 \text{ lb}$

5-4. GORDON-RANKINE FORMULA

The preceding paragraphs have presented two empirical methods for estimating the reduction in strength of a column as the slenderness ratio increased. The straight-line formulas reduced the allowable stress in direct proportion to the slenderness ratio while the parabolic formulas reduced the stress by some squared value of the slenderness ratio. Another method of reducing the allowable stress with increasing slenderness is to divide the base stress by a number larger than one, which increases as l/r increases. The Gordon-Rankine expression is a formula of this type, has a fairly rational derivation and gives results in fairly close agreement with actual tests for l/r values from 120 to 200. For a good many years the AISC used the Gordon-Rankine expression to follow for the design of secondary members for l/r values from 120 to 200. (A *main member* is defined as a member of primary importance whose failure would be expected to cause immediate collapse of all or a large part of a structure. A *secondary member* is a member of secondary importance whose failure would not be so serious. It is sometimes further defined as a member which theoretically does not support dead loads other than its own weight nor live loads other than those caused by wind.)

$$\frac{P}{A} = \frac{18,000}{1 + \left[(l/r)^2/18,000\right]}$$

During the years when the AISC required this formula to be used for the design of secondary members, they also permitted its use for main members in the same l/r range if the formula was multiplied by $1.6 - (l/r)/200$. This type of equation was developed by an English engineer, Lewis Gordon, but he used the square of the least side dimension instead of r^2. A famous Scottish engineer, W. J. M. Rankine, developed the equation in its present form.[3] For this reason it is commonly called the Rankine formula rather than the Gordon-Rankine formula. Example 5-5 illustrates the application of a Rankine formula to a bracing problem.

Example 5-5

A 15-ft wind-bracing member is to be designed for a load of 35,000 lb. It is to consist of a pair of angles with a $\frac{3}{8}$-in. gusset plate between them. The following Rankine formula is to be used in the design.

$$\frac{P}{A} = \frac{18,000}{1 + \left[(l/r)^2 / 18,000 \right]}$$

Capstone House, University of South Carolina, Columbia, S.C. (Courtesy of Owen Steel Co., Inc.)

[3] W. J. M. Rankine, *A Manual of Civil Engineering* (London: Charles Griffin and Company, 1883), pp. 522–523.

SOLUTION

Assume $l/r = 140$, P/A obtained by substituting into Rankine formula equals 8617 psi, and A reqd. $= 35,000/8617 = 4.06$ in.^{2}

Try 2Ls4×3×$\frac{5}{16}$ ($A = 4.18$, $r = 1.27$) long legs back-to-back

$$\frac{l}{r} = \frac{12 \times 15}{1.27} = 142$$

$$\text{allowable } \frac{P}{A} = \frac{18,000}{1 + \left[(142)^2 / 18,000 \right]} = 8,490 \text{ psi}$$

$$\text{allowable } P = (8490)(4.18) = 35,500 \text{ lb} > 35,000 \text{ lb} \qquad \text{OK}$$

Use 2Ls4×3×$\frac{5}{16}$

5-5. THE SECANT FORMULA

A rational type of formula which can be made to approximate test results for all values of l/r is the secant formula. Many investigators have worked with this type of column formula which is probably the most precise of all column expressions and which has fairly wide usage despite some difficulties of application. The results of all column formulas, no matter how complex, can be plotted on a graph or recorded in a table for practical use. On the other hand, the engineer may desire to substitute directly in the formula and for such cases the simpler formulas are probably more desirable.

The ASCE formed a committee for column research in 1923. In its reports this committee recommended the secant formula as being a very satisfactory expression for column design. The general form of the equation is as follows.

$$\frac{P}{A} = \frac{F_y}{1 + \frac{ec}{r^2} \sec\left(\sqrt{\frac{P}{AE}} \, \frac{l}{2r} \right)}$$

In this expression F_y is the yield point of the steel, c is the distance from the centroid to the extreme fiber, and e is the eccentricity of load application. This formula has a definite advantage in that it provides specifically for some eccentricity of load application. On the basis of studies of test results this committee recommended that a value for ec/r^2 (called the eccentric ratio) equal to 0.25 be used. This value supposedly provides for secondary stresses caused by bending and initial crookedness of columns.

The difficulty in directly using the secant formula is that while the goal is to obtain an allowable P/A stress for a particular column, P/A appears in both sides of the equation. The value of P/A, therefore, can be determined only with a trial-and-error process. The value finally obtained from the formula is actually the estimated load at which yielding begins in an eccentrically loaded column as its derivation made use of Hooke's law. The difference between the loads at which yielding begins and at which failure occurs may vary from only a few percent to approximately 50%. The variation depends on such items as the slenderness ratio, direction of bending, shape of section, etc.

5-6. THE AISC AND AASHTO FORMULAS

The AISC expressions were developed to incorporate the latest research information available concerning the behavior of steel columns. These formulas take into account the effect of residual stresses, the actual end restraint conditions of the columns, and the varying strengths of different steels. The important effect of residual stresses on the stress-strain curves was discussed in Section 4-2 and illustrated in Fig. 4-1. Different types of end restraints cause entirely different effective lengths and column strengths. A discussion of this subject was previously presented in Section 4-5 and is continued in Section 5-7 as it pertains to the AISC formulas.

The use of the AISC formulas results in more logical and economical designs than those given by the older expressions. Design by many other column formulas gives members that are appreciably overdesigned in the lower l/r range but the AISC formulas give fairly economical designs for all ranges of l/r. The expressions are rather complex to solve mathematically but as tables are available in the Steel Manual, designs can be made with little difficulty.

The AISC assumes that because of residual stresses the upper limit of elastic buckling is defined by an average stress equal to one-half of the yield point ($\frac{1}{2} F_y$). If this stress is equated to the Euler expression, the value of the slenderness ratio at this upper limit can be determined for a particular steel. This value is referred to as C_c, the slenderness ratio dividing elastic from inelastic buckling, and is determined as follows.

$$\frac{1}{2} F_y = \frac{\pi^2 E}{(l/r)^2}$$

$$= \frac{\pi^2 E}{C_c^{\,2}}$$

$$C_c = \sqrt{\frac{2\pi^2 E}{F_y}} \quad ^*$$

The values of C_c can be computed with little difficulty but the Steel Manual gives its values for each steel (126.1 for A36, 116.7 for the 42,000

*For SI units $C_c = 1987/\sqrt{F_y}$ with F_y in MPa.

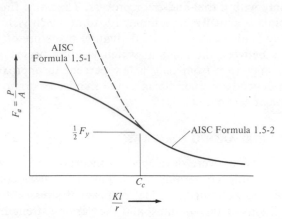

Figure 5-2

psi yield point steels, etc.). For slenderness ratios less than C_c a parabolic formula (AISC Formula 1.5-1) is used. This is the Structural Stability Research Council equation for the ultimate strength of a centrally loaded column with a factor of safety applied.[4] In this expression F_a is the allowable axial stress (P/A) and K is the factor to be multiplied by the unsupported column lengths to give their estimated effective lengths (see Section 5-7).

$$F_a = \frac{\left[1 - \frac{(Kl/r)^2}{2C_c^{\,2}}\right]F_y}{\frac{5}{3} + \frac{3(Kl/r)}{8C_c} - \frac{(Kl/r)^3}{8C_c^{\,3}}} \qquad \text{(AISC Formula 1.5-1)}$$

For values of Kl/r greater than C_c the Euler formula is used. With a factor of safety of 1.92 (or 23/12) the expression becomes

$$F_a = \frac{12\pi^2 E}{23(Kl/r)^2} \qquad \text{(AISC Formula 1.5-2)}$$

Figure 5-2 shows the ranges in which the two AISC expressions are used.

The denominator of Formula 1.5-1 is actually the factor of safety to be used and usually gives a value not much greater than the one used for axially loaded tension members. Tests have shown that short columns are not greatly affected by small eccentricities permitting the use of lower factors of safety. As columns become slenderer they become more sensitive to small imperfections and the factor of safety is increased up to 15%. It

[4]Structural Stability Research Council, *Guide to Stability Design Criteria for Metal Structures*, B. G. Johnson (ed.), 3d ed. (New York: Wiley, 1976).

might also be noticed that the medium length columns are those in which residual stresses and initial column crookedness become quite important. The F. S. expression is a quarter sine wave which equals $\frac{5}{3}$ when Kl/r is equal to zero and increases up to $\frac{23}{12}$ when Kl/r equals C_c.

A third equation for compression members is given by the AISC for axially loaded bracing and secondary members when l/r exceeds 120.

$$F_{as} = \frac{F_a \,(\text{by Formula 1.5-1 or 1.5-2})}{1.6 - l/200r} \qquad (\text{AISC Formula 1.5-3})$$

The AISC values given for bracing and secondary members are quite liberal because of the relative unimportance of these members. Another reason for the high allowable stresses available for the design of bracing and secondary members is that the persons writing the specifications took

Central Laboratory and Office Building, Baltimore, Md. (Courtesy of Owen Steel Co., Inc.)

into account the appreciable end restraint that secondary members usually have. In other words, effective lengths have already been considered and *the designer should use the actual lengths of secondary and bracing members and not their effective lengths*. This discussion means that the secondary AISC column formulas should be used only for members that are fairly well braced against rotation and translation at the points where they are connected.

The 1977 AASHTO Specifications have the same two types of column formulas as does the AISC Specification. The two categories, as for the AISC, are for those columns that have low slenderness ratios and fail due to inelastic buckling and those columns that have high slenderness ratios and fail by elastic buckling. For the inelastic range the AASHTO requires the use of the following parabolic formula

$$F_a = \frac{F_y}{\text{F.S.}} \left[1 - \frac{\left(\frac{Kl}{r}\right)^2 F_y}{4\pi^2 E} \right]$$

and for the elastic range (that is, where $Kl/r > C_c$) the Euler formula is to be used

$$F_a = \frac{\pi^2 E}{\text{F.S.}(Kl/r)^2}$$

For both cases a constant factor of safety (F.S.) equal to 2.12 is used.

5-7. EFFECTIVE COLUMN LENGTHS

The value of Kl used in the AISC and AASHTO Specifications is the *effective length* of the column which has previously been defined as the distance between the inflection points of the column. This distance was found to vary for different columns depending upon their types of end restraint. Should a column be perfectly fixed at each end [see Fig. 4-5(b)] inflection points would occur at the quarter points when it is laterally deflected giving an effective length of $l/2$. Should the column be connected with frictionless pins [see Fig. 4-5(a)] the effective length would equal the actual length of the column.

Actually there are no perfect pin connections nor any perfect fixed ends, and the usual column falls in between the two extremes. This discussion would seem to indicate that column effective lengths always vary from an absolute minimum of $l/2$ to an absolute maximum of l, but there are exceptions to this rule. An example is given in Fig. 5-3 where a simple bent is shown. The base of each of the columns is pinned and the other end is free to rotate and move laterally (called sidesway). Examination of this figure will show that the effective length will exceed the actual length of the column as the elastic curve will theoretically take the shape of the curve of a pinned end column of twice its length. Table C1.8.1 in the

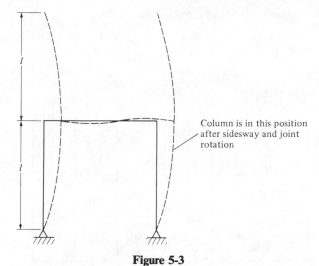

Column is in this position after sidesway and joint rotation

Figure 5-3

A heavy column showing base plate with combination welded and high-strength bolted connection. (Courtesy of Bethlehem Steel Company.)

Table 5-1   Effective Lengths for Main Members Only

	(a)	(b)	(c)	(d)	(e)	(f)
Buckled shape of column is shown by dashed line						
Theoretical K value	0.5	0.7	1.0	1.0	2.0	2.0
Recommended design value when ideal conditions are approximated	0.65	0.80	1.2	1.0	2.10	2.0

End condition code		
		Rotation fixed and translation fixed
		Rotation free and translation fixed
		Rotation fixed and translation free
		Rotation free and translation free

SOURCE: *Specification for the Design, Fabrication and Erection of Structural Steel for Buildings* November *1, 1978.* (*New York*: *AISC*, 1978), *p.* 5–124 *in the AISC Manual.*

"Commentary on AISC Specification" in the Steel Handbook gives recommended effective lengths when ideal conditions are approximated. This table is reproduced here as Table 5-1 with the permission of the AISC. K is the theoretical value to be multiplied times the column length to give its theoretical effective length; however, more conservative values are recommended in the table for actual design practices. The use of these latter values will be illustrated in several design problems in the pages to follow and Kl will be referred to as the effective length of the column.

The selection of the K value to be used for a particular column has given practicing engineers as much trouble as any other part of the AISC Specification. On many occasions K values have been selected which are too large, resulting in overdesigned columns with resulting economy losses. The difficulty seems to lie in distinguishing symmetrical buckling from sidesway buckling. When sidesway buckling occurs a smaller load can be supported.

When translation of the tops of the columns is clearly prevented as by diagonal bracing, shear walls, attachment to adjacent buildings, etc. (read Section 1.8.2 of the AISC Specification), symmetrical buckling will occur and the structure is referred to as a *braced frame*. For such cases as these the column effective lengths can be no greater than their actual lengths. Values of K equal to 1.0 can be conservatively assumed or lesser values estimated from parts (a), (b), or (d) of Table 5-1 (AISC Table C.1.8.1) or by charts prepared for this purpose such as the "sidesway prevented" chart given by the Structural Stability Research Council in their publication *Guide to Stability Design Criteria for Metal Structures*. A large percentage of the columns designed by the usual structural designer fall into this class.

It should be realized that most of the columns which the structural engineer has to design serve as members of frames and have effective lengths that are controlled by the amounts of restraint applied to their ends by the other members of the frame and by the walls of the structure itself. As no column ends are completely fixed or perfectly pinned, the designer may wish to interpolate between the values given in the table, the interpolation to be based on his judgment of the actual restraint conditions. Another reasonable method for estimating effective lengths involves the careful sketching of the anticipated deflected shape of a particular column and the measuring of the distance between the inflection points on the sketch.

For most buildings, sidesway is substantially eliminated by masonry walls, but for buildings built with light curtain walls and large column spacings or for tall buildings built without a positive system of lateral bracing, sidesway is appreciable. Such frames are referred to as *unbraced frames*. For such cases the bending stiffness of the structural frames provides most of the lateral support.[5] The effective lengths of columns for such laterally unsupported continuous frames must always be greater than 1.0 because of sidesway. This topic is more applicable to beam columns and is thus continued in Chapter 8 which deals with those types of members.

5-8. MAXIMUM SLENDERNESS RATIOS

The AISC Specification (1.8.4) states that the slenderness ratio Kl/r of all compression members may not exceed 200. The AASHTO Specifications (1.7.5) states that the ratio may not exceed 120 for main members. For secondary members whose purpose is to brace the structure against lateral or longitudinal forces or to reduce the unbraced length of other members the ratio may not exceed 140.

5-9. DESIGN OF COLUMNS WITH AISC FORMULAS

Examples 5-6 through 5-10 illustrate the design of different columns using the AISC expressions. In part (a) of Example 5-6 the solution is made by substituting into the formulas but in the remainder of this example and in the other examples the AISC Manual tables are used to simplify the calculations. It should be remembered that the yield points of several

[5]*Specification for the Design, Fabrication and Erection of Structural Steel for Buildings November 1, 1978* (New York: AISC, 1978), pp. s-123 through s-127 AISC Manual.

high-strength steels vary for different thicknesses (unless specially alloyed to keep values equal for thicker members); therefore, when these steels are involved it is necessary to carefully determine their yield stresses as described on pages 1-5 and 1-6 of the AISC Manual.

Example 5-6

Select a W section for the column and load shown in Fig. 5-4 using (a) A36 steel and (b) A441 steel.

SOLUTION

(a) Using A36 steel: Assume $Kl/r = 40$. Thus $F_a = 19.19$ ksi from the AISC Manual and A reqd. $= \underline{600/19.19 = 31.27}$ in.[2]

$$\text{Try W14} \times 109 \ (A = 32.0, \ r = 3.73)$$

Effective length from Table 5-1 (or Table C. 1.8.1. in AISC Specification commentary)

$$Kl = (0.65)(15) = 9.75 \text{ ft}$$

$$\frac{Kl}{r} = \frac{(12)(9.75)}{3.73} = 31.4$$

The value of F_a is taken from Table 3-36 in Section 5 of the AISC Specification as there is no need to laboriously substitute into AISC

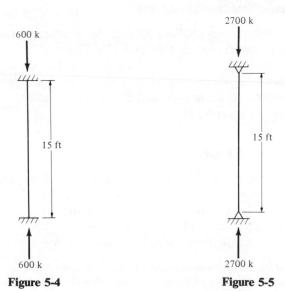

| Figure 5-4 | Figure 5-5 |

Formula 1.5-1.

$$F_a = 19.84 \text{ ksi}$$

$$\text{allowable } P = (19.84)(32.0) = 635 \text{ k} > 600 \text{ k} \qquad \text{OK}$$

Use W14×109

(b) Using A441 steel: Assume that $Kl/r=40$ and $F_y=50$ ksi. Thus $F_a = 25.83$ ksi from AISC Manual and the area required is $600/25.83 = 23.23$ in².

Try W14×82 ($A=24.1$ in.², $r=2.48$, Group 2 AISC; therefore, $F_Y = 50,000$)

$$\frac{Kl}{r} = \frac{12 \times 9.75}{2.48} = 47.2$$

$$\text{allowable } F_a = 24.78 \text{ ksi (from table)}$$

$$\text{allowable } P = (24.78)(24.1) = 597 \text{ k} < 600 \text{ k}$$

Use W14×82 (slightly overstressed)

Example 5-7

(a) Select a W section for the load and column shown in Fig. 5-5 using A441 steel.

(b) Using A36 steel and assuming a W14×311 is on hand select PLs so the section will be satisfactory for the conditions of Fig. 5-5.

SOLUTION

(a) Using A441 steel: Assume that the allowable F_a is 22 ksi and that the area required is $2700/22 = 122.7$ in.²

Try W14×426 ($A=125$, $r=4.34$, group 4 AISC; therefore, $F_y = 42,000$)

$$Kl = (1.0)(15) = 15 \text{ ft}$$

$$\frac{Kl}{r} = \frac{12 \times 15}{4.34} = 41.4$$

$$\text{allowable } F_a = 21.94 \text{ ksi (by substitution into AISC formula 1.5-1)}$$

$$\text{allowable } P = (21.94)(125) = 2742 \text{ k} > 2700 \text{ k} \qquad \text{OK}$$

Use W14×426

(b) Using A36 steel: Assume that the allowable F_a is 19 ksi and the area required is $2700/19 = 142$ in.²

Try W14×311 ($A=91.4$ in.^{2})

With 1 cover PL20×$1\frac{1}{4}$ each flange $=50.00$ in.^{2}

$$\text{total area} = 141.4 \text{ in.}^2$$
$$I_x = 4330 + (2)(25)(9.185)^2 = 8548 \text{ in.}^4$$
$$I_y = 1610 + \left(\tfrac{1}{12}\right)(2.50)(20)^3 = 3277 \text{ in.}^4$$
$$r=\sqrt{\frac{3277}{141.4}} = 4.81$$
$$Kl = (1.0)(15) = 15 \text{ ft}$$
$$\frac{Kl}{r} = \frac{(12)(15)}{4.81} = 37.42$$

allowable $F_a = 19.39$ ksi (from table)

allowable $P = (19.39)(141.4) = 2742 \text{ k} > 2700 \text{ k}$

Use W14×311 with 1 cover PL20×$1\frac{1}{4}$ each flange

Example 5-8

Select a pair of unequal leg angles with a $\frac{3}{8}$-in. gusset plate between them using A36 steel and the AISC Specification to support a 32-k load. The member is to be 19 ft long and is assumed to be a secondary member.

SOLUTION
Assume $l/r=$ about 180 and allowable F_a is 6.5 ksi and that the area required is $32/6.5=4.92$ in.^{2}

Try 2Ls4×3×$\frac{3}{8}$ ($A=4.97, r=1.26$) long legs back to back

$$\frac{l}{r} = \frac{12\times19}{1.26} = 181$$

Using actual length and not the effective length for a secondary member.

allowable $F_a = 6.56$ ksi

allowable $P = (6.56)(4.97) = 32.6 \text{ k} > 32 \text{ k}$ OK

Use 2Ls4×3×$\frac{3}{8}$

Example 5-9

Select a pair of channels for the column and load shown in Fig. 5-6. A36 steel and the AISC Specification are to be used. For connection purposes the back-to-back distance of the channels is to be 12 in.

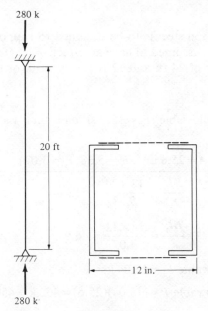

280 k

20 ft

280 k

12 in.

Figure 5-6

SOLUTION

Assume that the allowable F_a is 17.5 ksi and that the area required is $280/17.5 = 16.0$ in.^{2}

Try 2C12×30s ($A = 17.64$, $r_x = 4.29$)

$$I_y = (2)(5.14) + (17.64)(5.33)^2 = 511 \text{ in.}^4$$

$$r_y = \sqrt{\frac{511}{17.64}} = 5.38$$

$$Kl = (1)(20) = 20 \text{ ft}$$

$$\frac{Kl}{r} = \frac{(12)(20)}{4.29} = 56$$

allowable $F_a = 17.81$ ksi

allowable $P = (17.81)(17.64) = 314 \text{ k} > 280 \text{ k}$ OK

Use 2C12×30s

Columns may often be supported at different intervals in one direction than in the other, as where girts running parallel to the exterior walls of a building frame into the sides of the columns. For such cases the columns are probably turned so that their stronger axes face the larger unsupported lengths. The design will be governed by the larger of the two slenderness ratios. Example 5-10 illustrates the design of a column which is braced at different intervals about its x and y axes. For another example the student is referred to Example 1 in the column section of the AISC Manual.

Example 5-10

A W12 column of A36 steel is to be designed to support an axial load of 450 k. The column is assumed to have an effective length with respect to its strong or major axis of 24 ft, and 12 ft for its weak or minor axis.

SOLUTION
Assume that the allowable $F_a = 18$ ksi and that the area required is $450/18 = 25$ in.^{2}

$$\underline{\text{Try W12} \times 87 \left(A = 25.6 \text{ in.}^2, r_x = 5.38, r_y = 3.07 \right)}$$

$$\frac{Kl_x}{r_x} = \frac{12 \times 24}{5.38} = 53.5 \leftarrow$$

$$\frac{Kl_y}{r_y} = \frac{12 \times 12}{3.07} = 46.9$$

allowable $F_a = 18.04$ ksi

allowable $P = (18.04)(25.6) = 462 \text{ k} > 450 \text{ k}$ OK

$\underline{\text{Use W12} \times 87}$

5-10. LACING AND TIE PLATES

The necessity for built-up compression members in large buildings and bridges has previously been discussed in Chapter 4. When members are built up from more than one section it is necessary for them to be connected or laced together across their open sides. The purpose of lacing or lattice work is to hold the various parts parallel and the correct distance apart and to equalize the stress distribution between the various parts. The student will understand the necessity for lacing if she considers a built-up member consisting of several sections (such as the four-angle member of Fig. 4-2) which supports a heavy compressive load. Each of the parts will tend to buckle laterally unless they are tied together to act as a unit in supporting the load. In addition to lacing it is necessary to have tie plates (also called stay plates or batten plates) as near the ends of the member as possible and at intermediate points if the lacing is interrupted. Parts (a) and (b) of Fig. 5-7 show arrangements of tie plates and lacing. Other possibilities are shown in parts (c) and (d) of the same figure.

The failure of several structures in the past has been attributed to inadequate lacing of built-up compression members. Perhaps the best-known example was the failure of the Quebec Bridge in 1907. Following its collapse the general opinion was that the lattice work of the compression chords was too weak and resulted in failure. This disaster brought home to the engineering profession the importance of carefully designed lacing.

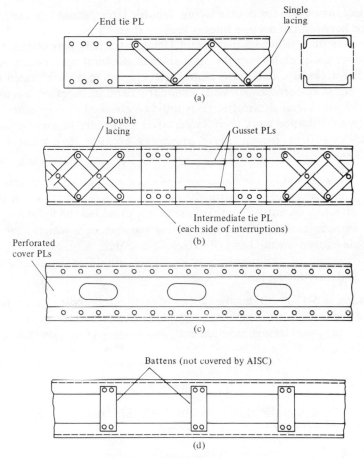

Figure 5-7

Dimensions of tie plates and lacing are usually controlled by the specifications. The AISC (Section 1.18.2) states that the tie plates shall have a thickness at least equal to $\frac{1}{50}$ the distance between the connection lines of rivets, bolts, or welds and shall have a length parallel to the axis of the main member at least equal to the distance between the connection lines.

Lacing usually consists of flat bars but may occasionally consist of angles, perforated cover plates, channels, or other rolled sections. These pieces must be so spaced that the individual parts being connected will not have l/r values between connections which exceed the governing value for the entire built-up member. Lacing is assumed to be subjected to a shearing stress normal to the member equal to not less than 2% of the total compression in the member. The AISC column formulas are used to design the lacing in the usual manner. Slenderness ratios are limited to 140 for single lacing and 200 for double lacing. Double lacing should be used if the distance between connection lines is greater than 15 in.

Rather than using tie plates and lacing bars it is permissible to use continuous cover plates over the open sides of the built-up sections. Access holes are necessary, and these plates are referred to as perforated cover plates. Stress concentrations and secondary bending stresses are usually neglected but lateral shearing stresses must be checked as they must be for other types of lattice work. Perforated cover plates are becoming popular with an increasing percentage of engineers because they possess these advantages.

1. They are easily fabricated with modern gas cutting methods.
2. Some specifications permit the inclusion of their net areas in the effective section of the main members, provided the holes are made in accordance with their empirical requirements which have been developed on the basis of extensive research.
3. Painting of the members is probably simplified as compared to ordinary lacing bars.

Example 5-11 illustrates the design of lacing and end tie plates for the built-up column of Example 5-9. Bridge specifications are somewhat different in their lacing requirements from the AISC but the design procedure is much the same.

Example 5-11

Using the AISC Specification design bolted single lacing for the column of Example 5-9. Reference is made to Fig. 5-8.

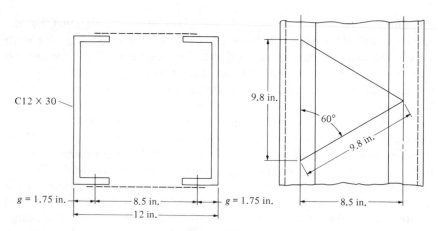

Figure 5-8

Eversharp, Inc. building at Milford, Conn. (Courtesy of Bethlehem Steel Company.)

SOLUTION

Distance between lines of bolts is 8.5 in. < 15 in.; therefore, single lacing OK.

Assume an inclination of 60° with axis of member. Length of channels between lacing connections is $8.5/\cos 30° = 9.8$ in., and l/r of 1 channel between connections is $9.8/0.763 = 12.9 < 56$, which is l/r of main member.

Force on lacing bar:

$$V = 0.02 P = (0.02)(280) = 5.6 \text{ k}$$

$$\tfrac{1}{2} V = 2.8 \text{ k} = \text{shearing force on each plane of lacing}$$

$$\text{force on bar} = (9.8/8.5)(2.8) = 3.23 \text{ k}$$

Properties of a flat bar:

$$I = \tfrac{1}{12}bt^3$$

$$A = bt$$

$$r = \sqrt{\dfrac{\frac{1}{12}bt^3}{bt}}$$

$$r = 0.289t$$

Design of bar:

$$\max \frac{l}{r} = 140$$

$$\frac{9.8}{0.289t} = 140$$

$$t = 0.242 \text{ in. } \left(\text{try } \tfrac{1}{4}\text{-in. flat bar}\right)$$

$$\frac{l}{r} = \frac{9.8}{(0.289)(0.250)} = 136$$

$$\text{allowable } F_a = 8.78 \text{ ksi} \qquad \text{(for secondary members)}$$

$$\text{area reqd.} = \frac{3.23}{8.78} = 0.368 \text{ in.}^2 \left(1.47 \times \tfrac{1}{4}\right)$$

Minimum edge distance if $\tfrac{3}{4}$-in. bolt used $= 1\tfrac{1}{4}$ in.

<u>Use $\tfrac{1}{4} \times 2\tfrac{1}{2}$ bar</u>

Design of end tie plates:

$$\text{min length} = 8.5 \text{ in.}$$

With t not less than $\tfrac{1}{50}$ distance between bolt lines,

$$t = \left(\tfrac{1}{50}\right)(8.5) = 0.17 \text{ in.}$$

<u>Use $\tfrac{3}{16} \times 8\tfrac{1}{2} \times 11\tfrac{1}{2}$ end tie PLs</u>

5-11.   STIFFENED AND UNSTIFFENED ELEMENTS

Up to this point in the text the author has only considered the overall stability of members and yet it is entirely possible for the thin flanges or webs of a column or beam to buckle locally in compression well before the calculated buckling strength of the whole member is reached. When thin plates are used to carry compressive stresses they are particularly susceptible to buckling about their weak axes due to their small moments of inertia in those directions.

The AISC in its Section 1.9 provides limiting values for the width-thickness ratios of the individual parts of compression members and for the parts of beams in their compression regions. The student is quite well aware of the lack of stiffness of thin pieces of cardboard or metal or plastic with free edges. If, however, one of these elements is folded its stiffness is appreciably increased. For this reason two categories are listed in the AISC Manual, these being *stiffened elements* and *unstiffened elements.*

An unstiffened element is a projecting piece with one free edge parallel to the direction of the compression force while a stiffened element is supported along the two edges in that direction. These two types of elements are illustrated in Fig. 5-9. In each case the width, b, and the thickness, t, of the elements in question are shown.

If the width-thickness ratio (b/t) of the various elements of beams and columns exceed certain values local buckling will occur before the yield stresses are reached. For such cases it is necessary to reduce allowable stresses or to reduce effective areas of the elements in question.

The AISC provides maximum b/t ratios for both unstiffened and stiffened elements. These various ratios given in the next paragraph are calculated for several F_y values in Table 6 of Appendix A of the 1978 AISC Specification.

For unstiffened elements b/t is limited to $95/\sqrt{F_y}$ except for single angle struts and for double angle struts with separators where it is $76/\sqrt{F_y}$ and $127/\sqrt{F_y}$ for stems of tees. For stiffened elements b/t is limited to $253/\sqrt{F_y}$ unless it is the flange of a square or rectangular box section of uniform thickness where the ratio is reduced to $238/\sqrt{F_y}$.

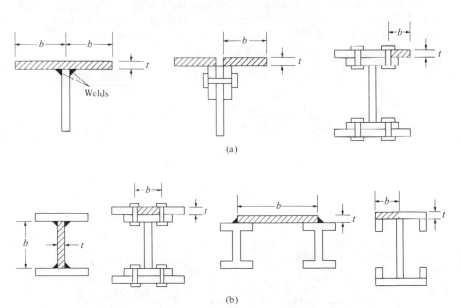

Figure 5-9 (a) Unstiffened elements. (b) Stiffened elements.

5-12. BASE PLATES FOR CONCENTRICALLY LOADED COLUMNS

The allowable compressive stress in a concrete or other type of masonry footing is much smaller than it is in a steel column. When a steel column is supported by a footing it is necessary for the column load to be spread over a sufficient area to keep the footing from being overstressed. Loads from steel columns are transferred through a steel base plate to a fairly large area of the footing below. (It will be noted that a footing performs a related function in that it spreads the load over an even larger area so that the underlying soil will not be overstressed.)

The base plates for steel columns can be welded directly to the columns or they can be fastened by means of some type of bolted or welded lug angles. These connection methods are illustrated in Fig. 5-10. A base plate welded directly to the column is shown in part (a) of the figure. For small columns these plates are probably shop-welded to the columns but for larger columns it may be necessary to ship the plates separately and set them to the correct elevations. The columns are then set and connected to the footing with anchor bolts which pass through the lug angles which have been shop-welded to the columns. This type of arrangement is shown in part (b) of the figure. Some designers like to use lug angles on both flanges and web.

The lengths and widths of column base plates are usually selected in multiples of even inches and their thicknesses in multiples of $\frac{1}{8}$ in. To make sure that column loads are spread uniformly over their base plates it is essential to have good contact between the two. To accomplish this objective it is necessary to straighten plates thicker than 2 in. up through 4 in. by pressing or planing. Plates thicker than 4 in. need to have their upper surfaces planed. The bottom of the plates will be in contact with cement grout and do not have to be planed.

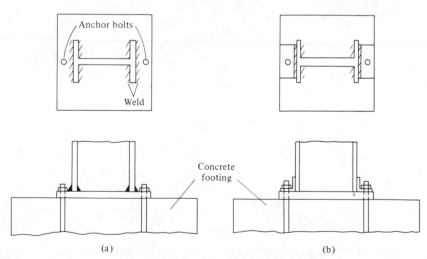

(a)   (b)

Figure 5-10 Column base plates.

The method of design suggested here is the one that is recommended by the Steel Manual. To analyze the base plate shown in Fig. 5-11 the column is assumed to apply a total load P to the base plate, and this load is assumed to be transmitted uniformly through the base plate to the footing below with a value of f_p psi (P/A). The footing will push back with a pressure of f_p psi and tend to curl up the cantilevered parts of the base plate outside the column as shown in the figure. This pressure also tends to push up the part of the base plate between the flanges of the column.

With reference to Fig. 5-11 the Steel Manual suggests that maximum moments in a base plate occur at distances approximately $0.80b_f$ and $0.95d$ apart. The bending moment is calculated at each of these sections and the larger value used to determine the plate thickness needed. This method of analysis is only a rough approximation of the true conditions because the actual plate stresses are caused by a combination of bending in two directions.

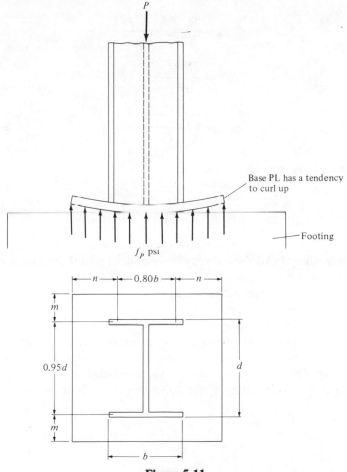

Figure 5-11

From Fig. 5-11 the following moment expressions can be written for the two critical sections considering in each case a 1-in. width of plate.

$$M = f_p n \frac{n}{2} = \frac{f_p n^2}{2}$$

$$M = f_p m \frac{m}{2} = \frac{f_p m^2}{2}$$

The section modulus of a 1-in. width of plate of t thickness is

$$S = \frac{I}{C} = \frac{\left(\frac{1}{12}\right)(1)(t^3)}{t/2} = \frac{t^2}{6}$$

Since the stress is $Mc/I = M/S$, the required thickness of the base plate can be determined.

$$F_b = \frac{M}{S} = \frac{f_p(m^2/2)}{t^2/6}$$

$$= \frac{3f_p m^2}{t^2}$$

$$t = \sqrt{\frac{3f_p m^2}{F_b}}$$

Similarly in the other direction,

$$t = \sqrt{\frac{3f_p n^2}{F_b}}$$

The reader of this section may have noticed that if the values of m and n for a particular base plate are close to zero, the plate thickness computed from the preceding expressions will also be close to zero. Such a value is not realistic. For such cases the highest stresses in the base plate actually occur at the face of the column web halfway between the insides of the flanges.

To keep the designer from having to make a complicated theoretical plate analysis for this situation, a simple procedure is provided in the AISC Manual. In Table C, page 3-100, a value labeled n' is provided. If the larger of m or n is less than n' the following plate thickness equation is to be used:

$$t = \sqrt{\frac{3f_p n'^2}{F_b}}$$

The values of n necessary to solve the problems of this chapter are (as taken from Table C) as follows: 2.77 in. for W8×24 through W8×28, 4.77 in. for W12×65 through W12×336, 4.43 in. for W14×61 through W14×82, and 5.64 in. for W14×90 through W14×132.

Knowing the values of m, n, and n', the thickness of plate required can be calculated. The example to follow illustrates the design of a column base plate. The allowable bearing stress in a steel base plate by the AISC is $0.75F_y$ (27,000 for A36 steel). The AISC Specification (Section 1.5.5) states that, in the absence of Code regulations, when the load is applied to the full area of the concrete support an allowable bearing pressure on the footing (F_p) equal to $0.35f_c'$ is to be used where f_c' is the specified compressive strength of the concrete. Should the base plate cover less than the full area of the supporting concrete the allowable bearing stress is to be determined by the following expression in which A_1 is the bearing area in square inches and A_2 is the entire supporting concrete area in square inches.

$$F_p = 0.35f_c'\sqrt{\frac{A_2}{A_1}} \leqslant 0.7f_c'$$

For the usual footing F_p will thus equal $0.7f_c'$. For bearing on brick in cement mortar a value of $F_p = 0.25$ ksi is specified while 0.40 ksi is permitted for bearing on sandstone or limestone.

Before the final dimensions of the base plates are selected reference should be made to the Steel Manual so that standard sizes will be used. In this way the designer will be insured of economy and promptness of delivery. Example 5-12 illustrates the design of a base plate by this procedure.

Example 5-12

Design a base plate with A36 steel for a W12×65 column and a load of 370 k. The column is to be supported by a concrete footing with an allowable bearing pressure of 1750 psi. The dimensions are as shown in Fig. 5-12.

SOLUTION

$$\text{area reqd.} = \frac{370,000}{1750} = 211.4 \text{ in.}^2 \quad (\text{say } 14 \times 16 = 224 \text{ in.}^2)$$

$$f_p = \frac{370,000}{224} = 1652 \text{ psi}$$

$$m = 2.24 \text{ in.}, \ n = 2.20 \text{ in.}$$

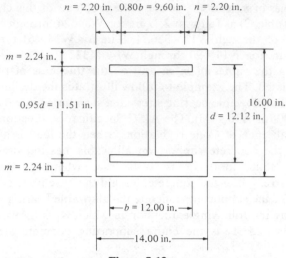

Figure 5-12

But $n' = 4.77$ in.

$$t = \sqrt{\frac{3 f_p n'^2}{F_b}} = \sqrt{\frac{(3)(1652)(4.77)^2}{27,000}}$$

$$t = 2.04 \text{ in.}$$

Use PL$2\frac{1}{8} \times 14 \times 1$ ft-4 in. ($54 \times 360 \times 410$ mm)

$$t = \sqrt{\frac{3 f_p n^2}{F_b}} = \sqrt{\frac{(3)(1652)(2.24)^2}{27,000}}$$

$$t = 0.960 \text{ in.}$$

Use PL$1 \times 14 \times 1$ ft-4 in. ($25 \times 360 \times 410$ mm)

The base plate discussion presented in this section is applicable to axially loaded columns only with simple supports with no moments to be resisted. In Chapter 23 entitled "Miscellaneous Topics" the design of base plates to resist bending moments as well as axial loads is presented.

Problems

5-1. Using the 1920 AREA formula ($P/A = 15,000 - 50 l/r$), with a maximum value of 12,500 psi and a maximum l/r of 120 determine the allowable axial compression load that each of these columns can support.
   (a) A 16-ft W14×53 (*Ans.* 156 k)
   (b) A 15-ft HP14×117 (*Ans.* 429.8 k)

(c) A 12-ft WT15×66 (*Ans.* 228.9 k)

(d) A 7-ft C12×30 (*Ans.* 83.7 k)

5-2. Using the AREA formula given in Prob. 5-1 determine the allowable axial compressive load that each of the following members can support.

(a) An 18-ft W12×79

(b) A 10-ft MC12×50

(c) An 8-ft L5×5×$\frac{7}{8}$

(d) A 15-ft HP12×53

5-3. Determine the allowable concentric load that each of the compression members shown in the accompanying illustration can support. Use the column formula given in Prob. 5-1. [*Ans.* (a) 517.9 k. (b) 180.7 k. (c) 167.6 k. (d) 364.3 k]

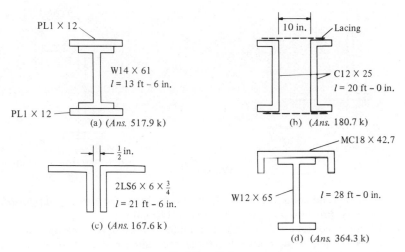

PL1 × 12

W14 × 61
l = 13 ft – 6 in.

PL1 × 12

(a) (*Ans.* 517.9 k)

10 in.   Lacing

C12 × 25
l = 20 ft – 0 in.

(h) (*Ans.* 180.7 k)

$\frac{1}{2}$ in.

2LS6 × 6 × $\frac{3}{4}$
l = 21 ft – 6 in.

(c) (*Ans.* 167.6 k)

MC18 × 42.7

W12 × 65   *l* = 28 ft – 0 in.

(d) (*Ans.* 364.3 k)

Problem 5-3

5-4. Select the lightest available W12 section to support a 200-k axial compression load for an unsupported length of 14 ft using the column formula of Prob. 5-1.

5-5. Select the lightest available W14 section to support a 450-k axial load for an unsupported length of 12 ft–0 in. using the column formula of Prob. 5-1. (*Ans.* W14×132)

5-6. Determine the allowable axial compressive load that can be supported by each of the sections shown in the accompanying illustration. Use the parabolic equation $P/A = 16,000 - \frac{1}{4}(l/r)^2$.

5-7. Select the lightest W14 section that will support an axial compressive load of 400 k for an unsupported length of 16 ft–0 in. using the column equation of Prob. 5-6. (*Ans.* W14×90)

5-8. A chord member of a bridge truss is to be 19 ft–0 in. long and is to support an axial compressive load equal to 360 k. Using the column equation $P/A = 18,000 - \frac{1}{2}(l/r)^2$ select the lightest available W12 section.

5-9. The following parabolic expression is to be used to select the most economical W14 column section to support a 300-k axial load, $P/A = 17,000 -$

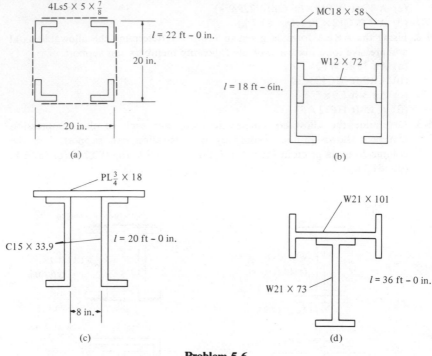

(a)

(b)

(c)

(d)

Problem 5-6

$0.4(l/r)^2$. The column is to be 28 ft long and is to be laterally supported in the weak direction at middepth. (*Ans.* W14×68)

5-10. Select the lightest available W12 section to support an axial compressive load of 100 k. The unsupported length of the member is 25 ft–0 in. and the Gordon-Rankine formula to follow is to be used.

$$\frac{P}{A} = \frac{18,000}{1 + \dfrac{(l/r)^2}{18,000}}$$

5-11. A W14×53 with a $\frac{3}{4}$×12 cover plate bolted to each flange is used for a 26 ft–0 in. column. What total allowable axial compressive load can it support using the Gordon-Rankine equation of Prob. 5-10? (*Ans.* 1 617 kN)

5-12. Using the AISC Specification determine the allowable axial compressive loads that the following columns can support.
    (a) A W8×35 with fixed ends, $l=16$ ft–6 in, A36 steel
    (b) A W12×96 with pinned ends, $l=20$ ft–0 in., A36 steel
    (c) A W10×68 with one end fixed and the other pinned, $l=24$ ft–6 in., $F_y=50$ ksi
    (d) A W14×193 with fixed ends, $l=22$ ft–0 in., $F_y=50$ ksi
    (e) A Pipe 10 Std. with pinned ends, $l=20$ ft–0 in., A36 steel
    (f) Two $8×8×\frac{3}{4}$ Ls separated $\frac{3}{8}$ in. (for gusset PL at ends), pinned ends, $l=24$ ft–6 in., A36 steel

5-13. Determine the allowable axial loads that the columns shown in the accompanying illustration can support. Use the AISC Specification and A36 steel.

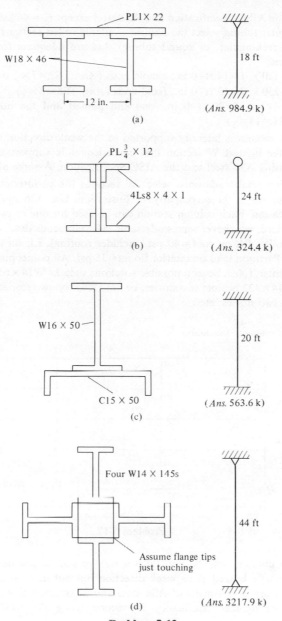

PL1 × 22

W18 × 46

12 in.

18 ft

(*Ans.* 984.9 k)

(a)

PL$\frac{3}{4}$ × 12

4Ls8 × 4 × 1

24 ft

(*Ans.* 324.4 k)

(b)

W16 × 50

C15 × 50

20 ft

(*Ans.* 563.6 k)

(c)

Four W14 × 145s

Assume flange tips
just touching

44 ft

(*Ans.* 3217.9 k)

(d)

Problem 5-13

5-14. Several building columns are to be designed using A36 steel and the AISC
Specification. Select the lightest available W sections for these columns that
are described as follows.
(a) $P = 300$ k, $l = 12$ ft–0 in., simple supports
(b) $P = 220$ k, $l = 14$ ft–0 in., fixed ends
(c) $P = 450$ k, $l = 16$ ft–6 in., fixed at bottom, pinned at top
(d) $P = 1200$ k, $l = 15$ ft–0 in., simple supports

5-15. Using the AISC Specification and A36 steel except $F_y = 46$ ksi for square and rectangular tubing select the lightest available rolled sections (W, M, S, HP, square, rectangular, or round tubing) that are adequate for the following situations.

(a) $P = 120$ k, $l = 14$ ft–0 in., simple ends (*Ans.* TS7×7×$\frac{1}{4}$ or TS8×6×$\frac{1}{4}$)

(b) $P = 250$ k, $l = 15$ ft–0 in., fixed ends (*Ans.* TS8×8×$\frac{3}{8}$)

(c) $P = 440$ k, $l = 20$ ft–0 in., one end pinned and the other fixed (*Ans.* TS14×14×$\frac{3}{8}$)

5-16. A 27-ft column is laterally supported in the weak direction at its middepth. Select the lightest W section that can adequately support an axial load of 300 k using A36 steel and the AISC Specification. Assume all $k_s = 1.0$.

5-17. Assuming axial loads only, select W sections for an interior column of the frame shown in the accompanying illustration. Use A36 steel and the AISC Specification. Each column section can be used for one or two stories before it is spliced, whichever seems advisable. Miscellaneous data: concrete weighs 150 lbs/ft^3. LL on roof = 40 psf (includes roofing). LL on interior floors = 80 psf. Partition load on interior floors = 15 psf. All points pinned. Frames 30 ft on center. (*Ans.* Several possible solutions such as W14×61 top two stories and W14×132 bottom two stories; or W12×65 top two stories and W12×136 bottom two stories; etc.)

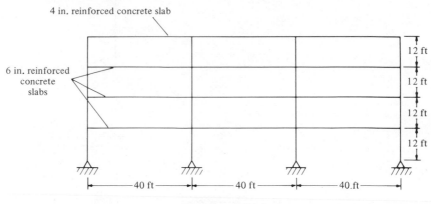

Problem 5-17

5-18. A 14-ft column is to be built into a wall in such a manner that it will be continuously braced in its weak direction but not in its strong direction. If the member is to consist of A36 steel and is assumed to have pinned ends select the lightest satisfactory W section using the AISC Specification. $P = 800$ k.

5-19. Repeat Problem 5-18 if A514 Grade 90 steel is used. (*Ans.* W14×61)

5-20. A W section is to be selected to support an axial compressive load of 300 k. The member, which is to be 24 ft long and is to be pinned top and bottom, has lateral support (pinned) supplied in the weak direction at middepth. Select the section using A36 steel and the AISC Specification.

5-21. Repeat Prob. 5-20 if lateral support (pinned) is supplied in the weak direction at the one-third points. (*Ans.* W12×58)

5-22. Twenty-seven-ft steel columns for a building are to be braced at the one-third points for the minor axis. If the columns are to support axial loads of 90 k each and are to consist of A36 steel, select a C or MC section for each using the AISC Specification. Assume top and bottom supports are fixed and lateral supports are pinned. (*Ans.* MC10×25.3)

5-23. Four $4 \times 4 \times \frac{1}{2}$ angles are used to form the member shown in the accompanying illustration. The member is 20 ft long, has pinned ends, and consists of A36 steel. Determine the maximum allowable axial compressive load that the member can support according to the AISC. Design single lacing and end tie plates assuming connections are made to the angles with $\frac{3}{4}$-in. bolts. (*Ans.* $P=298.8$ k. Use $\frac{1}{4} \times 13 \times 1$ ft–4 in. end tie plates and $\frac{15}{32} \times 2\frac{1}{2} \times 1$ ft–$8\frac{7}{8}$ in. single lacing at 45°)

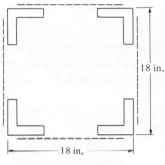

18 in.

18 in.

Problem 5-23

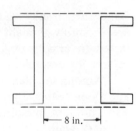

8 in.

Problem 5-24

5-24. Select a pair of channels to support an axial compressive load of 300 k. The member is to be 24 ft long with both ends pinned and is to be arranged as shown in the accompanying illustration. Use the AISC Specification, A36 steel, and design single lacing and end tie plates assuming $\frac{3}{4}$-in. bolts are to be used for connections.

5-25. A W12×65 supports an axial load of 330 k. Using the AISC Specification and A36 steel, design a base plate for the column if the supporting reinforced concrete footing has an allowable bearing pressure of 1125 psi. (*Ans.* One satisfactory answer is a PL1$\frac{3}{4}$ ×17×1 ft–6 in.)

5-26. Design a column base plate for a W14×120 supporting an axial load of 575 k. The supporting footing has an allowable bearing pressure of 1500 psi. Use A36 steel and the AISC Specification.

5-27. A W8×24 column supporting a 75-k axial load is bearing on a block wall with an allowable bearing pressure of 500 psi. Design a base plate for the column using A36 steel and the AISC Specification. (*Ans.* Several plates can be selected, one which is satisfactory is $\frac{13}{16} \times 12 \times 1$ ft–1 in.)

5-28. A W14×74 column has a 2×20×1 ft–8 in. base plate which rests on a reinforced concrete spread footing with an allowable unit bearing pressure of 1125 psi. What is the maximum allowable axial load that this bearing plate can support? Use A36 steel and the AISC Specification.

Chapter 6
Design of Beams

6-1. TYPES OF BEAMS

Beams are usually said to be members that support transverse loads. They are probably thought of as being used in horizontal positions and subjected to gravity or vertical loads; but there are frequent exceptions—rafters, for example.

Among the many types of beams are joists, lintels, spandrels, stringers, and floor beams. *Joists* are the closely spaced beams supporting the floors and roofs of buildings, while *lintels* are the beams over openings in masonry walls such as windows and doors. A *spandrel beam* supports the exterior walls of buildings and perhaps part of the floor and hallway loads. The discovery that steel beams as a part of a structural frame could support masonry walls is said to have permitted the construction of today's "skyscrapers." *Stringers* are the beams in bridge floors running parallel to the roadway, whereas *floor beams* are the larger beams in many bridge floors which are perpendicular to the roadway of the bridge and are used to transfer the floor loads from the stringers to the supporting girders or trusses. The term *girder* is rather loosely used but usually indicates a large beam and perhaps one into which smaller beams are framed. These and other types of beams are discussed in the paragraphs to follow.

6-2. THE FLEXURE FORMULA

Included in the items that need to be considered in beam design are moments, shears, crippling, buckling, lateral support, deflection, and perhaps fatigue. Beams will probably be selected which satisfactorily resist the bending moments and then checked to see if any of the other items are critical. To select a beam for a given situation the maximum moment is calculated for the assumed loading and a section having that much resisting moment is selected from the AISC Manual.

The resisting moment of a particular section can be computed with the flexure formula ($M_R = F_b I / c$). This expression and the P/A expression

Harrison Avenue Bridge, Beaumont, Tex. (Courtesy of Bethlehem Steel Company.)

are perhaps the most famous of all formulas to civil engineers. The frequency of their application is illustrated by the fact that some of the more modest structural designers often say, "All I know is P/A and Mc/I."

In the flexure formula f_b is the fiber stress in the outermost fiber a distance c from the neutral axis and I is the moment of inertia of the cross section. It should be remembered that this formula is limited to stress situations below the elastic limit because it is based on the usual elastic assumptions: a plane section before bending remains a plane section after bending; stress is proportional to strain, etc.

The value of I/c is constant for a particular section and is known as the section modulus. If a beam is to be designed for a particular bending moment M and for a certain allowable stress F_b the section modulus required to provide a beam of sufficient bending strength can be obtained from the flexure formula as follows.

$$\frac{M}{F_b} = \frac{I}{c} = S = \text{the section modulus}$$

6-3. SELECTION OF BEAMS

The W shapes will normally prove to be the most economical beam sections and they have largely replaced channels and S sections for beam usage. Channels are sometimes used for beams subjected to light loads, such as purlins, and in places where clearances available require narrow flanges. They have very little resistance to lateral forces and need to be braced as illustrated by the sag rod problem in Chapter 3. The W shapes have more steel concentrated in their flanges than do S beams and thus have larger section moduli values for the same weights. They are relatively wide and have appreciable lateral stiffness. (The small amount of space devoted to S beams in the AISC Manual clearly shows how much their use has decreased from former years. They are today used primarily for special situations as where narrow flange widths are desirable, or where shearing forces are very high, or where the greater flange thickness next to the web may be desirable where lateral bending occurs as perhaps with crane rails.)

Another common type of beam section is the open web joist or bar joist which is discussed at length in Chapter 13. This type of section which is commonly used to support floor and roof slabs is actually a light shop fabricated parallel chord truss. It is particularly economical for long spans and light loads.

It is rare for beams to be made from bars, angles, or T sections because these shapes have low section modulus values in comparison to their weights and thus little resistance to bending.

A table is given in the AISC Manual entitled "Allowable Stress Design Selection Table." From this table steel shapes having sufficient section moduli can be quickly selected. Two important items should be remembered in selecting shapes. These are as follows.

1. These steel sections cost so many cents per pound and it is therefore desirable to select the lightest possible shape having the required section modulus (assuming that the resulting section is one which will reasonably fit into the structure). The table has the sections arranged in various groups having certain ranges of section moduli. The heavily typed section at the top of each group is the lightest section in that group and the others are arranged successively in the order of their section moduli. Normally the deeper sections will have the lightest weights giving the required section moduli, and they will be generally selected unless their depth causes a problem in obtaining the desired headroom, in which case a shallower but heavier section will be selected.

2. The section moduli values in the table are given about the horizontal axes for beams in their upright positions. If a beam is to be turned on its side the proper section modulus can be found in the tables giving dimensions and properties of shapes in the AISC Manual. A W shape turned on its side may only be from 5 to 15% as strong as one in the upright position when subjected to gravity loads. In the same manner, the strength of a wood joist with the actual dimensions 2×10 in. turned on its side would only be 20% as strong as in the upright position (the percentage being based on the relative values of the section moduli).

The examples to follow illustrate the design of steel beams whose compression flanges have lateral support, thus permitting the use of the same allowable stresses in the tension and compression flanges. Beams without sufficient lateral support for the compression flanges are considered in Sections 6-6 and 6-7.

In each of these examples the weight of the beam to be selected must be included in the calculation of the bending moment to be resisted, as the beam must support itself as well as the external loads. The estimates of beam weight are very close here because the author was fortunately able to perform a little preliminary paperwork before making his estimate. The student is not expected to be able to glance at a problem and estimate exactly the weight of the beam required. Following the same procedure as did the author, however, he can do a little figuring on the side and make a very reasonable estimate. For instance, he could calculate the moment due to the external loads only, obtain the required section modulus, and select a beam having the required value. From this beam size he should be able to make a very good estimate of the weight of the final beam section which will often be a little larger than the trial beam size.

Example 6-1

Select a beam section for the span and loading shown in Fig. 6-1, assuming full lateral support is provided for the compression flange by the floor above. Allowable bending stresses are 24 ksi (165 MPa).

SOLUTION
Assume beam weight $= 62$ lb/ft.

$$M = \frac{wl^2}{8} = \frac{(4.362)(21)^2}{8} = 240.46 \text{ ft-k}$$

$$\text{required section modulus} = \frac{(12)(240.46)}{24} = 120.23 \text{ in.}^3$$

The possible solutions include the following: (a) A W14×82 is the section that has the closest S on the safe side. (b) But a W21×62 is the most economical solution. (c) Should depth be restricted, a W18×71 or even a W14×82 (very uneconomical) could be selected.

Use W21×62

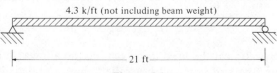

4.3 k/ft (not including beam weight)

21 ft

Figure 6-1

Example 6-2

A 5-in. reinforced concrete slab is to be supported with steel beams 8 ft 0 in. on centers. The beams, which will span 20 ft, are assumed to be simply supported. If the concrete slab is designed to support a live load of 100 psf, determine the lightest steel section required to support the slab. The compression flange of the beam will be incorporated in the concrete slab and is thus laterally supported. The concrete weighs 150 lb/ft^3 and the allowable bending stresses in the steel are 24 ksi.

SOLUTION

$$\text{DL}:\quad \text{slab} = (8)\left(\frac{5}{12}\right)(150) = 500 \text{ lb/ft}$$

$$\text{estimated beam weight} = 26$$

$$\text{LL}:\quad 8 \times 100 = \underline{800}$$

$$\text{total uniform load} = 1326 \text{ lb/ft}$$

$$M = \frac{(1.33)(20)^2}{8} = 66.5 \text{ ft-k}$$

$$S_{\text{reqd.}} = \frac{(12)(66.5)}{24} = 33.2 \text{ in.}^3$$

$$\underline{\text{Use W12} \times 26 \ \left(S_x = 33.4 \text{ in.}^3\right)}$$

The limiting moment permitted by the allowable stress method is the moment at which the stress in the outermost fibers first reaches the yield point. The true bending strength of a beam, however, is larger than this commonly used value because the beam will not fail at this condition. The outermost fibers will yield and the stress in the inner fibers will increase until they reach the yield stress, etc., until the whole section is plastified (see Section 21-3). It will be shown in Chapter 21 that before local failure will occur moments must be produced approximately 12% larger than those which will first produce the yield stress in the outermost fibers of W shapes.

The plastification process just mentioned is correct only if the beam remains stable in other ways: that is it must have sufficient lateral support to prevent lateral buckling of the compression flange (see Section 6-6) and it must have a sufficiently stocky profile to prevent local buckling.

The AISC Specification gives different allowable bending stresses for different conditions. For most cases the allowable bending stress is as follows.

$$F_b = 0.66 F_y$$

This expression can be used to determine the allowable bending stress in the extreme fibers of compact hot-rolled shapes and built-up members (not including members consisting of A514 steel which has $F_y = 90$ or

100 ksi nor girders built up with different yield stress steels called hybrid girders) which are symmetrical about and loaded in the plane of their minor axis and which meet the other requirements of Section 1.5.1.4.1 of the AISC Specification. One of these requirements is that the flange must be continuously attached to the web. A built-up section with its flanges intermittently welded to the web does not meet this requirement.

6-4. COMPACT SECTIONS

A compact section is one that is capable of developing its plastic moment capacity before any local buckling occurs. To qualify as compact a section must meet the requirements of subparagraphs 1 through 4 of Section 1.5.1.4.1. Nearly all W and S shapes of A36 steel and a large percentage of those shapes made from higher strength steels are compact.

For noncompact laterally supported sections the AISC requires a reduction in F_b below $0.66F_y$ while for laterally supported compact sections the allowable stress is equal to $0.66F_y$. The proportions necessary for a section to be classed as being compact are specified by the AISC and are summarized in the following paragraphs.

Flanges

Limitations are given by the AISC for the width/thickness ratios for both unstiffened and stiffened compression beam flanges. For the usual hot rolled sections such as W sections the flanges are unstiffened while they may very well be stiffened for certain built-up sections (see Fig. 5-9).

The AISC requires that the width of an unstiffened projecting element of a compression flange divided by its thickness (that is, $b_f/2t_f$) not exceed $65/\sqrt{F_y}$. For stiffened elements the width thickness ratio (b/t_f) may not be greater than $190/\sqrt{F_y}$ where b is the actual width of the stiffened element.

Webs

In addition to the flange requirements the depth thickness ratios (d/t) of compact sections are not permitted to exceed certain values. These values are $640/\sqrt{F_y} [1 - 3.74(f_a/F_y)]$ when $f_a/F_y \leqslant 0.16$ and $257/\sqrt{F_y}$ when $f_a/F_y > 0.16$. The term f_a represents the stress caused by a concurrent axial load (if any).

The limitations of web and flange sizes are calculated for different yield stress values and tabulated in Table 6-1. These values are as given in Table 6 of Appendix A of the 1978 AISC Specification. Nearly all W and S sections are compact when made of A36 steel while a large proportion of the same shapes are compact if F_y is 50 ksi.

Table 6-1   Maximum Width Thickness Ratios for Compact Sections

Yield stress F_y (ksi)		36	42	46	50	60	65
Unstiffened flanges $\dfrac{65}{\sqrt{F_y}}$		10.8	10.0	9.6	9.2	8.4	8.1
Stiffened flanges $\dfrac{190}{\sqrt{F_y}}$		31.7	29.3	28.0	26.9	24.5	23.6
Web	$\dfrac{640}{\sqrt{F_y}}$	106.7	98.8	94.4	90.5	82.6	79.4
	$\dfrac{257}{\sqrt{F_y}}$	42.8	39.7	37.9	36.3	33.2	31.9

It is obvious from the values shown in Table 6-1 that the higher the yield stress of a particular section the more likely it is to be noncompact. It is quite simple to determine the yield stress above which the flange of a particular section is noncompact as it is for the web. For instance if the maximum width thickness ratio of an unstiffened flange is equated to $b_f/2t_f$ and solved for F_y the result, which is referred to as F_y' is

$$\frac{65}{\sqrt{F_y}} = \frac{b_f}{2t_f}$$

$$F_y = F_y' = \left(\frac{65}{b_f/2t_f}\right)^2$$

A similar derivation for the web when it is subject to combined bending and axial stress with $f_a/F_a > 0.16$ follows.

$$\frac{257}{\sqrt{F_y}} = \frac{d}{t_w}$$

$$F_y = F_y''' = \left(\frac{257t_w}{d}\right)^2$$

If the yield stress in question is $> F_y'$ the flange is noncompact and if $> F_y'''$ the web is noncompact. The previously mentioned allowable stress selection table has the noncompact shapes clearly indicated by showing the value of F_y' for each section. The previous edition (seventh) of the AISC Manual contained F_y'' values computed for the depth-thickness ratio of beam webs when f_a was zero. The values of F_y'' are no longer shown because they are all higher than 70 ksi and plastic behavior is not recognized by the AISC for such steels. If $F_y > 70$ ksi the maximum value of F_b permitted is $0.60F_y$ which is the lower stress range for noncompact sections anyway.

If the web is noncompact the maximum allowable bending stress permitted by the AISC is $0.60F_y$. If, however, the web is compact and the flange has a $b_f/2t_f$ value $> 65/\sqrt{F_y}$ but less than $95/\sqrt{F_y}$ it is said to be *partially compact*. For partially compact sections a linear transition in F_b between $0.66F_y$ and $0.60F_y$ is provided by AISC Formula 1.5-5a given at the end of this paragraph. The purpose of this formula is to avoid some of the abruptness of the transition from an allowable stress of $0.66F_y$ to $0.60F_y$. The transition does not apply to members of A514 steel nor to hybrid girders. The partially compact section formula is

$$F_b = F_y\left[0.79 - 0.002\left(\frac{b_f}{2t_f}\right)\sqrt{F_y}\right]$$

Should a doubly-symmetric I or H shape be bent about its minor or y axis and should $b_f/2t_f > 65/\sqrt{F_y}$ but less than $95/\sqrt{F_y}$ the allowable bending stress is to be computed with AISC Formula 1.5-5b which follows.

$$F_b = F_y\left[1.075 - 0.005\left(\frac{b_f}{2t_f}\right)\sqrt{F_y}\right]$$

Example 6-3, which follows, illustrates the calculations necessary to determine the allowable bending stress and the resisting moment of a noncompact section. It will be noted that the allowable stress design selection table in the AISC Manual contains resisting moment values (M_R) for the commonly used beam sections with 36 and 50 ksi yield stress steels. Their values have been computed with the correct F_b values whether the sections are compact or noncompact. The values given in the other beam tables in the Manual have also accounted for reduced allowable bending stresses for noncompact sections.

Example 6-3

Compute the resisting moment of a W12×65 with (a) $F_y = 36$ ksi and (b) $F_y = 50$ ksi. Assume the section has full lateral support for its compression flange.

SOLUTION

(a) $F_y = 36$ ksi

Using a W12×65 ($d_w = 12.12$ in., $t_w = 0.390$ in., $b_f = 12.000$ in., $t_f = 0.605$

in., $S_x = 87.9$ in.^{3}) and checking "compact section" requirements,

$$\frac{b_f}{2t_f} = \frac{12.000}{(2)(0.605)} = 9.92 < 10.8 \qquad \text{OK}$$

$$\frac{d}{t_w} = \frac{12.12}{0.390} = 31.08 < 106.7 \qquad \text{OK}$$

Therefore

$$F_b = 0.66F_y = 24 \text{ ksi}$$

$$M_R = F_b S_x = (24)(87.9) = 2110 \text{ in.-k} = \underline{\underline{175.8}} \text{ ft-k}$$

(b) $F_y = 50$ ksi

Checking compact section requirements,

$$\frac{b_f}{2t_f} = \frac{12.000}{(2)(0.605)} = 9.92 > 9.2$$

Therefore flange is noncompact

$$\frac{d}{t_w} = \frac{12.12}{0.390} = 31.08 < 90.5 \qquad \text{OK}$$

Applying AISC Formula 1.5-5a,

$$F_b = 50\left[0.79 - 0.002\left(\frac{12.000}{2 \times 0.605}\right)\sqrt{50}\,\right] = 32.49 \text{ ksi}$$

$$M_R = F_b S_x = (32.49)(87.9) = 2856 \text{ in.-k} = \underline{\underline{238}} \text{ ft-k}$$

Note: The AISC Manual can be used to quickly determine if a section is compact as follows.

$$F_y' = 43.0 \text{ ksi} > 36 \text{ ksi} \quad \text{but} \quad < 50 \text{ ksi}$$

Therefore flange is noncompact for $F_y = 50$ ksi

$$F_y''' = \; > 70 \text{ ksi and is thus not listed}$$

Therefore web is compact for both steels

6-5. HOLES IN BEAMS

It is often necessary to have holes in steel beams. They are obviously required for the installation of bolts and rivets and sometimes for pipes, conduits, ducts, etc. If at all possible these latter type holes should be completely avoided. When absolutely necessary they should be placed through the web if the moment is large and through the flange if the shear is large. Cutting a hole through the web of a beam does not reduce its section modulus greatly or its resisting moment; but, as will be described in Section 7-1, a large hole in the web tremendously reduces the shearing strength of a steel section. When large holes are put in beam webs, extra plates are sometimes connected to the webs around the holes to serve as reinforcing against possible web buckling. An example design for such reinforcing is presented by Kussman and Copper.[1]

The presence of holes of any type in a beam certainly does not make it stronger and in all probability weakens it somewhat. The effect of holes has been a subject which has been argued back and forth for many years. The questions, "Is the neutral axis affected by the presence of holes?" and "Is it necessary to subtract holes from the compression flange which are going to be plugged with rivets or bolts?" are frequently asked.

The theory that the neutral axis might move from its normal position to the theoretical position of its net section when holes are present is quite questionable. A linear distribution of stress has been assumed in the preceding paragraphs of this chapter but when holes are present the situation is changed because there is considerable stress concentration around the holes.

Tests seem to show that flange holes for rivets or bolts do not appreciably change the location of the neutral axis. It is logical to assume that the location of the neutral axis will not follow the exact theoretical variation with its abrupt changes in position at rivet or bolt holes as shown in part (b) of Fig. 6-2. A more reasonable change in neutral axis location is shown in part (c) of this figure where it is assumed to have a more gradual variation in position.

It is interesting to note that flexure tests of steel beams seem to show that their failure is based on the strength of the compression flange even

---

[1]R. L. Kussman and P. B. Cooper, "Design Example for Beams with Web Openings," *Engineering Journal*, Second Quarter 1976, **13**, No 2 (New York: AISC), pp. 48–56.

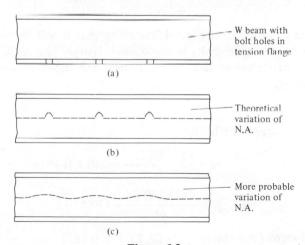

(a)

(b)

(c)

Figure 6-2

though there may be rivet or bolt holes in the tension flange. The presence of these holes does not seem to be as serious as might be thought particularly as compared to holes in a pure tension member. These tests show little difference in the strengths of beams with no holes and in beams with holes up to 15% of the gross area of either flange.

The AISC does not require the subtraction of holes in either flange provided the hole area in any one flange does not exceed 15% of the gross area of that flange, and then the deduction is only for the area in excess of 15%. Furthermore the AISC does not make a distinction between holes in the compression and tension flanges. Although the 15% value is permitted by the AISC, some specifications (notably the bridge ones) and a good many designers have not adopted the idea and follow the more conservative practice of deducting all holes.

The AASHTO and AREA require the calculation of two moments of inertia when holes are present. For compressive stresses the gross moment of inertia is to be used regardless of the presence of rivet or bolt holes. For tensile stresses the net moment of inertia is to be used. The neutral axis is assumed to remain at its normal position for both calculations. The effect of using the two different moments of inertia is to assume that rivet or bolt holes on the compression side of the beam have less effect than those on the tension side.

The usual practice is to subtract the same area of holes from both flanges whether they are present or not. For a section with two holes in the tension flange only, the properties of the section would be computed based on the subtraction of two holes from the tension flange and two holes from the compression flange. Example 6-4 illustrates this method. Again the author has made a few preliminary calculations in estimating the member size.

Example 6-4

Redesign the beam of Example 6-1 assuming that it will be necessary to punch holes for two $\frac{3}{4}$-in. bolts in the tension flange. The AISC reduction is not to be permitted in this beam. Figure 6-3 shows a sketch of the assumed section.

SOLUTION
Assume beam weight = 68 lb/ft.

$$M = \frac{(4.368)(21)^2}{8} = 240.8 \text{ ft-k}$$

$$\text{net } S_{\text{reqd.}} = \frac{(12)(240.8)}{24} = 120.4 \text{ in.}^3$$

Try W24×68 $\left(S = 154 \text{ in.}^3,\ d = 23.73,\ t_f = 0.585 \right)$

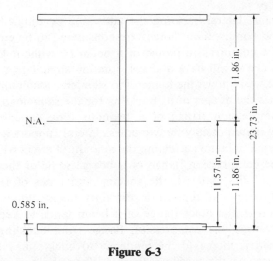

Figure 6-3

Assuming two holes in each flange, I of holes about N. A. is

$$(4)\left(\tfrac{7}{8}\right)(0.585)(11.57)^2 = 274 \text{ in.}^4$$

and S of holes is

$$\frac{274}{11.86} = 23.1 \text{ in.}^3$$

$$\text{actual net } S = 154 - 23.1 = 130.9 \text{ in.}^3 > 120.4 \text{ in.}^3 \qquad \text{OK}$$

Use W24×68

Should a hole be present in only one side of a flange of a W section, there will be no axis of symmetry for the net section of the shape. A correct theoretical elastic solution of the problem would involve the location of the principal axes, the calculation of the principal moments of inertia, etc. or substitution into the lengthy generalized equations for unsymmetrical bending presented in Section 7-5. Rather than following such lengthy procedures over a minor point it seems logical to assume holes in both sides of the flange. The results obtained will probably be just as satisfactory as those obtained by the more laborious methods mentioned.

6-6. LATERAL SUPPORT OF BEAMS

Probably the large majority of steel beams are used in such a manner that their compression flanges are restrained against lateral buckling. (Unfortunately, however, the percentage has not been quite as high as the design profession has assumed.) The upper flanges of beams used to support concrete building and bridge floors are often incorporated in these concrete floors. For situations of this type where the compression flanges are restrained against lateral buckling the allowable bending fiber stresses in the tension and compression flanges are considered to be equal.

Should the compression flange of a beam be without lateral support for some distance it will have a stress situation similar to that existing in columns. As is well known the longer and slenderer a column becomes the greater becomes the danger of its buckling for the same loading condition. When the compression flange of a beam is long enough and slender enough it may quite possibly buckle unless lateral support is provided.

There are many factors affecting the amount of stress which will cause buckling in the compression flange of a beam. Some of these factors are the properties of the material, the spacing and types of lateral support provided, the types of end support or restraints, the loading conditions, etc.

The tension in the other flange of a beam tends to keep that flange straight and restrain the compression flange from buckling; but as the bending moment is increased the tendency to buckle may become large enough to overcome the tensile restraint. When the compression flange does begin to buckle, twisting or torsion will occur, and the smaller the torsional strength of the beam the more rapid will be the failure. The W, S, and channel shapes so frequently used for beam sections do not have a great deal of resistance to lateral buckling and the resulting torsion. Some other shapes, notably the built-up box shapes, are tremendously stronger. These types of members have a great deal more torsional resistance than the W, S, and plate girder sections. Tests have shown that they will not buckle laterally until the strains developed are well in the plastic range. If box girders meet the requirements for compact shapes their allowable bending stresses are $0.66F_y$ and if they are not compact their allowable bending stresses are equal to $0.60F_y$ unless their depths are greater than six times their widths. For such cases lateral support requirements should be specially investigated (AISC 1.5.1.4.4.).

Some judgment needs to be used in deciding what does and what does not constitute satisfactory lateral support for a steel beam. Perhaps the most common question asked by practicing steel designers is "What is lateral support?" A beam that is wholly encased in concrete or that has its compression flange incorporated in a concrete slab is certainly well supported laterally. When a concrete slab rests on the top flange of a beam, the engineer must study the situation carefully before he counts on friction to provide full lateral support. Perhaps if the loads on the slab are fairly well fixed in position, they will contribute to the friction and it may be reasonable to assume full lateral support. If on the other hand there is much movement of the loads and appreciable vibration, the friction may well be reduced and full lateral support should not be assumed. Such situations occur in bridges due to traffic, and in buildings with vibrating machinery such as printing presses.

Should lateral support of the compression flange not be provided by a floor slab, it is possible that such support may be provided with connecting

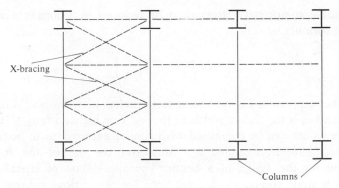

X-bracing

Columns

Figure 6-4

beams or with special members inserted for that purpose. Beams that frame into the sides of the beam or girder in question and are connected to the compression flange can usually be counted on to provide full lateral support at the connection. If the connection is made primarily to the tensile flange, little lateral support is provided to the compression flange. Before support is assumed from these beams the designer should note if they themselves are prevented from moving. The series of beams represented with horizontal dotted lines in Fig. 6-4 provide questionable lateral support for the main beams between columns. For a situation of this type some system of x-bracing may be desirable in one of the bays. Such a system is shown in Fig. 6-4. This one system will provide sufficient lateral support for the beams for several bays.

The corrugated sheet-metal roofs which are usually connected to the purlins with metal straps probably furnish only partial lateral support. A similar situation exists when wood flooring is bolted to supporting steel beams. At this time the student quite naturally asks, "If only partial support is available, what am I to consider to be the distance between points of lateral support?" The answer to this question is for him to use his judgment. As an illustration, a wood floor is assumed to be bolted every 4 ft to the supporting steel beams in such a manner that it is thought only partial lateral support is provided at those points. After studying the situation the engineer might well decide that the equivalent of full lateral support at 8-ft intervals is provided. Such a decision seems to be within the spirit of the specifications.

6-7. DESIGN OF LATERALLY UNSUPPORTED BEAMS

The general practice through the years has been to reduce the allowable stress in the fibers of the compression flange of a beam with little lateral support. As an illustration, the 1977 AASHTO Specifications permit bending stresses of $0.55F_y$ or 20,000 psi for A36 steel if full lateral support is provided by embedment in concrete. If full lateral support is not provided, the allowable compressive stress is to be reduced in accordance with the following formula.

$$F_b = 20,000 - 7.5\left(\frac{l}{b_f}\right)^2$$

In this expression l is the distance in inches between points of lateral support and b_f is the flange width in inches. For higher strength steels the allowable values can be computed with similar expressions to be found in those specifications. Example 6-5 illustrates the use of the AASHTO expression for the design of a beam. The application of lateral-support formulas is quite similar to that required for the various column expressions, and a section that can adequately resist the moment can quickly be found. It will be noted, however, that the determination of the lightest section that can adequately support the loads may involve quite a lengthy trial-and-error process, and for this reason curves such as those available in the AISC Manual and to be discussed later in this section are highly desirable.

Example 6-5

Select a steel section for the loads and span of Fig. 6-5. The beam is to be designed with A36 steel and the 1977 AASHTO Specifications. Lateral support is provided only at the beam ends.

SOLUTION
Assume beam weight = 70 lb/ft.

$$M = (20)(10) + \frac{(0.070)(20)^2}{8} = 203.5 \text{ ft-k}$$

Assume allowable fiber stress = 15 ksi

$$S_{\text{reqd.}} = \frac{12 \times 203.5}{15} = 163 \text{ in.}^3$$

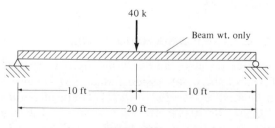

Figure 6-5

Try W24×76 ($b = 8.990$, $S = 176$)

$$F_b = 20,000 - 7.5\left(\frac{12 \times 20}{8.990}\right)^2 = 14.65 \text{ ksi}$$

resisting moment $= F_b S = \dfrac{(14.65)(176)}{12} = 215 \text{ ft-k} > 203.5 \text{ ft-k}$ OK

Use W24×76

The AISC Specification presents three expressions (1.5-6a, 1.5-6b and 1.5-7) for determining the allowable bending fiber stresses in beams for which continuous lateral support is not provided. The expressions are applicable to rolled shapes, plate girders, and built-up members having an axis of symmetry in the plane of the web. Depending upon the proportions of the member and the unbraced length the designer will substitute in Formulas 1.5-6a and 1.5-7 or into 1.5-6b and 1.5-7 and use the larger value so obtained provided the result is not greater than the maximum permissible value of $0.60F_y$.

The lateral buckling strength of a beam can be estimated by taking into account the torsional resistance of the beam about its longitudinal axis and the lateral bending resistance of the beam plus the resistance of the flange to torsion.[2-5] The resulting expression is, however, too complicated for practical engineering use.

For shallow thick-walled sections the resistance to torsion about the longitudinal axis and the lateral buckling resistance are the most important factors. For these cases, AISC Formula 1.5-7 is considered to give a reasonable approximation of an allowable buckling stress. In the expression that follows, l is the distance between points of lateral support, d is the beam depth, and A_f is the flange area. Should the beam under consideration be a cantilever the unsupported length can be conservatively assumed to equal the actual length.[6]

The allowable bending compression stresses permitted by the AISC Specification for sections loaded in the plane of their webs and having an axis of symmetry in their webs equals the larger value computed by Formulas 1.5-6a or 1.5-6b and 1.5-7 as described in the following paragraphs except not more than $0.60F_y$. This procedure for determining

[2] Karl De Vries, "Strength of Beams as Determined By Lateral Buckling," *Trans. ASCE* **112** (1947), pp. 1245–1271.

[3] G. Winter et al., "Discussion of 'Strength of Beams as Determined By Lateral Buckling,'" *Trans. ASCE* **112** (1947), pp. 1272–1320.

[4] G. G. Kubo, B. G. Johnston, and W. J. Eney, "Nonuniform Torsion of Plate Girders," *Trans. ASCE* **121** (1956), pp. 759–785.

[5] K. Basler and B. Thürlimann, "Strength of Plate Girders In Bending," *Proc. ASCE* **87**, no. 6 (August, 1961), pp. 153–181.

[6] S. Timoshenko and J. M. Gere, *Theory of Elastic Stability*, 2d ed. (New York: McGraw-Hill, 1961), pp. 257–262.

allowable stresses also applies to compression on extreme fibers of channels bent about their major axes. When

$$\sqrt{\frac{102 \times 10^3 C_b}{F_y}} \leqslant \frac{l}{r_T} \leqslant \sqrt{\frac{510 \times 10^3 C_b}{F_y}}$$

$$F_b = \left[\frac{2}{3} - \frac{F_y (l/r_T)^2}{1530 \times 10^3 C_b} \right] F_y \qquad \text{(AISC Formula 1.5-6a)}$$

When

$$\frac{l}{r_T} \geqslant \sqrt{\frac{510 \times 10^3 C_b}{F_y}}$$

$$F_b = \frac{170 \times 10^3 C_b}{(l/r_T)^2} \qquad \text{(AISC Formula 1.5-6b)}$$

or, when the compression flange is solid and approximately rectangular in cross section and its area is not less than that of the tension flange

$$F_b = \frac{12 \times 10^3 C_b}{ld/A_f} \qquad \text{(AISC Formula 1.5-7)}$$

Formula 1.5-7 gives allowable values that more nearly approach the shallow thick-walled sections but is somewhat conservative for a few deep thin W sections and nearly all plate girders. The designer could use this formula and ignore the other ones. Her designs would be perfectly safe although considerably overdesigned for the cases mentioned. For these members the resistance of the flange to torsion is the predominant factor and AISC formulas 1.5-6a and 1.5-6b more clearly estimate the effect of this item.

In these expressions l is the unbraced length of the compression flange; r_T is the radius of gyration of the compression flange plus one-third of the compression web area taken about an axis in the plane of the web; and C_b is a bending coefficient determined as follows.

$$C_b = 1.75 + 1.05 \left(\frac{M_1}{M_2} \right) + 0.3 \left(\frac{M_1}{M_2} \right)^2 \leqslant 2.3$$

The end restraint conditions and the loading conditions may be such as to provide appreciable resistance to lateral buckling. C_b is a modifier that is given to estimate the effect of these items. In this expression for C_b, M_1 is the smaller and M_2 the larger of the bending moments at the ends of the unbraced length taken about the strong axis of the member. Should the moment at any point within the unbraced length be larger than the end moments, C_b shall be taken as 1. The ratio M_1/M_2 is considered positive if M_1 and M_2 have the same sign (reverse curvature bending) and negative if they have opposite signs (single curvature bending).

The AISC Manual provides information that greatly simplifies the application of these seemingly complex lateral support equations. Of particular use are the L_c and L_u values which are provided in the beam and column sections of the manual. In subparagraph 5 of Section 1.5.1.4 the AISC states that for flanged beams the distances between points of lateral bracing should not exceed $76b_f/\sqrt{F_y}$ nor $20{,}000/[(d/A_f)F_y]$ if the members are to be assumed to have adequate lateral bracing. The least of these two values is referred to as L_c.

Should the distance between points of lateral bracing be greater than L_c the AISC says that the allowable bending stress must be reduced from $0.66F_y$, with the appropriate formula but in no case may it exceed $0.60F_y$. When these formulas are used, however, there is a range in which they give a value above $0.60F_y$. For each beam there is an unbraced length for which the controlling formula yields an allowable stress exactly equal to $0.60F_y$. This length is called L_u throughout the manual. Based on this information it is possible to make the following simplifying statements.

1. If the unbraced length $\leqslant L_c$, F_b equals $0.66F_y$ assuming the other requirements of AISC Section 1.5.1.4 are met.
2. If the unbraced length is $> L_c$ but $\ll L_u$, F_b equals $0.60F_y$.
3. If the unbraced length is $> L_c$ and $> L_u$, F_b will be less than $0.60F_y$ and can be determined by the appropriate formulas from Section 1.5.1.4.5 of the AISC Specification. The designer will often find that the curves given in the beam part of the AISC Manual entitled "Allowable Moments in Beams with Unbraced Length Greater Than L_u" and discussed later in this section will be helpful in this regard.

Example 6-6 illustrates the calculations necessary to determine the allowable stresses in a W section that has different unbraced lengths.

Example 6-6

Determine the allowable bending stresses in a W33×130 (see Fig. 6-6) for simple spans of 10, 13, 20, and 30 ft without lateral support. Use A36 steel and the AISC Specification.

SOLUTION
Computing properties of the section

$$A_f + \frac{1}{6}A_w = (11.510)(0.855) + \left(\frac{1}{3}\right)\left(\frac{31.38}{2}\right)(0.580) = 12.87 \text{ in.}^2$$

$$r_T = \sqrt{\frac{\frac{1}{2} \times 218}{12.87}} = 2.91$$

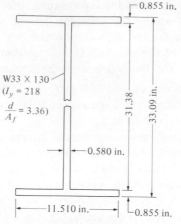

Figure 6-6

(Approximate I_y for tee section $= \frac{1}{2} I_y$ for whole section. More exact values of r_T are given in the Steel Manual with r_T given there as 2.88 for this section.) $C_b = 1.0$ since moment in span exceeds value at both ends.

(a) $l_{unbr.} = 10$ ft

$$L_c = 12.1 \text{ ft from manual} > 10\text{ft}$$

Therefore $F_b = 0.66 F_y = 24$ ksi.

(b) $l_{unbr.} = 13$ ft

$$L_u = 13.8 \text{ ft from manual}$$

$$l_{unbr.} > L_c < L_u$$

Therefore $F_b = 0.60 F_y = 22$ ksi.

(c) $l_{unbr.} = 20$ ft

$$l_{unbr.} > L_u$$

Therefore, one must use formulas

$$\sqrt{\frac{102 \times 10^3 \times 1.0}{36}} = 53 < \frac{l}{r_T} \quad \text{of} \quad \frac{12 \times 20}{2.88} = 83.3 < \sqrt{\frac{510 \times 10^3 \times 1.0}{36}}$$

$$= 119$$

Therefore, use Formulas 1.5-6a and 1.5-7.

$$F_b = \left[\frac{2}{3} - \frac{(36)(83.3)^2}{(1530)(10^3)(1)} \right] 36 = 18.1 \text{ ksi} \leftarrow$$

$$F_b = \frac{12 \times 10^3 \times 1}{12 \times 20 \times 3.36} = 14.9 \text{ ksi}$$

(d) $l_{\text{unbr.}} = 30$ ft

$$\sqrt{\frac{102 \times 10^3 \times 1.0}{36}} = 53 < \frac{l}{r_T} \quad \text{of} \quad \frac{12 \times 30}{2.88} = 125 > \sqrt{\frac{510 \times 10^3 \times 1.0}{36}}$$

$$= 119$$

Therefore, use Formulas 1.5-6b and 1.5-7.

$$F_b = \frac{170 \times 10^3 \times 1.0}{(125)^2} = 10.88 \text{ ksi} \leftarrow$$

$$F_b = \frac{12 \times 10^3 \times 1.0}{12 \times 30 \times 3.36} = 9.92 \text{ ksi}$$

Example 6-7 illustrates the design of a beam without full lateral support using the AISC Specification. The problem is very simple if the charts in the manual entitled "Allowable Moments in Beams with Unbraced Lengths Greater Than L_u" are used. In these charts the resisting moments of the sections commonly used as beams are plotted for different unbraced lengths and with C_b assumed to equal 1.0.

The designer enters the chart with unbraced length on the bottom scale and bending moment on the horizontal scale. She goes to the intersection of these two values and then moves up and to the right. Any section encountered in that direction will have a greater unbraced length and a greater resisting moment than needed. The first solid line encountered represents the most economical section available. Frequently a dashed line is encountered before a solid line is reached. The shape represented by the dashed line will have a sufficient resisting moment but will not be as economical.

Example 6-7

Select the lightest available steel section for the beam shown in Fig. 6-7 using the AISC Specification and A36 steel. Lateral support is provided at the ends only.

SOLUTION

Assume beam weight = 100 lb/ft.

$$M = \frac{(7.1)(20)^2}{8} = 355 \text{ ft-k}$$

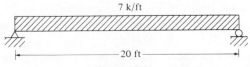

7 k/ft

20 ft

Figure 6-7

From Chart, noting $C_b = 1.0$,

<u>Use W30×99</u>

The charts are given for beams with C_b values equal to 1.0. Should the unbraced length be greater than L_u and the bending moment within that length smaller than at either end C_b will be larger than 1.0 and the allowable moment should be determined in accordance with the requirements of Section 1.5.14.5(2a) of the AISC Specification.

Should it be necessary to design a beam without full lateral support when the charts cannot be used (as where charts are not available for the steel being used, or where the particular conditions do not fall on the chart, etc.) it will be necessary to use a trial and error solution. An allowable stress can be assumed, the required section modulus computed, a trial beam selected, the allowable stress determined for the trial beam size, etc. There are, however, several variables in the AISC formulas and although it's not difficult to find a section that will adequately support the load it is quite difficult to find the absolutely lightest section by trial and error.

The student may quite logically ask the question, "What do I do if the beam in question has no compression flange?" Such a situation might very well occur in the bottom chord member of a truss subject to an intermediate load which causes bending in addition to the normal axial stress. Should the member consist of a pair of angles or a structural tee with the flanges on the tension side there will be no compression flange, only a web.

The formulas presented in this section are for members that are symmetrical about both x and y axes, as W or S shapes. For other shapes more complicated expressions are needed for estimating the allowable stresses. For reference the student is referred to *Guide to Stability Design Criteria for Metal Structures.*[7] A reasonable and very conservative practice is to assume $F_b = 0.60F_y$ for all such situations, as long as the local buckling requirements of AISC Section 1.9.1 are satisfied.

6-8. DESIGN OF CONTINUOUS MEMBERS

The AISC Specification for the elastic design of continuous members leans definitely toward the plastic design theories. Both theory and tests show clearly that continuous ductile steel members meeting the requirements for compact sections have the desirable ability of being able to redistribute the moments caused by overloads. (The student is again referred to the introductory paragraphs of Chapter 21 for a detailed discussion of this subject.)

The continuous beam of Fig. 6-8 is considered in this paragraph. The magnitude of the load applied to this beam can be increased until it

[7]Structural Stability Research Council, *Guide to Stability Design Criteria for Metal Structures,* B. G. Johnston, ed., 3rd ed (New York: Wiley, 1976).

Girders for all-welded Connecticut expressway bridge. (Courtesy of The Lincoln Electric Company.)

reaches a value above which there will be no increase in the support moments. (These moments, which are defined as plastic moments in Chapter 21, occur when the steel has been stressed to its yield point all the way through the section at some point.) Should the load be increased further the beam will act as though there is no continuity over the interior supports. These points will act as though they are hinges so far as load increases go but will continue to transfer the plastic moments. Increases in load will cause increases in the positive moments out in the "simple beams" between supports but no increases at the supports. The result is that the negative and positive moments in the beam tend to equalize as the

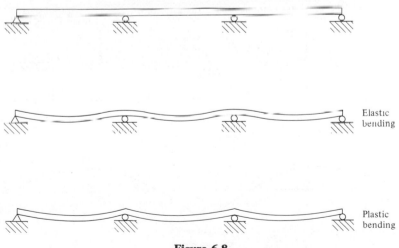

Figure 6-8

load is increased. The beam may be said to have a reserve of strength and will not fail until the load is decidedly increased above the value at which the outermost fibers of the steel were first stressed to its yield point.

The AISC says that for continuous compact sections the design may be made on the basis of $\frac{9}{10}$ of the maximum negative moments caused by gravity loads if the positive moments are increased by $\frac{1}{10}$ of the average negative moments at the adjacent supports. (The 0.9 factor is applicable only to gravity loads and not to lateral loads such as those caused by wind and earthquake. The factor can also be applied to columns which have axial stresses of less than $0.15F_y$.) This moment reduction does not apply to members consisting of A514 steel, hybrid girders nor to moments produced by loading on cantilevers. Example 6-8(a) illustrates the design of a two-span beam falling into this class. Part (b) of this example illustrates the design of the same beam if continuous lateral bracing is not provided.

Example 6-8

(a) A W18×55 consisting of a steel with $F_y = 45$ ksi is used for the span and loads of Fig. 6-9. Is the section satisfactory according to the AISC Specification if continuous lateral support is provided by a reinforced concrete slab?

(b) Is the section satisfactory if lateral support is provided only at beam supports and midspans?

SOLUTION

(a) Continuous lateral support:

$$F_b = 0.66F_y = 29.7 \text{ ksi}$$

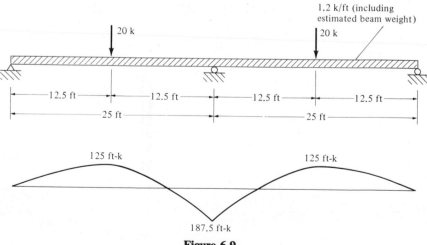

Figure 6-9

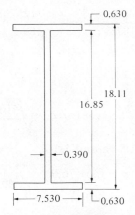

Figure 6-10

max negative M for design $= (0.9)(187.5) = 168.8$ ft-k

max positive M for design $= 125 + (0.10)\left(\dfrac{0 + 187.5}{2}\right) = 134.4$ ft-k

$$f_b = \frac{12 \times 168.8}{98.3} = 20.6 \text{ ksi} < 29.7 \text{ ksi} \qquad\qquad \text{OK}$$

(b) Lateral support at 12.5-ft intervals: Considering the 12.5 ft at ends of beam,

$$C_b = 1.75 + (1.05)\left(\frac{0}{125}\right) + (0.3)\left(\frac{0}{125}\right)^2 = 1.75$$

$$r_T = 1.95 \text{ from manual}$$

$$\sqrt{\frac{102 \times 10^3 \times 1.75}{45}} = 63.0 < \frac{12 \times 12.5}{1.95} \qquad \text{of}$$

$$76.9 < \sqrt{\frac{510 \times 10^3 \times 1.75}{45}} \qquad\qquad \text{of} \quad 140.8$$

Therefore use Formulas 1.5-6a and 1.5-7.

$$F_b = \left[\frac{2}{3} - \frac{45(76.9)^2}{1530 \times 10^3 \times 1.75}\right] 45 = 25.5 \text{ ksi}$$

$$F_b = \frac{12 \times 10^3 \times 1.75}{12 \times 12.5 \times 3.82} = 36.6 \text{ ksi} \leftarrow \text{use } 0.60F_y = 27 \text{ ksi}$$

$$f_b = \frac{12 \times 125}{98.3} = 15.30 \text{ ksi} < 27 \text{ ksi} \qquad\qquad \text{OK}$$

Considering the interior 12.5-ft section,

$$C_b = 1.75 + 1.05\left(+\frac{125}{187.5}\right) + 0.3\left(+\frac{125}{187.5}\right)^2 = 2.58 > 2.30$$

$$C_b = 2.30$$

$$\sqrt{\frac{102 \times 10^3 \times 2.30}{45}} = 72.2$$

$$< \frac{12 \times 12.5}{1.95} \quad \text{of} \quad 76.9 < \sqrt{\frac{510 \times 10^3 \times 2.30}{45}} \quad \text{of} \quad 161.5$$

Therefore use Formulas 1.5-6a and 1.5-7.

$$F_b = \left[\frac{2}{3} - \frac{(45)(76.9)^2}{1530 \times 10^3 \times 2.30}\right] 45 = 26.6 \text{ ksi}$$

$$F_b = \frac{12 \times 10^3 \times 2.30}{12 \times 12.5 \times 3.82} = 48.2 \text{ ksi} \leftarrow$$

Therefore, use $0.60F_y = 27$ ksi.

$$f_b = \frac{12 \times 168.8}{98.3} = 20.6 \text{ ksi} < 27 \text{ ksi} \qquad\qquad \text{OK}$$

Section is satisfactory

Problems*

6-1. Select the most economical section available for a 30-ft simple span if the beam is to support a uniform load of 3 k/ft. The beam is assumed to have full lateral support and an allowable bending stress of 20,000 psi. (*Ans.* W27×84)

6-2. to 6-6. Select the most economical section available for each of the beams shown in the accompanying illustrations assuming full lateral support is provided for the compression flanges. Use A36 steel and the AISC Specification.

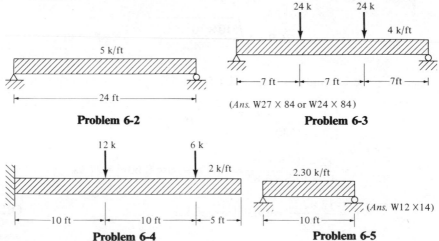

Problem 6-2

(*Ans.* W27 × 84 or W24 × 84)

Problem 6-3

Problem 6-4

(*Ans.* W12 ×14)

Problem 6-5

*NOTE: Consider moments only for the problems of this chapter.

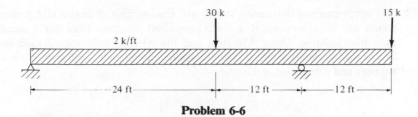

Problem 6-6

6-7. Select the most economical section for the beam shown in the accompanying illustration using a steel with an allowable bending stress of 30,000 psi. Full lateral support is assumed for the entire span. (*Ans.* W36×150)

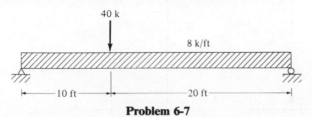

Problem 6-7

6-8. Select the most economical section available for a 32-ft simple span if the beam is to support a uniform load of 10 k/ft. Use A242 steel, the AISC Specification and assume that full lateral support is provided.

6-9. A W24×68 consisting of A36 steel is used for a simple span of 30 ft. Using the AISC Specification and assuming full lateral support determine the maximum allowable uniform load the beam can support in addition to its own weight. (*Ans.* $w=2.67$ k/ft or 38.97 kN/m)

6-10. A beam consists of a W16×40 with a $\frac{1}{2}$×12 cover plate welded to each flange. If the allowable bending stress is 24,000 psi, determine the allowable uniform load it can support in addition to its own weight for a 30-ft simple span.

6-11. Using the AISC Specification and A36 steel and assuming full lateral support determine the maximum allowable uniform load the beam shown in the accompanying illustration can support in addition to its own weight for a 24-ft simple span. (*Ans.* 49.18 kN/m)

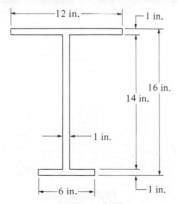

Problem 6-11

6-12. The accompanying illustration shows the arrangement of beams and girders which are used to support a 6-in. reinforced concrete floor for a small industrial building. Using A36 steel and the AISC Specification design the beams and girders assuming they are simply supported. Assume full lateral support and a live load of 120 psf.

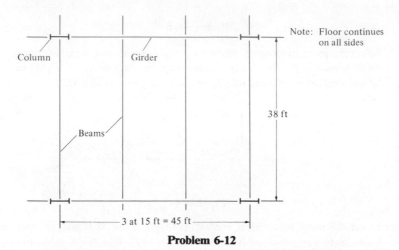

Note: Floor continues on all sides

Problem 6-12

6-13. A 40-ft simple beam is to support two movable 20-k loads a distance of 12 ft apart. If the beam is to have full lateral support select a section to resist the largest possible bending moment. AISC. A36. (*Ans.* W24×68)

6-14. Select the lightest C or MC section available to support a uniform load of 700 lbs/ft for 22-ft simple span. A36. AISC. Full lateral support.

6-15. A W27×94 with full lateral support is used for a 30-ft simple span and supports a 4 k/ft uniform load. There are assumed to be two holes for 1-in. bolts in each flange. Compute the maximum compressive stress using the gross section properties and the maximum tensile stress using the net section properties. Do not use the AISC Specification (Section 1.10.1) as regards the 15% rule of the gross flange area. (*Ans.* $f_c = 22.7$ ksi; $f_t = 27.6$ ksi)

6-16. The section shown in the accompanying illustration has two 1-in. bolts passing through each flange. Find the allowable uniform load this section can

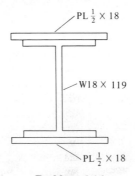

Problem 6-16

support in addition to its own weight for a 30-ft simple span. Allowable bending stress = 22 ksi. Not AISC Specification.

6-17. Rework Prob. 6-1 assuming two $\frac{3}{4}$-in. bolts pass through each flange of the section at the point of maximum moment. Not AISC. (*Ans.* W27×94)

6-18. Rework Prob. 6-15 using the AISC Specification as regards the 15% of the gross flange area rule (Section 1.10.1).

6-19. It is desired to select a section to support a 2.5 k/ft load for an 18-ft simple span. If full lateral support is assumed and two $\frac{3}{4}$-in. bolts are needed in each flange, select a W section using A36 steel and the AISC Specification. (*Ans.* W18×35)

6-20. Select a section for the span and loads shown in the accompanying illustration. Use an allowable $F_b = 20$ ksi and assume there are to be two holes for 1-in. bolts in the tension flange. Not AISC Specification.

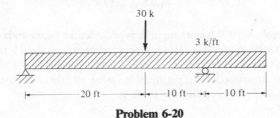

30 k

3 k/ft

|← 20 ft →|← 10 ft →|← 10 ft →|

Problem 6-20

6-21. Using the 1977 AASHTO expression for A36 steel $[F_b = 20{,}000 - 7.5(l/b_f)^2]$ select a W24 to support a 5 k/ft load for an 18-ft simple span if lateral support is provided at the beam ends only. (*Ans.* W24×76)

6.22. Compute the allowable bending stresses in a W21×57 for simple spans of 6, 12, and 25 ft if lateral support is provided at ends only. Use A36 steel and the AISC Specification.

6-23. A W36×150 consisting of A36 steel is used for a simple span of 24 ft, and has lateral support at its ends only. What is the largest concentrated load that can be placed at the beam center line? AISC Specification. (*Ans.* 111.2 k)

6.24. The beam shown in the accompanying illustration has lateral support provided at points *A* and *B* only. Using A36 steel and the AISC Specification design the beam.

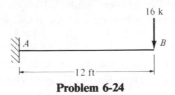

16 k

A   *B*

|← 12 ft →|

Problem 6-24

6-25. Using a steel with $F_y = 65$ ksi and the AISC Specification what uniformly distributed load can a simply supported W33×241 carry for a span of 40 ft when (a) the compression flange is braced laterally (*Ans.* 14.58 k/ft) and (b) the compression flange has lateral support only at its ends? (*Ans.* 5.36 k/ft)

6-26. What is the allowable uniform load that can be placed on a W30×99 (A242 steel) which has lateral support provided at its ends only? The beam is simply supported and has a span of 30 ft. AISC Specification.

6-27. A W33×130 is used to support the loads and moment shown in the accompanying illustration. Using A36 steel and the AISC Specification and neglecting the dead weight of the beam, find if the beam is overloaded
   (a) If full lateral support is provided (*Ans.* Beam is satisfactory, $M = 340$ ft-k, resisting moment $= 812$ ft-k)
   (b) If lateral support is provided at the ends only. (*Ans.* Beam is satisfactory, $M = 340$ ft-k, resisting moment $= 368.1$ ft-k)

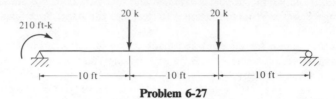

Problem 6-27

6-28. Repeat Prob. 6-19 if lateral support is provided at beam ends only.

6-29. The W30×99 shown has lateral support supplied at the 25 feet points only (beam ends and centerline). If F_y is 65 ksi and the AISC Specification is used determine the maximum permissible value of the concentrated load P. Neglect beam weight. (*Ans.* $P = 19.91$ k)

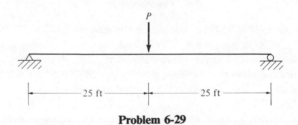

Problem 6-29

6-30. A W30×99 with lateral support supplied at its ends only supports a concentrated load at the center of its 30-ft simple span. Using A242 steel and the AISC Specification determine the maximum permissible value of the load.

6-31. The architects specify that a beam no greater than 13.00 in. in depth be designed to support a uniform load of 10 k/ft for an 18-ft simple span. Design the beam with A36 steel and the AISC Specification assuming full lateral support. (*Ans.* W10×112 with 1PL$\frac{3}{4}$×12 each flange—many other solutions possible)

6-32. Select the lightest W section that will be satisfactory for all three spans of the beam shown. Use A36 steel, the AISC Specification, and assume full lateral support is provided.

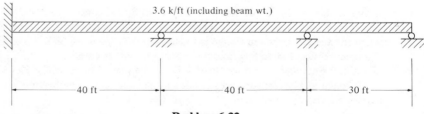

Problem 6-32

6-33. The member shown in the accompanying illustration is to have full lateral support and consist of A36 steel. If beam weight is neglected what is the lightest available W section that can be used according to the AISC Specification? (*Ans.*W27×84)

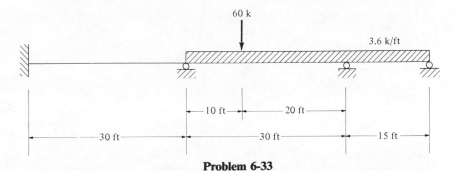

Problem 6-33

6-34. Using A36 steel and the AISC Specification select the lightest available section for the beam shown in the accompanying illustration. The section is provided with full lateral support.

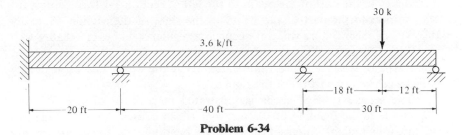

Problem 6-34

6-35. Three methods of supporting a roof are shown in the accompanying illustration. Design the beam for each case using A36 steel and the AISC Specification if a 2 k/ft uniform load is to be supported. Assume full lateral support. [*Ans.* (a) W24×76, (b) W21×62, (c) W21×62]

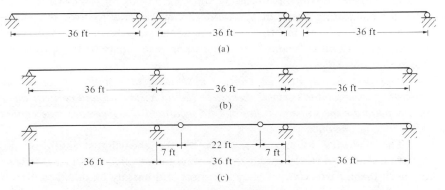

Problem 6-35

Chapter 7
Design of Beams
(Continued)

7-1. SHEAR

The beam supporting transverse loads, shown in Fig. 7-1(a), is sometimes said to be subjected to two types of shear—*transverse* and *longitudinal*. The first of these two shears is illustrated in part (b) of the figure where there is a tendency of the part of the beam to the left of section 1-1 to slide upward with respect to the part of the beam to the right of the section. Actually this type of shear failure will not occur in a regular steel beam because web crippling (discussed in next section) will occur first. Transverse shear, however, can feasibly cause failure directly if the beam has been deeply coped as shown in part (c) of the figure.

The second type of shear is illustrated in part (d) of Fig. 7-1 and occurs due to the bending of the member which causes changes in lengths of the longitudinal fibers. For positive bending the lower fibers are stretched and the upper fibers are shortened while somewhere in between there is a neutral axis where the fibers do not change in length. Due to these varying deformations a particular fiber has a tendency to slip on the fiber above or below. The largest value of longitudinal or horizontal shear occurs at the neutral axis.

If a wooden beam was made by stacking boards on top of each other and not connecting them they would obviously tend to take the shape shown in part (d) of the figure. The student may have observed short heavily loaded timber beams with large transverse shears which split along horizontal planes.

This presentation may be entirely misleading in seeming to completely separate horizontal and vertical shears. In reality horizontal and vertical shear at any point are the same and may not be separated. Furthermore, one cannot occur without the other.

The tendency to slip is resisted by the shearing strength of the material. Although steel beam sizes are rarely controlled by shear it is wise to check them, particularly if they are short and heavily loaded. Maximum

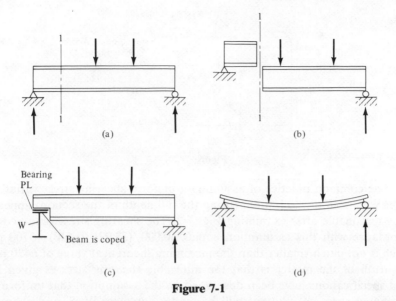

Figure 7-1

external shear usually occurs near the supports but shear is present throughout the beam. The average transverse shearing stress on the cross section of a beam at a certain point in the span equals the external shear divided by the cross-sectional area of the beam. The longitudinal shearing stress formula, however, shows that the shearing stress is not constant across a beam cross section but is zero at the outermost fibers and has its largest value at the neutral axis. The familiar formula, to follow, can be used to calculate the unit shearing stress at any point. (This formula applies to beams with open cross sections which are not subjected to torsion.)

$$f_v = \frac{VQ}{bI}$$

where

V = external shear at the section in question
Q = statical moment of that portion of the section lying outside (either above or below) the line on which f_v is desired, taken about the neutral axis
I = moment of inertia of the entire section about the neutral axis
b = width of the section where the unit shearing stress is desired

When this expression is used to calculate the shear across the face of a W, M, or S section the resulting values are very small in the flanges and quite large in the web. The values in the web are fairly uniform from top to bottom. Figure 7-2 shows the variation of shear on the cross section of a W24×76 which is resisting an external shear of 60,000 lb.

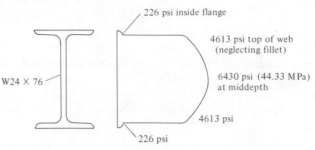

Figure 7-2

The common practice of assuming a uniform shearing stress variation in the web from top to bottom (using the full depth of the section) appears to be reasonable after examining Fig. 7-2. The shearing stress obtained in accordance with this assumption equals $60,000/(23.92)(0.440)=5700$ psi which is not much smaller than the maximum theoretical value of 6430 psi. The truth of the matter is that the allowable shearing stresses given by most specifications have been developed on the assumption that uniformly distributed shear calculations will be made. For most W, S, and channel sections having fairly large flanges and thin webs the results obtained from this assumption are reasonable. For other shapes a few calculations may be necessary to see if the percent error is appreciable.

A more conservative approach followed by some engineers is to use only the web depth in figuring the average shear. This method would give a result equal to $60,000/(22.56)(0.440)=6044$ psi in the case being considered. Although some designers use the total beam depth and others only the web depth or even the depth between the toes of the fillets for rolled sections, they nearly all agree to use only the web depth for plate girders.

The AISC Specification (1.5.1.2.1) states that the allowable shearing stress F_v is $0.40F_y$ on the cross-sectional area effective in resisting shear. They further state that for rolled and fabricated shapes this area equals the overall depth times the web thickness. In other words the shear stress is to be calculated as follows.

$$f_v = \frac{V}{dt_w}$$

The maximum permissible external shear which a particular beam can resist can be calculated by substituting the allowable shear stress F_v into the preceding expression and solving for V.

$$V = F_v \, dt_w$$

In the beam tables of the AISC Manual this value V is tabulated for each of the sections normally used as beams with yield stresses of 36 and 50 ksi. The designer can with a glance at these tables check the shear in a particular beam. If it is too high he can quickly select another section which has a satisfactory allowable V. On some occasions when short spans and high shears are present S beams with their rather heavy webs may prove to be economical.

Generally shear is not a problem in steel beams because the webs of rolled shapes are capable of resisting quite large shearing stresses. Perhaps it is well, however, to list here the most common situations where shear might be excessive. These are as follows.

1. Should large concentrated loads be placed near beam supports they will cause large external shears without corresponding increases in bending moments. A fairly common example of this type occurs in tall buildings where on a particular floor the upper columns are offset with respect to the columns below. The loads from the upper columns applied to the beams on the floor level in question will be quite large if there are many stories above.

2. Probably the most common shearing stress problem occurs where two members (as a beam and a column) are rigidly connected together so their webs lie in a common plane. In Section 1.5.1.2 of the *Commentary on the AISC Specification*, formulas are given which show when the webs are overstressed in shear and how much they need to be thickened or reinforced when overstressed. This situation, which commonly occurs at the junction of columns and beams (or rafters) in rigid frame structures, is discussed in Chapter 22.

3. Where beams are notched or coped as was shown in Fig. 7-1(c) shear can be a problem. For this case shear stresses can be calculated for

Combined welded and bolted joint, Transamerica Pyramid, San Francisco, Calif. (Courtesy of Kaiser Steel Corporation).

the remaining beam depth. A similar discussion can be made where holes are cut in beam webs for duct work or other items.

4. Theoretically, very heavily loaded short beams can have excessive shears but practically this does not occur too often unless it is like case 1.

5. Shear may very well be a problem even for ordinary loadings when very thin webs are used as in plate girders or in light gage cold formed steel members.

Should calculated shearing stresses exceed allowable values ($F_v =$ $0.40F_y$ in the AISC) shear plates can be connected to the webs in the zones of excessive stress.

7-2. WEB CRIPPLING

Beams that support heavy concentrated loads sometimes fail by web crippling or crushing unless the web is stiffened near the loads. Web crippling occurs due to the stress concentrations at the junction of the flange and the web, where the beam is trying to transfer compression in the relatively wide flange to the narrow web. Failure will occur when the metal begins to fail at the toe of the fillet in bearing and the flange and web have a tendency to fold over each other.

Most specifications assume the reaction or load spreads out from its place of application along a 45° plane (see Fig. 7-3). The toe of the fillet is the most dangerous location for failure because the resisting area has its smallest value there. The AISC does not permit the compression at this point in beams without web stiffeners to exceed $0.75F_y$. Should the allowable value be exceeded it is necessary to use stiffeners, or the bearing of the load or reaction must be spread over a greater length. The expressions for the length of bearing required are shown in Fig. 7-3. The top flanges of a beam must have lateral support at the reaction points or the web crippling strength is greatly reduced.

The minimum bearing length at the end of a certain beam is calcu-

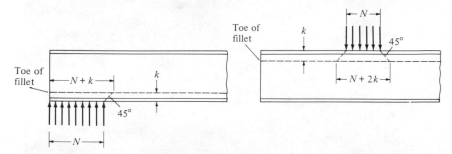

Figure 7-3 Left: Reaction (AISC Formula 1.10-9). $N =$ length of bearing, which cannot be less than k; $k =$ distance from toe of fillet to outside of flange; $t =$ thickness of web; $R =$ reaction; $R/t(N+k)$ may not exceed $0.75F_y$. Right: Concentrated load (AISC Formula 1.10-8). $R =$ concentrated load; $R/t(N+2k)$ may not exceed $0.75F_y$.

lated in Example 7-1. For many beams the theoretical bearing length determined is too small to be practical and may even be negative. Should this be the case, the designer will select a reasonable value that will fit in with the construction requirements. For steel beams bearing on masonry a minimum bearing length of probably $3\frac{1}{2}$ or 4 in. (90 or 100 mm) should be used. It is to be remembered that a fairly large percentage of structural failures have occurred, particularly during erection, due to insufficient bearing. Another fact to keep in mind is that the bearing area should be large enough to keep the bearing stress from exceeding the allowable value of the supporting material. This subject is considered in Section 7-10 of this chapter.

Example 7-1

A W33×130 consisting of A36 steel has been selected for the loading and span of Fig. 7-4. (a) Determine the minimum bearing length required at the reactions. (b) The 50-k concentrated loads are applied to the beam over a width of 6 in. Is this sufficient?

SOLUTION

Properties of a W33×130

$$\text{web } t = 0.580 \text{ in.}$$

$$k = 1\frac{11}{16} \text{ in.} = 1.688 \text{ in.}$$

(a) Minimum bearing length required at the reactions:

$$\frac{R}{t(N+k)} = 0.75F_y$$

$$\frac{80,000}{(0.580)(N+1.688)} = (0.75)(36,000)$$

$$N = 3.42 \text{ in.}$$

(b) Web-crippling stress under concentrated loads:

$$\frac{R}{t(N+2k)} \text{ not to exceed } 0.75F_y$$

$$\frac{50,000}{(0.580)(6+2\times1.688)} = 9194 \text{ psi} < 27,000 \text{ psi} \qquad \text{OK}$$

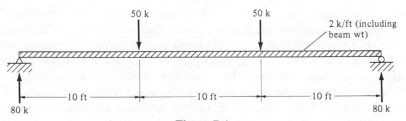

Figure 7-4

7-3. VERTICAL BUCKLING OF WEBS

The usual steel sections are rolled with proportions such that web crippling will occur before web buckling is possible. Although steel specifications at one time required the investigation of steel beams for web buckling, the requirement has been dropped; and they now only call for investigation of the possibility of web crippling. Built-up sections with high thin webs are a different matter and are considered in Chapter 16.

7-4. MAXIMUM DEFLECTIONS OF BEAMS

The deflections of steel beams are usually limited to certain maximum values. Among the several excellent reasons for deflection limitations are the following.

1. Excessive deflections may damage other materials attached to or supported by the beam in question. Plaster cracks caused by large ceiling joist deflections are one example.
2. The appearance of structures is often damaged by excessive deflections.
3. Extreme deflections do not inspire confidence in the persons using a structure although it may be completely safe from a strength standpoint.
4. It may be necessary for several different beams supporting the same loads to deflect equal amounts.

Standard American practice for buildings has been to limit live-load deflections to approximately $\frac{1}{360}$ of the span length. This deflection is supposedly the largest value that ceiling joists can deflect without causing cracks in underlying plaster and is the value permitted by the AISC for beams and girders supporting plastered ceilings. The $\frac{1}{360}$ deflection is only one of many maximum deflection values in use because of different loading situations, different engineers, and different specifications. For situations where precise and delicate machinery is supported maximum deflections may be limited to 1/1500 or 1/2000 of the span lengths. The 1977 AASHTO Specifications limit deflections in steel beams and girders due to live load and impact to $\frac{1}{800}$ of the span. (For bridges in urban areas which are partly used by pedestrians the AASHTO *recommends* a maximum value equal to 1/1000 of the span lengths.)

Before substituting blindly into a formula that will give the deflection of a beam for a certain loading condition the student should thoroughly understand the theoretical methods of calculating deflections. These methods include the moment area, conjugate beam, and virtual work procedures. From these methods the centerline deflection expression for a uniformly loaded simple beam can be determined. Example 7-2 illustrates the application of this expression. In this problem all units are changed into pound and inch units. The conversion of the uniform load given in kips per foot to pounds per inch should be particularly noted.

$$\delta_{\mathcal{L}} = \frac{5wl^4}{384EI}$$

Example 7-2

Determine the centerline deflection of the W24×62 beam used in Example 6-1. Does the resulting value exceed $\frac{1}{360}$ of the span length? Assume $E = 29 \times 10^6$ psi and $I = 1550$ in.^{4}.

SOLUTION

$$\delta_{\mathcal{L}} = \frac{(5)(4362/12)(21 \times 12)^4}{(384)(29 \times 10^6)(1550)}$$

$$\delta_{\mathcal{L}} = 0.425 \text{ in.}$$

$$\text{max allowable } \delta = \left(\frac{1}{360}\right)(12 \times 21) = 0.700 \text{ in.} > 0.425 \text{ in.} \qquad \text{OK}$$

Another deflection expression frequently used is the one giving the deflection at the centerline of a beam loaded with a concentrated load at the centerline.

$$\delta_{\mathcal{L}} = \frac{Pl^3}{48EI}$$

Some specifications handle the deflection problem by requiring certain minimum depth-span ratios. For example the AASHTO suggests the depth-span ratio be limited to a minimum value of 1/25. A shallower section is permitted but it should have sufficient stiffness to prevent a deflection greater than would have occurred if the 1/25 ratio had been used.

A steel beam can be cold-bent or cambered an amount equal to the deflection caused by dead load or the deflection caused by dead load plus some percentage of the live load. Approximately 25% of the camber so produced is elastic and will disappear when the cambering operation is completed. Detailed information for particular shapes is given in the Steel Handbook in the section entitled "Standard Mill Practice." It should be remembered that a beam which is bent upward looks much stronger and safer than one which sags downward (even a very small distance).

Other than the $\frac{1}{360}$ plaster limitation the AISC Specification does not specify exact maximum permissible deflections. There are so many different materials, types of structures, and loadings that no one single set of deflection limitations is acceptable for all cases. Thus limitations must be set by the individual designer on the basis of experience and judgment.

The AISC Commentary (1.13.1 and 1.13.2) does suggest minimum beam depths for a few cases as follows.

1. For fully stressed floor beams and girders the depth is recommended to be at least equal to $F_y/800$ times the span length where F_y is in kips per square inch. Should members of lesser depth be used the Commentary states that the allowable stress should be decreased in the same proportion as the depth is decreased from the recommended value.

2. Depths of fully stressed roof purlins are recommended to be not less than $F_y/1000$ times their span lengths except for flat roofs which are discussed under the subject of ponding in this chapter and in Chapter 23.

3. For steel beams supporting large open floor areas without partitions or other features that might cause damping, depths should preferably be not less than $\frac{1}{20}$ times span lengths so that vibrations due to pedestrian traffic are kept within acceptable values.

Ponding

If water on a flat roof accumulates faster than it runs off, the result is called *ponding* because the increased load causes the roof to deflect into a dish shape that can hold more water, which causes greater deflections, etc. This process continues until equilibrium is reached or until collapse occurs. Ponding is a serious matter as illustrated by the large number of flat roof failures which occur every year in the United States.

Ponding will occur on almost any flat roof to a certain degree even though roof drains are present. Drains may very well be used but they may be inadequate during severe storms or they may become stopped up and, furthermore, they are often placed along the beam lines which are actually the high points of the roof. The best method of preventing ponding is to have an appreciable slope on the roof ($\frac{1}{4}$ in./ft or more) together with good drainage facilities. It has been estimated that probably two-thirds of the flat roofs in the United States have slopes less than this value which is the minimum recommended by the National Roofing Contractors Association (NRCA). It costs approximately 3 to 6% more to construct a roof with this desired slope than building with no slope.[1]

When a very large flat roof (perhaps an acre or more) is being considered the effect of wind on water depth may be quite important. The problem of ponding will logically occur during a heavy rainstorm. Such a storm will frequently be accompanied by heavy winds. When a large quantity of water is present on the roof a strong wind may very well push a great deal of water to one end creating a dangerous depth of water as

[1]Gary Van Ryzin, "Roof Design: Avoid Ponding by Sloping to Drain," *Civil Engineering* (New York: American Society of Civil Engineers, January, 1980), pp. 77–81.

regards the load in pounds per square foot applied to the roof. For such situations *scuppers* are sometimes used. These are large holes or tubes in the walls or parapets which enable water above a certain depth to quickly drain off the roof.

Ponding failures will be prevented if the roof system (consisting of the roof deck and supporting beams and girders) has sufficient stiffness. The AISC Specification (1.13.3) describes a minimum stiffness to be achieved if ponding failures are to be prevented. If this minimum stiffness is not provided it is necessary to make other investigations to be sure that a ponding failure is not possible.

Theoretical calculations for ponding are very complicated. The AISC requirements are based on work by F. J. Marino[2] in which he considered the interaction of a two-way system of secondary or submembers supported by a primary system of main members or girders. In Chapter 23 of this text entitled "Miscellaneous Topics" the subject of ponding is continued. The discussion includes numerical applications.

7-5. UNSYMMETRICAL BENDING

From mechanics of materials it is remembered that each beam cross section has a pair of mutually perpendicular axes known as the principal axes for which the product of inertia is zero. Bending that occurs about any axis other than one of the principal axes is said to be unsymmetrical bending. When the external loads are not in a plane with either of the principal axes or when loads are simultaneously applied to the beam from two or more directions unsymmetrical bending is the result.

When the external loads are not in the principal plane the stresses can be determined by breaking the loads into components perpendicular to the principal axes, calculating the moments about each axis, and determining the maximum stresses caused by a combination of the two moments. The following expression can be written for the stress at any point in a beam subjected to unsymmetrical bending.

$$f_b = \frac{M_x y}{I_x} \pm \frac{M_y x}{I_y} = \frac{M_x}{S_x} \pm \frac{M_y}{S_y}$$

It might be noted that the longitudinal shearing stresses for a beam of this type can be calculated with a similar expression.

$$f_v = \frac{V_x Q_x}{b I_x} \pm \frac{V_y Q_y}{b I_y}$$

When a section has one axis of symmetry that axis is one of the principal axes and the calculations necessary for determining stress are

[2]F. J. Marino, "Ponding of Two-Way Roof System," *Engineering Journal* (New York: AISC, July, 1966), pp. 93–100.

quite simple. For this reason unsymmetrical bending is not difficult to handle in the usual beam section, which is probably a W, S, M, or C. Each of these sections has at least one axis of symmetry and the calculations are appreciably reduced. A further simplifying factor is that the loads are usually gravity loads and probably perpendicular to the x axis. For a case of this type the formula for stress is simply $f_b = Mc/I$.

Among the beams that must resist unsymmetrical bending are crane girders in industrial buildings and purlins for ordinary roof trusses. The x axes of purlins are parallel to the sloping roof surfaces while the large percentage of their loads (roofing, snow, etc.) are gravity loads. These loads do not lie in a plane with either of the principal axes of the inclined purlins and the result is unsymmetrical bending. Wind loads are generally considered to act perpendicular to the roof surface and thus perpendicular to the x axes of the purlins with the result that they are not considered to cause unsymmetrical bending. The x axes of crane girders are usually horizontal but the girders are subjected to lateral thrust loads from the moving cranes as well as to gravity loads.

Should a compact, laterally supported beam be bent about its x axis only the allowable stress F_b will be $0.66F_y$ using the AISC Specification. Should a similar shape be bent about its y axis only, F_b equals $0.75F_y$. Which allowable stress is to be used when both types of bending occur simultaneously? The AISC uses an allowable stress which is in effect a combination of the two allowable values. This is done by using the interaction equation to follow

$$\frac{f_{bx}}{F_{bx}} + \frac{f_{by}}{F_{by}} \lessgtr 1.0$$

For compact laterally supported shapes the equation becomes

$$\frac{f_{bx}}{0.66F_y} + \frac{f_{by}}{0.75F_y} \lessgtr 1.0$$

Examples 7-3 and 7-4 illustrate the design of beams subjected to unsymmetrical bending. To illustrate the trial-and-error nature of the problem the author did not do quite as much scratchwork as in the first examples. The first design problems of this type which the student attempts may quite well take him several trials. Consideration needs to be given to the question of lateral support for the compression flange. Should the lateral support be of questionable nature the engineer should reduce the allowable compressive stresses by means of one of the expressions previously given for that purpose.

Example 7-3

A certain beam is estimated to have a vertical bending moment of 120 ft-k and a lateral bending moment of 25 ft-k. These moments include the effect of the estimated beam weight. The loads are assumed to pass through the centroid of the section and the entire section modulus value is therefore available about each axis. Select a W shape that can resist these moments. Use A36 steel and the AISC Specification and assume full lateral support for the compression flange.

SOLUTION
First trial: The S required for vertical bending moments alone is $12 \times 120/24 = 60$ in.^{3} Try a section with a larger section modulus.

Try W24×62 $(S_x = 131, S_y = 9.80)$

$$f_{bx} = \frac{12 \times 120}{131} = 10.99 \text{ ksi}$$

$$F_{bx} = 0.66F_y = 24 \text{ ksi}$$

$$f_{by} = \frac{12 \times 25}{9.80} = 30.61 \text{ ksi}$$

$$F_{by} = 0.75F_y = 27 \text{ ksi}$$

$$\frac{f_{bx}}{F_{bx}} + \frac{f_{by}}{F_{by}} = \frac{10.99}{24} + \frac{30.61}{27} = 1.592 > 1.00 \qquad \text{NG}$$

Final trial:

Try W14×74 $(S_x = 112, S_y = 26.6)$

$$f_{bx} = \frac{12 \times 120}{112} = 12.86 \text{ ksi}$$

$$f_{by} = \frac{12 \times 25}{26.6} = 11.28 \text{ ksi}$$

$$\frac{f_{bx}}{F_{bx}} + \frac{f_{by}}{F_{by}} = \frac{12.86}{24} + \frac{11.28}{27} = 0.954 < 1.00 \qquad \text{OK}$$

Use W14×74

It will be noticed in the solution for Example 7-3 that though the procedure used will yield a section that will adequately support the moments the selection of the absolutely lightest satisfactory section listed in the AISC Manual could be quite lengthy. In the steel textbook by Gaylord and Gaylord[3] and in the one by Salmon and Johnson[4] practical methods are presented to assist the designer in selecting I-shaped sections for biaxial bending.

[3] E. H. Gaylord, Jr. and C. N. Gaylord, *Design of Steel Structures*, 2nd ed. (New York: McGraw-Hill, 1972), Chapter 5.
[4] C. G. Salmon and J. E. Johnson, *Steel Structures Design and Behavior*, 2nd. ed. (New York: Harper & Row, 1980), Chapter 7.

A further complication in the design of beams subject to unsymmetrical bending is caused by the fact that the loads often do not pass through the centroid of the section. For instance, the loads applied to a purlin are generally applied to its top flange and the result is a twisting moment or torsion in the inclined purlin. Rather than become involved in torsion calculations, a common design practice is to assume the lateral loads are carried only by the top flange of the beam. Only one-half of the section modulus about the y axis is considered effective and the value of f_{by} becomes $M_y / \frac{1}{2} S_y$. This rather poor design practice is used in Example 7-4.

Example 7-4

Redesign the beam of Example 7-3 if the lateral loads are assumed to be applied to the top flange of the beam and thus do not pass through the centroid of the section. Reduce the effective modulus for the y axis by 50%.

SOLUTION

Try W14×90 ($S_x = 143$, $S_y = 49.9$)

$$f_{bx} = \frac{12 \times 120}{143} = 10.07 \text{ ksi}$$

$$f_{by} = \frac{12 \times 25}{\frac{1}{2} \times 49.9} = 12.02 \text{ ksi}$$

$$\frac{f_{bx}}{F_{bx}} + \frac{f_{by}}{F_{by}} = \frac{10.07}{24} + \frac{12.02}{27} = 0.865 < 1.0 \qquad \text{OK}$$

Use W14×90

In this example (7-4) the lateral moment was estimated to cause a stress of 12.02 ksi. Of this amount 6.01 ksi was thrown in to approximate the torsion effect. (Is this a precise engineering calculation?) This subject of torsion is continued in Section 7-8 of this chapter.

When a section does not have an axis of symmetry, the work required to locate the neutral axis, to calculate the principal moments of inertia, and to determine the necessary distances for a particular point are quite tedious. The expressions (from mechanics of materials) at the end of this paragraph can be used to locate the principal axes and calculate the principal moments of inertia. In these expressions I_{xy} is the product of inertia while the various other symbols used are shown in Fig. 7-5.

$$\tan 2\phi = \frac{2I_{xy}}{I_y - I_x}$$

$$I_{x1} = \cos^2 \phi I_x + \sin^2 \phi I_y - 2 \sin \phi \cos \phi I_{xy}$$

$$I_{y1} = \cos^2 \phi I_y + \sin^2 \phi I_x + 2 \sin \phi \cos \phi I_{xy}$$

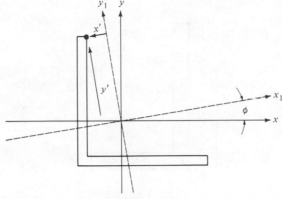

Figure 7-5

After the principal moments of inertia are determined the stresses can be obtained by breaking the loads into components perpendicular to each of the principal axes, calculating the moments about each, and substituting in the unsymmetrical bending expression. It will, however, be noted that the remaining calculations are still tedious particularly those involved in computing the perpendicular distances to each point (x and y in Fig. 7-5) from the principal axes.

The equation to follow may be more useful when the section in question has no axis of symmetry.

$$f = \frac{M_y I_x - M_x I_{xy}}{I_x I_y - I_{xy}{}^2}\, x + \frac{M_x I_y - M_y I_{xy}}{I_x I_y - I_{xy}{}^2}\, y$$

This general expression is applicable to any straight beam of constant cross section regardless of the shape and regardless of whether the section is open or closed. The x and y axes can be assumed in any convenient direction as long as they are perpendicular to each other and pass through the centroid of the cross section.

Example 7-5 illustrates the application of this expression. For substitution in the formula (as well as in the calculations for I_{xy}), x is considered to be positive for points to the right of the y axis and minus to the left, while y is positive for points above the x axis and minus below. A plus sign for the resulting stress indicates compression, and a minus sign indicates tension.

Example 7-5

Determine the stress at points A, B, and C in the $6 \times 4 \times \frac{3}{4}$-in. angle shown in Fig. 7-6 due to gravity loads which cause a moment of 5 ft-k.

Figure 7-6

SOLUTION

Properties of section:

$$I_x = 24.5 \text{ in.}^4$$

$$I_y = 8.7 \text{ in.}^4$$

$$I_{xy} = (3.00)(+0.92)(-1.71) + (3.96)(-0.71)(+1.30)$$

$$= -8.37 \text{ in.}^4$$

Stress at A:

$$f = \frac{M_y I_x - M_x I_{xy}}{I_x I_y - I_{xy}{}^2} x + \frac{M_x I_y - M_y I_{xy}}{I_x I_y - I_{xy}{}^2} y$$

$$f_A = \frac{0 - (60,000)(-8.37)}{(24.5)(8.7) - (-8.37)^2}(-1.08) + \frac{(60,000)(8.7) - 0}{(24.5)(8.7) - (-8.37)^2}(+3.92)$$

$$= +10,510 \text{ psi} \quad \text{(compression)}$$

Stress at B:

$$f_B = \frac{0 - (60,000)(-8.37)}{(24.5)(8.7) - (-8.37)^2}(-1.08) + \frac{(60,000)(8.7) - 0}{(24.5)(8.7) - (-8.37)^2}(-2.08)$$

$$= -11,378 \text{ psi} \quad \text{(tension)}$$

Stress at C:

$$f_C = \frac{0 - (60,000)(-8.37)}{(24.5)(8.7) - (-8.37)^2}(+2.92) + \frac{(60,000)(8.7) - 0}{(24.5)(8.7) - (-8.37)^2}(-2.08)$$

$$= +2660 \text{ psi} \quad \text{(compression)}$$

7-6. DESIGN OF PURLINS

To avoid bending in the top chords of roof trusses, it is theoretically desirable to place purlins only at panel points. For large trusses, however, it is more economical to space them at closer intervals. If this practice is not followed for large trusses the purlin sizes may become so large as to be impractical. When intermediate purlins are used the top chords of the truss

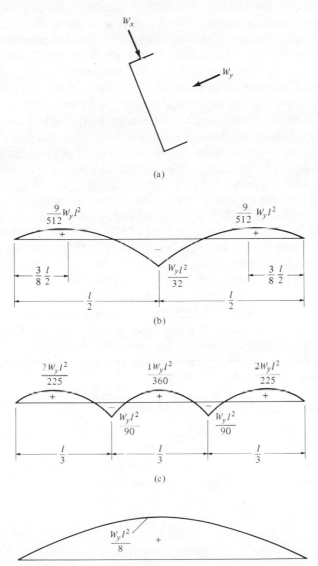

Figure 7-7 (a) Channel purlin, (b) Moment about web axis of purlins—sag rods at midspan. (c) Moment about web axis of purlins—sag rods at one-third points. (d) Moment about X axis of purlin.

should be designed for bending as well as axial stress. Purlins are usually spaced from 2 to 6 ft apart depending on loading conditions, while their most desirable depth to span ratios are probably in the neighborhood of 1/24. Channels or S sections are the most frequently used sections but on some occasions other shapes may be convenient.

As previously described in Chapter 3, the channel and S sections are very weak about their web axes and sag rods are often necessary to reduce the span lengths for bending about those axes. Sag rods, in effect, make the purlins continuous sections for their y axes and the moments about these axes are greatly reduced, as shown in Fig. 7-7. These moment diagrams are developed on the assumption that the changes in length of the sag rods are negligible. It is further assumed that the purlins are simply supported at the trusses. This assumption is on the conservative side since they are often continuous over two or more trusses and appreciable continuity may be achieved at their splices. The student can easily reproduce these diagrams from her knowledge of moment distribution or other analysis methods. In the diagrams l is the distance between trusses, w_y is the load component *perpendicular* to the web axis of the purlin, and w_x is the load component *parallel* to the web axis.

If sag rods were not used the maximum moment about the web axis of a purlin would be $w_y l^2/8$. When sag rods are used at midspan this moment is reduced to a maximum of $w_y l^2/32$ (a 75% reduction) and when used at one-third points is reduced to a maximum of $2w_y l^2/175$ (a 91% reduction). In the example problem to follow (Example 7-6), sag rods are used at the midpoints and the purlins are designed for a moment of $w_x l^2/8$ parallel to the web axis and $w_y l^2/32$ perpendicular to the web axis.

In addition to being of advantage in reducing moments about the web axes of purlins, sag rods can serve other useful purposes. First, they can provide lateral support for the purlins; secondly, they are useful in keeping the purlins in proper alignment during erection until the roof deck is installed and connected to the purlins.

Example 7-6

Select a W6 purlin for the roof shown in Fig. 7-8. The trusses are 15 ft–0 in. on center and sag rods are used at the midpoints between trusses. Full lateral support is assumed to be supplied from the roof above. Use A36 steel and the AISC Specification and assume there is no torsion. Thus the full value of S_y is to be used. Loads are as follows in terms of pounds per square foot of roof surface.

$$\text{snow} = 30 \text{ psf}$$

$$\text{roofing} = 6 \text{ psf}$$

$$\text{estimated purlin weight} = \underline{3 \text{ psf}}$$

$$\text{total} = 39 \text{ psf}$$

$$\text{wind pressure} = 15 \text{ psf} \perp \text{to roof surface}$$

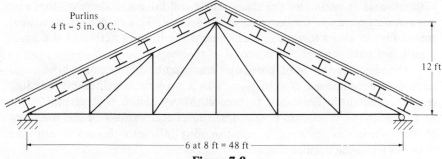

Purlins
4 ft – 5 in. O.C.

12 ft

←————————— 6 at 8 ft = 48 ft —————————→

Figure 7-8

SOLUTION

$$w_{\text{gravity}} = (4.42)(39) = 172.4 \text{ lb/ft}$$

$$w_{\text{wind}} = (4.42)(15) = 66.3 \text{ lb/ft}$$

$$w_x = 66.3 + \left(\frac{2}{\sqrt{5}}\right)(172.4) = 220.5 \text{ lb/ft}$$

$$w_y = \left(\frac{1}{\sqrt{5}}\right)(172.4) = 77.1 \text{ lb/ft}$$

$$M_x = \frac{(0.2205)(15)^2}{8} = 6.20 \text{ ft-k}$$

$$M_y = \frac{(0.0771)(15)^2}{32} = 0.542 \text{ ft-k}$$

Try W6×9 $(S_x = 5.56, S_y = 1.11)$

$$f_{bx} = \frac{(12)(6.20)}{5.56} = 13.38 \text{ ksi}$$

$$f_{by} = \frac{(12)(0.542)}{1.11} = 5.86 \text{ ksi}$$

$$\frac{f_{bx}}{F_{bx}} + \frac{f_{by}}{F_{by}} = \frac{13.38}{24} + \frac{5.86}{27} = 0.775 < 1.0 \qquad \text{OK}$$

Use W6×9

7-7. THE SHEAR CENTER

The shear center is defined as the point on the cross section of a beam
through which the resultant of the transverse loads must pass so that the
stresses in the beam may be calculated only from the theories of pure
bending and transverse shear. Should the resultant pass through this point,
it is unnecessary to analyze the beam for torsional moments. For a beam
with two axes of symmetry the shear center will fall at the intersection of
the two axes, thus coinciding with the centroid of the section. For a beam

with one axis of symmetry the shear center will fall somewhere on that axis but not necessarily at the centroid of the section. This surprising statement means that to avoid torsion in some beams the lines of action of the loads should not pass through the centroid of the section.

The shear center is of particular importance for beams whose cross sections are composed of thin parts which provide considerable bending resistance but little resistance to torsion. Many common structural members such as the W, S, and C sections, angles and various beams made up of thin plates (as in aircraft construction) fall into this class and the problem has wide application.

The location of shear centers for several open sections are shown in Fig. 7-9. Sections such as these are relatively weak in torsion and for them and similar shapes the location of the resultant of the external loads can be a very serious matter. A previous discussion has indicated that the addition of one or more webs to these sections so they are changed into box shapes greatly increases their torsional resistance.

The average designer probably does not take the time to go through the sometimes tedious computations involved in locating the shear center and calculating the effect of twisting. He may instead just ignore the situation, or he may make a rough estimate of the effect of torsion on the bending stresses. An illustration of a rough estimate of the torsion effect in purlins was given in Example 7-4, where only one-half of S_y was considered effective when lateral loads were applied to the top flange. The large percentage of the total stress due to the rough estimate made in that example should keep the designer on the lookout for cases where his estimates might be too inaccurate.

Shear centers can be located quickly for beams with open cross sections and relatively thin webs. For other beams the shear centers can

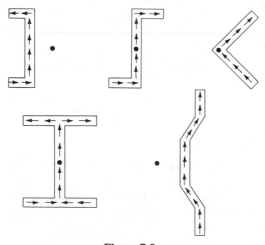

Figure 7-9

probably be found but only with considerable difficulty. The term *shear flow* is often used when speaking of thin-wall members although there is really no flowing involved. It refers to the shear per inch of the cross section and equals the unit shearing stress times the thickness of the member. (The unit shearing stress has been determined by the expression VQ/bI and the shear flow can be determined by VQ/I if the shearing stress is assumed to be constant across the thickness of the section.) The shear flow acts parallel to the sides of each element of a member.

The channel section of Fig. 7-10(a) will be considered for this discussion. In this figure the shear flow is shown with the small arrows and in part (b) the values are totaled for each component of the shape and labeled H and V. The two H values are in equilibrium horizontally and the internal V value balances the external shear at the section. Although the horizontal and vertical forces are in equilibrium, the same cannot be said for the moment forces unless the lines of action of the resultant of the external forces passes through a certain point called the shear center. The horizontal H forces in part (b) of the figure can be seen to form a couple. The moment produced by this couple must be opposed by an equal and opposite moment which can only be produced by the two V values. The location of the shear center is a problem in equilibrium; therefore moments should be taken about a point that eliminates the largest number of forces possible.

From this information the following equation can be written from which the shear center can be located.

$$Ve = Hh \qquad \text{(moments taken about c.g. of web)}$$

Examples 7-7 and 7-8 illustrate the calculations involved in locating the shear center for two shapes. It will be noted that the location of the shear center is independent of the value of the external shear although V values are given in these examples.

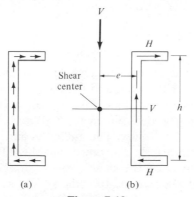

Figure 7-10

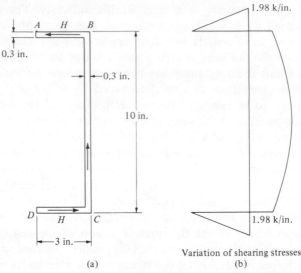

(a)   (b)

Variation of shearing stresses

Figure 7-11

Example 7-7

The channel section shown in Fig. 7-11(a) is subjected to an external shear of 30 k. Locate the shear center.

SOLUTION
Properties of section:

$$I_x = \left(\tfrac{1}{12}\right)(3)(10)^3 - \left(\tfrac{1}{12}\right)(2.7)(9.4)^3 = 63 \text{ in.}^4$$

$$f_v \text{ at } B = \frac{(30)(2.85 \times 0.3 \times 4.85)}{63} = 1.97 \text{ k/in.}$$

$$\text{Total } H = \left(\tfrac{1}{2}\right)(2.85)(1.97) = 2.81 \text{ k}$$

Location of shear center:

$$Ve = Hh$$

$$30e = (2.81)(9.7)$$

$$e = 0.91 \text{ in. from } \text{\textcentoblique} \text{ of web}$$

The student should clearly understand that shear stress variation across the corners, where the webs and flanges join, cannot be determined correctly with the mechanics of materials expression (VQ/bI or VQ/I for shear flow) and cannot be determined too well even after a complicated study with the theory of elasticity. The author as an approximation in the two examples presented here has assumed that shear flow continues up to the middle of the corners on the same pattern (straight line variation for horizontal members and parabolic for others). The values of Q are computed for the corresponding dimensions. Other assumptions could have been made such as assuming a shear flow variation continuing vertically for the full depth of webs and only for the protruding parts of flanges horizontally, or vice versa. It is rather disturbing to find that whichever assumption is used the values do not check out perfectly. For instance in Example 7-8 which follows the sum of the vertical shear flow values does not check out very well with the external shear.

Example 7-8

The open section of Fig. 7-12 is subjected to an external shear of 20 k. Locate the shear center.

SOLUTION
Properties of section:

$$I_x = \left(\tfrac{1}{12}\right)(0.25)(16)^3 + (2)(3.75 \times 0.25)(4.87)^2 = 129.8 \text{ in.}^4$$

Values of shear flow (labeled q):

$$q_A = 0$$

$$q_B = \frac{(20)(3.12 \times 0.25 \times 6.44)}{129.8} = 0.775 \text{ k/in.}$$

$$q_C = q_B + \frac{(20)(3.75 \times 0.25 \times 4.87)}{129.8} = 1.48 \text{ k/in.}$$

$$q_{\mathcal{L}} = q_C + \frac{(20)(4.87 \times 0.25 \times 2.44)}{129.8} = 1.94 \text{ k/in.}$$

These shear flow values are shown in Fig. 7-13(a) and the summation for each part of the member is given in part (b) of the figure.

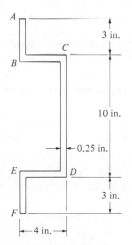

Figure 7-12

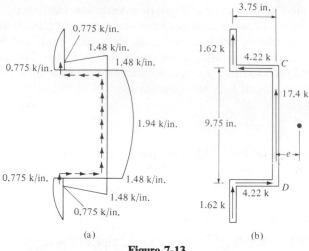

(a)   (b)

Figure 7-13

Taking moments about the center line of CD

$$-(4.22)(9.75)+(2)(1.62)(3.75)+20e=0$$

$$e=1.45 \text{ in.}$$

The theory of the shear center is a very useful one in design but it has certain limitations which should be clearly understood. For instance, the approximate analysis given in this section is only valid for thin sections. In addition steel beams often have variable cross sections along their spans with the result that the locus of the shear centers are not straight lines along those spans. The result is that if the resultant of the loads passes through the shear center at one cross section it might very well not do so at other cross sections.

7-8. TORSION

Torsion occurs when the line of action of a load does not pass through the shear center of the beam. When this happens the resulting torque or twisting moment equals the product of the load times the perpendicular distance from its line of action to the shear center. It is, of course, desirable to avoid situations where members subject to moments and shears (and perhaps axial loads) are also subject to torsion.

When torsion cannot be avoided it is desirable to use some type of box or round or tube shaped section. The most efficient of these as to torsional resistance is the tubular section. For rather short stocky members a W section may occasionally be satisfactory for resisting bending and torsion.

It is beyond the scope of this textbook to present a detailed theoretical discussion of torsion. For such information the reader should refer to an advanced text on mechanics of materials. The author's purpose here is to describe the nature of the problem and to introduce a fairly good approximate design method for handling torsion. It is to be remembered that a very poor approximate method was illustrated in Example 7-4 where the $1/2S_y$ value was used.

If a round bar or tube is subjected to torsion each cross section rotates in its own plane without warping and the resistance to torsion is supplied by the shear stresses. In other words a plane section before twisting remains a plane section after twisting. These shear stresses are proportional to the distance from the centroid and can be determined with the following expression in which T is the torsional moment, r is the radius, and J the polar moment of inertia of the cross section.

$$f_v = \frac{Tr}{J}$$

Should a member with a noncircular cross section be subjected to torsion the individual cross sections along the member will not only rotate but also will deform in a nonuniform manner in the longitudinal plane so that plane sections before twisting do not remain plane sections after twisting. This latter deformation is called *warping*. When such a member whose sections are unrestrained from warping is subjected to torsion its torsional resistance is due to distribution of shear stress which is called "St. Venant torsion" (named for the French engineer St. Venant who developed the theory that is the basis of present-day torsion analysis). In most closed cross sections such as tubes and box girders and some very compact open sections St. Venant torsion predominates.

The maximum shear stress due to torsion that occurs in a solid rectangular shape free to warp can be determined from the expression ($f_v = Tr/J$) with J equal to $bt^3/3$ if the rectangle is several times longer than its width. For I-shaped sections J can be estimated fairly accurately by adding the $bt^3/3$ values together for the three rectangles that make up the cross section. In its "Torsion Properties" tables the AISC Manual gives J values for standard shapes.

If warping is restrained, as by the type of end supports, other shear stresses act which together with the St. Venant shear stresses, resist the torsional moment. Most structural members with large open cross sections such as W sections resist torsion through a combination of St. Venant torsion and warping torsion.

Unless the elements of members with open cross sections such as W sections are quite thick they will have little torsional resistance. Such members, however, can be supported in a manner such that they can resist appreciable torsion. An example of this important fact is illustrated in Fig. 7-14. In part (a) of the figure the flanges of the section at the left end support are considered to be relatively free to rotate so that the member can be easily twisted. For this beam the torsional moment is resisted only by the St. Venant torsional strength of the member and the member is

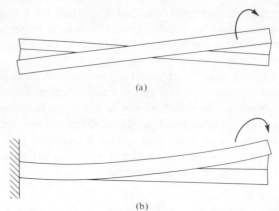

Figure 7-14 Torsion in beams. (a) Left support provides little resistance; member can be twisted with little torsional force, (b) Left end firmly restrained; torsional resistance greatly increased due to bending resistance of flanges.

subject to uniform warping. For such a beam the torsional resistance is not affected by any restraint of the warping.

In part (b) of the figure the left end of the member is assumed to be almost completely fixed as if it is embedded in concrete. For this case the torsional resistance is greatly increased. In the first case the flanges, though twisted, remain almost straight while in the second case they cannot be twisted laterally unless they are also bent out of shape. At the fixed end warping is prevented and St. Venant shear stresses are equal to zero. The resistance to torsion is provided solely by the shear stresses due to lateral bending of the flanges of the section. At other cross sections of the beam torsion is resisted by a combination of St. Venant torsion and warping torsion. Thus the flanges for the restrained end situation offer considerable resistance to torsion (particularly if they are wide) and they in effect act as horizontal beams bending about an axis lying in the center of the web.

The lateral bending moment in the flanges due to warping torsion can be conservatively calculated by replacing the torsional moment with a couple acting through the center of the flanges. This approximate method is illustrated in Fig. 7-15 where a vertical eccentric load is applied to the

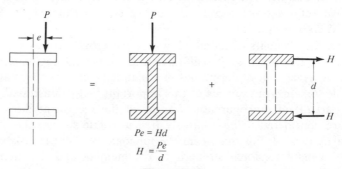

Figure 7-15

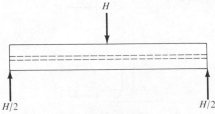

Figure 7-16

beam and the value of H calculated. It is then assumed as shown in Fig. 7-16 that the load H is applied laterally to the beam.

This approximate method is desirably applied only to short members because results will be overly conservative in long ones. The lateral bending stresses are calculated with the flexure formula

$$f_b = \frac{M_{\text{flange}}(b_f/2)}{(I_y/2)}$$

and the lateral shear stresses can be calculated with $f_v = VQ/I_y t$. The formula for f_b has been written as though the lateral moment due to H on one flange only is considered and so only one-half of I_y is used in the denominator.

This simple procedure for handling eccentric loading illustrated in Example 7-9, which follows, will give calculated bending stresses that are on the conservative side but it can be shown that the shearing stresses obtained are on the small side. As a result, if torsion is very large, the designer should use a more rigorous approach.

Example 7-9

A W24×104 section consisting of A36 steel has a 40-k gravity load applied eccentrically at the centerline of an 18-ft simple span as shown in Fig. 7-17. The flanges are restrained from warping at the ends only. Check the adequacy of the section according to the AISC Specification. Neglect beam weight.

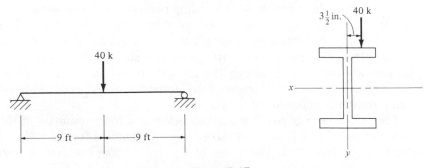

Figure 7-17

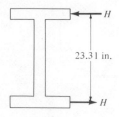

23.31 in.

SOLUTION

Using a W24×104 ($d=24.06$, $t_f=0.750$, $S_x=258$, $S_y=40.7$)

$$M_x = \frac{Pl}{4} = \frac{(40)(18)}{4} = 180 \text{ ft-k}$$

$$f_{bx} = \frac{(12)(180)}{258} = 8.37 \text{ ksi}$$

$$F_{bx} = 0.60F_y = 22 \text{ ksi} \qquad \text{since } l_{\text{unbr.}} > L_c < L_u$$

Replacing the eccentric moment ($40 \times 3.5 = 140$ in.-k) with two equal and opposite forces acting at centers of flanges:

$$\text{each flange force } H = \frac{140}{23.31} = 6.01 \text{ k}$$

$$\text{lateral moment acting on one flange} = \frac{Hl}{4} = \frac{(6.01)(18)}{4} = 27.045 \text{ ft-k}$$

$$f_{by} = \frac{(12)(27.045)}{(40.7/2)} = 15.95 \text{ ksi}$$

$$F_{by} = 0.75F_y = 27 \text{ ksi}$$

Checking interaction equation,

$$\frac{f_{bx}}{F_{bx}} + \frac{f_{by}}{F_{by}} = \frac{8.37}{22} + \frac{15.95}{27} = 0.971 < 1.0 \qquad\qquad \text{OK}$$

As has been described in the preceding paragraphs of this section the bending of the flanges is accompanied by a warping of the section. The greater the distance from a restraining support the greater the amount of warping. Beyond a certain distance from the support the warping is approximately the same as though the support offered no restraint. From that point on the torque is resisted by torsional action only with no bending resistance supplied by the flanges.

The *torsion bending constant a* given below is a rough estimate of the distance along a beam from a restrained location required for the effect of restraint to be dissipated. The value of C_w, the *warping constant*, is also tabulated in the AISC Manual for standard beam sections.

$$a = 1.61\sqrt{\frac{C_w}{J}}$$

7-9. LINTELS

Openings in brick or block walls which have relatively flat tops, such as windows and doors, require beams above them to support the masonry and other loads above. Although considerable arch action exists in the masonry after the mortar has set, it is still necessary to support the masonry while it is green to prevent settlement and possible destruction of the arch action before it can develop. These supporting beams, called lintels, may consist of any number of different shapes. For very small openings, flat bars or plates may be used, whereas angles are convenient for slightly larger openings. When the openings become quite large structural tees, channels, S beams, W sections, or even built-up sections may be necessary.

The amount of load that must be supported by a particular lintel is quite uncertain and a good estimate can only be made after a careful study is made of the particular situation. The window shown in Fig. 7-18(a) is assumed to be located in a brick wall which is solid and continuous above and around the sides of the window. It seems logical to assume that the lintel will carry the triangular shaped portion of the wall above the window. This loading situation appears even more reasonable when it is known that examinations of old brick buildings often reveal cracks above the windows in approximately the shape of a triangle when the lintels have settled. In part (b) of the figure an expression is developed for the maximum moment in this type of lintel.

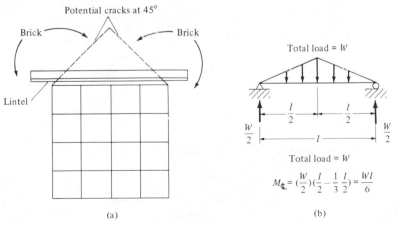

(a)   (b)

Figure 7-18

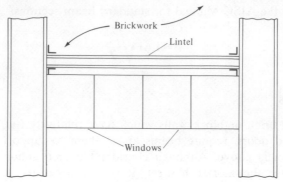

Figure 7-19

Figure 7-19 shows a group of windows located between two columns. A spandrel wall is located above the windows and as the figure shows the lintel will in all probability have to support the entire overhead wall.

Another loading condition for lintels which is often encountered in small buildings is shown in Fig. 7-20. The lintel shown there, which is placed above a series of windows on one floor, supports not only the walls above but also the reactions from the beams under the floor above.

The lintel selected for a particular wall must have dimensions that fit in reasonably well so that the brick or blocks can be placed without unusual difficulty. As it is desirable for the lintel not to be visible from one or both sides of a wall, it is usually buried in the wall. For this reason an inverted structural tee (⊥) or a pair of angles back to back (⅃L) are very satisfactory in many situations.

For many walls that consist of more than one thickness of brick or blocks (or combinations of bricks and blocks), the lintel may consist of two or more angles or other shapes or some combination of shapes one in each

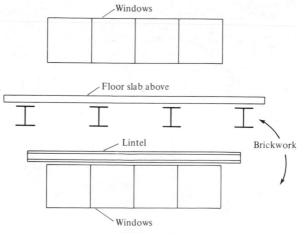

Figure 7-20

thickness which are not connected to each other. Should nonconnected shapes be used, an effort should be made to keep the deflections in each part equal. It is true that the window and door headers below might prevent the separate parts of the lintel from deflecting differently. Even if this is true, additional loads will probably be thrown on the lintel, which tends to have the smallest deflection, and cause it to be overstressed.

There is a great deal of information available as to the size of lintels that should be used for different size openings. One table suggests a $3\frac{1}{2} \times \frac{3}{8}$-in. flat bar for each 4-in. thickness of wall for openings of less than 2 ft; a $3\frac{1}{2} \times 3\frac{1}{2} \times \frac{5}{16}$-in. angle for each 4-in. thickness of wall for openings of 2 to 5 ft, etc. Despite the availability of this excellent information concerning lintel sizes consideration should be given to the particular loading conditions and member arrangements in the building under study to ensure the selection of satisfactory sizes.

7-10. BEAM-BEARING PLATES

When the ends of beams are supported by direct bearing on concrete or other masonry construction it is frequently necessary to distribute the beam reactions over the masonry by means of beam bearing plates. The reaction is assumed to be spread uniformly through the bearing plate to the masonry and the masonry is assumed to push up against the plate with a uniform pressure equal to f_p psf. This pressure tends to curl up the plate and the bottom flange of the beam. The AISC Manual recommends that the bearing plate be considered to take the entire bending moment produced and that the critical section for moment be assumed to be a distance k from the centerline of the beam (see Fig. 7-21). The distance k is the same as the distance from the outer face of the flange to the web toe of the fillet given in the tables for each section (or it equals the flange thickness plus the fillet radius).

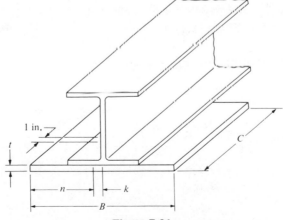

Figure 7-21

The determination of the true pressure distribution in a beam bearing plate is a very formidable task and the uniform pressure distribution assumption is usually made. This assumption is probably on the conservative side as the pressure is probably larger at the center of the beam than at the edges. The outer edges of the plate and flange tend to bend upward and the center of the beam tends to go down, concentrating the pressure there.

The required thickness of a 1-in.-wide strip of plate can be determined as follows:

$$M = f_p n \frac{n}{2} = \frac{f_p n^2}{2}$$

$$S = \frac{I}{c} = \frac{\left(\frac{1}{12}\right)(1)(t)^3}{t/2} = \frac{t^2}{6}$$

$$\frac{M}{f} = S$$

$$\frac{f_p n^2}{2f} = \frac{t^2}{6}$$

$$t = \sqrt{\frac{3f_p n^2}{F_b}}$$

Example 7-10 illustrates the calculations involved in designing a beam bearing plate. Notice that the distances B and C are desirably taken to the nearest full inch.

Example 7-10

A W18×71 beam ($b=7.635$ $k=1\frac{1}{2}$ in. $t_w=0.495$) is to be end supported by a masonry wall which has an allowable bearing strength of 250 psi. Design a bearing plate for the beam with A36 steel and the AISC Specification. The end reaction is 40 k and the maximum length of bearing available is 12 in.

SOLUTION
Minimum length of bearing for web crippling:

$$\frac{R}{t_w(N+k)} = 0.75 F_y$$

$$\frac{40,000}{0.495\left(N+1\frac{1}{2}\right)} = (0.75)(36,000)$$

$$N = 1.49 \text{ in.} \qquad \text{(minimum permissible value of } C\text{)}$$

$$\text{area required} = \frac{40,000}{250} = 160 \text{ in.}^2$$

$$B = \frac{160}{12} = 13.33 \text{ in.}$$

Try 12×14 in. PL

$$n = 7.00 - 1\tfrac{1}{2} = 5.50 \text{ in.}$$

$$f_p = \frac{40{,}000}{12 \times 14} = 238 \text{ psi}$$

$$t = \sqrt{\frac{3 f_p n^2}{F_b}} = \sqrt{\frac{(3)(238)(5.50)^2}{27{,}000}}$$

$$= 0.894 \text{ in.}$$

Use $\tfrac{15}{16} \times 12 \times 14$ in. PL

On some occasions the beam flanges alone probably provide sufficient bearing area, but bearing plates are nevertheless recommended as they are useful in erection and ensure an even bearing surface for the beam. They can be placed separately from the beams and carefully leveled to the proper elevations. When the ends of steel beams are enclosed by the concrete or masonry walls it is considered desirable to use some type of wall anchor to prevent the beam from moving longitudinally with respect to the wall. The usual anchor consists of a bent steel bar passing through the web of the beam and running parallel to the wall. These are called government anchors and details of their sizes are given in the AISC Manual. Occasionally clip angles attached to the web are used instead of government anchors. Should longitudinal loads of considerable size be anticipated, regular vertical anchor bolts may be used at the beam ends.

Problems*

7-1. A $W24 \times 68$ has a maximum external shear of 75 k.

    (a) Calculate the maximum theoretical shearing stress in the section. (*Ans.* 8.60 ksi)

    (b) Calculate the average shearing stress in the section using the full depth of the section. (*Ans.* 7.62 ksi)

7-2. A $W18 \times 46$ has a maximum end shear of 40 k. Calculate the average shearing stress in the section (using the total depth of the section) and the actual maximum theoretical shearing stress.

7-3. Using A36 steel and the AISC Specification select the lightest section for the span and loading shown in the accompanying illustration. Neglect beam weight in all calculations. (*Ans.* W30 × 116)

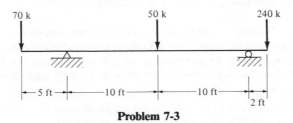

70 k   50 k   240 k

$\leftarrow$ 5 ft $\rightarrow\!\mid\leftarrow$ 10 ft $\longrightarrow\!\mid\leftarrow$ 10 ft $\longrightarrow\!\mid$
                                        2 ft

Problem 7-3

*Note: Full lateral support is provided for all sections unless noted otherwise.

7-4. Using the AISC Specification and A36 steel determine the maximum uniform load that can be placed on a W14×34, which has a simple span of 6 ft-6 in.

7-5. A W16×40 consisting of A36 steel is used as a simple beam for a span of 6 ft. Using the AISC Specification determine the maximum uniform load the beam can support. (*Ans.* 23.56 k/ft)

7-6. A W36×245 consisting of A36 steel is used as a simple beam for a span of 15 ft. Using the AISC Specification determine the maximum uniform load the beam can support.

7-7. A W33×221 is required for a certain span but, due to a strike in the steel mills, cannot be obtained on time; however an extra W33×118 of sufficient length is available along with a good supply of plates. Select $\frac{7}{8}$-in. thick cover plates to be welded to the flanges of the W33×118 to provide a satisfactory substitute for the original beam. Use A36 steel and the AISC Specification. (*Ans.* W33×118 with one $\frac{7}{8}$×16 PL each flange)

7-8. A W30×173 section has been specified for use on a certain job. By mistake, a W30×124 section was shipped to the field. This beam must be erected today. Assuming that plates are obtainable immediately, select cover plates to be welded to the flanges to obtain the necessary section. Use A36 steel and the AISC Specification.

7-9. Repeat Prob. 7-7 if the original beam was to have been of a steel with $F_y =$ 50 ksi and the substitute material is available only in A36 steel. Assume also that only $\frac{15}{16}$-in. plates are available. (*Ans.* W33×118 with $\frac{15}{16}$×24 cover PL each flange)

7-10. Design a beam of A36 steel with a depth no greater than 12.00 in. to support a uniform load of 8 k/ft for a 20-ft simple span. Use AISC Specification.

7-11. A simply supported W24×146 beam is 6 ft long and supports a concentrated load of 300 k 2 ft from the left end. If the length of bearing at the left support is 8 in. and at the concentrated load is 12 in., check the beam for bending, shear, and web crippling. Use A36 steel and the AISC Specification. (*Ans.* Section is unsatisfactory as to web crippling)

7-12. A 30-ft beam is to support a moving concentrated load of 36 k. Using A36 steel and the AISC Specification select the most economical section. Also determine the maximum shearing stress in the section selected and compute the length of bearing required at each support from the standpoint of web crippling.

7-13. A 20-ft simple beam which supports a concentrated load of 20 k at midspan, is laterally unsupported except at its ends and is limited to a maximum total deflection of $l/1000$. Select the most economical W section available using A36 steel and the AISC Specification. (*Ans.* W18×55)

7-14. Design a beam for a 24-ft simple span to support a 4 k/ft uniform load. The maximum permissible deflection is 1/1200th of the span and the allowable bending stress is 20 ksi.

7-15. Select the lightest available section for the span and loads shown in the accompanying illustration if the maximum permissible deflection is 1/1500th of the span. Neglect beam weight in all calculations and use A36 steel and the AISC Specification. (*Ans.* W36×300 with $\frac{1}{4}$×7 PL each flange)

7-16. Select the lightest section available for the span and loads shown in the accompanying illustration if the maximum permissible deflection is 1/1000th of the span. Use A36 steel and the AISC Specification.

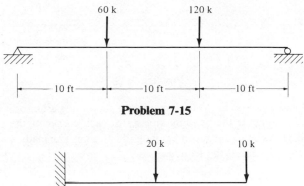

Problem 7-15

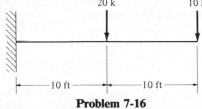

Problem 7-16

7-17. The maximum permissible deflection in the beam shown in the accompanying illustration is 1/1200th of the span. Select the lightest available section using A36 steel and the AISC Specification. (*Ans.* W36×194)

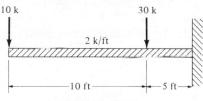

Problem 7-17

7-18. Select the lightest available W section for the beam shown in the accompanying illustration. Lateral support is provided only at the 12-ft points and a maximum deflection equal to 1/1500th of the 24-ft span is permitted. Neglect beam weight in all calculations and use A36 steel and the AISC Specification.

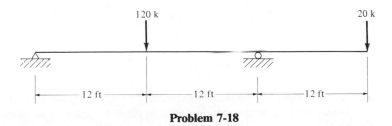

Problem 7-18

7-19. Select the lightest W section (A36 steel) that can support the loads shown in the accompanying illustration so the deflection will not exceed 1/1200th of the span. Use the AISC Specification. (*Ans.* W30×116)

6 k/ft

|← 30 ft →|← 30 ft →|

Problem 7-19

7-20. A W21×62 serves as an 18-ft simple beam supporting a uniform gravity load of 2 k/ft. Compute the bending stresses at each corner of the beam if it is inclined as shown in the accompanying illustration. The loads are assumed to act through the c.g. of the section.

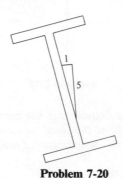

1

5

Problem 7-20

7-21. Find the shear center for the section shown in the accompanying illustration. (*Ans. e* = 0.93 in. from web ℄)

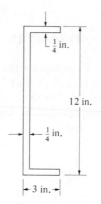

$\frac{1}{4}$ in.

12 in.

$\frac{1}{4}$ in.

|← 3 in. →|

Problem 7-21

7-22. Find the shear center for the section shown in the accompanying illustration.

7-23. Find the shear center for the section shown in the accompanying illustration. (*Ans. e* = 1.03 in.)

7-24. Select a W12 section to serve as a purlin between roof trusses 24 ft on centers. The roof is assumed to support a dead load of 20 psf of roof surface and a snow load of 20 psf of horizontal roof surface projection. The slope of

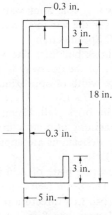

Problem 7-22

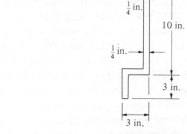

Problem 7-23

the roof truss is 1 vertically to 2 horizontally and the purlins are to be spaced 12 ft on centers. Use A36 steel and the AISC Specification and assume loads pass through c.g. of section. Sag rods are assumed to be placed halfway between trusses.

7-25. Repeat Prob. 7-24 assuming lateral support for the purlins is provided at the 12-ft points only. (*Ans.* W12×26)

7-26. The load shown in the accompanying illustration is assumed to be 4 in. off center and full torsional fixity is assumed for torsional restraint at the ends. Select the lightest available W24 section that will be satisfactory. Lateral support for the compression flange is provided at the ends and centerline. Use A36 steel and the AISC Specification.

Problem 7-26

7-27. Select the most economical W18 section available to support the load shown in the accompanying illustration which is 3.5 in. off center. Use A36 steel and the AISC Specification. Lateral support is provided at the fixed end only. The fixed end provides full torsional restraint. (*Ans.* W18×97)

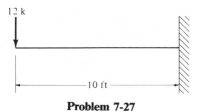

Problem 7-27

7-28. Design an angle lintel to go over a 5-ft window in a 4-in. brick wall. The wall, which is assumed to exist on the sides and over the window, is assumed to weigh 120 lb/ft.^{3} Use A36 steel and the AISC Specification.

7-29. A lintel is to be located over a doorway in a brick building. The wall is 8 in. thick and weighs 125 lb/ft.^{3} The clear opening of the doorway is 12 ft–4 in. Select an inverted structural tee with a flange width of approximately 7 in. Use AISC Specification and A36 steel. (*Ans.* WT7×15)

7-30. A 14-ft opening is to be constructed in a 12-in. brick wall. Using A36 steel and the AISC Specification design a lintel consisting of a structural tee and an angle. The parts should be selected so their deflections are approximately equal. The wall weighs 120 lb/ft^3 (19 kN/m^3).

7-31. Using the AISC Specification design a steel bearing plate of A36 steel for a W21×68 beam with an end reaction of 62 k. The beam bears on a masonry wall having an allowable bearing stress of 500 psi. In a direction perpendicular to the wall the bearing plate may not be longer than 8 in. (*Ans.* Several sizes possible: one is a PL1$\frac{9}{16}$×8×1 ft 4 in.)

7-32. Design a steel bearing plate of A36 steel for a W30×116 beam supported by a concrete wall. The allowable bearing on the concrete is assumed to be 750 psi and the maximum beam reaction is 105 k. Use the AISC Specification and assume the width of the plate perpendicular to the wall may not exceed 6 in.

7-33. Repeat Prob. 7-32 using a steel with $F_y = 45$ ksi steel for the beam and the bearing plate. Assume the allowable bearing pressure on the concrete is 1125 psi. (*Ans.* Several sizes possible: one is a PL2×6×1 ft. 4 in.)

Chapter 8
Bending and
Axial Stress

8-1. OCCURRENCE

Structural members that are subjected to a combination of bending and axial stress are far more common than the student may realize. This section is devoted to listing a few of the more obvious cases. Columns that are part of a steel building frame must nearly always resist sizable bending moments in addition to the usual compressive loads. It is almost impossible to erect and center loads exactly on columns even in a testing lab and in an actual building the student can see that it is even more impossible. Even if building loads could be perfectly centered at one time they would not stay in one place. Furthermore, columns may be initially crooked or have other flaws with the result that lateral bending is produced. The beams framing into columns are quite commonly supported with framing angles or brackets on the sides of the columns. These eccentrically applied loads produce moments. Wind and other lateral loads cause columns to bend laterally, and the columns in rigid frame buildings are subjected to moments even when the frame is supporting gravity loads alone. The members of bridge portals must resist combined stresses as do building columns. Among the causes of the combined stresses are heavy lateral wind loads, vertical traffic loads whether symmetrical or not, and the centrifugal effect of traffic on curved bridges.

The previous experience of the student has probably been to assume truss members axially loaded only. Purlins for roof trusses, however, are frequently placed in between truss joints, causing the top chords to bend. Similarly the bottom chords may be bent by the hanging of light fixtures, ductwork, and other items between the truss joints. All horizontal and inclined truss members have moments caused by their own weights, while all truss members whether vertical or not are subjected to secondary bending stresses. Secondary stresses are developed because the members are not connected with frictionless pins as assumed in the usual analysis, the member centers of gravities or those of their connectors do not exactly coincide at the joints, etc.

Inryco Building, Chicago, Ill. (Courtesy of Inryco, Inc.)

Moments in tension members are not as serious as those in compression members because tension tends to reduce lateral deflections while compression increases them. Increased lateral deflection in turn results in larger moments, which result in larger lateral deflections, etc. It is hoped that members in such situations are quite stiff so as to keep the additional lateral deflections from becoming excessive.

8-2. CALCULATION OF STRESSES

Sometimes members are encountered which have as their most important loadings transverse bending moments but which are also subjected to axial loads. However, this is not nearly as common a situation as where the major loading is axial with some transverse bending occurring at the same time. The name *beam-column* is often given to members that have appreciable amounts of both axial compression and bending stresses. The stresses in members subjected to a combination of axial load and bending are difficult to obtain exactly and the stresses calculated in this chapter are truly approximate.

The stress at any point in a member subject to bending and direct stress is usually obtained from the familiar expression to follow. (This expression is said to be approximate because it does not include the effect of increased lateral deflections caused by moments and their subsequent effect on the moments and thus the stresses.)

$$f = \frac{P}{A} \pm \frac{Mc}{I}$$

Frequently bending occurs about some axis other than the x or y axis; that is, it occurs about both axes simultaneously. The corner columns of buildings may fall into this class. Stresses for members subjected to axial loads and bending about both axes are usually determined by the following expression.

$$f = \frac{P}{A} \pm \frac{M_x y}{I_x} \pm \frac{M_y x}{I_y}$$

Example 8-1 shows the calculations involved in computing the total stress in a member of a bridge truss due to its computed axial stress plus the flexure stress caused by its own weight. The calculations here are quite straightforward, but a few comments will probably be of value. First, it will be noted that although this is a bolted member the gross area of the section is used, because the bolt holes are at the ends of the member while the maximum bending is assumed to occur at the centerline. Secondly, the flexure stresses due to the weight of the member in this example are not very large, and as a matter of fact they are not very large in the average truss member. As truss members become unusually long and heavy the moments due to their own weights may become of appreciable magnitude.

The engineers who prepare the usual specifications used for the design of truss members have reduced the allowable axial stresses by approximately one-fourth to take into account the so-called secondary stresses. A similar discussion can be made for the magnitude of column flexure stresses due to column imperfections, slightly off-centered loads, and moments due to lateral deflections. This discussion shows that if the engineer went overboard theoretically, she would design all members, whether columns or beams or truss members, for axial load and bending. It is probable, however, that the decreased allowable stresses provide a sufficient margin to cover the usual case of off-center loads or otherwise imperfect columns and secondary truss member stresses. Unless the situation is severe the designer will probably not include the moments in his design. Specifications rarely mention the matter and it is up to the individual designer to use his own judgment. This discussion is not intended to apply to truss members that have transverse loads applied to them between the truss joints nor to columns that have appreciable moments applied as a part of a rigid frame structure.

Example 8-1

A bottom chord of a bolted bridge truss is 22 ft long, has a maximum tensile force of 310 k, and is made up of 4Ls $6 \times 4 \times \frac{1}{2}$, arranged as

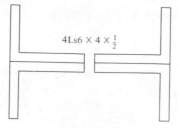

Figure 8-1

shown in Fig. 8-1. Determine the maximum stress in the member due to the axial tension force plus the flexure stress due to the member's own weight.

SOLUTION

$$\{A_{\text{gross}} = 19 \text{ in.}^2, \; I_x = 2[12.5 + 9.50 \, (0.987)^2] = 43.5 \text{ in.}^4\}$$

$$\text{weight of Ls} = \left(\frac{19}{144}\right)(490) = 64.8 \text{ lb/ft}$$

$$M = \frac{wl^2}{8} = \frac{(64.8)(22)^2}{8} = 3{,}920 \text{ ft-lb}$$

$$f = \frac{P}{A} + \frac{Mc}{I} = \frac{310{,}000}{19} + \frac{(12)(3920)(4.00)}{43.5}$$

$$f = 16{,}315 + 4325 = 20{,}640 \text{ psi} \qquad (142 \text{ MPa})$$

$$\text{percent stress due to member's weight} = \frac{4325}{20{,}640} = 21\%$$

8-3. SPECIFICATIONS FOR COMBINED STRESSES

The preceding paragraphs have shown how stresses due to flexure and axial load can be combined. As a matter of fact the calculations were quite simple, but the problem of establishing an allowable combined stress is a much more difficult situation. There are allowable stresses for pure bending and other allowable stresses for pure axial loads, but the two values given in one specification are probably appreciably different. What is the value to be used when the two types of stresses occur simultaneously? Most organizations use an allowable stress which is some combination of the two individual allowables. Expressions of this type are referred to as *interaction equations*. One interaction equation used by many specifications is as follows.

$$\frac{f_a}{F_a} + \frac{f_b}{F_b} \leqslant 1.0$$

In this expression f_a is the axial stress (P/A), F_a is the allowable stress if only axial stresses occurred, f_b is the flexure stress (Mc/I), and F_b is the allowable flexure stress if only bending stresses were present. For many years the AISC also used this expression for members subjected to a combination of the two types of stresses. Today the AISC permits its use only for certain conditions (to be described later in this chapter). This expression might be thought of as a percentage formula. For instance, if 60% of the allowable axial stress is used up by f_a/F_a, only 40% of the allowable flexure stress remains for the Mc/I stress. This expression has the effect of giving an allowable combined stress which will fall proportionately between the allowable individual values as each type of stress is to its allowable value. When the flexure stress is large with respect to the axial stress, the allowable combined stress will approach very nearly the allowable flexure stress. Similarly, when the axial stress is large in comparison to the flexure stress the allowable combined stress will be close to the allowable axial stress. Example 8-2 presents an illustration of the combination of the two types of stresses according to this commonly used expression.

Should bending occur about both axes, the following expression is used to consider the combined stress situation.

$$\frac{f_a}{F_a} + \frac{f_{bx}}{F_{bx}} + \frac{f_{by}}{F_{by}} \leqslant 1.0$$

Example 8-2

A W10×49 ($A = 14.40$ in.^{2}, $S_x = 54.6$, $r = 2.54$ in.) is subjected to a moment of 40 ft-k and an axial compressive load of 100 k. The member is 15 ft long and has an allowable bending stress of 18 ksi and an allowable axial compressive stress to be determined from the formula $F_a = 15,000 - \frac{1}{4}(l/r)^2$. Is the member overstressed according to the $f_a/F_a + f_b/F_b \leqslant 1$ expression?

SOLUTION

$$f_a = \frac{100}{14.4} = 6.94 \text{ ksi}$$

$$F_a = 15,000 - \frac{1}{4}\left(\frac{12 \times 15}{2.54}\right)^2 = 13.75 \text{ ksi}$$

$$f_b = \frac{12 \times 40}{54.6} = 8.8 \text{ ksi}$$

$$F_b = 18 \text{ ksi}$$

$$\frac{f_a}{F_a} + \frac{f_b}{F_b} = \frac{6.94}{13.75} + \frac{8.8}{18} = 0.994 < 1 \qquad \text{OK}$$

8-4. DESIGN FOR AXIAL COMPRESSION AND BENDING

When the average designer is faced with the problem of selecting a member to resist combined stresses, she will probably estimate the size member required, after which she will check its combined stress in accordance with the requirements of the specification being used. Example 8-3 illustrates the design of a member of this type. To arrive at the assumed section in this example it was assumed that roughly one-half of each allowable stress (F_a and F_b) was caused by the loads. The resulting axial stress value was divided into the total axial load to estimate the cross-sectional area required, and the flexure stress value was divided into the bending moment to estimate the required section modulus. A section was selected which had approximately these proportions. Although the section selected by the method was satisfactory in this example it is usually necessary to make another trial or two to obtain an economical solution. The division of stresses was purely an estimate and some other division might be better. For instance if the axial load seems quite large in proportion to the bending moment, the designer might decide to estimate that 75% of the axial stress allowable is used and 25% of the bending stress allowable; or some other division. (Much more convenient methods are available for designing beam-columns, such as the one presented in Section 8-6 for the AISC Specification.)

Example 8-3

Select a W10 beam-column to support an axial load of 120 k and a bending moment of 50 ft-k. The member has an unsupported length of 18 ft, has an allowable flexure stress of 18 ksi, and an allowable compressive stress of $F_a = 15,000 - \frac{1}{4}(l/r)^2$. The $f_a/F_a + f_b/F_b \leqslant 1.0$ expression is to be used in the design.

SOLUTION
Estimated axial stress = 7 ksi.

$$A \text{ reqd.} = \frac{120}{7} = 17.1 \text{ in.}^2$$

Estimated flexure stress = 9 ksi.

$$S \text{ reqd.} = \frac{12 \times 50}{9} = 66.7 \text{ in.}^3$$

Try W10×60 ($A = 17.6$ in.^{2}, $S_x = 66.7$, $r = 2.57$)

$$f_a = \frac{120,000}{17.6} = 6818 \text{ psi}$$

$$F_a = 15,000 - \frac{1}{4}\left(\frac{12 \times 18}{2.57}\right)^2 = 13,230 \text{ psi}$$

$$f_b = \frac{(12)(50,000)}{66.7} = 8996 \text{ psi}$$

$$F_b = 18,000 \text{ psi}$$

$$\frac{f_a}{F_a} + \frac{f_b}{F_b} = \frac{6818}{13,230} + \frac{8996}{18,000}$$

$$= 0.515 + 0.500 = 1.015 \qquad \text{(slightly overstressed)} \qquad \text{OK}$$

8-5. AISC REQUIREMENTS

For many years the AISC used the previously discussed $f_a/F_a + f_b/F_b \leqslant 1.0$ formula, but today they permit its use only under certain conditions. The present expressions are more conservative than the older expressions because a factor has been introduced in the equations to estimate the effect of additional moments caused by lateral displacements. As previously indicated a member subjected to a moment deflects laterally and an additional moment equal to the axial load times the lateral deflection is developed.

The present AISC Specification says that members subjected to a combination of axial compression and flexure stress must be proportioned to meet the requirements of both of the following expressions.[1]

$$\frac{f_a}{F_a} + \frac{C_{mx}f_{bx}}{\left(1 - \dfrac{f_a}{F'_{ex}}\right)F_{bx}} + \frac{C_{my}f_{by}}{\left(1 - \dfrac{f_a}{F'_{ey}}\right)F_{by}} \leqslant 1.0 \qquad \text{(AISC Formula 1.6-1a)}$$

$$\frac{f_a}{0.60F_y} + \frac{f_{bx}}{F_{bx}} + \frac{f_{by}}{F_{by}} \leqslant 1.0 \qquad \text{(AISC Formula 1.6-1b)}$$

Should f_a/F_a be 0.15 or less, the influence of the $[1-(f_a/F'_e)]$ factor is so small that the following formula can be used in lieu of the preceding ones.

$$\frac{f_a}{F_a} + \frac{f_{bx}}{F_{bx}} + \frac{f_{by}}{F_{by}} \leqslant 1.0 \qquad \text{(AISC Formula 1.6-2)}$$

Should bending occur about one axis only the appropriate term for the other axis will drop out of these equations.

In these expressions f_a, f_b, F_a, and F_b are the same values, which have been previously defined. F'_e is the Euler stress divided by a safety factor of $\frac{23}{12}$. Its value is given by the expression to follow in which l_b is the *actual*

---

[1] For development of these formulas the student is referred to pages 364–369 in *Structural Steel Design* written by Beedle et al. (New York: Ronald Press, 1964).

unsupported length in the plane of bending, r_b is the corresponding radius of gyration, and K is the effective length factor in the plane of bending.

$$F_e' = \frac{12\pi^2 E}{23(Kl_b/r_b)^2}$$

The value of F_e' can be increased by one-third for wind and seismic stresses according to the AISC Specification provided the section used is not overstressed (not counting the one-third increase) by the gravity dead, live, and impact loads. In the first of the two AISC expressions $1 - f_a/F_e'$ is the *amplification factor* which has as its purpose the estimation of the increased moments caused by lateral deflection. The amplification factor is in the direction of greater conservatism as compared to the old $f_a/F_a + f_b/F_b$ expression. However, for some conditions (depending on the actual slenderness ratios, axial loads, and moments involved) it is too conservative; and the formula has a *modification* or *reduction factor* to keep the estimated moments due to deflection from being too large.

There are three separate categories of C_m as described in Section 1.6.1 of the AISC Specification. These are as follows.

1. In Category 1 the columns are considered to be parts of frames that depend upon the bending stiffnesses of their members for lateral stiffness. These members are subject to joint translation or sidesway and C_m is considered to equal 0.85.

2. In Category 2 the members are prevented from joint translation or sidesway and they are not subject to transverse loading between their ends. For such members the modification factor is to be determined as follows.

$$C_m = 0.6 - 0.4\frac{M_1}{M_2} \geqslant 0.4$$

In this expression M_1/M_2 is the ratio of the smaller moment to the larger moment at the ends of the unbraced length. The ratio is negative if the moments cause the member to bend in single curvature ($)$ or $()$) and positive if they bend the member in reverse curvature (ς or $\wr$). It should be obvious that a member in single curvature has larger lateral deflections than a similar member bent in reverse curvature. With larger lateral deflections the moments due to the axial loads and thus the stresses will be greater. When a column is bent in single curvature substitution in the AISC formula will, therefore, give smaller allowable stresses.

3. Category 3 applies to members which are subjected to transverse loading between the joints and which are braced against joint translation or sidesway in the plane of loading. The compression chord of a truss with a purlin load between its joints is a typical example of this category. The AISC Specification states that the value of C_m may be taken as follows.

   a. For members with restrained ends $C_m = 0.85$.
   b. For members with unrestrained ends $C_m = 1.0$.

Table 8-1

Case	ψ	C_m
(a)	0	1.0
(b)	-0.4	$1-0.4\dfrac{f_a}{F_e'}$
(c)	-0.4	$1-0.4\dfrac{f_a}{F_e'}$
(d)	-0.2	$1-0.2\dfrac{f_a}{F_e'}$
(e)	-0.3	$1-0.3\dfrac{f_a}{F_e'}$
(f)	-0.2	$1-0.2\dfrac{f_a}{F_e'}$

Instead of these values C_m may be determined for various end conditions and loads by the values given in Table 8-1 which is a reproduction of Table C1.6.1 of the AISC Commentary.

Most of the members encountered in actual practice which are subject to appreciable amounts of combined bending and axial stress are parts of a rigid frame structure, and the other members rigidly connected to the member in question have some appreciable effect on that member. This means that to determine the allowable axial stress the effective length of the member needs to be determined as previously described. From that previous discussion it is remembered that if sidesway is possible the effective length can be greater than the actual length, but if sidesway is prevented as by fairly heavy masonry walls the effective length will be less than the actual length.

In category 1 the effective lengths of the members are used in calculating F_a and it can never be less than the unbraced length and may even be greater. The effective length in the direction of bending is used in calculating F'_e. For the moment computations the actual unbraced length is used. In category 2 the columns have no sidesway nor transverse loading and the effective length is used for calculating F_a. It cannot be greater than the unbraced length and may be less. Again the effective length in the direction of bending is used in calculating F'_e. For the moment calculations the actual unbraced length is used.

Examples 8-4 and 8-5 illustrate the design of members subjected to combined bending and axial stress in accordance with the AISC Specification. It will be noted in these examples that f_b is the bending stress at the point of the member being considered. When there are no transverse loads the stress will be computed for the largest of the moments at the ends of the unbraced length. When a transverse load is applied the largest moment between the points of lateral support is used to compute f_b for substitution in the first of the two AISC expressions. For substitution in the second expression the largest moment at either of the supported points is used.

The joints of a truss are restrained from translation. For this reason it might seem reasonable to use an effective length for the compression members of a truss somewhat less than the actual lengths. Section 1.8 of the AISC Commentary, however, suggests that the use of $K=1.0$ is wise when the ultimate load situation is considered. Should all of the members of a truss reach their ultimate load capacity at the same time the restraints against translation mentioned would be drastically reduced or eliminated.[2]

Example 8-4

For the truss shown in Fig. 8-2(a) a W8×31 is used as a continuous top chord member from joint L_0 to joint U_3. If the member consists of A36 steel, does it have sufficient strength to resist the loads shown in part (b) of the figure using the AISC Specification? Part (b) shows the portion of the chord from L_0 to U_1 and the 10-k load represents the effect of a purlin.

SOLUTION

Try W8×31 $\left(A=9.13 \text{ in.}^2, b_f/2t_f=9.2, F'_y=50.0, S_x=27.5, r_x=3.47, r_y=2.02, d/A_f=2.30 \right)$

$$f_a = \frac{125}{9.13} = 13.69 \text{ ksi}$$

Assuming that the member is fixed at L_0 and U_1 for the purpose of estimating moment,

$$M = \frac{Pl}{8} = \frac{(10)(13)}{8} = 16.25 \text{ ft-k}$$

$$f_b = \frac{12 \times 16.25}{27.5} = 7.09 \text{ ksi}$$

[2] "Commentary on the AISC Specification," *Manual of Steel Construction* (New York: AISC, Inc., 1978) pp. 5–127.

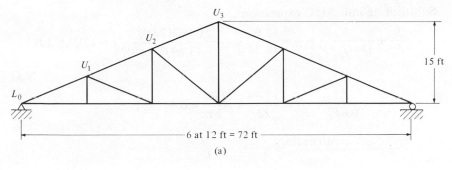

(a)

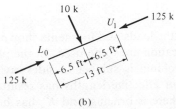

(b)

Figure 8-2

Lateral support (pinned) is provided at 6.5-ft intervals.

$L_c = 8.4$ ft (from AISC Manual) > 6.5 ft $< L_u = 20.1$ ft and $F_y < F_y'$
$\qquad = 50.0$ ksi

Therefore,

$$F_b = 0.66F_y = 24 \text{ ksi}.$$

$$F_e' = \frac{(12)(\pi)^2(29 \times 10^3)}{23\left[\left(\dfrac{1.00 \times 12 \times 13}{3.47}\right)\right]^2} = 73.89 \text{ ksi}$$

(can be looked up in AISC
Manual, pp. 5–79)

The member falls into category 3.

$$C_m = 1 - 0.2\frac{f_a}{F_e'} = 1 - 0.2\frac{13.69}{73.89} = 0.963$$

Determining allowable value of F_a,

$$\frac{K_y l_y}{r_y} = \frac{(1.0)(12)(6.5)}{2.02} = 38.61$$

$$\frac{K_x l_x}{r_x} = \frac{(1.0)(12)(13)}{3.47} = 44.96 \leftarrow$$

$$F_a = 18.78 \text{ ksi} \qquad \text{(from AISC Manual)}$$

Substituting into AISC expressions,

$$\frac{f_a}{F_a} + \frac{C_m f_b}{(1-f_a/F_e')F_b} = \frac{13.69}{18.78} + \frac{0.963(7.09)}{(1-13.69/73.89)24} = 1.078 > 1.0$$

N.G.

$$\frac{f_a}{0.6F_y} + \frac{f_b}{F_b} = \frac{13.69}{22} + \frac{7.09}{24} = 0.917 < 1$$

Section is unsatisfactory

Example 8-5

Select a W14 to resist the loads and moments shown in Fig. 8-3 if A36 steel and the AISC Specification are to be used. In the plane of loading there is no bracing and the column is subject to sidesway but has no transverse loading. An analysis of effective lengths has resulted in $K_x = 1.92$. In the perpendicular plane there is bracing and K_y has been estimated to equal 0.80.

SOLUTION

Try W14×159 ($A = 46.7$, $S_x = 254$, $r_x = 6.38$, $r_y = 4.00$, $L_c = 16.4$)

$$f_a = \frac{500}{46.7} = 10.71 \text{ ksi}$$

$$f_b = \frac{(12)(220)}{254} = 10.39 \text{ ksi}$$

$$\frac{K_x l_x}{r_x} = \frac{(1.92)(12)(14)}{6.38} = 50.6 \leftarrow$$

$$\frac{K_y l_y}{r_y} = \frac{(0.80)(12)(14)}{4.00} = 33.6$$

$$F_a = 18.30 \text{ ksi}$$

$$L_c = 16.4 \text{ ft} > 14 \text{ ft}$$

Therefore,

$$F_b = 0.66 \, F_y = 24 \text{ ksi}$$

and

$$F_e' = \frac{(12)(\pi)^2(29 \times 10^3)}{(23)\left(\dfrac{1.92 \times 12 \times 14}{6.38}\right)^2} = 58.42 \text{ ksi}$$

Member falls into category 1, therefore

$$C_m = 0.85$$

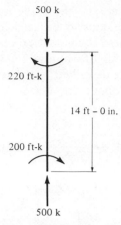

Figure 8-3

Substituting into AISC expression,

$$\frac{f_a}{F_a} + \frac{C_m f_b}{(1-f_a/F'_e)F_b} = \frac{10.71}{18.30} + \frac{(0.85)(10.39)}{(1-10.71/58.42)24} = 1.036 > 1.0$$

N.G.

$$\frac{f_a}{0.6F_y} + \frac{f_b}{F_b} = \frac{10.71}{22} + \frac{10.39}{24} = 0.920 < 1.0$$

Section is unsatisfactory since it does not satisfy both AISC formulas.

8-6.   DESIGN OF BEAM-COLUMNS (AISC)

A common method used for selecting sections to resist both moments and axial loads is the *equivalent axial load method*. In this method the eccentric load is replaced with a concentric load of such a magnitude that the maximum stress it causes is the same as the maximum stress produced by the eccentric load.

It is assumed here that it is desired to select the most economical section to resist both a moment and an axial load (say 140 ft-k and 400 k, respectively). By a trial and error procedure it is possible to eventually select the lightest section. Somewhere, however, there is a fictitious axial load which will require the same section size as that required for the actual moment and the actual axial load. This fictitious load is called the *equivalent axial load*.

Equations are used to convert the bending moment into an estimated equivalent axial load P' which is added to the axial load P. The total of $P + P'$ is called the equivalent axial load P_{eq} and it is used to enter the concentric column tables for choice of a trial section.

Space is not taken here to show the complete derivation of the equivalent axial load expressions to be used in place of AISC Formulas 1.6-1a, 1.6-1b and 1.6-2. These equivalent expressions which are referred to as Modified Formulas 1.6-1a, 1.6-1b, and 1.6-2 by the AISC Manual can be derived by the procedure described in the several paragraphs to follow.

If bending is present about the x axis only Formula 1.6-1a can be written as follows.

$$\frac{f_a}{F_a} + \frac{C_m f_b}{(1 - f_a / F_e') F_b} \leqslant 1.0$$

Replacing f_a with P/A and f_b with M/S the formula becomes

$$\frac{P}{AF_a} + \frac{M}{SF_b} \left(\frac{C_m}{1 - P/AF_e'} \right) = 1.0$$

Multiplying both sides of the equation by AF_a gives

$$P + \left(\frac{MA}{S} \right) \left(\frac{F_a}{F_b} \right) \left(\frac{C_m}{1 - P/AF_e'} \right) = AF_a$$

If both numerator and denominator of the $C_m / (1 - P/AF_e')$ part of the formula is multiplied by F_e' the result is

$$\frac{C_m}{1 - P/AF_e'} \left(\frac{F_e'}{F_e'} \right) = \frac{C_m F_e'}{F_e' - P/A}$$

Substituting into this expression the value of $F_e' = \pi^2 E / (Kl_b / r_b)^2$ including the AISC value of $E = 29{,}000$ ksi the expression becomes

$$\frac{C_m F_e'}{F_e' - P/A} = \frac{C_m 149{,}000 A r^2}{149{,}000 A r^2 - P(Kl)^2}$$

Then this value is put into the expression for bending about the x axis only and in addition A/S is replaced with B, the so-called bending factor, and $149{,}000 A r_x^{\,2}$ is replaced with the letters a_x. The values of B and a can easily be computed for each shape and they are given in the concentric column tables of the AISC Manual. The resulting equation is

$$P_{eq} = P + \left[BMC_m \left(\frac{F_a}{F_b} \right) \left(\frac{a_x}{a_x - P(Kl)^2} \right) \right]$$

In a similar fashion the equivalent or modified formulas can be written for all three of the AISC formulas including bending about both x

and y axes with the following results.

$$P + P_x' + P_y' = P + \left[B_x M_x C_{mx} \left(\frac{F_a}{F_{bx}} \right) \left(\frac{a_x}{a_x - P(Kl)^2} \right) \right]$$

$$+ \left[B_y M_y C_{my} \left(\frac{F_a}{F_{by}} \right) \left(\frac{a_y}{a_y - P(Kl)^2} \right) \right] \quad \text{(Modified Formula 1.6-1a)}$$

$$P + P_x' + P_y' = P \left(\frac{F_a}{0.6 F_y} \right) + \left[B_x M_x \left(\frac{F_a}{F_{bx}} \right) \right]$$

$$+ \left[B_y M_y \left(\frac{F_a}{F_{by}} \right) \right] \quad \text{(Modified Formula 1.6-1b)}$$

AISC Formula 1.6-2 is to be used in lieu of Formulas 1.6-1a and 1.6-1b when

$$\frac{f_a}{F_a} \leqslant 0.15$$

$$P + P_x' + P_y' = P + \left[B_x M_x \left(\frac{F_a}{F_{bx}} \right) \right] + \left[B_y M_y \left(\frac{F_a}{F_{by}} \right) \right] \quad \begin{array}{l} \text{(Modified Formula} \\ \text{1.6-2)} \end{array}$$

These equations which are very useful for preliminary design yield columns that are on the conservative size. Should bending occur about one axis only the equations are reduced by deleting the appropriate terms.

In these expressions B_x and B_y are the bending factors which are respectively equal to A/S_x and A/S_y. Their values, which are tabulated in the Steel Manual, vary from 0.168 to 1.133 for I-shaped sections for B_x while those for B_y vary from 0.408 to 4.722. The larger values occur with the smaller sections. a_x is a component of the amplification factor when bending is about the x axis and equals $0.149 A r_x^2 \times 10^6$. Similarly a_y is $0.149 A r_y^2 \times 10^6$ and both a_x and a_y are given for sections normally used as columns in the column section of the AISC Manual. In these equations the moments M_x and M_y must be used in in.-kips.

Example 8-6 illustrates the design of a beam-column using the equivalent axial load procedure. The design of such members involves a trial and error process but the AISC Manual provides such a quick and simple procedure for selecting an economical trial section that the process is greatly abbreviated. An approximate equivalent axial load is calculated with the expression to follow in which $P_{\text{eff}} = P_{\text{eq}}$ the equivalent axial load, P_o is the actual axial load, m is a factor given in Table B on pages 3–10 of the AISC Manual (8th ed.) and U is a factor provided in the column tables. In applying the equation the moments M_x and M_y must be used in ft-kips.

$$P_{\text{eff}} = P_o + M_x m + M_y m U$$

To apply the expression a value of m is taken from the 1st Approximation section of Table B and a value of U is assumed equal to 3. The equation is solved for P_{eff} and a column is selected from the column tables for that load. Then the equation for P_{eff} is solved again using a revised value of m from the Subsequent Approximations part of Table B and the value of U is taken from the column tables for the column initially selected. Another shape is selected and the process is continued until m and U stabilize. The properties of the shape so selected are then used in the modified formulas 1.6-1a, 1.6-1b, and 1.6-2 as applicable to determine the final equivalent axial load.

Example 8-6

Select a W14 for the column shown in Fig. 8-4 using A36 steel and the AISC Specification. Assume $K = 1.0$ and $C_{mx} = 0.85$.

SOLUTION

For $kl = 15$ ft select a value of $m = 2.2$ from the 1st Approximation part of Table B

$$P_{\text{eff}} = P_o + M_x m + M_y mU$$

$$P_{\text{eff}} = 400 + (140)(2.2) = 708 \text{ k}$$

From column tables select W14 × 132 ($m = 1.7$ from the Subsequent Approximations part of Table B)

$$P_{\text{eff}} = 400 + (140)(1.7) = 638 \text{ k}$$

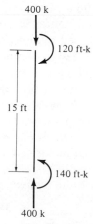

400 k

120 ft-k

15 ft

140 ft-k

400 k

Figure 8-4

From column tables select W14 × 120 ($m = 1.7$)
Try W14 × 120 (allowable axial load = 654 k, A = 35.3, r_y = 3.74,
L_c = 15.5 ft., therefore F_b = 24 ksi, B_x = 0.186, a_x = 204.8 × 10^6)

$$f_a = \frac{400}{35.3} = 11.33 \text{ ksi}$$

$$\frac{Kl}{r_y} = \frac{(1.0)(12 \times 15)}{3.74} = 48.13$$

$$F_a = 18.52 \text{ ksi}$$

$$\frac{f_a}{F_a} = \frac{11.33}{18.52} = 0.612 > 0.15$$

Therefore, we must use Modified Formulas 1.6-1a and 1.6-1b

Formula 1.6-1a

$$P + P'_x = 400 + \left[(0.186)(12 \times 140)(0.85)\left(\frac{18.52}{24}\right) \right.$$

$$\left. \left(\frac{204.8 \times 10^6}{204.8 \times 10^6 - (400)(12 \times 15)^2} \right) \right]$$

$$= 618.8 \text{ k} \leftarrow$$

Formula 1.6-1b

$$P + P'_x = (400)\left(\frac{18.52}{22}\right) + \left[(0.186)(12 \times 140)\left(\frac{18.52}{24}\right) \right] = 577.9 \text{ k}$$

Reenter tables with equivalent axial load of 618.8 k and select W14 × 120

In the AISC Manual example designs are presented for beam-columns in which lateral wind loads are involved. In these examples the lateral loads are multiplied by $\frac{3}{4}$ to take into account the provision of Section 1.5.6 of the AISC Specification which permits an increase in allowable stresses of $\frac{1}{3}$ for wind or seismic loads or for wind or seismic loads acting in combination with the design dead and live loads. This reduction is permissible only if the resulting sections are as large as they would have been if the member were selected for the design dead and live loads and impact without the stress increase and without the wind or seismic loads applied.

There seems to be some question concerning the justification of this allowable stress increase for lateral loads. Although it is true that wind or seismic loads are of short duration, steel does not exhibit large increases in strength for short-term loads as does timber. One logical explanation for the stress increase is that it is quite unlikely for maximum wind and seismic loads to occur simultaneously with the other maximum design loads. An interesting article on this stress increase appeared in the *Engineering Journal.*[3]

Limitations of AISC Tables

The tables in the AISC Manual which have been described here work very well unless the moment becomes quite large in proportion to the axial load. For such cases the members selected with the tables will be capable of supporting the loads and moments but may very well not provide a very economical solution. The tables are limited to the W14s, W12s, and shallower sections but when the moment is large in proportion to the axial load there will often be a much deeper section such as a W27 or W30 which may be more economical.

Example 8-7 illustrates this situation. The method previously presented in Example 8-3 is used for estimating the size and then the regular AISC expressions are used to check the section. It will be noted that in this problem the estimated area required is 47.78 in.^{2} (which corresponds to about a W30×173) and the estimated section modulus required is 273 in.^{3} (which corresponds approximately to a W30×99). The author interpolated between these two sections and tried a W30×124 which proved to be satisfactory. Subsequent checks were made on W33s, W27s, and W24s. The W30×124 was the lightest one which was satisfactory. If the column tables in the AISC Manual were used together with the equivalent axial load method appreciably heavier sections would be the result such as the W14×159 or the W12×170.

Example 8-7

Using the AISC Specification and A36 steel select a W section to support an axial load $P=430$ k (1 913 kN) and a bending moment about the x axis $M_x=250$ ft-k. The member, which has pinned ends, is to be 18 ft long and is to be laterally braced (pinned) in the weak direction at middepth. Assume $C_m=0.85$.

SOLUTION
Assume

$$\frac{f_a}{F_a} + \frac{f_b}{F_b} = 0.5 + 0.5 = 1.0$$

Assume

$$F_a = 18 \text{ ksi}$$
$$\frac{f_a}{F_a} = \frac{430/A}{18} = 0.5$$

---

[3] D. S. Ellifritt, "The Mysterious $\frac{1}{3}$ Stress Increase," *Engineering Journal* **14**, No. 4, October 1977 (New York:AISC), pp. 138–140.

Approximate A required $=47.78$ in.^{2}
   Assume
$$F_b = 22 \text{ ksi.}$$
$$\frac{f_b}{F_b} = \frac{(12)(250)/S_x}{22} = 0.5$$

Approximate S_x required $=273$ in.^{3}
   From this information several possible estimated deep sections includ-
ing the W30×124, W27×146, W24×131, and others may be tried.
   Try W30×124 ($A = 36.5$ in.^{2}, $S_x = 355$ in.^{3}, $r_x = 12.1$, $r_y = 2.23$,
$L_c = 11.1$, $L_u = 15.0$)

$$f_a = \frac{430}{36.5} = 11.78 \text{ ksi}$$

$$\frac{Kl_x}{r_x} = \frac{(1)(12 \times 18)}{12.1} = 17.85$$

$$\frac{Kl_y}{r_y} = \frac{(1)(12 \times 9)}{2.23} = 48.43 \leftarrow$$

$$F_a = 18.49 \text{ ksi}$$

$$\frac{Kl_b}{r_b} = \frac{(1)(12 \times 18)}{12.1} = 17.85$$

$$F_e' = \frac{(12)(\pi)^2(29,000)}{(23)(17.85)^2} = 468.7 \text{ ksi}$$

$$f_{bx} = \frac{(12)(250)}{355} = 8.45 \text{ ksi}$$

$$F_b - 24 \text{ ksi}$$

assuming unbraced length of compression flange $=9$ ft.
   Substituting into Formulas 1.6-1a and 1.6-1b since $f_a/F_a = 11.78/18.49 > 0.15$,

$$\frac{11.78}{18.49} + \frac{(0.85)(8.45)}{(1 - 11.78/468.7)(24)} = 0.944 < 1.00 \qquad \text{OK}$$

$$\frac{11.78}{22} + \frac{8.45}{24} = 0.888 < 1.00 \qquad \text{OK}$$

   Use W30×124

8-7. COMBINED AXIAL TENSION AND BENDING

The foregoing paragraphs and example problems have all pertained to
members subjected to a combination of axial compression and bending. A

less critical case but one which can occasionally be quite important occurs when a member is subjected to simultaneous axial tension and bending. Various specifications have different design requirements. For instance the AREA says that members subject to both axial tension and bending stress shall be so proportioned that the total axial tension plus the bending tension shall not exceed the allowable tension for axially loaded tensile members. (In addition they say that the compression caused by such bending may not exceed the allowable value for axially loaded compression members.)

The AISC requires a member subject to a combination of axial tension and bending to satisfy the expression which follows. In this expression f_b is the maximum tensile bending stress and F_b is the allowable bending tensile stress.

$$\frac{f_a}{0.60F_y} + \frac{f_{bx}}{F_{bx}} + \frac{f_{by}}{F_{by}} \leqslant 1.0$$

A further requirement is that the computed bending compressive stress may not exceed the applicable value for allowable bending compressive stress.

8-8. FURTHER DISCUSSION OF EFFECTIVE LENGTHS OF COLUMNS

The subject of effective lengths was previously discussed in Chapters 4 and 5. In fact Table 5-1 was presented to provide suggested effective lengths for columns with different degrees of end restraint. It is to be remembered that this table was developed for certain idealized conditions which might be entirely different from practical design conditions. This method is quite satisfactory for preliminary designs and for situations where sidesway is prevented by bracing.

Truthfully the effective length of a column is a property of the whole structure of which the column is a part. In the large majority of existing buildings it is probable that the masonry walls provide sufficient lateral support to prevent sidesway. Many modern buildings, however, are subject to sidesway. When light curtain walls are used, as they often are in modern buildings, there is probably little resistance to sidesway. Sidesway is also present in tall buildings in appreciable amounts unless a definite diagonal bracing system is used. For these cases it seems logical to assume that resistance to sidesway is primarily provided by the lateral stiffness of the frame alone.

Perhaps a few explanatory remarks should be made at this point concerning the definition of sidesway as it pertains to effective lengths. In this discussion sidesway refers to a sidesway type of buckling. This not only includes sidesway as used in the analysis of indeterminate frames (where the frames deflect laterally due to the presence of lateral loads or unsymmetrical vertical loads) but also to columns whose ends could move transversely if the columns were loaded until buckling occurred.

For structures with sidesway the designer may make her own mathematical analysis of the structure to determine its effective length (although too lengthy to be practical for the average designer and design problem); she may on the basis of her judgment of the particular conditions interpolate between the values given in Table 5-1; or she may obtain effective lengths from the chart in the AISC Manual entitled "Alignment Chart for Effective Length of Columns in Continuous Frames." This chart is reproduced in Fig. 8-5 with the permission of the AISC.

The alignment chart is to be used to estimate the effective lengths of columns in frames where resistance to lateral movement is provided by the stiffness of the members of the frames. To use the chart it is necessary to obtain the sizes of the girders and columns framing into the column in question before its effective length can be determined. In other words, before the chart can be used a trial design has to be made of each of the members.

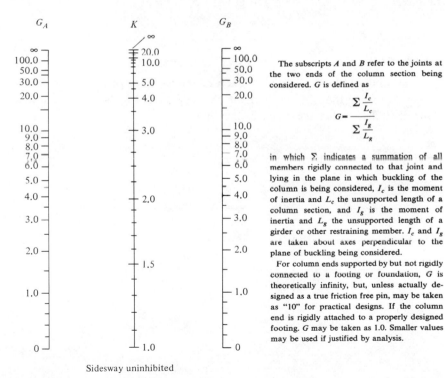

The subscripts A and B refer to the joints at the two ends of the column section being considered. G is defined as

$$G = \frac{\Sigma \dfrac{I_c}{L_c}}{\Sigma \dfrac{I_g}{L_g}}$$

in which Σ indicates a summation of all members rigidly connected to that joint and lying in the plane in which buckling of the column is being considered, I_c is the moment of inertia and L_c the unsupported length of a column section, and I_g is the moment of inertia and L_g the unsupported length of a girder or other restraining member. I_c and I_g are taken about axes perpendicular to the plane of buckling being considered.

For column ends supported by but not rigidly connected to a footing or foundation, G is theoretically infinity, but, unless actually designed as a true friction free pin, may be taken as "10" for practical designs. If the column end is rigidly attached to a properly designed footing, G may be taken as 1.0. Smaller values may be used if justified by analysis.

Sidesway uninhibited

Figure 8-5 Alignment chart for effective length of columns in continuous frames.

The effective lengths of each of the columns of a frame are estimated with the alignment chart in Example 8-8. (When sidesway is possible it will be found that the effective lengths are always greater than the actual lengths as is illustrated in this example. When frames are braced in such a manner that sidesway is not possible K will be less than 1.0.) An initial design has provided preliminary sizes for each of the members in the frame of Example 8-8. After the effective lengths are determined each column can be redesigned. Should the sizes change appreciably, new effective lengths can be determined, the column designs repeated, etc. Several tables are used in the solution of this example. These should be self-explanatory after the clear directions given by the AISC on the alignment chart are examined.

For most buildings the values of K_x and K_y should be examined separately. The reason for such individual study lies in the different possible framing conditions in the two directions. Many multistory frames consist of rigid frames in one direction and conventionally connected frames with sway bracing in the other. In addition the points of lateral support may often be entirely different in the two planes. An examination of Example 8-6 will show that it is highly possible for the column action to be governed by K_y and for K_x to govern F_e'.

Example 8-8

Determine the effective lengths of each of the columns of the frame shown in Fig. 8-6 using the AISC alignment chart given in Fig. 8-5. The tentative sizes of each member are given in the figure.

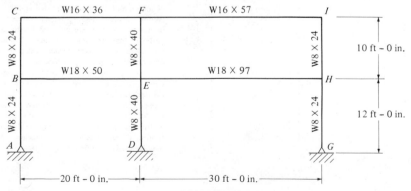

Figure 8-6

SOLUTION

Stiffness factors:

Member	Shape	I	L	I/L
AB	W8×24	82.8	144	0.575
BC	W8×24	82.8	120	0.690
DE	W8×40	146	144	1.014
EF	W8×40	146	120	1.217
GH	W8×24	82.8	144	0.575
HI	W8×24	82.8	120	0.690
BE	W18×50	800	240	3.333
CF	W16×36	448	240	1.867
EH	W18×97	1,750	360	4.861
FI	W16×57	758	360	2.106

G factors for each joint:

Joint	$\sum(I_c/L_c)/\sum(I_g/L_g)$	G
A	See Fig. 8-5	10.0
B	$\dfrac{0.575+0.690}{3.333}$	0.380
C	$\dfrac{0.690}{1.867}$	0.370
D	See Fig. 8-5	10.0
E	$\dfrac{1.014+1.217}{3.333+4.861}$	0.272
F	$\dfrac{1.217}{1.867+2.106}$	0.306
G	See Fig. 8-5	10.0
H	$\dfrac{0.575+0.690}{4.861}$	0.260
I	$\dfrac{0.690}{2.106}$	0.328

Column K factors from chart (Fig. 8-5).

Column	G Values at Column Ends		K
AB	10.0	0.380	1.72
BC	0.380	0.370	1.12
DE	10.0	0.272	1.70
EF	0.272	0.306	1.09
GH	10.0	0.260	1.70
HI	0.260	0.328	1.10

Another chart is available[4] where sidesway is prevented. This chart is presented in Fig. 8-7. Examples 8-9 and 8-10 present two more beam-column problems and in these problems use is made of these effective length charts. These alignment charts are applicable to elastic columns. For inelastic columns the G values used may be reduced by multiplying them by a stiffness reduction factor with resulting smaller K values. This procedure described on pages 3-6 through 3-8 of the AISC Manual is based on work by Yura.[5]

Example 8-9

Check the adequacy of the lower W8×40 column shown in Fig. 8-6 if it is assumed to have no bracing against sidesway in the plane of the frame. The member is assumed to be subjected to an axial load of 100 k and to

[4] Structural Stability Research Council, *Guide to Stability Design Criteria for Metal Structures*, B. G. Johnson, ed., 3rd ed. (New York: Wiley, 1976).
[5] J. A. Yura, "The Effective Length of Columns in Unbraced Frames," *Engineering Journal*, AISC 8, No. 2, April 1971, pp. 37–42.

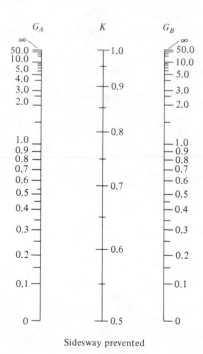

Sidesway prevented

Figure 8-7

20 ft-k

30 ft-k

Figure 8-8

have the moment diagram shown in Fig. 8-8. Assume $K = 1.0$ for the y direction and use A36 steel and the AISC Specification.

SOLUTION

Using W8×40 ($A = 11.7$, $S_x = 35.5$, $r_x = 3.53$, $r_y = 2.04$)

$$f_a = \frac{100}{11.7} = 8.55 \text{ ksi}$$

$$\frac{Kl_x}{r_x} = \frac{(1.70)(12 \times 12)}{3.53} = 69.3$$

(The value of $K = 1.70$ was obtained in Example 8-8. The stiffness reduction for G is 1.0 here and does not affect K.)

$$\frac{Kl_y}{r_y} = \frac{(1.00)(12 \times 12)}{2.04} = 70.6 \leftarrow$$

$$F_a = 16.37 \text{ ksi}$$

$$\frac{f_a}{F_a} = \frac{8.55}{16.37} = 0.522 > 0.15$$

Therefore, must use Formulas 1.6-1a and 1.6-1b.

$$f_b = \frac{(12)(30)}{35.5} = 10.14 \text{ ksi}$$

$L_c = 8.5$ ft and $L_u = 25.3$ ft,   therefore $F_b = 0.60 F_y = 22$ ksi

$$C_m = 0.85$$

$$F_e' = 31.1 \text{ ksi} \quad \text{for} \quad \frac{Kl_x}{r_x} = 69.3$$

$$\frac{8.55}{16.37} + \frac{(0.85)(10.14)}{(1-8.55/31.1)22} = 1.062 > 1.0 \qquad \text{(Formula 1.6-1a)} \qquad \text{N.G.}$$

$$\frac{8.55}{22} + \frac{10.14}{22} = 0.850 < 1.0 \qquad \text{(Formula 1.6-1b)} \qquad \text{N.G.}$$

Section is unsatisfactory

Example 8-10

The columns in the frame shown in Fig. 8-9 are braced against sidesway in both directions. Check one of the upper columns which is subjected to an axial load of 180 k and the moments shown in the figure. Use A36 steel and the AISC Specification.

SOLUTION

Using a W12×65 ($A = 19.1$, $I_x = 533$, $S_x = 87.9$, $r_x = 5.28$, $r_y = 3.02$, $L_c = 12.7$ ft, $L_u = 27.7$ ft)

$$f_a = \frac{180}{19.1} = 9.42 \text{ ksi}$$

$$\frac{Kl}{r} = \frac{(1.0)(12 \times 15)}{3.02} = 59.6$$

$$F_a = 17.47 \text{ ksi}$$

$$\frac{f_a}{F_a} = \frac{9.42}{17.47} = 0.539 > 0.15$$

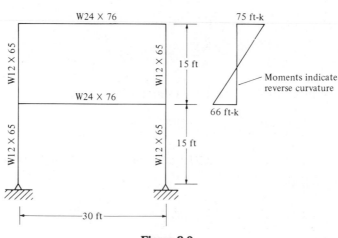

Figure 8-9

Therefore, must use Formulas 1.6-1a and 1.6-1b.

$$f_b = \frac{(12)(75)}{87.9} = 10.24 \text{ ksi}$$

$$F_b = 22 \text{ ksi}$$

$$C_m = \left[0.60 - (0.40)\left(+ \frac{66}{75} \right) \right] = 0.248 < 0.4$$

Use 0.4

Effective length factor in plane of bending determined as follows (Fig. 8-10). The stiffness reduction factor is again equal to 1.0.

$$\frac{Kl}{r_x} = \frac{(0.74)(12 \times 15)}{5.28} = 25.23$$

$$F'_e = 234.8 \text{ ksi}$$

$$\frac{9.42}{17.47} + \frac{(0.40)(10.24)}{(1 - 9.42/234.8)(22)} = 0.733 < 1.0 \qquad \text{(Formula 1.6-1a) OK}$$

$$\frac{9.42}{22} + \frac{10.24}{22} = 0.894 < 1.0 \qquad \text{(Formula 1.6-1b) OK}$$

Section is satisfactory

Information has been presented here only for prismatic columns used in standard building beam-column construction or for trusses. There is, however, considerable information available in engineering literature con-

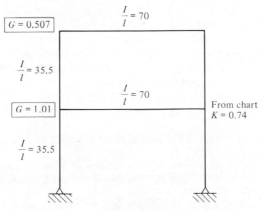

Figure 8-10

cerning K factors for other cases such as gabled frames, stepped columns, and columns with axial loads applied at intermediate points.[6, 7, 8]

Problems

8-1. Determine the maximum stresses in the compression chord of a roof truss caused by its own weight. The member is 24 ft–0 in. long and consists of two C12×20.7s with a top cover plate $\frac{1}{2}$×12 in. Assume the lacing and tie plates weigh 200 lb. (*Ans.* $f_t = +1.18$ ksi, $f_c = -0.647$ ksi)

8-2. The top chord of a roof truss is shown in the accompanying illustration. Determine the maximum stresses at the extreme fibers of this member if it is subjected to a moment of 90 ft-k and an axial compression of 450 k.

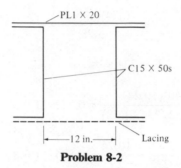

PL1 × 20

C15 × 50s

├────12 in.────┤   Lacing

Problem 8-2

8-3. An 18 ft W14×90 is used to support an axial load of 300 k and a bending moment about its major axis of 60 ft-k. Is it adequate according to the following interaction formula and allowable stress values?

$$F_a = 15,000 - \frac{1}{4}\left(\frac{l}{r}\right)^2$$

$$F_b = 18,000 \text{ psi}$$

$$\frac{f_a}{F_a} + \frac{f_b}{F_b} \leqslant 1.0$$

(*Ans.* Section unsatisfactory; interaction equation gives 1.08)

8-4. A column has a laterally unsupported length of 20 ft, an axial compression of 150 k and a bending moment of 60 ft-k. Select a W section for this member using the interaction formula and allowable stress values of Prob. 8-3.

8-5. A 12-ft column in a rigid frame building is subjected to an axial load of 300 k and to a moment of 80 ft-k. Using the interaction formula and the allowable stresses of Example 8-3 select a W shape. (*Ans.* W14×109)

[6]Le-Wu-Lu, "Effective Length of Columns in Gable Frames," *Engineering Journal* **2**, No. 1, January, 1965 (New York: AISC), pp. 6–7.

[7]J. P. Anderson and J. H. Woodward, "Calculation of Effective Lengths and Effective Slenderness Ratios of Stepped Columns," *Engineering Journal* **9**, No. 3, October, 1972 (New York: AISC), pp. 157–166.

[8]B. S. Sandhu, "Effective Length of Columns with Intermediate Axial Loads," *Engineering Journal* **9**, No. 3, October, 1972 (New York: AISC), pp. 154–156.

8-6. The column shown in Prob. 8-6 is to be designed to support an eccentric load of 150 k, located as shown in the figure. Select the lightest available W section using the design information of Prob. 8-3. Length = 18 ft.

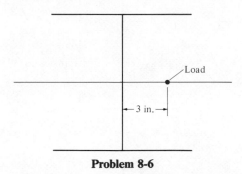

Problem 8-6

8-7. Same question as for Prob. 8-6 except the load is 100 k and is applied as shown in the accompanying illustration. (*Ans.* W12×72)

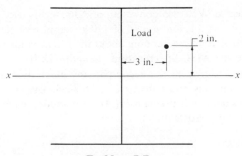

Problem 8-7

8-8. A column has a laterally unsupported length of 16 ft, an axial compression of 350 k and a bending moment of 100 ft-k. Using the design formulas given in Prob. 8-3 select the most economical W section.

8-9. A column with an unsupported length of 28 ft must support an axial compressive load of 65 k and a moment of 65 ft-k. Select a W section using the design formulas of Prob. 8-3. (*Ans.* W12×53 slightly overstressed)

8-10. A W10×49 pin-connected beam-column, which is subjected to sidesway, is 15-ft long. A 150-k load is applied to the column at its upper end with an eccentricity of 2 in. so as to cause bending about the major axis of the section. Check the adequacy of the member with the AISC specification if it consists of A36 steel.

8-11. Sidesway is prevented for the beam-column shown in the accompanying illustration. If bending is about the major axis, is the member satisfactory? Use A36 steel and the AISC Specification. (*Ans.* Interaction equations yield 0.870 and 0.774, therefore section is satisfactory)

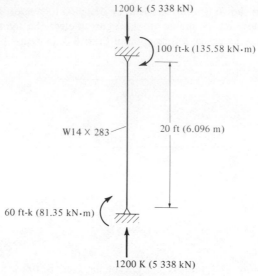

Problem 8-11

8-12. A pin-connected W10×45 consisting of A36 steel is used as a beam-column. If sidesway is possible and a load of 170 k is applied 1.50 in. off center at the upper end so as to cause bending about the *y* axis, is the member satisfactory according to the AISC Specification? Length = 12 ft.

8-13. Is the member shown in the accompanying illustration adequate? It is bent about its major axis and consists of A36 steel. Use the AISC Specification and assume sidesway is prevented. (*Ans.* Interaction equations yield 1.01 and 0.733 so probably satisfactory)

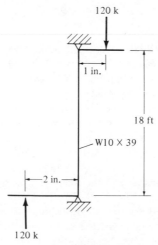

Problem 8-13

8-14. Select a W14 for the situation shown in the accompanying illustration. Use A36 steel, the AISC Specification, and assume sidesway occurs.

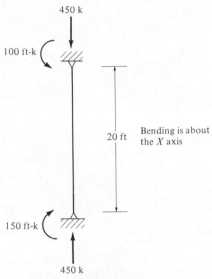

450 k

100 ft-k

20 ft   Bending is about the X axis

150 ft-k

450 k

Problem 8-14

8-15. Repeat Prob. 8-8 using the AISC Specification and A36 steel. The column is assumed to be pinned at both ends, to have no sidesway, and to have no transverse loading applied. The moments are maximum at the column ends and tend to bend the column in reverse curvature. (*Ans.* W14×90)

8-16. Repeat Prob. 8-9 using the AISC Specification and A36 steel. The member is fixed at its ends and has no sidesway or transverse loading. The moments are maximum at the column ends and tend to bend the column in single curvature.

8-17. A 16-ft pinned-end column is subjected to a moment of 100 ft-k at one end and 120 ft-k at the other end such that it is bent in single curvature. There are no transverse loads, and joint translation is prevented. If the axial load is 150 k, select a W14 section using the AISC Specification and A36 steel. (*Ans.* W14×82, W14×74 slightly overstressed)

8-18. A 15-ft pinned end column is subject to a bending moment of 300 ft-k and an axial load of 300 k. If A36 steel is used and sidesway is not prevented select the lightest satisfactory W section.

8-19. The accompanying diagram represents a highway sign board. Is a W10×45 of A36 steel satisfactory for the column labeled AB in the figure if the AISC Specification is used? Neglect member weights in all calculations. (*Ans.* section is satisfactory)

8-20. The W12×58 beam column shown is subject to an axial load of 170 k and to a lateral wind load perpendicular to the x axis of the member. Using A36 steel and the AISC specification determine the maximum value of w in kips per foot.

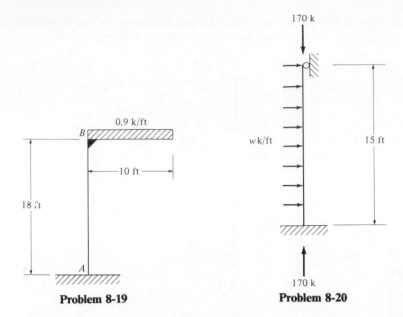

0.9 k/ft

B

10 ft

18 ft

A

Problem 8-19

170 k

w k/ft

15 ft

170 k

Problem 8-20

8-21. Using the AISC Specification and A36 steel select a WT8 for a 10-ft horizontal truss member with a 65-k axial compression load and a transverse uniform load of 75 lb/ft. The ends of the member are assumed to be pinned and joint translation is prevented. (*Ans.* WT8 × 20 slightly overstressed)

8-22. Repeat Prob. 8-21 if the member ends are assumed to be fixed.

8-23. Repeat Prob. 8-21 if one member end is assumed to be fixed and the other pinned. (*Ans.* WT8 × 20)

8-24. Repeat Prob. 8-7 using A36 steel and the AISC Specification. Assume the member is fixed at its ends and has no sidesway or transverse loading. The moments are maximum at the column ends and tend to bend it in single curvature.

8-25. For the frame shown in the accompanying illustration select tentative beam and column sizes assuming moments at beam ends of $wl^2/10$ and the same

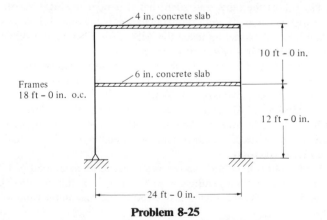

4 in. concrete slab

10 ft – 0 in.

6 in. concrete slab

Frames
18 ft – 0 in. o.c.

12 ft – 0 in.

24 ft – 0 in.

Problem 8-25

values in the columns. Use the AISC Specification and A36 steel. The roof is to be designed to support a built-up roof weighing 6 psf plus a live load of 30 psf and the interior floor slab is to support a live load of 125 psf. Using the member sizes selected determine the column K values from the chart given in Fig. 8-5.

8-26. Select a W14 beam-column of A36 steel for an axial load of 200 k, $M_x = 50$ ft-k, and $M_y = 30$ ft-k. Use AISC Specification and assume $Kl_x = Kl_y = 18$ ft and C_{mx} and $C_{my} = 0.85$.

8-27. The frame of Prob. 8-25 is to be designed as a rigid frame structure. Use the tentative member sizes selected and distribute the moments. Select new member sizes using the K values determined in Prob. 8-25 in the column design. Determine the new column K values with the AISC chart.

Chapter 9
Bolted
Connections

9-1. INTRODUCTION

For many years the accepted method of connecting the members of a steel structure was by riveting. In recent years, however, the use of rivets has declined rapidly due to the tremendous upsurge of welding and more recently high-strength bolting. This chapter is devoted almost entirely to high-strength bolts while a very brief discussion of rivets is presented in Chapter 10 together with some additional information concerning bolts.

Bolting of steel structures is a very rapid field erection process which requires less skilled labor than does riveting or welding. These facts give bolting a distinct economic advantage over the other connection methods in the United States where labor costs are so very high. Even though the purchase price of a high-strength bolt is nearly three times that of a rivet the overall cost of bolted construction is cheaper than that for riveted construction because of reduced labor and equipment costs and the smaller number of bolts required to resist the same loads.

9-2. TYPES OF BOLTS

There are several types of bolts which can be used for connecting structural steel members. These include unfinished bolts, turned bolts, ribbed bolts, and high-strength bolts. A few descriptive comments are presented in the following paragraphs concerning each of these types.

1. *Unfinished bolts* (also called common, machine, ordinary or rough bolts) are made of a low carbon steel. The ASTM does not specify a minimum yield point for these bolts which they classify as A307. These bolts, however, do exhibit a well-defined yield stress equal to approximately 55 ksi (379 MPa) and they have a minimum tensile strength of about 74 ksi (510 MPa). For some light structures they may very well provide the cheapest connections available. For larger structures, however, they are usually not economical because so many of them are required in a

connection as compared to high-strength bolts. Furthermore, they are satisfactory for static loadings only.

A307 bolts generally have square heads and nuts to reduce costs, but hexagonal heads are sometimes used because they have a little more attractive appearance, are easier to turn and easier to hold with the wrenches, and they require less turning space. As they have relatively large tolerances in shank and thread dimensions, their allowable stresses are appreciably smaller than those for rivets or turned bolts. They are primarily used in light structures subjected to static loads and for secondary members (such as purlins, girts, bracing, platforms, small trusses, and so forth).

Designers are often guilty of specifying high-strength bolts for connections when common bolts would be satisfactory. *The strength and advantages of common bolts have usually been greatly underrated in the past.* The analysis and design of A307 bolted connections are handled exactly as are riveted connections in every way except that the allowable stresses are different.

2. *Turned bolts* are probably formed from hexagonal stock and their threads cut with a die. They are machined to close tolerances to provide close fits in their holes, usually within $\frac{1}{50}$ in. Because of these close tolerances the holes must be very accurately made, probably necessitating reaming or drilling. With such close fitting they are more satisfactory in resisting shear than are unfinished bolts; however, they are used very infrequently for structural connections today due to their high cost.

3. *Ribbed bolts* are those which have standard rivet heads and raised fins or ribs spaced evenly around their shanks. The outside diameters of the ribs are a little larger than the hole diameters and they must be driven. The ribs cut grooves into the connected members ensuring tight fits but making the bolts rather difficult to install—particularly when they are to pass through several thicknesses of steel because of the difficulty of perfectly lining up the holes. An appreciable amount of field reaming is probably the result. Ribbed bolts are usually assumed to have a strength equal to that of standard rivets of the same size.

A variation of the ribbed bolt is the *interference bolt*, which is made in accordance with the high-strength bolt specifications but which is also of the ribbed type. These bolts have knurled configurations with diameters a little larger than the holes in which they are to be placed. The usual tensioning methods with impact wrenches are not required, thus permitting their installation with ordinary spud wrenches. Because they can be installed with hand wrenches they are very useful in steel erection where the normal equipment for tightening is difficult to handle (such as for TV and transmission towers). Both the ribbed and interference bolts are rather difficult to insert in the holes when several plates are being connected. They do fit snugly in the holes however and the nuts may be tightened

High-strength bolt. (Courtesy of Bethlehem Steel Company.)

without simultaneously holding the heads as is required for the usual smooth bolts which fit loosely in the holes.

4. *High-strength bolts* are made from medium carbon heat-treated steel and from alloy steel and have tensile strengths several times those of ordinary bolts. As described in Section 9-8 of this chapter they are of two basic types, the A325 bolts (made from a heat treated medium carbon steel) and the higher strength A490 bolts (also heat treated but made from an alloy steel). Although they are a relatively new type of fastener for structural steel they are today the most popular field connection method. High-strength bolts are being used for all types of structures, from small buildings to "skyscrapers" and monumental bridges. These bolts were developed to overcome the weaknesses of rivets—primarily insufficient tension in their shanks after cooling. The resulting rivet tensions may not be large enough to hold them in place during the application of severe impactive and vibrating loads. The result is that they may become loose and vibrate and may eventually have to be replaced. High-strength bolts are tightened until they have very high tensile stresses. The connected parts are clamped tightly together between the bolt and nut heads permitting loads to be transferred primarily by friction.

9-3. HISTORY OF HIGH-STRENGTH BOLTS

The joints obtained using high-strength bolts are superior to riveted joints in performance and economy and they have become the leading field method of fastening structural steel members. Structural steel connections made with high-strength bolts are a relatively new development but their acceptance has been nothing short of astounding. C. Batho and E. H. Bateman first claimed in 1934 that high-strength bolts could satisfactorily be used for the assembly of steel structures[1] but it was not until 1947 that the Research Council on Riveted and Bolted Structural Joints of the Engineering Foundation was established. This group issued their first

[1] C. Batho and E. H. Bateman, "Investigations on Bolts and Bolted Joints," H. M. Stationery Office (London, 1934).

specifications in 1951 and high-strength bolts were adopted by both building and bridge engineers for both static and dynamic loadings with amazing speed. They not only quickly became the leading method of making field connections, but they also were found to have many applications for shop connections. The construction of the Mackinac Bridge in Michigan involved the use of more than one million high-strength bolts.

Connections that were formerly made with ordinary bolts and nuts were not too satisfactory when they were subjected to vibratory loads because the nuts frequently became loose. For many years this problem was dealt with by using some type of locknut, but the modern high-strength bolts furnish a far superior solution.

9-4. ADVANTAGES OF HIGH-STRENGTH BOLTS

Among the many advantages of high-strength bolts, partly explaining their great success, are the following.

1. Smaller crews are involved as compared to riveting. Two two-person bolting crews can easily turn out over twice as many bolts in a day as the number of rivets driven by the standard four-person riveting crew. The result is quicker steel erection.
2. In comparison to rivets, fewer bolts are needed to provide the same strength.
3. Good bolted joints can be made by people with a great deal less training and experience than is necessary to produce welded and riveted connections of equal quality. The proper installation of high-strength bolts can be learned in a matter of hours.
4. No erection bolts are required which may have to be later removed (depending on specifications) as in welded joints.
5. There is little noise as compared to riveting.
6. Cheaper equipment is used to make bolted connections.
7. No fire hazard is present and no danger present from the tossing of hot rivets.
8. Tests on riveted joints and bolted joints under identical conditions definitely show that bolted joints have a higher fatigue strength. Their fatigue strength is also equal to or greater than that obtained with equivalent welded joints.
9. Where structures are to be later altered or disassembled changes in connections are quite simple because of the ease of bolt removal.

9-5. INSTALLATION OF HIGH-STRENGTH BOLTS

High-strength bolts are usually tightened to a minimum tension equal to the proofload or pretension load of the bolt. The proofload is actually obtained by multiplying a stress equal to a lower boundary of the proportional or elastic limit of the steel (equal to approximately 70% of the specified minimum tensile strength of the bolts) by their tensile stress area A_s. The tensile stress area is determined from the following expression in which n is the number of threads per inch and D is the diameter of the shank of the bolt. This area falls between the gross area of a bolt and its area at the root of the threads.

$$A_s = 0.7854\left[D - \left(\frac{0.9743}{n} \right) \right]^2$$

Table 9-1, which is a reproduction of Table 1.23.5 of the AISC Specification, gives the proofloads for various sizes of A325 and A490 bolts.

Although many engineers felt that there would be some slippage as compared to rivets (because of the fact that the hot driven rivets more nearly filled the holes) the results show that there is less slippage in high-strength bolted joints than in riveted joints under similar conditions.

It is interesting to note that the nuts used for high-strength bolts need no special provisions for locking. Once these bolts are installed and sufficiently tightened to produce the tension required, there is no tendency for the nuts to become loose even after millions of cycles of loading. In fact, there apparently is no case on record where nuts have worked loose on high-strength bolts that were properly installed.

During the first several years of appreciable use of high-strength bolts several different methods of trying to ensure proper bolt tension were attempted. These methods included (1) applying a certain torque to each size bolt, (2) the use of automatic wrenches that stalled at a certain torque, and (3) turning the nut one full turn from the finger-tight position.

Great success in obtaining the proper tension was not obtained with any of these methods. One trouble with the methods of applying certain torques to the bolts was that different torques are required to tighten a bolt

Table 9-1 Minimum Bolt Tension, Kips

Bolt size (in.)	A325 Bolts	A490 Bolts
$\frac{1}{2}$	12	15
$\frac{5}{8}$	19	24
$\frac{3}{4}$	28	35
$\frac{7}{8}$	39	49
1	51	64
$1\frac{1}{8}$	56	80
$1\frac{1}{4}$	71	102
$1\frac{3}{8}$	85	121
$1\frac{1}{2}$	103	148

SOURCE: *Specification for the Design, Fabrication, and Erection of Structural Steel for Buildings*, Nov. 1, 1978, AISC, Inc.

from the head than from the nut. For a time specifications called for bolts to be tightened to 90% of the proofload by using a certain torque. This specification was not completely successful because many users tried very hard not to go on the high side of the 90% figure. They were afraid that if they put the bolts into the plastic range, failure would occur. The result of their carefulness was that the bolts were frequently not tightened as high as the 90% figure desired. Actually, overtightening the bolts does not seem to be a problem. If they are overtightened, they usually break and another one is installed.

In the "one turn of the nut" method slight differences in the nut and bolt threads, grit or dents in the threads, etc., made the so-called finger-tight position quite variable. Furthermore, the workers standing around with torque wrenches did not like to use their hands to tighten the nuts. (It has been said that this method was probably first developed for maintenance of, say, railroad bridges in remote areas where it was necessary to put in a few bolts here or there. The workers had a fairly good rule to go by to obtain adequate tightening and they could do it with whatever equipment they might have had with them.)

Today there are more satisfactory procedures for applying certain torques and a different "turn of the nut" method is used. Both methods are recommended without preference by the Research Council. In the calibrated wrench method at least three of each lot of bolts to be used on the job are tightened in a calibrating device and their tensions read. The wrenches are set to stall at the torque producing the desired tension. The calibration must be checked at frequent intervals to make sure proper tightening is being obtained.

In today's "turn of the nut" method, the nuts are spun on with an impact wrench until they are snug, the snug point being defined as the point at which the wrench has impacted a few times. (Supposedly this snug-tight position also corresponds to the tightening a person can achieve with full strength with a spud wrench for $\frac{3}{4}$ in. or larger bolts. That would be too much for smaller bolts because they may be deformed into the inelastic range.) From the snug point the nuts are turned from $\frac{1}{2}$ to $\frac{2}{3}$ turns, depending on the lengths, diameters, and types of bolts. At the snug point a bolt will have a stress of quite a few thousand pounds per square inch, and after turning the nut, a stress of at least 70% of the minimum required tensile strength of the bolt will be achieved.

If the "turn of the nut" procedure is followed for A325 bolts no washers are required, but if the torque method is used a hardened washer is required under the part being turned to minimize the variation in friction between the gripped material and the underside of the turned surface. For the A490 bolt a hardened washer must be used under the end being tightened regardless of the tightening method. Washers are necessary at each end of A490 bolts when the material being connected has a specified minimum yield point less than 40 ksi. The reason for using a washer is to

Torquing the nut for a high-strength bolt with an air-driven impact wrench. (Courtesy of Bethlehem Steel Company.)

provide as uniform a friction as possible under the twisted ends of the various bolts.

For a while it was thought that the use of washers at each end would reduce the amount of "bolt relaxation" occurring due to the high-stress concentrations under the head or nut. Tests, however, have shown that the actual losses without washers are less than approximately 5% and that about the same losses occur when washers are used at each end. Another reason for specifying washers formerly was to prevent galling, which is injury to the metal during twisting due to friction, etc. It has been found, however, that any galling that takes place is not detrimental to the strength of the joints, whether static or fatigue loads are involved.

There is another method which may be used for ensuring that high-strength bolts are properly tightened. It is called the *direct tension indicator method*. In this method a hardened washer with protrusions on one face is inserted between the bolt head and the material being gripped. These protrusions obviously cause a gap or opening. When the bolt is tightened, the protrusions are flattened somewhat and the gap becomes smaller. The tension in the bolt is determined by measuring the remaining gap. If the bolt is tightened correctly the gap[2] will be about 0.015 in.

The Research council does not specify a maximum bolt tension. This means that a bolt can be tightened to the highest load that will not break it

[2] J. H. A. Struik, A. O. Oyeledun, and J. W. Fisher, "Bolt Tension Control with a Direct Tension Indicator," *Engineering Journal* **10**, No. 1, 1973 (New York: AISC), pp. 1–5.

and the bolt will still do the job. Should the bolt break, another one is put in with no damage done. It might be noted that the nuts are stronger than the bolt and the bolt will break before the nut strips.

The surfaces of joints including the area adjacent to washers need to be free of loose scale, dirt, burs, and other defects which might prevent the parts from solid seating. It is necessary for the surface of the parts to be connected to have slopes of not more than 1 to 20 with respect to the bolt axis unless beveled washers are used. For friction-type joints the contact surfaces must also be free from oil, paint, lacquer, or galvanizing. (There is now much evidence available which indicates that this prohibition of galvanization is not necessary in such structures as bridges.)[3, 4] The 1977 AASHTO Specifications permit hot dip galvanization if the coated surfaces are scored with wire brushes or sand blasted after the galvanization and before steel erection.

About the only successful method other than the direct tension indicator washers of checking the tightness of high-strength bolts is with a torque wrench. The desired procedure is to check one or two of the bolts in a small connection and perhaps as many as 10% of those in a large connection. Should the torque required to tighten one or more of the bolts be less than the value required for installation, it will probably be necessary to check all of the bolts in that connection.

9-6. LOAD TRANSFER AND TYPES OF JOINTS

The following paragraphs present a few of the elementary types of bolted or riveted joints subjected to axial forces (that is, the loads are assumed to pass through the centers of gravities of the groups of connectors). For each of these joint types some comments are made about the methods of load transfer. Eccentrically loaded connections are discussed in later sections of this chapter.

For this initial discussion the reader is referred to part (a) of Figure 9-1. It is assumed that the plates shown are connected with a group of loosely fitting bolts. In other words, the bolts are not tightened sufficiently so as to significantly squeeze the plates together. If there is assumed to be little friction between the plates they will slip a little due to the applied loads shown. As a result the loads in the plates will tend to shear the connectors off on the plane between the plates and press or bear against the sides of the bolts as shown in the figure. These connectors are said to be in *single shear and bearing* (also called unenclosed bearing). They must

[3]J. R. Hall, "The Case for the Galvanized Bridge," *Civil Engineering*, November, 1964, pp. 31–34.

[4]P. C. Birkemoe, and D. C. Herrschaft, "Bolted Galvanized Bridges—Engineering Acceptance Near," *Civil Engineering*, April, 1970, pp. 42–46.

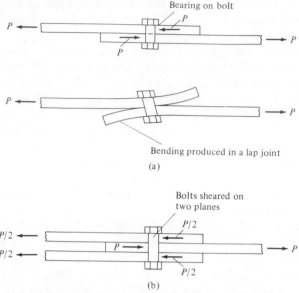

Figure 9-1 (a) Lap joint. (b) Butt joint.

have sufficient strength to satisfactorily resist these forces and the members forming the joint must be sufficiently strong to prevent the connectors from tearing through.

Should rivets be used instead of the loosely fitting bolts the situation is somewhat different because hot-driven rivets cool and shrink and squeeze or clamp the connected pieces together with sizable forces which greatly increase the friction between the pieces. As a result a large portion of the loads being transferred between the members are transferred by friction. The clamping forces produced in riveted joints, however, are generally not considered to be dependable and as a result specifications normally consider the connection to be loose with no frictional resistance. The same assumption is made for A307 common bolts which are not tightened so as to have large dependable tensions.

High-strength bolts are in a different class altogether. Using the tightening methods previously described a very dependable tension is obtained in the bolts resulting in large clamping forces and large dependable amounts of frictional resistance to slipping. Unless the loads to be transferred are larger than the frictional resistance the entire forces are resisted by friction and the bolts are not really placed in shear or bearing. If the load exceeds the friction there will be slippage with the result that the bolts will be placed in shear and bearing.

The Lap Joint

The joint shown in part (a) of Figure 9-1 is referred to as a lap joint. This type of joint has a disadvantage in that the center of gravity of the force in one member is not in line with the center of gravity of the force in the other member. A couple is present which causes an undesirable bending in the connection as shown in the figure. For this reason the lap joint, which is desirably used only for minor connections, should be designed with at least two fasteners in each line parallel to the length of the member to minimize the possibility of a bending failure.

The Butt Joint

A butt joint is formed when three members are connected as shown in Figure 9-1(b). If the friction between the members is negligible the members will slip a little and tend to shear off the bolts simultaneously on the two planes of contact between the members. Again the members are bearing against the bolts and the bolts are said to be in *double shear and bearing* (also called enclosed bearing). The butt joint is more desirable than the lap joint for two main reasons. These are:

1. The members are arranged so that the total shearing force *P* is split into two parts, causing the force on each plane to be only about one-half of what it would be on a single plane if a lap joint were used. From a shear standpoint, therefore, the load-carrying ability of a group of bolts in double shear is theoretically twice as great as the same number of bolts in single shear.
2. A more symmetrical loading condition is provided. (In fact the butt joint does provide a symmetrical situation if the outside members are the same thickness and have the same stress values.) The result is a reduction or elimination of the bending described for a lap joint.

Double-Plane Connections

The double-plane connection is one in which the bolts are subjected to single shear and bearing but in which bending moment is prevented. This type of connection, which is shown for a hanger in Fig. 9-2(a), subjects the bolts to single shear on two different planes.

Miscellaneous

Bolted connections generally consist of lap or butt joints or some combination of them but there are other cases. For instance there are joints in which more than three members are being connected and the bolts are in multiple shear as shown in Fig. 9-2(b). Several other types of bolted connections are discussed in this chapter. These include bolts in tension, bolts in shear and tension, etc.

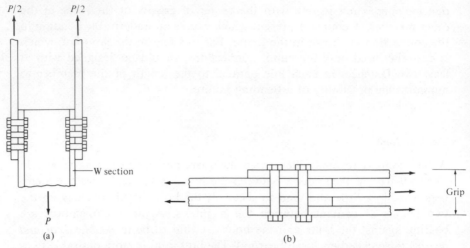

Figure 9-2 (a) Hanger connection. (b) Bolts in multiple shear.

9-7. FAILURE OF BOLTED JOINTS

Figure 9-3 shows several ways in which failure of bolted joints can occur. To be able to satisfactorily design bolted joints it is necessary to understand these possibilities. These are described as follows.

1. The possibility of failure in a lap joint by shearing of the bolt on the plane between the members (single shear) is shown in part (a).

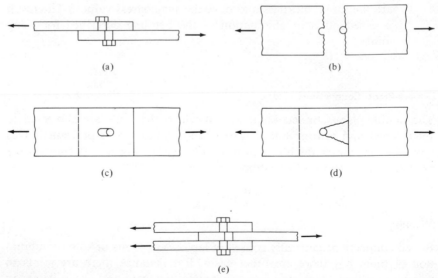

Figure 9-3 (a) Failure by shearing of bolt. (b) Tension failure of plate. (c) Crushing failure of plate. (d) Shear failure of plate behind bolt. (e) Double shear failure of a butt joint.

2. In part (b) the possibility of a tension failure of one of the plates through a bolt hole is shown.
3. A possible failure of the bolts and/or plates by bearing between the two is given in (c).
4. Part (d) shows another possibility in the shearing out of part of the member.
5. A butt joint is shown in (e) with the possibility of a shear failure of the bolts along two planes (double shear).

9-8. SPECIFICATIONS FOR HIGH-STRENGTH BOLTS

High-strength bolts are of the A325 or the A490 types as classified by the ASTM. The A325 bolt is made from a medium-carbon steel and gains its strength from heat treating by quenching and tempering while the occasionally used A490 is a quenched and tempered alloy steel bolt. When bolts larger than $1\frac{1}{2}$ in. and up to 3 in. in diameter are desired A449 bolts may be used. These are threaded bar stocks of medium-carbon steel with the same mechanical properties as the A325 bolts.

Both the A325 and A490 bolts are available as Types 1, 2, and 3. The A325 Type 1 bolt is the standard medium-carbon steel bolt while the A490 Type 1 is the standard alloy steel bolt. If the type is not specified in an order the buyer will receive Type 1 bolts. Both the A325 Type 2 and the A490 Type 2 are low-carbon martensite bolts which may be used for applications at atmospheric temperatures. If high temperatures are to be encountered Type 1 bolts must be used. Finally the A325 Type 3 and the A490 Type 3 are weathering steel bolts having corrosion resistance approximately equal to that of the A588 weathering steels.

The allowable shearing and bearing stresses permitted by the AISC Specification for A325 and A490 bolts are presented in this section. These values, which are for buildings, are appreciably higher than the ones permitted by the AASHTO and AREA Specifications for bridges. The AISC allowable bearing stresses are as for rivets equal to 1.5 times F_u, the specified minimum tensile strength of the connected parts.

The AISC has two basic types of allowable shearing stresses, one for the *friction-type connections* and one for the *bearing-type connections* in which the factor of safety is substantially reduced. When high-strength bolts are used they clamp the plates being connected so tightly together that a great deal of friction is created between them. The shearing stress given for the friction-type connection is the maximum allowable shear before the friction is assumed to be overcome. This type of connection is to be used for structures where there is a great deal of impact and vibration with resulting stress variations or reversals or where any slippage is undesirable. Under ordinary circumstances no slippage is really expected in either type of connection.

The bearing allowable stress is considerably higher because of the smaller factor of safety against slippage. The higher allowable stress values for this case clearly show that it is the bearing-type connection which makes use of the high strengths of these bolts and gives the best economy.

It should be realized that under the present (1978) AISC Specification high-strength bolts are to be tightened in the same manner whether they are to be used in friction-type or in bearing-type connections. The quality control provisions specified in the manufacture of the A325 and A490 bolts are more stringent than those for the A449 bolts. As a result, despite the method of tightening, the A449 bolts may not be used in friction-type connections.

If the bolt threads are not in the shear plane for bearing-type connections the full shank area of the bolts, A_b, is to be multiplied by the allowable shear stress, F_v (30 ksi for A325 bolts and 40 ksi for A490 bolts) to obtain the allowable load. If the threads are in the same shear plane a reduced area called the tensile stress area (see Section 9-5) is to be multiplied by the allowable stress. Because the tensile stress area is about $0.7A_b$ the AISC Specification has just multiplied the allowable stress by 0.7 (21 ksi for A325 bolts and 28 ksi for A490 bolts) and says to use the full bolt area A_b.

The allowable stresses for high-strength bolts given in Table 1.5.2.1 of the AISC Specification are shown in Table 9-2 of this chapter for friction and bearing type connections for bolts placed in standard size holes $\frac{1}{16}$ in. larger than the bolt diameters and for normal surface conditions (that is, untreated mill scale surfaces). In addition to these allowable shear stresses the AISC provides different values for different sizes and shapes of holes and for different kinds of surface treatments of the connected parts. In Table 9-2 values are given for normal surface conditions for (a) standard size holes, (b) oversized and short-slotted holes, and (c) long-slotted holes.

Table 9-2 Recommended Allowable Stresses for High-Strength Bolts for Buildings

| | | Allowable shear (F_v) | | | |
| | | Friction-type connections | | | |
Description of fasteners	Allowable tension (F_t)	Standard size holes	Oversized and short-slotted holes	Long-slotted holes	Bearing-type connections
A325 bolts, when threads are not excluded from shear planes	44.0	17.5	15.0	12.5	21.0
A325 bolts, when threads are excluded from shear planes	44.0	17.5	15.0	12.5	30.0
A490 bolts, when threads are not excluded from shear planes	54.0	22.0	19.0	16.0	28.0
A490 bolts, when threads are excluded from shear planes	54.0	22.0	19.0	16.0	40.0

SOURCE: *Specification for the Design, Fabrication, and Erection of Structural Steel for Buildings*, November 1, 1978, AISC, Inc.

In Appendix E of the AISC Specification allowable shear stresses are also given for each of these types of holes and for nine different surface conditions of the parts being connected. Test results have shown that these other surfaces (which have been subject to some kind of treatment such as sandblasting) provide superior shear resistance as compared to the normal mill scale surfaces. All of the allowable values mentioned here may have to be modified for fatigue loadings and for wind and earthquake loadings. For information on dimensions of oversized, short, and long slotted holes, the reader is referred to Section 1.23.4 of the AISC Specification.

There are other important items pertaining to high-strength bolts covered by the AISC Specification. For new construction, high-strength bolts that are installed in a friction-type connection prior to welding may be considered as sharing the stress with the weld. Other types of bolts are not considered to share the stresses including high-strength bolts in bearing type connections. Similarly, in new work, rivets and high-stength bolts may be considered to share the stresses resulting from dead and live loads, provided they are friction-type connections.

9-9. SPACING AND EDGE DISTANCES OF BOLTS

Before minimum spacings and edge distances can be discussed it is necessary for a few terms to be explained. The following definitions are given for a group of bolts in a connection and are shown in Fig. 9-4.

> *Pitch* is the center-to-center distance of bolts in a direction parallel to the axis of the member.
> *Gage* is the center-to-center distance of bolt lines perpendicular to the axis of the member.
> The *edge distance* is the distance from the center of a bolt to the adjacent edge of a member.
> The *distance between bolts* is the shortest distance between fasteners on the same or different gage lines.

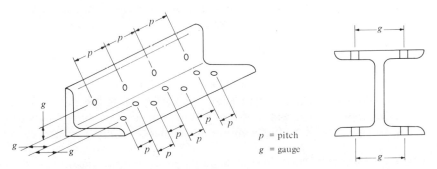

p = pitch
g = gauge

Figure 9-4

Minimum Edge Distances

Bolts should not be placed too near the edges of a member for two major reasons. The punching of holes too close to the edges may cause the steel opposite the hole to bulge out or even crack. The second reason applies to the ends of members where there is danger of the fastener tearing through the metal. The usual practice is to place the fastener a minimum distance from the edge of the plates equal to about 1.5 to 2.0 times the fastener diameter so the metal there will have a shearing strength at least equal to that of the fasteners. For more exact information it is necessary to refer to the specification being used. The AISC Specification in its Table 1.16.5.1 has a set of minimum permissible edge distances. The values given depend on bolt sizes, stress conditions, whether material edges are rolled or sheared, etc. (The distances required are greater for sheared edges than for the smoother edges of rolled shapes.) The values so obtained may have to be increased along a line of transmitted force if the expression at the end of the next paragraph so requires.

The minimum edge distance needed to prevent a plate from splitting in the direction of a transmitted force can be estimated by equating the force transferred by the end bolt to the force that would cause a shear failure in the material behind the fastener. This latter shearing strength can be expressed in terms of the thickness of the material, the material length outside of the hole, material strength, etc.[5] The result is that the AISC says that in the direction of the transmitted force the edge distance may not be less than the quantity obtained from the following expression:

$$\text{minimum edge distance in the direction of transmitted force} = \frac{2P}{F_u t}$$

In this expression P is the force in kips transmitted by one fastener to the critical connected part, F_u is the specified minimum tensile strength of the critical connected part in kips per square inch, and t is the thickness of the critical connected part in inches. Should the holes be oversized or slotted the minimum edge distance may not be less than the value required for a standard hole multiplied by an increment C_2 given in Table 1.16.5.4 of the AISC Specification. Another minimum edge distance value is provided in the AISC for end connections bolted to beam webs and designed for beam shear reactions only.

Minimum spacings

Bolts should be placed a sufficient distance apart to permit efficient installation and to prevent tension failures of the members between fasteners. The AISC specifies a minimum center-to-center distance equal

[5]J. W. Fisher and J. H. A. Struik, *Guide to Design Criteria for Bolted and Riveted Joints* (New York: Wiley, 1974), pp. 108–112.

to $2\frac{2}{3}$ times the nominal diameter (with a preferred minimum of 3 diameters). The value so obtained may have to be increased along a line of transmitted force if the expression at the end of this paragraph so requires. In this expression d is the nominal diameter of the fastener in inches.

$$\text{minimum distance center-to-center} = \frac{2P}{F_u t} + \frac{d}{2}$$

Should the holes be oversized or slotted the minimum required spacing of the fasteners is multiplied by an increment C_1 given in Table 1.16.4.2 of the AISC Specification.

Maximum Edge Distances

Many specifications give maximum distances bolts can be placed from the edge of a connection. The AISC maximum is 12 times the plate thickness but not to exceed 6 in. If bolts are too far from the edges, openings may develop between the members being connected. Maximum spacings for bolts may also be given for compression members so that buckling will not occur between bolts.

Holes cannot be punched very close to the web of a beam or the leg of an angle. They can be drilled but this rather expensive practice should not be followed unless there is an unusual situation. Even if the holes are drilled in these locations there may be considerable difficulty in placing and tightening them in the limited space.

9-10. BEARING-TYPE CONNECTIONS—LOADS PASSING THROUGH CENTER OF GRAVITY OF CONNECTIONS

In bearing-type connections it is assumed that the loads to be transferred are larger than the friction caused by tightening the bolts with the result that the members slip a little on each other, putting the bolts in shear and bearing. The strength of a bolt in bearing equals the allowable bearing unit stress of the bolt times the diameter of the shank of the bolt times the thickness of the member that bears on the bolt. The strength of a bolt in single shear is the allowable shearing unit stress times the cross-sectional area of the shank of the bolt. Should a bolt be in double shear its shearing strength is considered to be twice its single shear value.

Tests of bolted joints have shown that neither the bolts nor the metal in contact with the bolts actually fails in bearing. However, these tests have shown that the efficiency of the connected parts in tension and compression is affected by the magnitude of the bearing stress. Therefore, bearing stresses are given by the Specifications above which they feel the strength of the connected parts is impaired. In other words these apparently quite high allowable bearing stresses are not really allowable bearing stresses at all but rather indexes of the efficiency of the connected members. The

AISC Specification (Section 1.5.1.5.3) gives an allowable bearing stress $F_p = 1.5F_u$ where F_u is the specified minimum tensile strength of the connected parts.

Example 9-1 illustrates the calculations involved in determining the strength of a bearing-type connection. Using a similar procedure, the number of bolts required for a certain loading condition is calculated in Example 9-2.

Example 9-1

Determine the allowable tensile force P which can be applied to the plates shown in Fig. 9-5. The AISC Specification A36 steel and $\frac{7}{8}$-in. A325 bolts placed in standard holes with threads excluded from the shear plane are used in a bearing-type connection.

SOLUTION
Allowable tensile load on plates:

$$A_g = \left(\tfrac{1}{2}\right)(12) = 6.00 \text{ in.}^2$$

$$A_n = \left[\left(\tfrac{1}{2}\right)(12) - (2)(1.00)\left(\tfrac{1}{2}\right)\right] = 5.00 \text{ in.}^2$$

$$P = 0.60F_y A_g = (0.60)(36)(6.00) = 129.6 \text{ k}, \qquad \text{or}$$

$$P = 0.50F_y A_e \quad \text{where} \quad A_e = A_n$$

$$= (0.50)(58)(5.00) = 145 \text{ k}$$

Bolts in single shear and bearing on $\frac{1}{2}$ in.:

$$SS = (4)(0.6)(30) = 72 \text{ k} \leftarrow$$

$$Bear = (4)(0.5)\left(\tfrac{7}{8}\right)(1.5 \times 58) = 152.2 \text{ k}$$

Allow $P = 72$ k (320 kN)

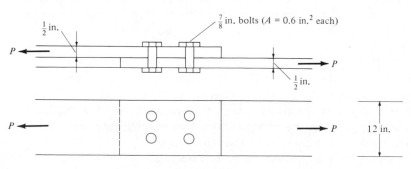

Figure 9-5

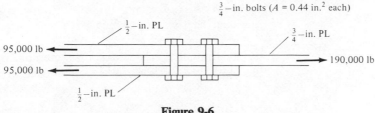

Figure 9-6

Example 9-2

How many $\frac{3}{4}$-in A325 bolts in standard size holes with threads excluded from the shear plane are required for the bearing-type connection shown in Fig. 9-6? Use A36 steel and the AISC Specification.

SOLUTION

Bolts in double shear and bearing on $\frac{3}{4}$ in.:

$$\text{allowable shear per bolt} = (2)(0.44)(30) = 26.4 \text{ k} \leftarrow$$

$$\text{allowable bearing per bolt} = \left(\tfrac{3}{4}\right)\left(\tfrac{3}{4}\right)(1.5 \times 58) = 48.9 \text{ k}$$

$$\text{no. of bolts reqd.} = \frac{190}{26.4} = 7 \ +$$

Use 8 or 9 bolts (depending on arrangement)

The assumption has been made that the loads applied to a bearing-type connection are equally divided between the bolts. For this distribution to be correct the plates must be perfectly rigid and the bolts perfectly elastic, but actually the plates being connected are elastic too and have deformations which decidedly affect the bolt stresses. The effect of these deformations is to cause a very complex distribution of load in the elastic range.

Should the plates be assumed to be completely rigid and nondeforming, all bolts would be deformed equally and have equal stresses. This situation is shown in part (a) of Fig. 9-7. Actually the load resisted by the bolts of a group are probably never equal (in the elastic range) when there are more than two bolts in a line. Should the plates be deformable, the plate stresses and thus the deformations will decrease from the ends of the connection to the middle as shown in part (b) of Fig. 9-7. The result is that the highest stressed elements of the top plate will be over the lowest stressed elements of the lower plate, and vice versa. The slip will be greatest at the end bolts and smallest at the middle bolts. The bolts at the ends will then have stresses much greater than those in the inside bolts.

The greater the spacing of bolts in a connection the greater will be the variation in bolt stresses due to plate deformation; therefore, the use of compact joints is very desirable as they will tend to reduce the variation in bolt stresses. It might be interesting to consider a theoretical (although not practical) method of roughly equalizing bolt stresses. The theory would

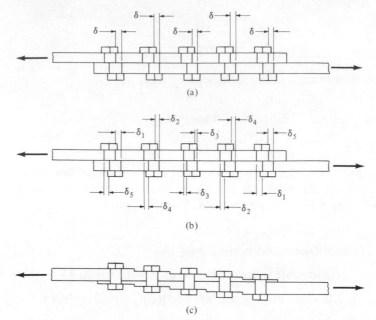

Figure 9-7 (a) Assuming nondeforming plates. (b) Assuming deformable plates. (c) Stepped joint (impractical).

involve the reduction of the thickness of the plate toward its end in proportion to the reduced stresses by stepping. This procedure, which is shown in Fig. 9-7(c), would tend to equalize the deformations of the plate and thus the bolt stresses. A similar procedure would be to scarf the overlapping plates.

The calculation of the theoretically correct elastic stresses in a bolted group based on plate deformations is a tedious problem and is rarely handled in the design office. On the other hand the analysis of a bolted joint based on the plastic theory is a very simple problem. In this theory the end bolts are assumed to be stressed to their yield point. Should the total load on the connection be increased, the end bolts will deform without resisting additional load, the next bolts in the line will have their stresses increased until they too are at the yield point, and so forth. Plastic analysis seems to justify to a certain extent the assumption of rigid plates and equal bolt stresses which is usually made in design practice. This assumption is used in the example problems of this chapter.

When there are only a few bolts in a line the plastic theory of equal stresses seems to be borne out very well but when there are a large number of bolts in a line the situation changes. Tests have clearly shown that the end bolts will fail before the full redistribution takes place.[6]

For load carrying bolted joints it is common for specifications to require a minimum of two or three fasteners. The feeling is that a single connector may fail to live up to its specified strength because of improper installation, material weakness, etc., but if several fasteners are used the effects of one bad fastener in the group will be overcome.

9-11.   FRICTION-TYPE CONNECTIONS—LOADS PASSING THROUGH CENTER OF GRAVITY OF CONNECTION

For fatigue situations where members are subjected to constantly fluctuating loads the friction-type connection may be very desirable. In this type connection the force to be carried is assumed to be less than the permissible friction between the members produced by tightening the bolts.

If bolts are tightened to their minimum tensile values there is very little chance of their bearing against the plates which they are connecting. In fact, tests show that there is very little chance of slip occurring unless there is a calculated shear of at least 50% of the bolt tension. This means that in the usual friction-type connection the bolts are not stressed in shear; however, the AISC gives an allowable bolt shear so that the connections may be proportioned by the same methods used for proportioning bearing-type bolted connections. It will be noticed that in a

Three high-strength bolted structures in Constitution Plaza Complex, Hartford, Conn., using approximately 195,000 bolts. (Courtesy of Bethlehem Steel Company.)

friction-type joint where there is no slip the bolts theoretically are not in shear nor are they in bearing.

The preceding discussion does not present the whole story because during erection the joints may be assembled with bolts, and as the members are erected their weights will often push the bolts against the side of the holes before they are tightened and put them in bearing and shear. Example 9-3 presents the calculations involved in determining the strength of a friction-type joint.

Example 9-3

Repeat Example 9-1 if a friction-type connection is used.

SOLUTION

Allowable tensile load on plates from solution to Example 9-1 = 129.6 k. Bolts in "single shear" and no bearing

$$\text{"SS"} = (4)(0.6)(17.5) = 42 \text{ k} \leftarrow$$

<u>Allowable $P = 42$ k</u>

Where cover plates are bolted to the flanges of W sections the bolts must carry the longitudinal shear on the plane between the plates and the flanges. With reference to the cover-plated beam of Fig. 9-8 the unit longitudinal shearing stress between a cover plate and the W flange can be determined with the VQ/bI expression. The total shear across the flange for a 1-in.-length of the beam equals VQ/I.

The spacing of pairs of bolts in Fig. 9-8 can be determined by dividing the shear per inch at a particular section into the strength of the two bolts.

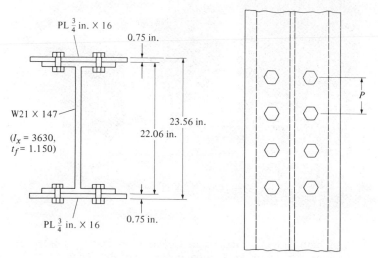

$\text{PL } \frac{3}{4} \text{ in.} \times 16$

0.75 in.

$\text{W21} \times 147$

$(I_x = 3630,$
$t_f = 1.150)$

23.56 in.
22.06 in.

P

$\text{PL } \frac{3}{4} \text{ in.} \times 16$

0.75 in.

Figure 9-8

The theoretical spacings will vary as the external shear varies along the beam. Example 9-4 illustrates the calculations involved in determining bolt spacing for a cover-plated beam.

Example 9-4

At a certain section in the cover-plated beam of Fig. 9-8 the external shear is 190 k. Determine the required spacing of $\frac{7}{8}$-in. A325 bolts used in a friction-type connection if the beam is rolled from A36 steel and the AISC Specification is to be used.

SOLUTION

$$I_x = 3630 + (2)\left(16 \times \tfrac{3}{4}\right)(11.405)^2 = 6752 \text{ in.}^4$$

$$f_v = \frac{VQ}{I} = \frac{(190,000)\left(16 \times \tfrac{3}{4} \times 11.405\right)}{6752} = 3851 \text{ lb/in.}$$

$$\text{shear} = (2)(0.6)(17,500) = 21,000 \text{ lb for 2 bolts}$$

$$\text{bearing} = (2)\left(\tfrac{7}{8}\right)\left(\tfrac{3}{4}\right)(87,000) = 114,188 \text{ lb for 2 bolts}$$

$$p = \frac{21,000}{3851} = 5.45 \text{ in.} \qquad \left(\text{say } 5\tfrac{1}{2} \text{ in.}\right)$$

9-12. BOLTS SUBJECTED TO ECCENTRIC SHEAR

Eccentrically loaded bolt groups are subjected to shears and bending moments. The student may feel such situations are rare but the truth is that they are much more common than he suspects. For instance, in a truss it is desirable to have the center of gravity of a member lined up exactly with the center of gravity of the bolts at its end connections. This feat is not quite as easy to accomplish as it may seem and connections are often subjected to moments.

Eccentricity is quite obvious in Fig. 9-9(a) where a beam is connected to a column. In part (b) of the figure a beam is connected to a column with a pair of web angles. For this connection it is obvious that even though no end moment is being considered the connection must resist some moment because the center of gravity of the load from the beam does not coincide with the reaction from the column.

The following several paragraphs are concerned with developing a method of calculating the forces in the bolts of an eccentric connection. For this discussion the bolts of Fig. 9-10(a) are assumed to be subjected to a load P which has an eccentricity of e from the c.g. (center of gravity) of the bolt group. To consider the force situation in the bolts an upward and downward force, each equal to P, is assumed to act at the c.g. of the bolt group. This situation, shown in part (b) of the figure, in no way changes the bolt force. The force in a particular bolt will, therefore, equal P divided

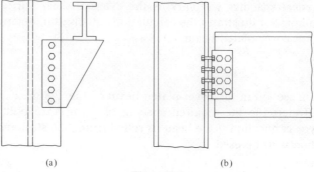

(a)   (b)

Figure 9-9

by the number of bolts in the group as seen in part (c), plus the force due to the moment caused by the couple shown in part (d) of the figure.

The magnitude of the forces in the bolts due to the moment Pe will now be considered. The distances of each bolt from the c.g. of the group are represented by the values d_1, d_2, etc. in Fig. 9-11. The moment produced by the couple causes the plate to rotate about the c.g. of the bolt connection with the amount of rotation or strain at a particular bolt being proportional to its distance from the c.g. (For this derivation the gusset plates are again assumed to be perfectly rigid and the bolts are assumed to be perfectly elastic.) Rotation is greatest at the bolt which is the greatest distance from the c.g. as will be the stress since stress is proportional to strain in the elastic range.

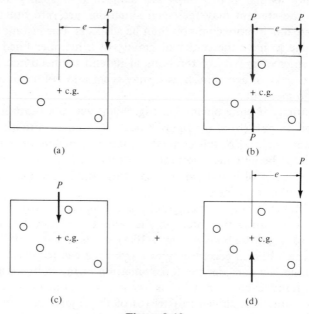

Figure 9-10

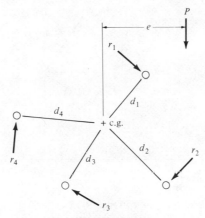

Figure 9-11

The rotation is assumed to produce forces of r_1, r_2, r_3, and r_4 respectively on the bolts in the figure. The moment transferred to the bolts must be balanced by resisting moments of the bolts as follows.

$$M_{c.g.} = Pe = r_1 d_1 + r_2 d_2 + r_3 d_3 + r_4 d_4 \qquad (1)$$

As the force caused on each bolt is directly proportional to the distance from the c.g. the following expression can be written.

$$\frac{r_1}{d_1} = \frac{r_2}{d_2} = \frac{r_3}{d_3} = \frac{r_4}{d_4}$$

and writing each r value in terms of r_1 and d_1

$$r_1 = \frac{r_1 d_1}{d_1} \qquad r_2 = \frac{r_1 d_2}{d_1} \qquad r_3 = \frac{r_1 d_3}{d_1} \qquad r_4 = \frac{r_1 d_4}{d_1}$$

Substituting these values into equation (1) and simplifying,

$$M = \frac{r_1 d_1{}^2}{d_1} + \frac{r_1 d_2{}^2}{d_1} + \frac{r_1 d_3{}^2}{d_1} + \frac{r_1 d_4{}^2}{d_1}$$

$$= \frac{r_1}{d_1} \left(d_1{}^2 + d_2{}^2 + d_3{}^2 + d_4{}^2 \right)$$

Therefore

$$M = \frac{r_1 \sum d^2}{d_1}$$

The force on each bolt can now be written as follows.

$$r_1 = \frac{M d_1}{\sum d^2} \qquad r_2 = \frac{d_2}{d_1} r_1 = \frac{M d_2}{\sum d^2} \qquad r_3 = \frac{M d_3}{\sum d^2} \qquad r_4 = \frac{M d_4}{\sum d^2}$$

Each value of r is perpendicular to the line drawn from the c.g. to the particular bolt. It is usually more convenient to break these down into

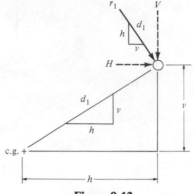

Figure 9-12

vertical and horizontal components. In accomplishing this purpose reference is made to Fig. 9-12.

The horizontal and vertical components of the distance d_1 are represented by h and v respectively and the horizontal and vertical components of force r_1, are represented by H and V respectively in this figure. It is now possible to write the following ratio from which H can be obtained.

$$\frac{r_1}{d_1} = \frac{H}{v}$$

$$H = \frac{r_1 v}{d_1} = \left(\frac{Md_1}{\sum d^2}\right)\left(\frac{v}{d_1}\right)$$

Therefore,

$$H = \frac{Mv}{\sum d^2}$$

By a similar procedure V is found to equal

$$V = \frac{Mh}{\sum d^2}$$

Example 9-5

Determine the force in the most stressed bolt of the group shown in Fig. 9-13.

SOLUTION

A sketch of each bolt and the forces applied to it by the direct load and the clockwise moment are shown in Fig. 9-14. From this sketch the student

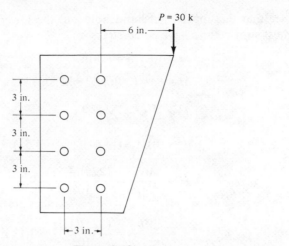

Figure 9-13

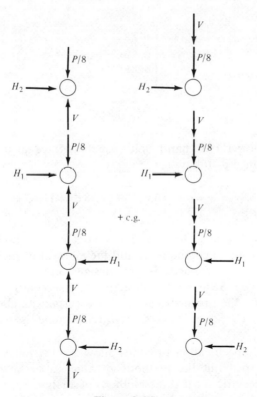

Figure 9-14

will see that the upper right-hand bolt and the lower right-hand bolt are the most stressed and have equal stresses.

$$e = 6 + 1.5 = 7.5 \text{ in.}$$

$$M = Pe = (30)(7.5) = 225 \text{ in.-k}$$

$$\sum d^2 = \sum h^2 + \sum v^2$$

$$\sum d^2 = (8)(1.5)^2 + (4)(1.5^2 + 4.5^2) = 108$$

$$H = \frac{Mv}{\sum d^2} = \frac{(225)(4.5)}{108} = 9.38 \text{ k}$$

$$V = \frac{Mh}{\sum d^2} = \frac{(225)(1.5)}{108} = 3.13 \text{ k}$$

$$\frac{P}{8} = \frac{30}{8} = 3.75 \text{ k}$$

3.13 k

3.75 k

9.38 k

Force in lower right-hand bolt (equal to force in upper right-hand bolt) is determined as follows.

$$r = \sqrt{(6.88)^2 + (9.38)^2} = 11.63 \text{ k}$$

Should the eccentric load be inclined it can be broken down into vertical and horizontal components and the moment of each about the c.g. of the bolt group determined. Several design formulas can be developed which will enable the engineer to directly design eccentric connections but in all probability the process of assuming a certain number and arrangement of bolts, checking stresses, and redesigning is probably just as satisfactory.

The AISC Specification provides allowable capacities for connectors but does not spell out the methods of analysis to be used. For bolts subjected to eccentric shear three methods of analysis are reasonable. First there is the method described in this section in which friction between the plates is neglected and plates are assumed to be rigid. This type of analysis has been commonly used since at least 1870.[7, 8] Another method that might be used is the one where friction is considered and an empirically reduced eccentricity is used.[9] If the allowable shear stresses of the 1978 AISC Specification are used this effective eccentricity method will not provide the desired safety factor of 2.5 and should not be used with that specification. Thirdly, an ultimate strength approach may be used.[10, 11]

The method described in this chapter for eccentrically loaded fastener groups has been based on the assumption that the behavior of the fasteners is elastic. It can, however, be shown that the load-deformation relationship of fasteners does not have well-defined shear yield stresses. The result is that, although the elastic procedure does provide a simple and conservative solution for the problem, it does not result in a constant safety factor, and in fact its solution may be extremely conservative.

With the plastic or ultimate strength approach the fact that the eccentric moment causes a rotation as well as a translation effect on the connectors is recognized. In effect this is equivalent to pure rotation about a single point, which is called the *instantaneous center of rotation*.

Studies by Crawford and Kulak[10] show that the shear force in a single fastener R can be obtained from the following relationship. R_{ult} is the ultimate shear load for a single fastener equaling 74 k for an A325 bolt, e is the base of the natural logarithm (2.718), and Δ is equal to the total deformation of a bolt experimentally determined as equal to 0.34 in. The coefficients 10.0 and 0.55 were also experimentally obtained.[10]

$$R = R_{ult}(1 - e^{-10\Delta})^{0.55}$$

Although the development of this plastic method of analysis was actually based on bearing-type connections where slip may occur, both theory and load tests have shown that it may conservatively be applied to friction-type or slip-resistant connections.[11] The actual application of the method involves a trial and error process.[12]

The numerical values given in the AISC Manual for eccentrically loaded connections (bolted and welded) are based on plastic analysis.

9-13. TENSION LOADS ON BOLTED JOINTS

Bolted and riveted connections subjected to pure tensile loads have been avoided as much as possible in the past by designers. The use of tensile

[7]W. McGuire, *Steel Structures* (Englewood Cliffs, N.J.: Prentice-Hall, 1968), p. 813.

[8]C. Reilly, "Studies of Iron Girder Bridges," *Proc. Inst. Civil Engrs.* 29 (London, 1870).

[9]T. R. Higgins, "New Formulas for Fasteners Loaded Off Center," *Engr. News Record* (May 21, 1964).

[10]S. F. Crawford and G. L. Kulak, "Eccentrically Loaded Bolted Connections," *Journal of Structural Division, ASCE* 97, ST3 (March, 1971), pp. 765–783.

[11]G. L. Kulak, "Eccentrically Loaded Slip–Resistant Connections," *Engineering Journal, AISC*, 12, No. 2 (2nd Quarter, 1975), pp. 52–55.

[12]C. G. Salmon and J. E. Johnson, *Steel Structures Design and Behavior*, 2nd ed. (New York: Harper & Row, 1980), pp.134–140.

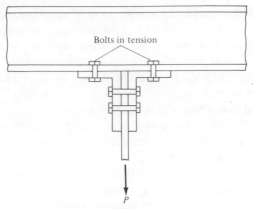

Figure 9-15

connections was probably "forced on" them more often for wind-bracing systems in tall buildings than for any other situation. There are some other locations where they have been used, however, such as hanger connections for bridges, flange connections for piping systems, etc. Figure 9-15 shows a bolted connection where some of the bolts are subjected to tensile loads.

Hot-driven rivets and tightened high-strength bolts are not free to shorten, with the result that large tensile forces are produced in them during their installation. These initial tensions are actually close to the yield points. There has always been considerable reluctance among designers to apply tensile loads to connectors of this type for fear that the external loads might easily increase their already present tensile stresses and cause them to fail. The truth of the matter, however, is that when ordinary external tensile loads are applied to connections of this type the connectors will experience little if any change in stress.

Hot-driven rivets which have cooled and shrunk or highly tightened bolts actually prestress the joints in which they are used against tensile loads. (To follow this discussion the student may like to think of a prestressed concrete beam which has external compressive loads applied at each end.) The tensile stresses in the connectors squeeze together the members being connected. If a tensile load is applied to this connection at the contact surface, it cannot exert any additional load on the bolts or rivets until the members are pulled apart and additional strains put on the bolts or rivets. The members cannot be pulled apart until a load is applied which is larger than the total tension in the connectors of the connection. This statement means that the joint is prestressed against tensile forces by the amount of stress initially put in the shanks of the connectors.

Another way of saying this is that if a tensile load P is applied at the contact surface it tends to reduce the thickness of the plates somewhat but at the same time the contact pressure between the plates will be correspondingly reduced and the plates will tend to expand by the same amount. The theoretical result then is no change in plate thickness and no change in connector tension. This situation continues until P equals the connector tension. At this time an increase in P will result in separation of the plates and thereafter the tension in the connector will equal P.

Should the load be applied to the outer surfaces there will be some immediate strain increase in the connector. This increase will be accompanied by an expansion of the plates even though the load does not exceed the prestress but the increase will be very slight because the load will go to the plate and connectors roughly in proportion to their stiffness. As the plate is much the stiffer it will receive most of the load. An expression can be developed for the elongation of the bolt based on the bolt area and the assumed contact area between the plates. Depending on the contact area assumed, it will be found that unless P is greater than the bolt tension its stress increase will be in the range of 10%. Should the load exceed the prestress the bolt stress will rise appreciably.

The preceding rather lengthy discussion is truthfully approximate but should explain why an ordinary tensile load applied to a riveted or bolted joint will not change the stress situation very much.

Recognizing the prestressing effect of high-strength bolts the Research Council and the AISC permit calculated tensile loads, independent of the tightening forces, to equal approximately twice the allowable tensile stress given in the specifications.

9-14.   PRYING ACTION

A further consideration that should often be given to tensile connections is the possibility of prying action. A tensile connection is shown in Fig. 9-16(a) which is subjected to prying action as illustrated in part (b) of the same figure. Should the flanges of the connection be quite thick and stiff

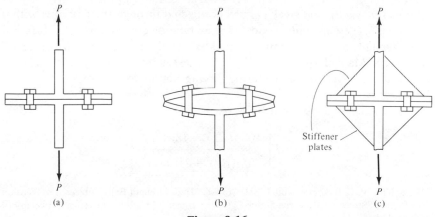

Figure 9-16

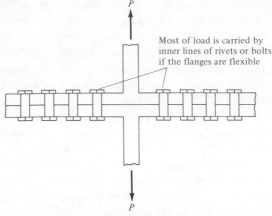

Figure 9-17

or have stiffener plates as shown in Fig. 9-16(c), the prying action will probably be negligible but not so if they are thin and flexible.

It is usually desirable to limit the number of rows of rivets or bolts in a tensile connection because a large percentage of the load is carried by the inner rows of multirow connections even at ultimate load. The tensile connection shown in Fig. 9-17 illustrates this point as the prying action will throw most of the load to the inner connectors, particularly if the plates are thin and flexible. For connections subjected to pure tensile loads estimates should be made of possible prying action and its magnitude.

The additional force in the bolts resulting from prying action should be added to the tensile force resulting directly from the applied forces. The actual determination of prying forces is quite complex and research on the subject is still being conducted. Several empirical formulas have been developed to approximate test results such as the ones given at the end of this paragraph. These particular expressions, which were given in the 7th edition of the AISC Manual, were based on study at the University of Illinois for two carbon steel T-stubs connected through their flanges with 4 in. bolts.[13] They are only good for that type of connection. For different arrangements other equations are needed. In the present or 8th edition of the AICS Manual (pp. 4-88 through 4-93) a very lengthy and detailed procedure is given for considering prying action for tee or double-angle hangars.

For A325 high-strength bolts

$$Q = F \left[\frac{100b(d_b)^2 - 18w(t_f)^2}{70a(d_b)^2 + 21w(t_f)^2} \right]$$

[13]R. S. Nair, P. C. Birkemoe, and W. H. Munse, "High Strength Bolts Subjected to Tension and Prying," Structural Research Series 353, Department of Civil Engineering, University of Illinois, Urbana, September 1969.

For A490 high-strength bolts

$$Q = F\left[\frac{100b(d_b)^2 - 14w(t_f)^2}{62a(d_b)^2 + 21w(t_f)^2}\right]$$

In these expressions

Q = the prying force per fastener, in kips
F = externally applied load per fastener, in kips
b = distance from center line of fastener to nearest face of outstanding leg of angle or web of tee minus $\frac{1}{16}$ in., in inches
d_b = bolt diameter, in inches
w = length of flange tributary to each fastener, in kips
t_f = thickness of angle of tee flange, in inches
a = distance from fastener line to edge of flange, not to exceed $2t_f$, in inches

Not only should the bolts be designed for the combined values of F and Q but also the bending moments in the flange of the tee or angles M_1 and M_2 illustrated in Fig. 9-18 need to be checked as well. In the figure the moment M_1 is equal to Qa while the moment M_2 is equal to $(F+Q)(b) - Q(a+b)$. The bending stresses resulting must be less than F_b which by the AISC is $0.75F_y$.

Example 9-6 which follows gives a brief example for a hanger type connection. Prying action is also a factor in the moment resisting connections discussed in Chapter 12.

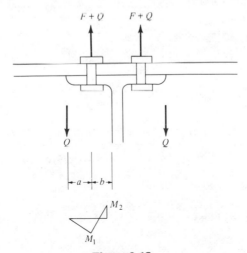

Figure 9-18

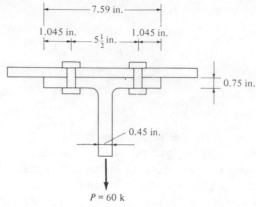

Figure 9-19

Example 9-6

Estimate the prying force and the bending moment in the flange of the tee stub cut from a W18×65 of A36 steel which is shown in Fig. 9-19. There are a total of six $\frac{7}{8}$ A325 bolts in the connection. They are spaced 4.50 in. on center longitudinally.

SOLUTION

$$F = \frac{60}{6} = 10 \text{ k}$$

$$b = 2.75 - \frac{0.45}{2} - \frac{1}{16} = 2.46 \text{ in.}$$

$$d_b = 0.875 \text{ in.}$$

$$w = 4.50 \text{ in.}$$

$$t_f = 0.75 \text{ in.}$$

$$a = 1.045 \text{ in.} < 2t_f, \qquad \text{therefore use } 1.045 \text{ in.}$$

Prying force

$$Q = 10 \left[\frac{(100)(2.46)(0.875)^2 - (18)(4.50)(0.75)^2}{(70)(1.045)(0.875)^2 + (21)(4.50)(0.75)^2} \right]$$

$$= 13.08 \text{ k}$$

total tensile force applied to bolts $= 10 + 13.08 = 23.08 \text{ k} < (0.6)(44) = 26.4 \text{ k}$

allowable tensile load   **OK**

Bending stress

$M_1 =$ moment at bolts $= (13.08)(1.045) = 13.67$ in.-k

$M_2 =$ moment $\frac{1}{16}$ in. from face of web

$= (10 + 13.08)(2.46) - (13.08)(1.045 + 2.46) = 10.93$ in.-k

$$S = \frac{\frac{1}{12}wt_f^3}{t_f/2} = \frac{wt_f^2}{6} = \frac{(4.50)(0.75)^2}{6} = 0.422 \text{ in.}^3$$

$$f_b = \frac{13.67}{0.422} = 32.39 \text{ ksi} > 0.75F_y = 27 \text{ ksi}$$ NG

9-15. BOLTS SUBJECTED TO COMBINED SHEAR AND TENSION

The bolts used for many structural steel connections are subjected to a combination of shear and tension. The combination is present even in the standard beam connections with the common web angles, to be described in Chapter 12. This stress condition is present in brackets and various types of moment-resisting connections, where the upper bolts are subjected to a vertical shear plus a tension caused by the fact that the beam end is trying to rotate downward and tends to pull the top portion of the connection away from the column or other member to which it is connected.

Tests on bolts subjected to combined shear and tension show that their ultimate strengths can be represented with an elliptical interaction curve. The curve has limits of F_t as the allowable stress if the bolt is stressed in tension alone and F_v if stressed in shear alone. Such a curve is shown in Fig. C1.6.3 of the AISC Commentary.

A high-strength bolted connection for a tension member is reviewed in Example 9-7. The applied load in this example is subjecting the bolts to a combination of tension and shear. In Section 1.6.3 of the AISC Specification allowable tensile stresses are given for A325 and A490 bolts in bearing-type connections when the threads are excluded and when they are not excluded from the shear plane. The following is the allowable for an A325 bolt when the threads are excluded from the shear plane.

$$F_t = 55 - 1.4f_v \leqslant 44$$

In this expression f_v is the calculated shearing stress produced by the applied loads. When a friction-type connection is used to resist an axially applied tensile force the clamping force is obviously reduced and the allowable shear stress F_v must be reduced in some proportion to the loss in clamping or pretension. This is accomplished in the AISC Specification by requiring that the maximum allowable shear stress be multiplied by the reduction factor

$$1 - \frac{f_t A_b}{T_b}$$

in which f_t is the calculated tensile stress due to the applied loads, T_b is the minimum bolt tension or proofload of the bolt (see Table 9-1) and A_b is the bolt's cross-sectional area. Thus for bolts in a friction-type connection with standard holes and threads excluded from the shear plane the allowable shearing stresses are

$$F_v \leqslant \left(1 - \frac{f_t A_b}{T_b}\right)(17.5) \qquad \text{for A325 bolts}$$

$$F_v \leqslant \left(1 - \frac{f_t A_b}{T_b}\right)(22.0) \qquad \text{for A490 bolts}$$

Sometimes a friction-type connection is subjected to combined shear and tension at the contact surface between a beam connection and the supporting member. If for such a case the fastener tension f_t is caused by moment in the plane of the beam web the shearing component of stress may be neglected in proportioning the fasteners for tension. The AISC Commentary says that for such a situation the part of the shear assigned to the fasteners subject to direct tensile stress is balanced by the increase in compression force on the compression side with the result that no actual shear force is applied to the fasteners in tension.[14]

Example 9-7

The tension member shown in Fig. 9-20 is connected to the column with eight $\frac{7}{8}$-in. A325 high-strength bolts in a bearing-type connection with the threads excluded from the shear plane and standard size holes. Is this a sufficient number of bolts to resist the applied load according to the AISC Specification?

SOLUTION
Shearing stress:

$$f_v = \frac{67.1}{(8)(0.6)} = 13.98 \text{ ksi} < 30.0 \text{ ksi} \qquad \text{OK}$$

Tension stress:

$$\text{allowable } F_t = 55 - (1.4)(13.98) = 35.43 \text{ ksi} \qquad \text{OK}$$

$$\text{actual } f_t = \frac{134.2}{(8)(0.6)} = 27.96 \text{ ksi} < 35.43 \text{ ksi} \qquad \text{OK}$$

<u>Connection is satisfactory (perhaps overdesigned)</u>

[14] Commentary on the AISC Specification (New York: AISC, 1978), p. 5–121.

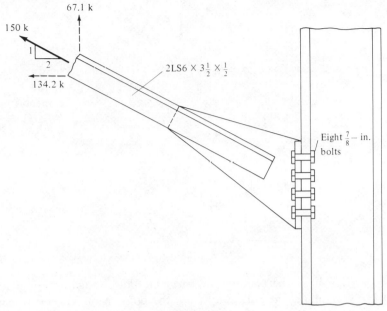

67.1 k

150 k

1

2

134.2 k

$2LS6 \times 3\frac{1}{2} \times \frac{1}{2}$

Eight $\frac{7}{8}$ – in. bolts

Figure 9-20

The problem of Example 9-7 is repeated in Example 9-8 except a friction-type connection is used. Although the connection is acting in shear and tension, friction prevents slippage of the bolt and consequent application of direct shear to the connectors.

Example 9-8

Rework Example 9-7 assuming that a friction-type connection is being used because no slip is permissible.

SOLUTION

$$f_t = \frac{134.2}{(8)(0.6)} = 27.96 \text{ ksi} < 44 \text{ ksi} \qquad\qquad \text{OK}$$

$$\text{allowable } F_v = \left(1 - \frac{27.96 \times 0.6}{39}\right)(17.5) = 9.98 \text{ ksi}$$

$$\text{actual } f_v = \frac{67.1}{(8)(0.6)} = 13.98 \text{ ksi} > 9.98 \text{ ksi} \qquad\qquad \text{NG}$$

Connection is not satisfactory as friction type

Example 9-9 illustrates the calculations involved in determining the stress situation in the high-strength bolts in the bracket connection of Fig. 9-21. The eccentric load on the bracket tends to bend the top of the

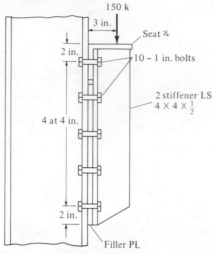

Figure 9-21

bracket away from the column and push the bottom against the column. Unless the tendency to pull away from the column on the top is greater than the initial tension in the bolts there will be almost no change in bolt tension, as described in Section 9-13. It is felt that the calculations in this example problem are self-explanatory.

Example 9-9

Determine the stresses in the 1-in. A325 high-strength bolts shown in the eccentrically loaded connection of Fig. 9-22.

SOLUTION
Stresses due to bolt tightening bearing the bracket against the column:

Proofload of 1-in. bolts is 51 k

Pressure on PLs when bolts tightened $= \dfrac{(10)(51)}{8 \times 20} = 3.19$ ksi

This stress diagram is shown in part (a) of Fig. 9-22
Stresses due to eccentric load:

$$f = \frac{Mc}{I} = \frac{(150)(3)(10)}{\left(\frac{1}{12}\right)(8)(20)^3} = 0.84 \text{ ksi}$$

This stress diagram is shown in part (b) of Fig. 9-22.

Note: The bolts going into this column cannot have an increase in stress unless the members start pulling apart, and the plates cannot pull apart unless the force applied exceeds the force bearing the pieces together.

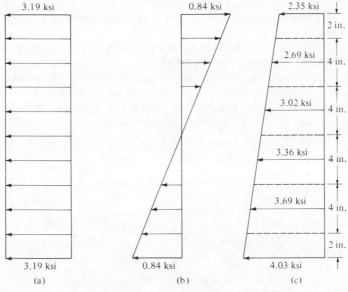

Figure 9-22

Portion of the bolt tension caused by the moment (top bolts):

$$T \text{ due to moment} = 51 - \left(\frac{2.35 + 2.69}{2} \right)(4 \times 4) = 10.70$$

or

$$\frac{10.70}{0.785} = 13.62 \text{ ksi}$$

Problems*

9-1. The plates used in the lap joint shown in the accompanying illustration are each $\frac{1}{2} \times 12$ in. and consist of A36 steel. Determine the maximum allowable

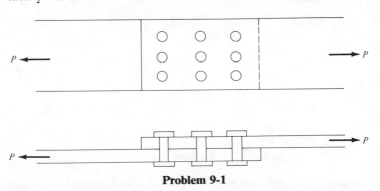

Problem 9-1

Note: For each of the problems listed the following information is to be used *unless otherwise indicated*: (a) AISC Specification; (b) standard sized holes; (c) threads of bolts excluded from shear plane; (d) members have clean mill scale surfaces (Class A).

load P which can be resisted if $\frac{7}{8}$-in. A325 high-strength bolts are used in (a) a friction-type connection and (b) a bearing-type connection. [*Ans*. (a) 94.5 k, (b) 129.6 k]

9-2. Rework Prob. 9-1 if 1-in. A490 bolts are used in (a) a friction-type connection and (b) a bearing-type connection.

9-3. If $\frac{7}{8}$-in A325 bolts are used in the connection shown in the accompanying illustration determine the tensile capacity of the joint (a) as a friction-type connection and (b) as a bearing-type connection. The plates are rolled from A36 steel and are part of a double lap splice connection. [*Ans*. (a) $P = 189$ k, (b) $P = 226.8$ k]

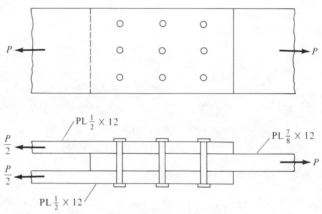

Problem 9-3

9-4. The truss member shown in the accompanying illustration consists of two C12×25s (A36 steel) connected to a 1-in. gusset plate. How many $\frac{7}{8}$-in. A325 bolts are required to develop the full tensile capacity of the member when it is used (a) as a friction-type connection and (b) as a bearing-type connection?

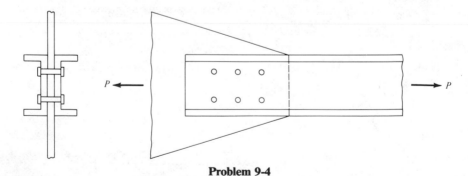

Problem 9-4

9-5. Rework Prob. 9-4 if 1-in. A490 bolts and A242 steel are used. [*Ans*. (a) 11.17 say 12, (b) 6.14 say 8]

9-6. For the connection shown in the accompanying illustration P equals 460 k. Determine the number of 1-in. A325 bolts required as (a) a friction-type connection and (b) as a bearing-type connection. A36 steel.

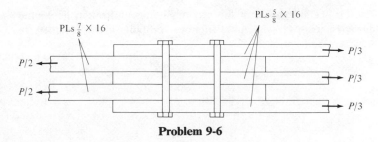

Problem 9-6

9-7. Rework Prob. 9-6 using $\frac{7}{8}$-in. A490 bolts. [*Ans.* (a) 8.71 say 10 (b) 4.79 say 6]

9-8. For the beam shown in the accompanying illustration what is the required spacing of $\frac{7}{8}$-in. A325 bolts in a bearing-type connection at a section where the external shear is 205 k? A36.

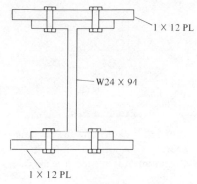

Problem 9-8

9-9. The cover-plated section shown in the accompanying illustration is used to support a 14.3 k/ft uniform load (including beam weight) for an 18-ft simple span. If $\frac{7}{8}$-in. A325 friction-type bolts are used arranged as shown, work out a spacing diagram for the entire span. Assume the cover plates extend for the full span. A3o. (*Ans.* Use 6 in. spacing entire span per AISC Section 1.18.2.3.)

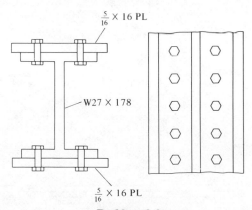

Problem 9-9

9-10. For the section shown in the accompanying illustration determine the required spacing of $\frac{3}{4}$-in. A490 friction-type bolts if the member consists of A572 Grade 60 steel. $V = 220$ k.

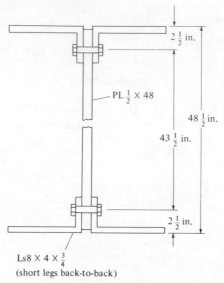

Ls8 × 4 × $\frac{3}{4}$
(short legs back-to-back)

Problem 9-10

9-11. For an external shear of 380 k determine the spacing required for $\frac{3}{4}$-in. A325 web bolts used in a bearing-type connection for the built-up section shown in the accompanying illustration. A36 steel. (*Ans.* 8.64 in. say $8\frac{1}{2}$ in.)

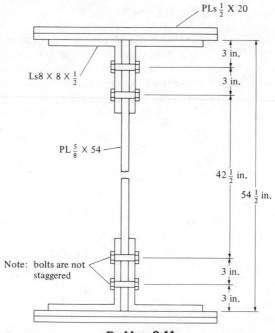

Problem 9-11

9-12. Determine the resultant load on the most stressed bolt in the eccentrically loaded connection shown in the accompanying illustration.

9-13. Determine the resultant load on the most stressed bolt in the eccentrically loaded connection shown in the accompanying illustration. (*Ans.* 20.96 k)

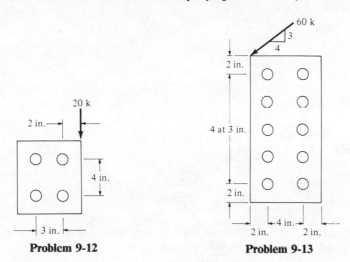

Problem 9-12   **Problem 9-13**

9-14. Determine the most stressed bolt and the magnitude of its force for the bolt group shown in the accompanying illustration. Determine also the size of A325 bearing-type bolts needed to resist that force. A36 steel.

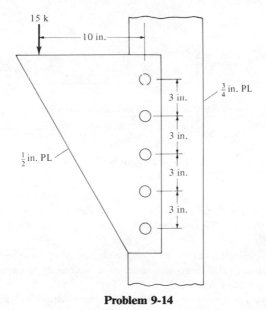

Problem 9-14

9-15. Find the most stressed bolt and the magnitude of the force on that bolt for the connection shown in the accompanying illustration (*Ans.* 6.15 k)

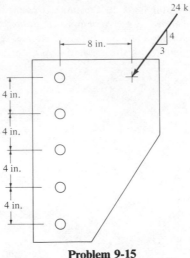

Problem 9-15

9-16. Determine the most stressed bolt and the magnitude of the total force on that bolt for the connection shown in the accompanying illustration.

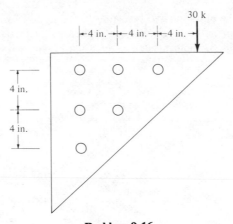

Problem 9-16

9-17. The 1-in A325 high-strength bolts shown in the accompanying illustration are used in a bearing-type connection to connect a $\frac{3}{4}$-in. plate to a $\frac{7}{8}$-in. plate. Are the bolts of sufficient size? A36 steel. (*Ans.* Yes, $R = 16.59$ k; allowable $R = 23.55$ k)

9-18. Determine the bolt size required for the connection shown in the accompanying illustration. A325 bolts are used in a bearing-type connection and the members consist of A36 steel.

9-19. For the bracket plate shown in the accompanying illustration $\frac{7}{8}$-in. A490 bolts are used in a bearing-type connection. Are they satisfactory? A36 steel. (*Ans.* Yes. R-15.63 k; allowable $R = 24$ k)

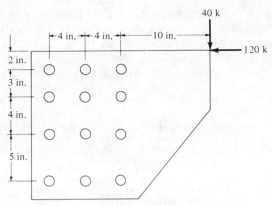

Problem 9-17

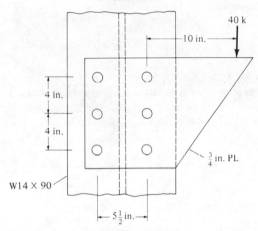

Problem 9-18

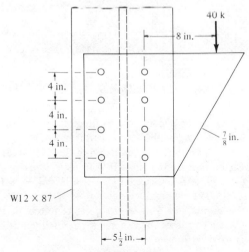

Problem 9-19

9-20. Rework Prob. 9-19 using 1-in. A325 bolts in a bearing-type connection.

9-21. Are the bolts shown in the tee stub satisfactory to resist direct force and prying and is the bending stress in the tee flange satisfactory? Use the AISC Specification and prying equation. There are eight 1-in. A325 bolts and they are spaced 4.00 in. on center longitudinally. (*Ans.* Yes)

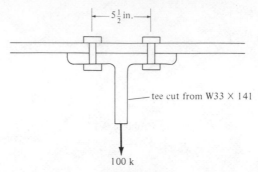

$5\frac{1}{2}$ in.

tee cut from W33 × 141

100 k

Problem 9-21

9-22. Repeat Prob. 9-21 if the tee is cut from a W30×99.

9-23. Is the connection shown in the accompanying illustration sufficient to resist the 80-k load which passes through the center of gravity of the bolt group? A36. (*Ans.* No. $f_v = 13.64$ ksi $> F_v = 12.50$ ksi)

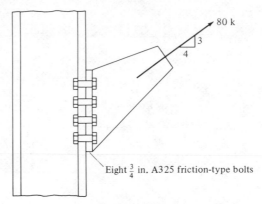

80 k

3
4

Eight $\frac{3}{4}$ in. A325 friction-type bolts

Problem 9-23

9-24. If the load shown in the accompanying illustration passes through the center of gravity of the rivet group, how large can it be? A36.

9-25. The connectors shown in the accompanying illustration are 1-in. A325 bolts used in a bearing-type connection. If A36 steel is used are the bolts sufficient? The $\frac{3}{4}$-in. plate is assumed to be connected to a 1-in. plate behind. (*Ans.* Yes. $R = 19.04$ k while allowable $R = 23.55$ k)

9-26. Determine the number of $\frac{3}{4}$-in. A325 bolts required in the angles and the flange of the W shape shown in the accompanying illustration if a bearing-type connection is used. Use A36 steel.

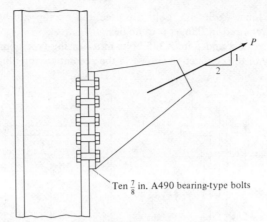

Ten $\frac{7}{8}$ in. A490 bearing-type bolts

Problem 9-24

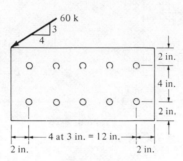

60 k

3

4

2 in.

4 in.

2 in.

|← 4 at 3 in. = 12 in. →|

2 in.   2 in.

Problem 9-25

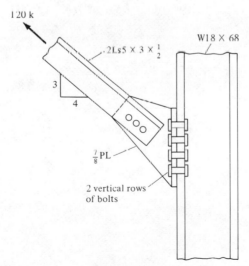

120 k

$2Ls 5 \times 3 \times \frac{1}{2}$

W18 × 68

3

4

$\frac{7}{8}$ PL

2 vertical rows
of bolts

Problem 9-26

9-27. Repeat Prob. 9-26 if $\frac{7}{8}$-in. bolts are used and no slip can be tolerated. (*Ans.* 10 bolts required in flange, 6 in angles)

9-28. Using A36 steel and $\frac{7}{8}$-in. A325 bolts in a bearing-type connection check the adequacy of the connection shown in the accompanying illustration.

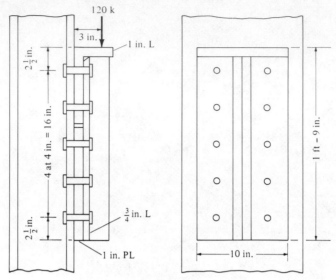

Problem 9-28

Chapter 10
Historical Notes
on Riveted
Connections

10-1. GENERAL

For many years rivets were the accepted method for connecting the members of steel structures. Today, however, they no longer provide the most economical connections. They are still occasionally used for fasteners but their use has declined to such a degree that many large steel fabricators in the United States have discontinued riveting altogether. It is, however, desirable for the designer to be familiar with rivets even though he will seldom if ever design riveted structures. He may have to analyze an existing riveted structure for new loads or for an expansion of the structure. The purpose of this brief chapter is to present only the differences in bolt and rivet theory.

The rivets used in construction work are usually made of a soft grade of steel which will not become brittle when heated and hammered with a riveting gun to form the head. The usual rivet consists of a cylindrical shank of steel with a rounded head on one end. It is heated in the field to a cherry-red color (approximately 1800°F), inserted in the hole, and a head is formed on the other end probably with a portable rivet gun operated by compressed air. The rivet gun, which has a depression in its head to give the rivet head the proper shape, applies a rapid succession of blows to the rivet.

For riveting done in the shop the rivets are probably heated to a light cherry-red color and driven with a pressure type riveter. This type of riveter, usually called a "bull" riveter, squeezes the rivet with a pressure of perhaps as high as 50 to 80 tons (445 to 712 kN) and drives the rivet with one stroke. Because of this great pressure the rivet in its soft state is forced to fill the hole very satisfactorily. This type of riveting is much to be preferred over that done with the pneumatic hammer but no greater allowable stresses are allowed by riveting specifications. The bull riveters were built for much faster operation than were the portable hand riveters

Handling hot rivets, Chicago, Ill. (Courtesy of Inryco, Inc.)

but the latter riveters are needed for places that are not easily accessible (i.e., field erection).

As the rivet cools it shrinks or contracts and squeezes together the parts being connected. The squeezing effect actually causes considerable transfer of stress between the parts being connected to take place by friction. The amount of friction is not dependable, however, and the specifications do not permit its inclusion in the strength of a connection. Rivets shrink diametrically as well as lengthwise and actually become somewhat smaller than the holes which they are assumed to fill. (Permissible stresses for rivets are actually given in terms of the nominal cross-sectional areas of the rivets before driving.)

Some shop rivets are driven cold with tremendous pressures. Obviously the cold-driving process works better for the smaller size rivets probably $\frac{3}{4}$ in. in diameter or less although larger ones have been successfully used. Cold-driven rivets fill the holes better, eliminate the cost of heating, and are stronger due to the fact that the steel is cold worked. There is, however, a reduction of clamping force since the rivets do not shrink after driving.

10-2. TYPES OF RIVETS

The sizes of rivets used in ordinary construction work are $\frac{3}{4}$ in. and $\frac{7}{8}$ in. in diameter but they can be obtained in standard sizes from $\frac{1}{2}$ in. to $1\frac{1}{2}$ in. in

$\frac{1}{8}$-in. increments. (The smaller sizes were used for small roof trusses, signs, small towers, etc., while the larger sizes were used for very large bridges or towers and very tall buildings.) The use of more than one or two sizes of rivets or bolts on a single job is usually undesirable because it is expensive and inconvenient to punch different-size holes in a member in the shop, and the installation of different-size rivets or bolts in the field may be confusing. Some cases arise where it is absolutely necessary to have different sizes, as where smaller rivets or bolts are needed for keeping the proper edge distance in certain sections, but these situations should be avoided if possible.

Rivet heads are usually round in shape, called button heads; but if clearance requirements dictate, the head can be flattened or even counter-sunk and chipped flush. These situations are shown in Fig. 10-1.

The countersunk and chipped-flush rivets do not have a head large enough to develop full strength and should be discounted 50% in design. A rivet with a flattened head is to be preferred to a countersunk rivet but if a smooth surface is required the countersunk and chipped-flush rivet is necessary. This latter type of rivet is appreciably more expensive than the button head type in addition to being weaker and should not be used unless absolutely necessary.

There are three ASTM classifications for rivets for structural steel applications as described in the following paragraphs.

ASTM Specification A502, Grade 1

These rivets are used for most structural work (when rivets are used). They have a low carbon content of about 0.80%, are weaker than the ordinary structural carbon steel, and have a higher ductility. The fact that these rivets are easier to drive than the higher strength rivets is probably the main reason that when rivets are used they probably are A502, Grade 1 regardless of the strength of the steel used in the structural members.

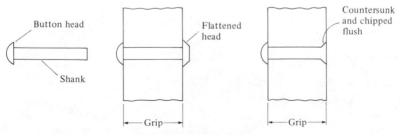

Figure 10-1 Types of rivets.

ASTM Specification A502, Grade 2

These carbon-manganese rivets have higher strengths than the Grade 1 rivets and were developed for the higher strength steels. Their higher strength permits the designer to use fewer rivets in a connection and thus smaller gusset plates.

ASTM Specification A502, Grade 3

These rivets have the same allowable stresses as the Grade 2 rivets but they have much higher resistance to atmospheric corrosion equal to approximately four times that of carbon steel without copper.

10-3. STRENGTH OF RIVETED CONNECTIONS—RIVETS IN SHEAR

The factors determining the strength of a rivet are the rivet diameter and the thickness and arrangement of the pieces being connected. The actual distribution of stress around a rivet hole is difficult to determine, if it can be determined at all; and to simplify the calculations it is assumed to vary uniformly over a rectangular area equal to the diameter of the rivet times the thickness of the plate.

The strength of a rivet in bearing equals the allowable bearing unit stress of the rivet times the diameter of the shank of the rivet times the thickness of the member that bears on the rivet. The strength of a rivet in single shear is the allowable shearing unit stress times the cross-sectional area of the shank of the rivet. Should a rivet be in double shear its shearing strength is considered to be twice its single-shear value. The allowable bearing stress for rivets, F_p, is the same as for bolts. It equals $1.5F_u$ where F_u is the specified minimum tensile strength of the steel used in the connected parts.

The allowable tension and shear stresses for rivets and A307 bolts permitted by the AISC are shown in Table 10-1. These values are taken from Table 1.5.2.1 of the AISC Specification. The allowable tension values are for static loads only while the allowable shear stress permitted in A307 bolts is not affected if the bolt threads are in the shear plane.

Table 10-1 AISC Allowable Tensions and Shears For Rivets and A307 Bolts

Fastener type	Allowable tension (F_t)	Allowable shear (F_v) bearing-type connections
A502, Grade 1, hot-driven rivets	23.0 ksi	17.5 ksi
A502, Grades 2 and 3, hot-driven rivets	29.0 ksi	22.0 ksi
A307 bolts	20.0 ksi	10.0 ksi

SOURCE: *Specification for the Design, Fabrication, and Erection of Structural Steel for Buildings*, November 1, 1978, Table 1.5.2.1. AISC, Inc.

For many years it was common for steel specifications to give different allowable bearing stresses depending on whether single shear or double shear was present. The feeling was that with single shear and bearing [see Fig. 9-1(a)] the eccentricity caused less uniform bearing stresses. Tests in recent years have shown that a distinction in the allowable bearing stresses between the two cases is not necessary. Even if there is a difference between the stress distributions (for the two cases) at small loads, the plastic behavior of the rivets at ultimate load makes a distinction unnecessary.

Examples 10-1 and 10-2 illustrate the simple calculations involved in determining the strength of riveted connections and the number of rivets required for a certain loading condition.

Example 10-1

Determine the allowable force P which can be applied to the plates shown in Fig. 10-2. The AISC Specification, A36 steel and A502, Grade 1 rivets are used in the connection. (Allowable shear in rivets 17.5 ksi and the allowable bearing in rivets is $1.5 \times 58 = 87$ ksi.)

SOLUTION
Allowable tensile load on plates

$$A_g = \left(\tfrac{1}{2}\right)(10) = 5.00 \text{ in.}^2$$

$$A_n = \left[\left(\tfrac{1}{2}\right)(10) - (2)\left(\tfrac{7}{8}\right)\left(\tfrac{1}{2}\right)\right] = 4.125 \text{ in.}^2$$

$$P = 0.60 F_y A_g = (0.60)(36)(5.00) = 108 \text{ k}$$

or

$$P = 0.50 F_u A_e \quad \text{where } A_e = A_n$$
$$= (0.50)(58)(4.125) = 119.6 \text{ k}$$

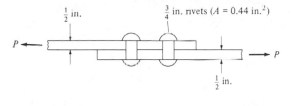

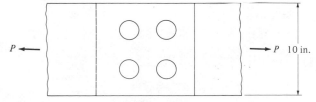

Figure 10-2

Rivets in single shear and bearing on $\frac{1}{2}$ in.:

$$\text{allowable shear in rivets} = (4)(0.44)(17.5) = 30.8 \text{ k}\leftarrow$$

$$\text{allowable bearing in rivets} = (4)\left(\tfrac{3}{4}\right)\left(\tfrac{1}{2}\right)(87) = 130.5 \text{ k}$$

Maximum permissible load $P = 30.8$ k

Example 10-2

How many $\frac{7}{8}$-in. A502, Grade 1 rivets are required for the connection shown in Fig. 10-3 if the plates are made from A36 steel? Use AISC Specification.

SOLUTION
Rivets in double shear and bearing on $\frac{1}{2}$ in.:

$$\text{allowable shear per rivet} = (2)(0.6)(17.5) = 21 \text{ k}$$

$$\text{allowable bearing per rivet} = \left(\tfrac{7}{8}\right)\left(\tfrac{1}{2}\right)(87.0) = 38.1 \text{ k}$$

$$\text{no. of rivets reqd.} = \frac{114.0}{21.0} = 5.43$$

Use six $\frac{7}{8}$-in. rivets

Little comment is made herein concerning A307 bolts. The reason is that all of the calculations for these fasteners are made exactly as they are for rivets except that the allowable stresses are different. Only one brief example (10-3) is considered necessary.

Example 10-3

Repeat Example 10-2 using $\frac{7}{8}$-in. A307 bolts.

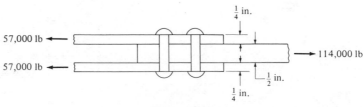

Figure 10-3

SOLUTION

Bolts in double shear and bearing on $\frac{1}{2}$ in.:

$$\text{allowable shear per bolt} = (2)(0.6)(10) = 12 \text{ k} \leftarrow$$

$$\text{allowable bearing per bolt} = \left(\tfrac{7}{8}\right)\left(\tfrac{1}{2}\right)(87) = 38.1 \text{ k}$$

$$\text{no. reqd.} = \frac{114}{12} = 9^{+}$$

Use ten $\frac{7}{8}$-in. A307 bolts

10-4. RIVETS SUBJECTED TO ECCENTRIC SHEAR

The theory required for considering eccentrically loaded riveted joints is exactly the same as for eccentrically loaded bolted connections and was previously described in Section 9-12 and thus is not repeated in this chapter.

10-5. RIVETS SUBJECTED TO COMBINED SHEAR AND TENSION

In the majority of framed connections for beam to beam and beam to column the rivets are subjected to a combination of shear and tension. This combination is true for the rivets in the bracket of Fig. 10-4 as well as for those in the connection shown in Figs. 12-1(a) and (b) and 12-3(a) and (b). In each of these connections the load being transferred tends to cause

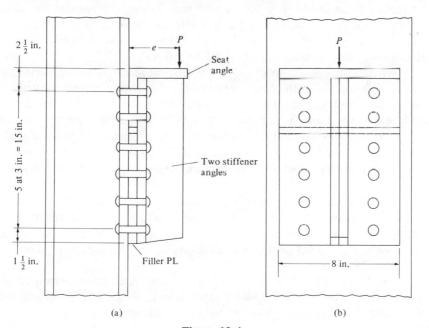

(a)   (b)

Figure 10-4

the connection to slide downward and this tendency is resisted by the shearing strength of the rivets. At the same time, the load is tending to rotate the connection, pulling the top rivets away from the column and pushing the lower part of the connection against the column.

The vertical shearing stress can be easily determined by dividing the vertical load by the cross-sectional area of the rivets involved; however, the stresses caused by bending are a little more difficult to compute. For this discussion the bracket connection of Fig. 10-4 with an eccentric load P is considered. The downward load and the upward shear in the rivets produce a couple. The distance between the two forces is assumed to be e and the moment produced Pe. An equal and opposite couple must be produced by the tension in the upper rivets and by the compression in the lower part of the connection.

An exact analysis of the rivet stresses produced by this situation is almost impossible; therefore, several different approximate methods have been developed by various designers. Perhaps the most common method is the one described in the paragraphs to follow.

In this method the initial tension in the rivets is neglected and the top of the connection is assumed to have pulled away from the column. Below the opening there is assumed to be a neutral axis above which the rivets are in tension and below which the connection angles are bearing against the column. It is assumed in this discussion that the bearing is spread over the full width of the angles. (Because of deformation of these angles the pressure at the outer edges is probably appreciably smaller than the pressure on the inside. For this reason many designers assume the compression is spread over some smaller width. It should be noted, however, that the varying widths used by different designers have little effect on the final tensile stresses determined in the top rivets.)

With the full-width assumption the neutral axis will usually lie about one-sixth to one-seventh of the depth of the connection above the bottom of the connection. From this information the location of the neutral axis can be estimated and its accuracy checked by taking moments about that location. After moments are taken it may be necessary to make a slight adjustment in the location. When the neutral axis is located the moment of inertia can be calculated and the stresses obtained by the flexure formula.

Tests on rivets subjected to combined shear and tension show that their ultimate strengths can be represented with an elliptical interaction curve. The curve has limits of F_t as the allowable stress if the rivet is stressed in tension alone and of F_v if stressed in shear alone. Such a curve is shown in Fig. C1.6.3 in the "Commentary on the AISC Specification" in the AISC Manual.

For A502, Grade 1 rivets the allowable combined stress permitted by the AISC is given by the following expression in which f_t is the calculated tensile stress and f_v is the calculated shearing stress.

$$F_t = 30 - 1.3 f_v \leqslant 23$$

In effect, this expression replaces the test result curve with three straight lines (as shown in Fig. C1.6.3 of the AISC Manual) and divides the values with appropriate factors of safety. It will be noted that the maximum permissible shear for Grade 1 rivets is 17.5 ksi while the maximum permissible tension is 23 ksi. This combined shear and tension formula says that if the calculated shear stress f_v is equal to or less than 5.384 ksi the permissible tension stress F_t will equal 23 ksi but if f_v is greater than 5.384 ksi the value of F_t will be less than 23 ksi. A further study of the formula will show that if f_t is equal to or less than 7.25 ksi the permissible shear F_v will be 17.5 ksi but if f_t is greater than 7.25 ksi F_v will be less than 17.5 ksi.

Example 10-4

Determine the maximum combined tensile and shearing stresses in the top rivets of the connection shown in Fig. 10-4. The stiffener angles are each $4 \times 4 \times \frac{1}{2}$ in.; the eccentric load is 75 k; the eccentricity is 3 in., and $\frac{7}{8}$-in. rivets are used. Does the resulting value exceed the AISC allowable for A502, Grade 1 rivets?

SOLUTION

Assume neutral axis one-sixth of the distance from bottom of connection.

$$\left(\tfrac{1}{6}\right)(19 \text{ in.}) = 3.166 \text{ in.} \qquad \text{(say 3 in.) (see Fig. 10-5)}$$

Check location of neutral axis:

$$(8)(3)(1.5) = (2)(0.6)(1.5 + 4.5 + 7.5 + 10.5 + 13.5)$$

$$36 = 45 \qquad\qquad\qquad \text{NG}$$

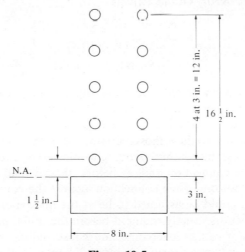

Figure 10-5

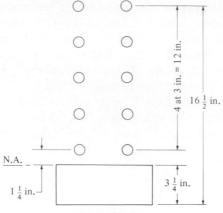

Figure 10-6

Assume neutral axis 3.25 in. from bottom of connection (see Fig. 10-6).

$$(8)(3\tfrac{1}{4})(1.625) = (2)(0.6)(1.25 + 4.25 + 7.25 + 10.25 + 13.25)$$

$$42.2 = 43.5 \qquad\qquad\qquad \text{OK}$$

$$I = (\tfrac{1}{3})(8)(3.25)^3 + (2)(0.6)(1.25^2 + 4.25^2 + 7.25^2 + 10.25^2 + 13.25^2)$$

$$= 91.4 + 422$$

$$= 513 \text{ in.}^4$$

$$f_t = \frac{Mc}{I} = \frac{(225)(13.25)}{513} = 5.8 \text{ ksi}$$

$$f_v = \frac{75}{(12)(0.6)} = 10.4 \text{ ksi}$$

Allowable F_t for A502, Grade 1 rivets:

$$F_t = 30 - 1.3f_v \leqslant 23$$

$$= 30 - (1.3)(10.4)$$

$$= 16.48 \text{ ksi} > 5.8 \text{ ksi} \qquad\qquad \text{OK}$$

As previously indicated, there are other methods of estimating the stresses in rivets subjected to shear and tension. The method described here neglected any initial tension present in the rivets. If the rivets did have an initial tension (as they most certainly do) they would squeeze the connection against the column. Bending would serve to reduce the pressure at the top of the connection and, as before, there would be an increase in pressure at the bottom of the connection against the column. For this case the neutral axis could be assumed to lie at the middepth of the section. A very large moment is usually required before the initial pressure at the top of the connection can be overcome and cause pulling of the connection away from the column, causing in turn a change in the top rivet stress. This subject was discussed at length in Chapter 9 for high-strength bolts.

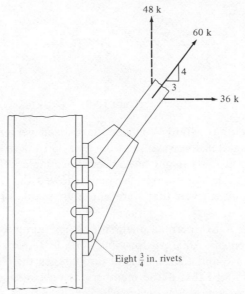

48 k

60 k

4

3

36 k

Eight $\frac{3}{4}$ in. rivets

Figure 10-7

Another case where rivets are subject to a combination of shear and tension is shown in Fig. 10-7 and illustrated by the calculations of Example 10-5.

Example 10-5

The tension member shown in Fig. 10-7 is connected to the column with eight $\frac{3}{4}$-in. A502, Grade 1 rivets. Is the connection satisfactory? Use A36 steel and the AISC Specification.

SOLUTION

$$f_v = \frac{48}{(8)(0.44)} = 13.64 \text{ ksi}$$

$$f_t = \frac{36}{(8)(0.44)} = 10.23 \text{ ksi}$$

$$F_t = 30 - 1.3 f_v \leqslant 23$$

$$= 30 - (1.3)(13.64)$$

$$= 12.27 \text{ ksi} > 10.23 \text{ ksi} \qquad \qquad \text{OK}$$

Therefore connection is satisfactory

10-6. RIVETS IN TENSION

Figure 10-8(a) shows a connection in which the rivets are in tension. Perhaps one of the most common cases of tension rivets occurs in wind

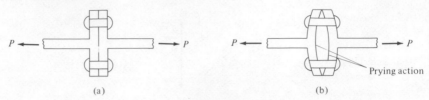

Figure 10-8

bracing for buildings. Many engineers have been very much opposed to the use of rivets in tension because of the fear that the rivet heads might be pulled off. This attitude might have been reasonable before 1930 but the results of tests published during that year[1] showed that improved riveting techniques produced rivets that were entirely capable of resisting tension loads.

Today the AISC permits their use in building design (giving an allowable tension of 23,000 psi for A502, Grade 1 rivets); but the AREA prohibits their use and the AASHTO discourages their use. It should be noted in Fig. 10-8(b) that if the members being connected are not fairly stiff, some prying action will occur, which will increase the tension in the rivets. Should the members be fairly flexible, the designer should attempt to estimate the stresses caused by prying. A more detailed discussion of the action of tensile connectors including prying was presented in Section 9-14.

Problems

10-1. Determine the maximum allowable load P which can be resisted by the lap joint shown in the accompanying illustration if $\frac{7}{8}$-in. A502, Grade 1 rivets and the AISC Specification are used. Each of the plates is $\frac{5}{8} \times 12$ and consists of A36 steel. (*Ans. P* = 126 k)

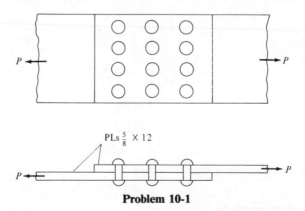

Problem 10-1

[1]W. M. Wilson and W. A. Oliver, *Tension Tests of Rivets*, University of Illinois Experiment Station, Bulletin 210, 1930.

10-2. Repeat Prob. 10-1 if the AASHTO allowable stresses are used. Allowable shear on rivets = 14,000 psi, allowable bearing on rivets = 40,000 psi and allowable F_t on net area of plates = 20,000 psi.

10-3. The truss tension member shown in the accompanying illustration consists of a single angle $5 \times 3 \times \frac{5}{16}$ in. and is made from A36 steel. The member is connected to a $\frac{1}{2}$-in. gusset plate with the five $\frac{7}{8}$-in. A502, Grade 1 rivets shown. Using the AISC Specification determine the allowable value of P. (*Ans. P* = 51.6 k)

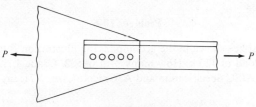

Problem 10-3

10-4. How many $\frac{3}{4}$-in. A502, Grade 1 rivets are needed to carry the load shown in the accompanying illustration? A36. AISC.

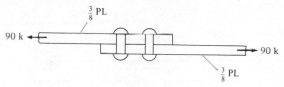

Problem 10-4

10-5. Using the allowable stresses given in Prob. 10-2 determine the allowable value of P in the butt joint shown in the accompanying illustration if 1-in. rivets are used. (*Ans.* 150 k)

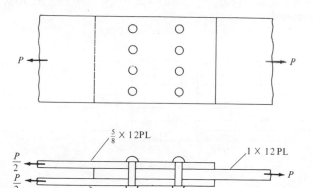

Problem 10-5

10-6. A502, Grade 1 rivets with 1-in. diameters are to be used for the butt joint shown. How many rivets are needed if A36 steel and the AISC Specification are to be used?

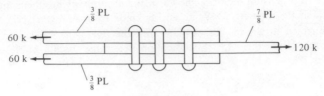

$\frac{3}{8}$ PL $\frac{7}{8}$ PL

60 k

60 k   120 k

$\frac{3}{8}$ PL

Problem 10-6

10-7. The butt splice shown in the accompanying illustration is to be designed for a tensile load of 120 k. How many $\frac{7}{8}$-in. A502, Grade 1 rivets are required, using the AISC Specification and A36 steel? (*Ans.* 5.71, say 6)

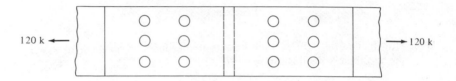

120 k   120 k

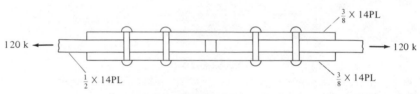

$\frac{3}{8}$ × 14PL

120 k   120 k

$\frac{1}{2}$ × 14PL $\frac{3}{8}$ × 14PL

Problem 10-7

10-8. How many $\frac{7}{8}$-in. A502, Grade 1 rivets are required for the connection shown in the accompanying illustration if A36 steel and the AISC Specification are used?

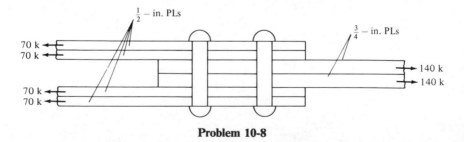

$\frac{1}{2}$ – in. PLs

$\frac{3}{4}$ – in. PLs

70 k
70 k

140 k

140 k

70 k
70 k

Problem 10-8

10-9. For the connection shown in the accompanying illustration P equals 475 k. Determine the number of $\frac{7}{8}$-in. A502, Grade 2 rivets required. Use A36 steel and the AISC Specification. (*Ans.* 8.99, say 9)

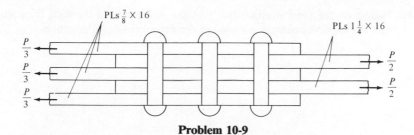

Problem 10-9

10-10. Determine the maximum allowable value of P for the connection shown in the accompanying illustration if 1-in. A502, Grade 2 rivets and the AISC Specification are used. The plates consist of A36 steel.

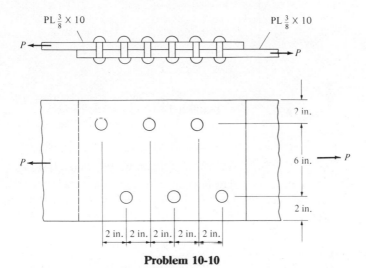

Problem 10-10

10-11. For the beam shown in the accompanying illustration what is the required spacing of $\frac{3}{4}$-in. A502, Grade 1 rivets at a section where the external shear is 106 k? (*Ans.* $p = 5.37$ in., say $5\frac{1}{4}$ in.)

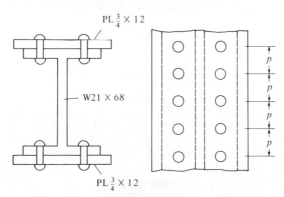

Problem 10-11

10-12. Determine the most stressed rivet and the magnitude of the total force on that rivet for the connection shown in the accompanying illustration.

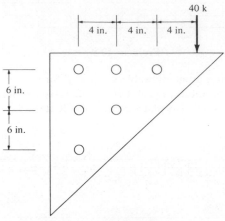

Problem 10-12

10-13. The MC18×42.7 shown in the accompanying illustration consists of A36 steel. Using the AISC Specification determine the maximum permissible value of the eccentric load P if $\frac{7}{8}$-in. A502, Grade 2 rivets are used. (*Ans.* 68.6 k)

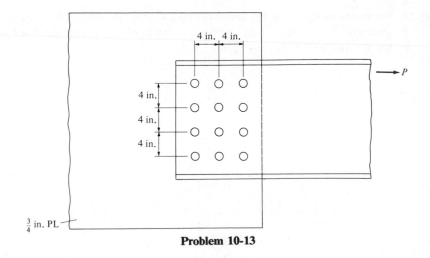

Problem 10-13

10-14. If the maximum allowable load on any one rivet in the rivet group shown in the accompanying illustration is 12 k determine the maximum permissible value of P.

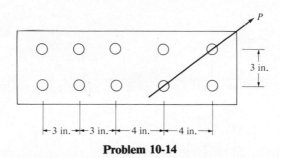

Problem 10-14

10-15. Using 1-in. A502, Grade 1 rivets determine if the connection shown in the accompanying illustration is satisfactory. A36 steel. AISC Specification. ($Ans. f_t = 5.38$ ksi $< F_t = 18.17$ ksi; OK)

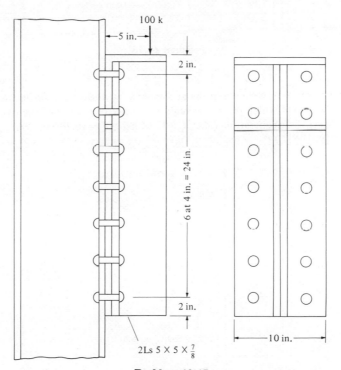

Problem 10-15

10-16. Using the AISC Specification, A36 steel and $\frac{7}{8}$-in. A502, Grade 1 rivets determine the permissible value of P for the connection shown in the accompanying illustration.

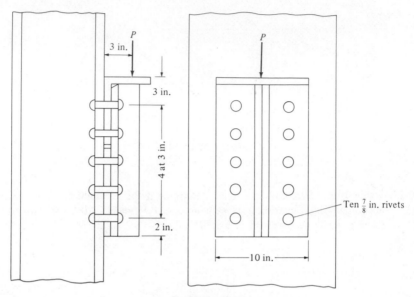

Problem 10-16

10-17. Is the connection shown in the accompanying illustration sufficient to resist the 60-k load which passes through the center of gravity of the rivet group? A36. AISC Specification. (*Ans.* $f_t = 13.64$ ksi $< F_t = 16.70$ ksi; OK)

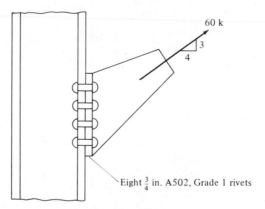

Problem 10-17

10-18. If the load shown in the accompanying illustration passes through the center of gravity of the rivet group, how large can it be? A36. AISC Specification.

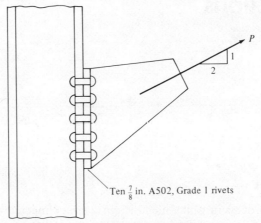

Ten $\frac{7}{8}$ in. A502, Grade 1 rivets

Problem 10-18

Chapter 11
Welded
Connections

11-1. GENERAL

Welding is a process in which metallic parts are connected by heating their surfaces to a plastic or fluid state and allowing the parts to flow together and join with or without the addition of other molten metal. It is impossible to determine when welding originated but it was several thousand years ago. Metalworking, including welding, was quite an art in ancient Greece at least three thousand years ago, but welding had undoubtedly been performed for many centuries before those days. Ancient welding was probably a forging process in which the metals were heated to a certain temperature (not to the melting stage) and hammered together.

Although modern welding has been available for a good many years, it has only come into its own in the last few decades for the building and bridge phases of structural engineering. The adoption of structural welding was quite slow for several decades because many engineers thought that welding had two great disadvantages—(1) that welds had reduced fatigue strength as compared to riveted connections and (2) that it was impossible to ensure a high quality of welding without unreasonably extensive and costly inspection.

These negative feelings persisted for many years, although tests seemed to indicate that neither reason was valid. Regardless of the validity of these fears they were widely held and undoubtedly slowed down the use of welding, particularly for highway bridges and to an even greater extent for railroad bridges. Today most engineers agree that there is little difference between the fatigue strength of bolted and welded joints. They will also admit that the rules governing the qualification of welders, the better techniques applied, and the excellent workmanship requirements of the AWS specifications make the inspection of welding a much less difficult problem. Consequently welding is today permitted for almost all structural work other than for some bridges.

On the subject of fear of welding it is interesting to consider welded ships. Ships are subjected to severe impactive loadings which are difficult

Fabrication of plate girders for Connecticut expressway bridge. (Courtesy of The Lincoln Electric Company.)

to predict and yet naval architects use all welded ships with great success. A similar discussion can be made for airplanes and aeronautical engineers. The slowest adoption of structural welding has been for railroad bridges. These bridges are undoubtedly subjected to heavier live loads than high-way bridges, larger vibrations, and more stress reversals; but are their stress situations as serious and as difficult to predict as those for ships and planes?

11-2. ADVANTAGES OF WELDING

Today it is possible to make use of the many advantages which welding offers, since the fatigue and inspection fears have been largely eliminated. Several of the many welding advantages are discussed in the following paragraphs.

1. To most persons the first advantage is in the area of economy, because the use of welding permits large savings in pounds of steel used. Welded structures allow the elimination of a large percentage of the gusset and splice plates necessary for riveted or bolted structures as well as the elimination of rivet or bolt heads. In some bridge trusses it may be possible to save up to 15% or more of the steel weight by using welding. Welding also requires appreciably less labor than does riveting because one welder can replace the standard four-person riveting crew.

2. Welding has a much wider range of application than riveting or bolting. Consider a steel pipe column and the difficulties of connecting it

to other steel members by riveting or bolting. A riveted or bolted connection may be virtually impossible but a welded connection will present no difficulties whatsoever. The student can visualize many other similar situations where welding has a decided advantage.

3. Welded structures are more rigid structures because the members are often welded directly to each other. The connections for riveted or bolted structures are often made through connection angles or plates which deflect due to load transfer, making the entire structure more flexible. On the other hand, greater rigidity can be a disadvantage where simple end connections with little moment resistance are desired. For such cases designers must be careful as to the type of joint they specify.

4. The process of fusing pieces together gives the most truly continuous structures. It results in one-piece construction and because welded joints are as strong or stronger than the base metal no restrictions have to be placed on the joints. This continuity advantage has permitted the erection of countless slender and graceful steel statically indeterminate frames through the world. Some of the more outspoken proponents of welding have referred to riveted and bolted structures, with their heavy plates and large number of rivets or bolts, as looking like tanks or armored cars when compared to the clean, smooth lines of welded structures. For a graphic illustration of this advantage the student should compare the moment resisting connections of Fig. 12-4.

5. It is easier to make changes in design and to correct errors during erection (and at less expense) if welding is used. A closely related advantage has certainly been illustrated in military engagements during the past few decades by the quick welding repairs made to military equipment under battle conditions.

6. Another item that is often important is the relative silence of welding. Imagine the importance of this fact when working near hospitals or schools or when making additions to existing buildings. Anyone with close to normal hearing who has attempted to work in an office within several hundred feet of a riveting job will go along with this advantage.

7. Fewer safety precautions are required for the public in congested areas as compared to those necessary for a riveted structure where tossing of hot rivets is a necessity.

8. Fewer pieces are used and as a result time is saved in detailing, fabrication and field erection.

11-3. TYPES OF WELDING

Although both gas and arc welding are available almost all structural welding is arc welding. Sir Humphry Davy discovered in 1801 how to create an electric arc by bringing close together two terminals of an electric circuit of relatively high voltage. Although he is generally given credit for the development of modern welding, a good many years actually elapsed after his discovery before welding was actually performed with the electric arc. (His work was of the greatest importance to the modern structural world but it is interesting to know that many people say his greatest discovery was not the electric arc but rather a laboratory assistant whose name was Michael Faraday.) Several Europeans formed welds of one type or another in the 1880s with the electric arc, while in the United States the first patent for arc welding was given to Charles Coffin of Detroit in 1889.[1]

The figures to follow in this chapter show the necessity of supplying additional metal to the joints being welded to give satisfactory connections. In electric-arc welding the metallic rod, which is used as the electrode, melts off into the joint as it is being made. When gas welding is used it is necessary to introduce a metal rod known as a *filler* or *welding rod*.

In gas welding a mixture of oxygen and some suitable type of gas is burned at the tip of a torch or blowpipe held in the welder's hand or by an automatic machine. The gas used in structural welding is probably acetylene and the process is called oxyacteylene welding. The flame produced can be used for flame cutting of metals as well as for welding. Gas welding is rather easy to learn and the equipment used is rather inexpensive. It is, however, a rather slow process as compared to other means of welding, and is normally used for repair and maintenance work and not for the fabrication and erection of large steel structures.

In arc welding an electric arc is formed between the pieces being welded and an electrode held in the operator's hand with some type of holder or by an automatic machine. The arc is a continuous spark which upon contact brings the electrode and the pieces being welded to the melting point. The resistance of the air or gas between the electrode and the pieces being welded changes the electrical energy into heat. A temperature of somewhere between 6000 and 10,000°F is produced in the arc. As the end of the electrode melts small droplets or globules of the molten metal are formed and are actually forced by the arc across to the pieces being connected penetrating the molten metal to become a part of the weld. The amount of penetration can actually be controlled by the amount of current consumed. Since the molten droplets of the electrodes are actually propelled to the weld, arc welding can be successfully used for overhead work.

A pool of molten steel can hold a fairly large amount of gases in solution and if not protected from the surrounding air will chemically combine with oxygen and nitrogen. After cooling the welds will be relatively porous due to the little pockets formed by the gases. Such welds are relatively brittle and have much less resistance to corrosion. A weld can be shielded by using an electrode coated with certain mineral compounds.

[1]Lincoln Electric Company, *Procedure Handbook of Arc Welding Design and Practice*, 11th ed., 1957, Part I.

The electric arc causes the coating to melt and creates an inert gas or vapor around the area being welded. The vapor acts as a shield around the molten metal and keeps it from coming freely in contact with the surrounding air. It also deposits a slag in the molten metal, which has less density than the base metal and comes to the surface to protect the weld from the air while the weld cools. After cooling, the slag can easily be removed by peening and wire brushing (such removal being absolutely necessary before painting or application of another weld layer). A picture showing the elements of the shielded arc welding process is shown in Fig. 11-1. This figure is taken from the *Procedure Handbook of Arc Welding Design and Practice* published by the Lincoln Electric Company. Shielded metal arc welding is frequently abbreviated herein with the letters SMAW.

The type of welding electrode used is very important as it decidedly affects the weld properties such as strength, ductility, and corrosion resistance. There are quite a number of different types of electrodes manufactured, the type to be used for a certain job being dependent upon the type of metal being welded, the amount of material which needs to be added, the position of the work, etc. The electrodes fall into two general classes—the *lightly coated electrodes* and the *heavily coated electrodes*.

The heavily coated electrodes are normally used in structural welding because the melting of their coatings produces very satisfactory vapor shields around the work as well as slag in the weld. The resulting welds are stronger, more resistant to corrosion, and more ductile than are those produced with lightly coated electrodes. When the lightly coated electrodes are used, no attempt is made to prevent oxidation and no slag is formed. The electrodes are lightly coated with some arc stabilizing chemical such as lime.

Another type of welding is the submerged (or hidden) arc welding. In this method (frequently labeled SAW herein) the arc is covered with a

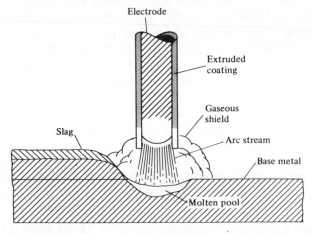

Figure 11-1 Elements of the shielded metal arc welding process (SMAW).

Lincoln ML-3 Squirtwelder mounted on a self-propelled trackless trailer deposits this $\frac{1}{4}$-in. web-to-flange weld at 28 in./min. (Courtesy of The Lincoln Electric Company.)

mound of granular fusible material and thus hidden from view. A bare metal electrode is fed from a reel and melted and deposited as filler material. SAW welds are quickly and efficiently made and are of high quality exhibiting high impact strength and corrosion resistance and good ductility. Furthermore they provide deeper penetration with the result that the area effective in resisting loads is larger.

11-4. WELDING INSPECTION

Three steps must be taken to insure good welding for a particular job. These are: (1) establishing good welding procedures, (2) use of prequalified welders, and (3) employment of competent inspectors in shop and field.

When the procedures established by the AWS and AISC for good welding are followed and when welders are used who have previously been required to prove their ability, good results will probably be obtained. However, to make absolutely sure, well-qualified inspectors are needed.

Good welding procedure involves the selection of proper electrodes, current, and voltage; the properties of base metal and filler; and the position of welding, to name only a few factors. The usual practice for large jobs is to employ welders who have certificates showing their qualifications. In addition, it is not a bad practice to have each person make an identifying mark on each weld so that those persons frequently doing poor work can be located. This practice probably improves the general quality of work performed.

Visual Inspection

Another factor that will cause welders to perform better work is just the presence of an inspector whom they feel knows good welding when he sees it. For a person to make a good inspector it is desirable for him to have done welding himself and to have spent much time observing the work of good welders. From this experience he should be able to know if a welder is obtaining satisfactory fusion and penetration. He should be able to recognize good welds as to their shape, size, and general appearance. For instance the metal in a good weld should approximate its original color after it has cooled. If it has been overheated it may have a rusty and reddish-looking color. He can use various scales and gages to check the sizes and shapes of welds.

Visual inspection by a good person will probably give a good indication of the quality of welds but is not a perfect source of information as to the subsurface condition of the weld. There are several methods for determining the internal soundness of a weld. These include the use of penetrating dyes and magnetic particles, ultrasonic testing, and radiographic procedures. These methods can be used to detect internal defects such as porosity, weld penetration, and presence of slag.

Liquid Penetrants

Various types of dyes can be spread over weld surfaces and they will penetrate into surface cracks of the weld. After the dye has penetrated into the crack the excess surface material is wiped off and a powdery developer is used to draw the dye out of the cracks. The outlines of the cracks can then be seen with the eye. Several variations of this method are used to improve the visibility of the defects such as the use of fluorescent dyes. After the dye is drawn from the cracks the cracks are made to stand out brightly by examination under black light.[2]

Magnetic Particles

In this method the weld being inspected is magnetized electrically. Cracks that are at or near the surface of the weld cause north and south poles to form on each side of the cracks. Dry iron powdered filings or a liquid suspension of particles are placed on the weld. The patterns of these

[2] James Hughes, "It's Superinspector," *Steelways* **25**, no. 4, (New York: American Iron and Steel Institute, September/October, 1969), pp. 19–21.

particles formed when many of them cling to the cracks show the locations of cracks and indicate their size and shape. A disadvantage of this method is that if multilayer welds are used the method has to be applied to each layer.

Ultrasonic Testing

In recent years the steel industry has applied ultrasonics to the manufacture of steel. Although the equipment is expensive the method is quite useful in welding inspection as well. Sound waves are sent through the material being tested and are reflected from the opposite side of the material. These reflections are shown on a cathode ray tube. Defects in the weld will affect the time of the sound transmission. The operator can read the picture on the tube and then locate flaws and learn how severe they are.

Radiographic Procedures

The more expensive radiographic methods can be used to check occasional welds in important structures. From these tests it is possible to make good estimates of the percentage of bad welds in a structure. The use of portable x-ray machines where access is not a problem and the use of radium or radioactive cobalt for making pictures are excellent but expensive methods of testing welds. These methods are satisfactory for butt welds (such as for the welding of important stainless steel piping at nuclear energy projects) but are not satisfactory for fillet welds because the pictures are difficult to interpret. A further disadvantage of these methods is the radioactive danger. Careful procedures have to be used to protect the technicians as well as nearby workers. On the average construction job this danger probably requires night inspection of welds when only a few workers are near the inspection area. (Normally a very large job would be required before the use of the extremely expensive radium or cobalt pills could be justified.)

A properly welded connection can always be made much stronger (perhaps $1\frac{1}{2}$ or 2 times) than the plates being connected. As a result, the factor of safety is much higher than is required by the specifications. The reasons for this extra strength are as follows: the electrode wire is made from premium steel, the metal is melted electrically (as is done in the manufacture of high quality steels) and the cooling rate is quite rapid. As a result of these facts it is probably a rare occasion for a welder to make a weld of less strength than required by the design.

11-5. CLASSIFICATION OF WELDS

There are three separate classifications of welds described in the following paragraphs. These classifications are based upon the types of welds made, positions of welds, and types of joints.

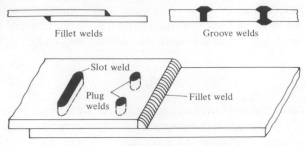

Figure 11-2

Type of Weld

The two main types of welds are the *fillet welds* and the *groove welds*. In addition there are plug and slot welds which are not as common in structural work. These four types of welds are shown in Fig. 11-2.

The fillet welds will be shown to be weaker than groove welds; however, most structural connections (about 80%) are made with fillet welds. Any person who has had experience in steel structures will understand why fillet welds are more common than are groove welds. Groove welds (which are welds made in grooves between the members to be joined) are used when the members to be connected are lined up in the same plane. To use them in every situation would mean that the members would have to fit almost perfectly and unfortunately the average steel structure does not fit together in that manner. Many students have seen steel workers pulling and ramming steel members to get them in position. When members are allowed to lap over each other larger tolerances are allowable in erection and fillet welds are the welds used. Nevertheless groove welds are quite common for many connections such as column splices, butting of beam flanges to columns, etc., and they comprise about 15% of structural welding.

A plug weld is a circular weld passing though one member to another and joining the two together. A slot weld is a weld formed in a slot or elongated hole which joins one member to the other member through the slot. The slot may be partly or fully filled with weld material. These two expensive types of welds may occasionally be used when members lap over each other and the desired length of fillet welds cannot be obtained. They may also be used to stitch together parts of a member as the fastening of cover plates to a built-up member.

Position

Welds are referred to as being flat, horizontal, vertical, and overhead. They are listed in the preceding sentence in order of their economy with the flat welds being the most economical and the overhead welds being the most expensive. Although the flat welds can often be made with an automatic

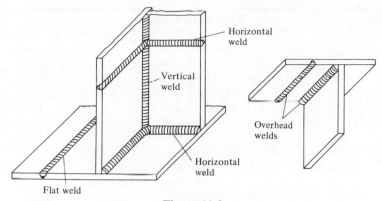

Figure 11-3

machine, most structural welding is done by hand. It has previously been indicated that the assistance of gravity is not necessary for the forming of good welds but it does speed up the process. The globules of the molten electrodes can be forced into the overhead welds against gravity and good welds will result but they are slow and expensive to make so it is desirable to avoid them whenever possible. These types of welds are shown in Fig. 11-3.

Type of Joint

Welds can be further classified according to the type of joint used as being: butt, lap, tee, edge, corner, etc. These joint types are shown in Fig. 11-4.

11-6. WELDING SYMBOLS

Figure 11-5 presents the method of making welding symbols, developed by the American Welding Society.[3] With this excellent shorthand system a

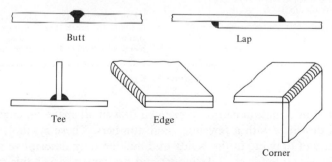

Figure 11-4

[3] Lincoln Electric Company, *Procedure Handbook of Arc Welding Design and Practice*, 11th ed., 1957. Courtesy of the publisher.

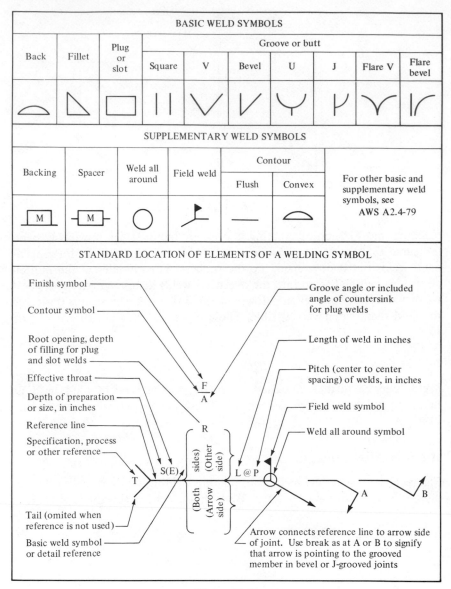

Figure 11-5

great deal of information can be presented in a small space on engineering plans and drawings with a few lines and numbers. These symbols remove the necessity of drawing in the welds and making long descriptive notes. It is certainly desirable for steel designers and draftsmen to use this standard system. Should most of the welds on a drawing be of the same size, a note to that effect can be given and the symbols omitted, except for the off-size welds.

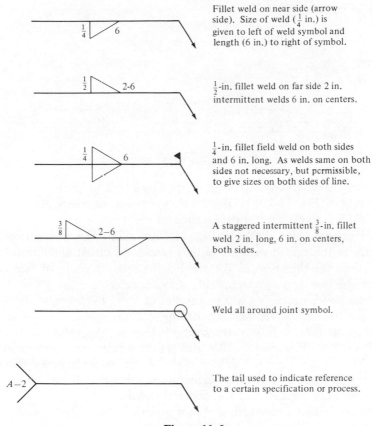

Fillet weld on near side (arrow side). Size of weld ($\frac{1}{4}$ in.) is given to left of weld symbol and length (6 in.) to right of symbol.

$\frac{1}{2}$-in. fillet weld on far side 2 in. intermittent welds 6 in. on centers.

$\frac{1}{4}$-in. fillet field weld on both sides and 6 in. long. As welds same on both sides not necessary, but permissible, to give sizes on both sides of line.

A staggered intermittent $\frac{3}{8}$-in. fillet weld 2 in. long, 6 in. on centers, both sides.

Weld all around joint symbol.

The tail used to indicate reference to a certain specification or process.

Figure 11-6

The purpose of this section is not to teach the student every possible type of symbol but rather to give him a general idea of the appearance of welding symbols and the information which they can show. He can easily refer to the detailed information published by the AWS and reprinted in many handbooks (including the AISC Manual). At first glance the information presented in Fig. 11-5 is probably quite confusing to the student. For this reason a few very common symbols for fillet welds are presented in Fig. 11-6, together with an explanation of each.

11-7. GROOVE WELDS

When complete penetration groove welds are subjected to axial tension or axial compression the weld stress is assumed to equal the load divided by the net area of the weld. Three types of groove welds are shown in Fig. 11-7. The square groove joint, shown in part (a) of the figure, is used to connect relatively thin material up to roughly $\frac{5}{16}$ in. (8 mm) thickness. As the material becomes thicker it is necessary to use the single-vee groove

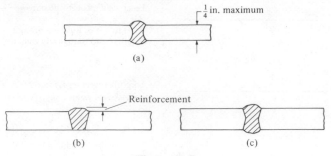

Figure 11-7

welds and the double-vee groove welds illustrated in parts (b) and (c) respectively of Fig. 11-7. For these two welds the members are bevelled before welding to permit full penetration of the weld.

The groove welds shown in Fig. 11-7 are said to have *reinforcement*. Reinforcement is added weld metal that causes the throat dimension to be greater than the thickness of the welded material. Because of reinforcement or the lack of it, groove welds may be referred to as 100%, 125%, 150%, etc., welds according to the amount of extra thickness at the weld. There are two major reasons for having reinforcement. These are: (1) reinforcement gives a little extra strength because the extra metal takes care of pits and other irregularities and (2) the welder can easily make the weld a little thicker than the welded material. He would have a difficult if not impossible task in making a perfectly smooth weld with no places that were thinner nor thicker than the material welded.

Reinforcement undoubtedly makes groove welds stronger and better when they are to be subjected to relatively static loads. When the connection is to be subjected to repeated and vibrating loads, however, reinforcement is not as satisfactory because stress concentrations seem to develop in the reinforcement and contribute to earlier failure. For such cases as these a common practice is to provide reinforcement and grind it off flush with the material being connected.

In Fig. 11-8 some of the edge preparations that may be necessary for groove welds are shown. In part (a) a bevel with a feathered edge is shown. When feathered edges are used there is a problem with burn-through. This may be lessened if a *land* is used such as shown in part (b) of the figure or a back-up strip as shown in part (c). The back-up strip is often a $\frac{1}{4}$-in. copper plate. Weld metal does not stick to copper and copper also has a very high conductivity which is useful in carrying away excess heat and reducing distortion. Sometimes steel back-up strips are used but they will become a part of the weld and are left in place. A land should not be used together with a back-up strip because there is a high possibility that a gas pocket might be formed, preventing full penetration. When double bevels are used as shown in part (d) of the figure spacers are sometimes provided to prevent burn-through. The spacers are removed after one side is welded.

Figure 11-8 Edge preparation for groove welds. (a) Bevel with feathered edge. (b) Bevel with a land. (c) Bevel with a back-up plate. (d) Double bevel with a spacer.

From the standpoints of strength, resistance to impact and stress repetition, and amount of filler metal required, groove welds are much to be preferred to fillet welds. From other standpoints, however, they are not so attractive and the vast majority of structural welding is fillet welding. Groove welds have higher residual stresses, and the preparations (such as the scarfing and veeing) of the edges of members for groove welds are expensive, but the major disadvantages probably lie in the problems involved in getting the pieces to fit together in the field. The advantages of fillet welds in this respect were described in Section 11-5. For these reasons field butt joints are not used very often, except on small jobs and where members may be fabricated a little long and cut in the field to the lengths necessary for precise fitting.

Sometimes connections are designed where the groove welds do not extend for the full thicknesses of the parts being connected. These welds are referred to as *partial-penetration groove welds*. Special design require-ments are given in specifications for these welds.

11-8. FILLET WELDS

Tests have shown that fillet welds are stronger in tension and compression than they are in shear, so the controlling fillet weld stresses given by the various specifications are shearing stresses. When practical, it is desirable to try to arrange welded connections so they will be subjected to shearing stresses only and not to a combination of shear and tension or shear and compression.

Fillet welds when tested to failure seem to fail by shear at angles of about 45° through the throat. Their strength is therefore assumed to equal the allowable shearing stress times the theoretical throat area of the weld. The theoretical throats of several fillet welds are shown in Fig. 11-9. The throat area equals the theoretical throat distance times the length of the weld. In this figure the root of the weld is the point where the faces of

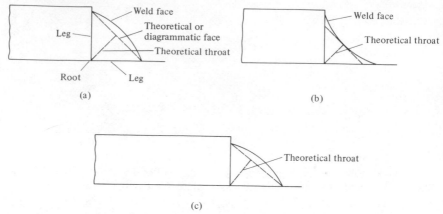

Figure 11-9 (a) Convex surface. (b) Concave surface. (c) Unequal leg fillet weld.

the original metal pieces intersect, and the theoretical throat of the weld is the shortest distance from the root of the weld to its diagrammatic face.

For the 45° or equal leg fillet the throat dimension is 0.707 times the leg of the weld, but it has a different value for fillet welds with unequal legs. The desirable fillet weld has a flat or slightly convex surface, although the convexity of the weld does not add to its calculated strength. At first glance the concave surface would appear to give the ideal fillet weld shape because stresses could apparently flow smoothly and evenly around the corner with little stress concentration. Years of experience, however, have shown that single-pass fillet welds of a concave shape have a greater tendency to crack upon cooling and this factor has proved to be of greater importance than the smoother stress distribution of convex types.

When a concave weld shrinks the surface is placed in tension tending to cause cracks, whereas when the surface of a convex weld shrinks it does not place the outer surface in tension. Rather as the face shortens it is placed in compression.

Another item of importance pertaining to the shape of fillet welds is the angle of the weld with the pieces being welded. The desirable value of this angle is in the vicinity of 45°. For 45° fillet welds the leg sizes are equal and such welds are referred to by the leg sizes (as a $\frac{1}{4}$-in. fillet weld). Should the leg sizes be different (not a 45° weld) both leg sizes are given in describing the weld (as a $\frac{3}{8}$ by $\frac{1}{2}$-in. fillet weld).

The automatic submerged arc welding method (SAW) provides a deeper penetration than does the usual shielded arc welding process. As a result the AISC permits the designer to use a larger throat area for welds made by this process. In Section 1.14.6.2 the AISC states that the effective throat thickness for SAW fillet welds with legs $\frac{3}{8}$ in. or less may equal the leg sizes. For larger leg sizes the effective throat thicknesses are considered to equal the actual throat thicknesses plus 0.11 in.

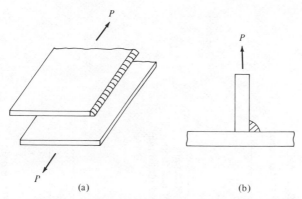

Figure 11-10 (a) Longitudinal fillet. (b) Transverse fillet.

11-9. STRENGTH OF WELDS

For this discussion reference is made to Fig. 11-10. As previously indicated the stress in a weld is considered to equal the load P divided by the effective throat area of the weld. This method of determining the strength of fillet welds is used regardless of the direction of load. Tests have shown that transverse fillets are without question about a third stronger than are longitudinal fillets, but this fact is not recognized by most specifications in order to simplify design calculations. One reason why transverse fillet welds are stronger is because they are more uniformly stressed for their entire length, while the longitudinal fillets are stressed unevenly due to varying deformations along the length of the weld. Another reason for their greater strength is given by tests which show failure occurs at an angle other than 45°, giving them a larger effective throat area.

11-10. AISC REQUIREMENTS

When welds are made the electrode material should have properties of the base metal. If the properties are comparable the weld metal is referred to as the "matching" base metal. Table 4.1.1 of AWS D1.1-77 gives details concerning "matching" weld metals.

Table 11-1 (which is reproduced from Table 1.5.3 of the AISC Specification) gives allowable stresses for various types of welds. Should a full penetration groove weld be used, the allowable stress in the weld is the same as in the base material. If the connection is in compression normal to the effective area of the weld metal a weld metal with a strength less than the "matching" weld metal may be used. If in tension, a "matching" weld metal must be used.

For fillet welds the allowable stresses for shear on the effective area of the welds are equal to 0.30 times the nominal tensile strength of the weld metal but the stress in the base material may not be greater than $0.60F_y$ for

Table 11-1 Allowable Stress on Welds

Type of weld and stress	Allowable stress	Required weld strength level
Complete-penetration groove welds		
Tension normal to effective area	Same as base metal	"Matching" weld metal must be used.
Compression normal to effective area	Same as base metal	
Tension or compression parallel to axis of weld	Same as base metal	Weld metal with a strength level equal to or less than "matching" weld metal may be used.
Shear on effective area	$0.30 \times$ nominal tensile strength of weld metal (ksi), except shear stress on base metal shall not exceed $0.40 \times$ yield stress of base metal	
Partial-penetration groove welds		
Compression normal to effective area	Same as base metal	Weld metal with a strength level equal to or less than "matching" weld metal may be used.
Tension or compression parallel to axis of weld	Same as base metal	

Kind of stress	Allowable stress	Required weld strength level
Shear parallel to axis of weld	0.30 × nominal tensile strength of weld metal (ksi), except shear stress on base metal shall not exceed 0.40 × yield stress of base metal	Weld metal with a strength level equal to or less than "matching" weld metal may be used
Tension normal to effective area	0.30 × nominal tensile strength of weld metal (ksi), except tensile stress on base metal shall not exceed 0.60 × yield stress of base metal	
Fillet welds		
Shear on effective area	0.30 × nominal tensile strength of weld metal (ksi), except shear stress on base metal shall not exceed 0.40 × yield stress of base metal	
Tension or compression parallel to axis of weld	Same as base metal	
Plug and slot welds		
Shear parallel to faying surfaces (on effective area)	0.30 × nominal tensile strength of weld metal (ksi), except shear stress on base metal shall not exceed 0.40 × yield stress of base metal	Weld metal with a strength level equal to or less than "matching" weld metal may be used

SOURCE: *Specification for the Design, Fabrication, and Erection of Structural Steel for Buildings,* November 1, 1973, AISC, Inc., Table 1.5.3.

tension nor $0.40F_y$ for shear. Values are also given in Table 11-1 for partial-penetration groove welds and for plug and slot welds.

The filler metal electrodes for shielded arc welding are listed as E60XX, E70XX, etc. In this classification the letter E represents an electrode while the first set of digits (as 60 or 70) indicates the minimum tensile strength of the weld in kips per square inch. The remaining digits designate the position, current, and other necessary information which are to be employed with a certain electrode.

In addition to the allowable stresses given in Table 11-1 there are several other AISC provisions applying to welding. Among the more important are the following.

1. The minimum length of a fillet weld may not be less than four times the nominal leg size of the weld. Should its length actually be less than this value the weld size considered effective must be reduced to one-quarter of the weld length.

2. The maximum size of a fillet weld along edges of material less than $\frac{1}{4}$ in. thick equals the material thickness. For thicker material it may not be larger than the material thickness less $\frac{1}{16}$ in., unless the weld is specially built out to give a full throat thickness.

3. The minimum size fillet welds are given in Table 1.17.2A of the AISC Specification and vary from $\frac{1}{8}$ in. for $\frac{1}{4}$ in. or less thickness of material up to $\frac{5}{16}$ in. for material over $\frac{3}{4}$ in. in thickness. The smallest practical weld size is about $\frac{1}{8}$ in. and the most economical size is probably about $\frac{5}{16}$ in. The $\frac{5}{16}$-in. weld is about the largest size that can be made in one pass with the shielded arc welded process (SMAW) and $\frac{1}{2}$ in. with the submerged arc process (SAW).

4. When practical, end returns (sometimes called boxing) should be made for fillet welds as shown in Fig. 11-11. The length of returns should not be less than twice the nominal size of the weld. When they are not used it is considered good practice by some designers to subtract two times the weld size from the effective weld length. End returns are useful in reducing the high stress concentrations which occur at the ends of welds, particularly for connections where there is considerable vibration and eccentricity

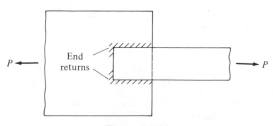

Figure 11-11

The all welded 56 story Toronto Dominion Bank Tower. (Courtesy of The Lincoln Electric Company.)

of load. The AISC says (Section 1.14.6.2) that the effective length of fillet welds includes the length of the end returns used.

 5. When longitudinal fillet welds are used alone for the connection of plates or bars, their length may not be less than the perpendicular distance between them. Furthermore, the distance between fillet welds may not be greater than 8 in. for end connections unless transverse bending in the weld is otherwise prevented.

11-11. DESIGN OF SIMPLE FILLET WELDS

Examples 11-1 through 11-3 illustrate the calculations necessary to determine the strength of various fillet welded connections while Example 11-4 presents the design of such a connection. In these and other problems weld lengths are selected no closer than to the nearest $\frac{1}{4}$ in., because closer work cannot be expected in shop or field.

Example 11-1

Determine the allowable shear force that may be applied to a 1 in. length of a $\frac{5}{16}$-in. fillet weld using (a) the shielded metal arc process (SMAW) and (b) the submerged arc process (SAW). Use E70 electrodes with a minimum tensile strength of 70 ksi and the AISC Specification.

SOLUTION

(a) Shielded metal arc process

effective throat thickness of weld $= (0.707)\left(\frac{5}{16}\right) = 0.221$ in.

allowable capacity $= (0.221)(0.30 \times 70)$

$$= 4.64 \text{ k/in. } (0.813 \text{ kN/mm})$$

(b) Submerged arc process

From AISC 1.14.6.2 the effective throat thickness of weld $= \frac{5}{16}$ in.

allowable capacity $= \left(\frac{5}{16}\right)(0.30 \times 70) = 6.56 \text{ k/in. } (1.149 \text{ kN/mm})$

Fillet welds may not be designed using an allowable stress that is greater than permitted on the adjacent members being connected. For instance, the allowable shear stress on the effective area of fillet welds is 0.30 times the tensile strength of the electrode but may not exceed the allowable stress in the base material ($0.60F_y$ in tension or $0.40F_y$ in shear). It should be remembered from the discussion earlier in this section that fillet welds are assumed in design to transmit loads by shear on the effective area regardless of the direction of the loads or of the position of the welds at the connections.

Examples 11-2 and 11-3 illustrate the calculations necessary to determine the allowable loads that can be applied to plates connected with SMAW and SAW fillet welds. In each of these examples the shearing strength per inch of the welds controls and is multiplied by the total length of the welds to give the total capacity of the connections. If the shear capacity of the plates ($0.40F_y t$) had controlled it, it also would have been multiplied by the total length of the welds (and not by the width of the plates) to give the total capacity.

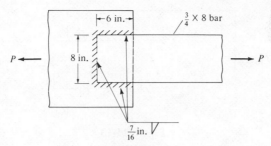

Figure 11-12

Example 11-2

What is the allowable capacity of the connection shown in Fig. 11-12 if A36 steel, the AISC Specification, and E70 electrodes are used? The $\frac{7}{16}$-in. fillet welds shown were made by the SMAW process.

SOLUTION

$$\text{effective throat thickness} = t_e = (0.707)\left(\tfrac{7}{16}\right) = 0.309 \text{ in.}$$

$$\text{shear capacity of weld/in.} = (0.309)(0.30 \times 70) = 6.489 \text{ k/in.} \leftarrow$$

$$\text{max shear capacity of plates} = 0.40 F_y t = (0.40)(36)\left(\tfrac{3}{4}\right) = 10.8 \text{ k/in.}$$

$$\text{allowable tensile capacity of weld} = (20)(6.489) = 129.8 \text{ k}$$

$$\text{allowable tensile capacity of PL} = \left(\tfrac{3}{4} \times 8\right)(0.60 \times 36) = 129.6 \text{ k} \leftarrow$$

$$\underline{\underline{P = 129.6 \text{ k}}}$$

Example 11-3

Repeat Example 11-2 if SAW welds are used.

SOLUTION

$$\text{effective throat thickness} = t_e = (0.707)\left(\tfrac{7}{16}\right) + 0.11 = 0.419 \text{ in.}$$

$$\text{shear capacity of weld/in.} = (0.419)(0.30 \times 70) = 8.799 \text{ k/in.} \leftarrow$$

$$\text{max shear capacity of plates} = (0.40)(36)\left(\tfrac{3}{4}\right) = 10.8 \text{ k/in.}$$

$$\text{allowable shear capacity of weld} = (20)(8.799) = 176 \text{ k}$$

$$\text{allowable tensile capacity of PL} = \left(\tfrac{3}{4} \times 8\right)(0.60 \times 36) = 129.6 \text{ k} \leftarrow$$

$$\underline{\underline{P = 129.6 \text{ k}}}$$

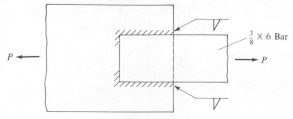

Figure 11-13

Example 11-4

Using A36 steel, the AISC Specification and E70 electrodes design SMAW fillet welds to resist a full capacity load on the $\frac{3}{8} \times 6$ in. member shown in Fig. 11-13.

SOLUTION

$$P = \left(\tfrac{3}{8}\right)(6)(22) = 49.5 \text{ k}$$

$$\text{max weld size} = \tfrac{3}{8} - \tfrac{1}{16} = \tfrac{5}{16} \text{ in.} \qquad \text{(AISC Section 1.17.3)}$$

$$\text{min weld size} = \tfrac{3}{16} \text{ in.} \qquad \text{(AISC Table 1.17.2A)}$$

Use $\tfrac{5}{16}$-in. weld

$$\text{effective thickness} = t_e = (0.707)\left(\tfrac{5}{16}\right) = 0.221 \text{ in.}$$

$$\text{capacity of weld/in.} = (0.221)(0.30 \times 70) = 4.64 \text{ k/in.} \leftarrow$$

$$\text{max shear capacity of PL} = (0.40)(36)\left(\tfrac{3}{8}\right) = 5.4 \text{ k/in.}$$

$$\text{length reqd.} = \frac{49.5}{4.64} = 10.67 \text{ in.}$$

use end return not less than $2 \times \tfrac{5}{16} = \tfrac{10}{16}$ $\left(\text{say 1 in.}\right)$

leaving $10.67 - 2 = 8.67$ in. or $4\tfrac{1}{2}$ in. each side

However use $6 - 1 = 5$-in. welds each side as required by AISC Section 1.17.4

On some occasions the lengths available for the usual longitudinal fillet welds are not sufficient for the load to be resisted. For the situation shown in Fig. 11-14 it may be possible to develop sufficient strength by welding along the back of the channel at the edge of the plate if sufficient space is available. The dashed lines shown in this figure show this weld.

Another possibility is the use of slot welds as illustrated in Example 11-5. There are several AISC requirements pertaining to slot welds which need to be mentioned here. This specification says that the width of a slot may not be less than the member thickness $+ \tfrac{5}{16}$ in. (rounded off to the next greater odd $\tfrac{1}{16}$ in. since structural punches are made in odd 16th's

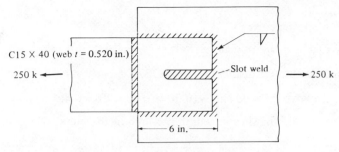

Figure 11-14

diameters) nor may it be greater than $2\frac{1}{4}$ times the weld thickness. For members up to $\frac{5}{8}$ in. thickness the weld thickness must equal the plate thickness, and for members greater than $\frac{5}{8}$ in. thickness the weld thickness may not be less than one-half the member thickness nor $\frac{5}{8}$ in. The maximum length permitted for slot welds is 10 times the weld thickness. The limitations given in specifications for the maximum sizes of plug or slot welds are caused by the detrimental shrinkage which occurs around these types of welds when they exceed certain sizes. Should holes or slots larger than those specified be used it is desirable to use fillet welds around the borders of the holes or slots rather than using a slot or plug weld. Slot and plug welds are normally used in conjunction with fillet welds in lap joints. Sometimes plug welds are used to fill in the holes temporarily used for erection bolts for beam and column connections. They may or may not be included in the calculated strength of these joints.

The strength of a plug or slot weld is equal to its allowable stress times its nominal area in the shearing plane. This area, which is commonly called the *faying surface*, is equal to the area of contact at the base of the plug or slot. The length of a slot weld can then be determined from the following.

$$L = \frac{\text{load}}{(\text{width})(\text{allow stress})}$$

Example 11-5 illustrates the design of the welds necessary to connect a channel to a plate. The calculations show that the ordinary side and end fillet welds do not provide sufficient strength. It is decided to use a slot weld to provide the remaining load resistance needed.

Example 11-5

Design SMAW fillet welds to connect a C15×40 to the plate shown in Fig. 11-14. The load to be resisted is 250 k and E70 electrodes are to be used. As shown in the figure the channel may lap over the plate by only 6 in. due to space limitations. See AISC Specification 1.17.9.

SOLUTION

Due to limited space use

$$\text{max weld size} = \text{web } t - \tfrac{1}{16} = \tfrac{1}{2} - \tfrac{1}{16} = \tfrac{7}{16} \text{ in.}$$

$$\text{effective throat thickness} = t_e = (0.707)\left(\tfrac{7}{16}\right) = 0.309 \text{ in.}$$

$$\text{capacity of weld/in.} = (0.309)(0.30 \times 70) = 6.489 \text{ k/in.} \leftarrow$$

but may not exceed the shear strength per inch of the channel $= (0.40)(36)(\tfrac{1}{2})$ $= 7.2$ k/in.

$$\text{length reqd.} = \frac{250}{6.489} = 38.53 \text{ in.} > 27 \text{ in. available}$$

Therefore use a slot weld

$$\text{min width of slot} = 0.520 + \tfrac{5}{16} = \tfrac{13}{16} \text{ in.}$$

$$\text{max width} = 2\tfrac{1}{4} \times \text{weld thickness}$$

$$= \left(2\tfrac{1}{4}\right)(\text{web } t \text{ of channel}) = \left(2\tfrac{1}{4}\right)\left(\tfrac{1}{2}\right)$$

$$= 1\tfrac{1}{8} \text{ say } \tfrac{17}{16} \left(\text{to odd } \tfrac{1}{16}\right)$$

Use $\tfrac{15}{16}$ in.

$$\text{capacity of } \tfrac{7}{16}\text{-in. fillet weld} = (6.489)\left(6 + 6 + 15 - \tfrac{15}{16}\right) = 169.1 \text{ k}$$

load to be resisted by slot weld $= 250 - 169.1 = 80.9$ k

$$\text{length reqd. for slot weld} = \frac{80.9}{\left(\tfrac{15}{16}\right)(0.30 \times 70)} = 4.11 \text{ in.}$$

Say $4\tfrac{1}{2}$ in.

$$\text{max length permitted by AISC} = (10)\left(\tfrac{1}{2}\right)$$

$$= 5.00 \text{ in.} > 4\tfrac{1}{2} \text{ in.} \qquad \text{OK}$$

Use $\tfrac{15}{16} \times 4\tfrac{1}{2}$ in. slot weld

ALTERNATE SOLUTION

Should space have been available on the back of the channel next to the plate a $\tfrac{7}{16}$-in. fillet weld would carry $(15)(6.489) = 97.3$ k > 80.9 k.

11-12.   DESIGN OF FILLET WELDS FOR TRUSS MEMBERS

Should the members of a welded truss consist of single angles, double angles, or similar shapes and be subjected to static axial loads only, the AISC permits their connections to be designed by the same procedures described in the preceding section. The designers can select the weld size, calculate the total length of the weld required, and place the welds around

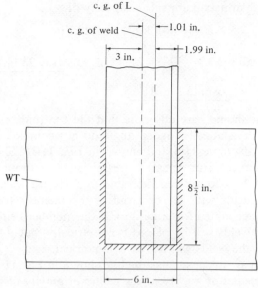

Figure 11-15

the member ends as they see fit.[4] (It would not be logical, of course, to place the weld all on one side of a member such as for the angle of Fig. 11-15 because of the rotation possibility.) Example 11-6 illustrates the simple calculations involved in designing the welds for the ends of a truss member.

Example 11-6

Using the AISC Specification, A36 steel, and the E70 electrodes design side and end fillet SMAW welds for the full capacity of a $6 \times 4 \times \frac{1}{2}$-in. angle tension member with the long leg connected. Assume static load.

SOLUTION

$$\text{tensile capacity of } L = (4.75)(22) = 104.5 \text{ k}$$

$$\text{max weld size} = \frac{1}{2} - \frac{1}{16} = \frac{7}{16} \text{ in.}$$

$$\text{min weld size} = \frac{3}{16} \text{ in.} \qquad \text{(from AISC Table 1.17.2A)}$$

Use $\frac{5}{16}$-in. weld as maximum size which can be made in one pass

$$\text{effective thickness} = t_e = (0.707)\left(\frac{5}{16}\right) = 0.221 \text{ in.}$$

$$\text{capacity of weld/in.} = (0.221)(0.30 \times 70) = 4.64 \text{ k/in.} \leftarrow$$

[4]*The Welding Journal*, January, 1942, pp. 44–45.

but not greater than maximum shear capacity of

$$L = (0.40)(36)\left(\tfrac{1}{2}\right) = 7.2 \text{ k/in.}$$

$$\text{length reqd.} = \frac{104.5}{4.64} = 22.52 \text{ in.} \qquad \underline{\text{(say 23 in.)}}$$

Place welds as shown in Fig. 11-15

The student should carefully note that the centroid of the welds and the centroid of the statically loaded angle do not coincide in the connection selected in Example 11-6 and shown in Fig. 11-15. Should a welded connection be subjected to repeated stresses (such as those occurring in a bridge member), it is considered desirable to place the welds so that their centroid will coincide with the centroid of the member (or the resulting torsion must be accounted for in design). If the member being connected is symmetrical the welds will be placed symmetrically, but if the member is not symmetrical the welds will not be symmetrical.

The force in an angle, such as the one shown in Fig. 11-16, is assumed to act along its center of gravity. If the center of gravity of weld resistance is to coincide with the angle force it must be asymmetrically placed, or in this figure L_1 must be longer than L_2. (When angles are connected by rivets or bolts there is usually an appreciable amount of eccentricity, but in a welded joint eccentricity can be fairly well eliminated.) The information necessary to handle this type of weld design can be easily expressed in equation form but only the theory behind the equations is presented here.

For the angle shown in Fig. 11-16 the force acting along line L_2 (designated here as P_2) can be determined by taking moments about L_1. The member force and the weld resistance are to coincide and the moments of the two about any point must be zero. If moments are taken about L_1 the force P_1 (which acts along line L_1) will be eliminated from the equation and P_2 can be determined. In a similar manner P_1 can be determined by taking moments along L_2 or by $\Sigma V = 0$. Example 11-7 illustrates the design of fillet welds of this type. A similar problem is handled in Example 11-8 except an end fillet weld is included. The center

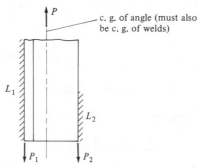

Figure 11-16

of gravity and resistance of the end weld are known and can be easily included in the moment equations.

There are other possible solutions for the design of the welds for the angle considered in these two examples. Although the $\frac{7}{16}$-in. weld is the largest one permitted at the rounded edges of the $\frac{1}{2}$-in. angle and at its end, a larger weld could be used on the other side next to the outstanding leg. From a practical point of view, however, the welds should be the same size because different size welds slow the welder down in that he has to change electrodes to make different sizes.

Example 11-7

Using the AISC Specification, A36 steel, E70 electrodes, and the SMAW process design side fillet welds for the full capacity of the $5 \times 3 \times \frac{1}{2}$-in. angle tension member shown in Fig. 11-17. Assume the member is subjected to repeated stress variations, making any connection eccentricity undesirable.

SOLUTION

$$\text{tensile capacity of } L = (22)(3.75) = 82.50 \text{ k}$$

$$\text{max weld size} = \tfrac{1}{2} - \tfrac{1}{16} = \tfrac{7}{16} \text{ in.}$$

Use $\frac{5}{16}$-in. weld

$$\text{effective throat thickness} = t_e = (0.707)\left(\tfrac{5}{16}\right) = 0.221 \text{ in.}$$

$$\text{capacity of weld/in.} = (0.221)(0.30 \times 70) = 4.641 \text{ k/in.} \leftarrow$$

but may not exceed $(0.40)(36)(\tfrac{1}{2}) = 7.2 \text{ k/in.}$

$$\text{total weld length reqd.} = \frac{82.50}{4.641} = 17.78 \text{ in.}$$

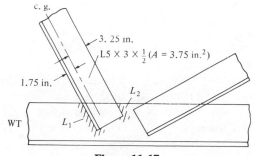

Figure 11-17

Taking moments about L_1 (see Fig. 11-16) to determine force P_2,

$$(82.50)(1.75) - 5.00P_2 = 0$$

$$P_2 = 28.88 \text{ k}$$

$$P_1 = P - P_1 = 82.50 - 28.88 = 53.62 \text{ k}$$

$$L_1 = \frac{53.62}{4.641} = 11.55 \text{ in.} \qquad \text{(say 12 in.)}$$

$$L_2 = \frac{28.88}{4.641} = 6.22 \text{ in.} \qquad \left(\text{say } 6\tfrac{1}{2} \text{ in.}\right)$$

Use end returns $2 \times \frac{5}{16} = \frac{10}{16}$ (say 1 in.). The end return lengths may be subtracted from the side weld lengths making them 11 in. and $5\frac{1}{2}$ in., respectively.

Example 11-8

Rework Example 11-8 using fillet welds along the sides and end of the angle.

SOLUTION

Assuming $\frac{5}{16}$-in. weld (capacity = 4.641 k/in. from Example 11-7).

$$\text{strength of end weld} = (5.00)(4.641) = 23.20 \text{ k}$$

Taking moments about L_1 to determine force P_2

$$(82.50)(1.75) - (2.50)(23.20) - 5.00P_2 = 0$$

$$P_2 = 17.28 \text{ k}$$

$$P_1 = 82.5 - 17.28 - 23.20 = 42.02 \text{ k}$$

$$L_1 = \frac{42.02}{4.641} = 9.05 \text{ in.} \qquad \text{say 9 in.}$$

$$L_2 = \frac{17.28}{4.641} = 3.72 \text{ in.} \qquad \text{say 4 in.}$$

11-13. SHEAR AND TORSION

Fillet welds are frequently loaded with eccentrically applied loads with the result that the welds are subjected to either shear and torsion or to shear and bending. Figure 11-18 is presented in an attempt to show the student the difference between the two situations. Shear and torsion, shown in part (a) of the figure, is the subject of this section while shear and bending, shown in part (b) of the figure, is the subject of Section 11-14.

For this discussion the welded bracket of part (a) of Fig. 11-18 is considered. The pieces being connected are assumed to be completely rigid

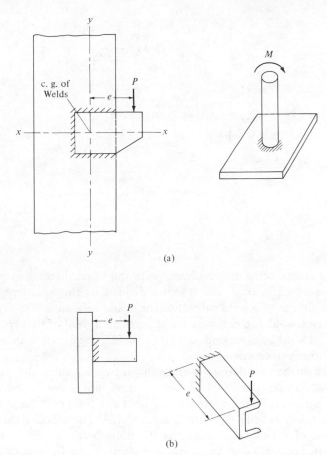

Figure 11-18 (a) Welds subjected to shear and torsion. (b) Welds subjected to shear and bending.

as they were in bolted connections. The effect of this assumption is that all deformation occurs in the weld. The weld is subjected to a combination of shear and torsion as was the eccentrically loaded bolt group considered in Section 9-12. The force caused by torsion can be computed from the following familiar expression.

$$f = \frac{Td}{J}$$

In this expression T is the torsion, d is the distance from the center of gravity of the weld to the point being considered, and J is the polar moment of inertia of the weld. It is usually more convenient to break the force down into its vertical and horizontal components. In the expressions to follow, h and v are the horizontal and vertical components of the distance d. (These formulas are almost identical to those used for determining stresses in bolt groups subject to torsion.)

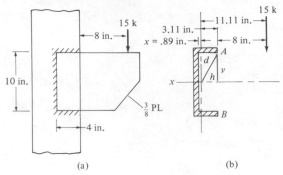

Figure 11-19

$$f_h = \frac{Tv}{J} \qquad f_v = \frac{Th}{J}$$

These components are combined with the usual direct shearing stress which is assumed to equal the reaction divided by the total length of the welds. For design of a weld subject to shear and torsion it is convenient to assume a 1-in. weld and compute the stresses on a weld of that size. Should the assumed weld be overstressed a larger weld is required, if understressed a smaller one is desirable.

Although the calculations will in all probability show the weld to be overstressed or understressed the math does not have to be repeated because a ratio can be set up to give the weld size for which the load would produce a stress exactly equal to the allowable. The student should note that the use of a 1-in. weld simplifies the units because 1 in. of length of weld is 1 in.^{2} of weld and the computed stresses can be said to be either kips per square inch or kips per inch of length. Should the calculations be based on some size other than a 1-in. weld the student will have to be very careful in keeping the units straight particularly in obtaining the final weld size. Example 11-9 illustrates the calculations involved in determining the weld size required for a connection subjected to a combination of shear and torsion. The tables in the AISC Manual for eccentrically loaded weld groups are based on an ultimate strength analysis as are their tables for eccentrically loaded bolted groups. For such an analysis there is assumed to be a relative rotation and translation of the parts connected by the welds with the rotation occurring about the *instantaneous center of rotation*.[5]

Example 11-9

For the bracket shown in Fig. 11-19(a), determine the fillet weld size required if E70 electrodes, the AISC Specification, and the SMAW process are used.

[5]L. J. Butler, S. Pal, and G. L. Kulak, "Eccentrically Loaded Welded Connections," *ASCE Journal of Structural Division* **98**, No. ST5, May 1972, pp. 989–1005.

SOLUTION

Assuming a 1-in. weld as shown in part (b) of Fig. 11-19,

$$A = 18 \text{ in.}^2$$

$$\bar{x} = \frac{(4)(2)(2)}{18} = 0.89 \text{ in.}$$

$$I_x = \left(\tfrac{1}{12}\right)(1)(10)^3 + (2)(4)(5)^2 = 283.3 \text{ in.}^4$$

$$I_y = (2)\left(\tfrac{1}{3}\right)(0.89^3 + 3.11^3) + (10)(0.89)^2 = 28.5 \text{ in.}^4$$

$$J = 283.3 + 28.5 = 311.8 \text{ in.}^4$$

Most stressed portions of weld are greatest distance from weld center of gravity [A and B in Fig. 11-19(b)].

$$f_h = \frac{Tv}{J} = \frac{(15 \times 11.11)(5)}{311.8} = 2.67 \text{ k/in.}^2$$

$$f_v = \frac{Th}{J} = \frac{(15 \times 11.11)(3.11)}{311.8} = 1.66 \text{ k/in.}^2$$

$$f_s = f_{shear} = \frac{15}{18} = 0.83 \text{ k/in.}^2$$

$$f_r = f_{resultant} = \sqrt{(1.66 + 0.83)^2 + (2.67)^2} = 3.65 \text{ k/in.}^2$$

allowable capacity of a 1-in. fillet weld (E70 electrode)

$$= (0.707)(1.00)(0.30 \times 70) = 14.85 \text{ k/in.}$$

$$\text{weld size reqd.} = \frac{3.65}{14.85} = 0.246 \text{ in.} \qquad \left(\text{say } \tfrac{1}{4} \text{ in.}\right)$$

checking shear capacity/in. for $PL = (0.40 \times 36)\left(\tfrac{3}{8}\right) = 5.4 \text{ k/in.}$

$$> (0.707)\left(\tfrac{1}{4}\right)(0.30 \times 70) - 3.71 \text{ k/in.} \quad \text{OK}$$

Use $\tfrac{1}{4}$-in. fillet weld

11-14. SHEAR AND BENDING

The welds shown in Fig. 11-18(b) and in Fig. 11-20 are subjected to a combination of shear and bending.

For short welds of this type the usual practice is to consider a uniform variation of shearing stress. If, however, the bending stress is assumed to be given by the flexure formula, the shear does not vary uniformly for vertical welds but as a parabola with a maximum value $1\tfrac{1}{2}$ times the average value. These stress and shear variations are shown in Fig. 11-21.

The student should carefully note that the maximum shearing stresses and the maximum bending stresses occur at different locations. It is, therefore, probably not necessary to combine the two stresses at any one

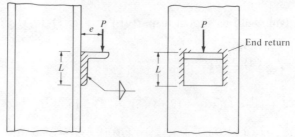

Figure 11-20

point. If the weld is capable of withstanding the worst shear and the worst moment individually it is probably satisfactory. In Example 11-10, however, a welded connection subjected to shear and bending is designed by the usual practice of assuming a uniform shear distribution in the weld and combining that value vectorially with the maximum bending stress.

Example 11-10

Using E70 electrodes, the SMAW process, and the AISC Specification determine the weld size required for the connection of Fig. 11-20 if $P = 30$ k, $e = 2\frac{1}{2}$ in., and $L = 8$ in. Assume that the member thicknesses do not control weld size.

SOLUTION

$$f_s = \frac{30}{(2)(8)} = 1.875 \text{ k/in.}$$

$$f = \frac{(30 \times 2.5)(4)}{\left(\frac{1}{12}\right)(1)(8)^3(2)} = 3.51 \text{ k/in.}$$

$$f_r = \sqrt{(1.875)^2 + (3.51)^2} = 3.98 \text{ k/in.}$$

$$\text{weld size reqd.} = \frac{3.98}{(0.707)(1)(0.30 \times 70)} = 0.268 \text{ in.} \quad \left(\text{say } \frac{5}{16} \text{ in.}\right)$$

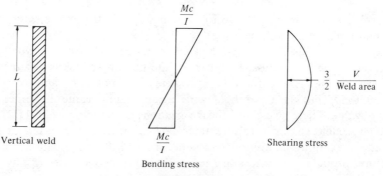

Figure 11-21

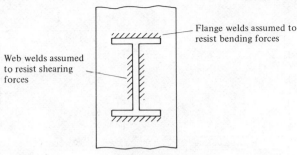

Figure 11-22

Should welds of the type shown in Fig. 11-22 be used for a W or S beam the web welds would probably be assumed to uniformly carry all of the shear and the flange welds all of the moment. Chapter 12 presents more information on this subject.

Problems

11-1. A $\frac{5}{16}$-in. fillet weld SMAW process is used to connect the members shown in the accompanying illustration. Determine the maximum load that can be applied to this connection according to the AISC if the steel is A36 and E70 electrodes are used. (*Ans.* $P = 64.8$ k)

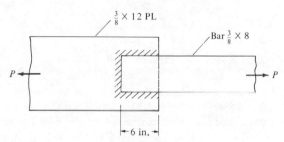

Problem 11-1

11-2. Rework Prob. 11-1 if the SAW process is used.

11-3. Rework Prob. 11-1 if A572 Grade 65 steel and E80 electrodes are used. (*Ans.* 106.1 k)

11-4. Design maximum size fillet welds to develop the full strength of the A36 bar shown in the accompanying illustration. Use E70 electrodes, the SMAW process, and the AISC Specification.

11-5. Rework Prob. 11-4 if the SAW process is used. (*Ans.* Use 8-in. welds each side as required by AISC Section 1.17.4 including end returns)

11-6. Rework Prob. 11-4 using side welds and a vertical end weld at the end of the $\frac{1}{2} \times 8$ bar. Also use A572 Grade 65 steel and E80 electrodes.

11-7. Rework Prob. 11-4 using side welds and welds at the end of the $\frac{1}{2} \times 8$ PL at the connection. (*Ans.* Total $L = 13.31$ in., say 14 in.)

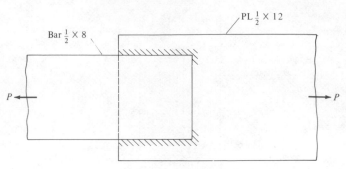

Problem 11-4

11-8. The $\frac{5}{8} \times 8$-in. PL shown in the accompanying illustration consists of A36 steel and is to be connected to a gusset plate with $\frac{5}{16}$-in. SMAW fillet welds. Determine the length L required to develop the full strength of the bar if E70 electrodes and the AISC Specification are used.

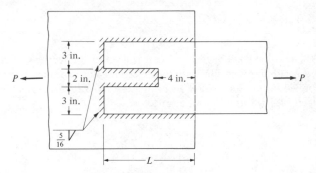

Problem 11-8

11-9. Design maximum size side SAW fillet welds to develop the full tensile strength of a $6 \times 4 \times \frac{1}{2} L$ using E70 electrodes, A36 steel, and the AISC Specification. The member is connected on the 6-in. leg and is subject to alternating loads. (*Ans.* Use 5 and 10 in. side welds)

11-10. Rework Prob. 11-9 using side welds and a weld at the end of the angle.

11-11. Rework Prob. 11-10 using A588 steel and E80 electrodes. (*Ans.* Use $6\frac{1}{2}$ and $9\frac{1}{2}$ in. side welds)

11-12. One leg of an $8 \times 8 \times \frac{3}{4}$ angle is to be connected with side welds and a weld at the end of the angle to a plate behind to develop the full capacity of the angle (A36). Balance the SMAW fillet welds around the center of gravity of the angle, use maximum weld size and assume the allowable stress in the welds is 11,500 psi.

11-13. It is desired to design SMAW fillet welds necessary to connect a C10×30 made from A36 steel to a $\frac{3}{8}$-in. gusset plate. End, side, and slot welds may be used to develop the full capacity of the channel. No welding is permitted on the back of the channel. Use E70 electrodes and the AISC Specification. It is assumed that due to space limitations the channel can lap over the gusset plate by a maximum of 10 in. (*Ans.* $\frac{5}{16}$-in. fillet welds and one $\frac{13}{16} \times 2\frac{1}{2}$ slot weld)

11-14. Rework prob. 11-13 using A572 Grade 60 steel and E70 electrodes.

11-15. Determine the maximum force per inch to be resisted by the fillet weld shown in the accompanying illustration. (*Ans.* 5.88 k/in.)

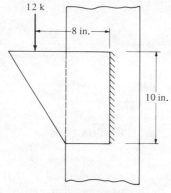

Problem 11-15

11-16. Determine the maximum force to be resisted per inch by the fillet weld shown in the accompanying illustration.

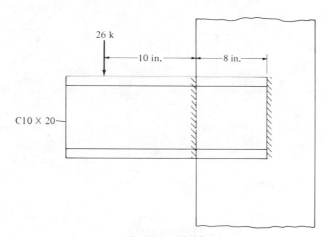

Problem 11-16

11-17. Rework Prob. 11-16 if welds are used on the sides of the channel in addition to those shown in the figure. (*Ans.* 2.90 k/in.)

11-18. Determine the maximum force per inch to be resisted by the fillet welds shown in the accompanying illustration. Also determine the required weld thickness using A36 steel, E70 electrode SMAW welds, and the AISC Specification.

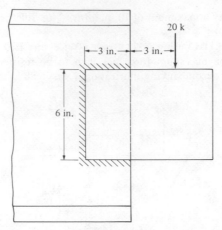

Problem 11-18

11-19. Determine the maximum eccentric load P that can be applied to the connection shown in the accompanying illustration if $\frac{1}{4}$-in. SMAW fillet welds are used. Assume plate thickness $= \frac{1}{2}$ in. E70. A36. AISC Specification. (*Ans.* 11.0 k)

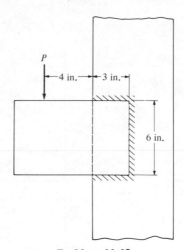

Problem 11-19

11-20. Rework Prob. 11-19 if $\frac{3}{8}$-in. fillet welds are used and the vertical weld is 10 in. high.

11-21. Determine the SMAW fillet weld size required for the connection of Prob. 11-15 if the load is increased to 30 k and the height of the weld is 12 in. A36. E70. AISC Specification. (*Ans.* 0.694 in., say $\frac{3}{4}$ in.)

11-22. Using E70 electrodes, the SMAW process, A36 steel, and the AISC specification determine the fillet weld size required for the bracket shown in the accompanying illustration. Assume member thicknesses $= \frac{1}{2}$ in.

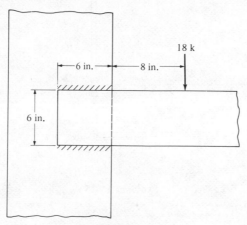

Problem 11-22

11-23. Rework Prob. 11-22 if the load is increased from 18 to 30 k and the weld lengths increased from 6 to 8 in. (*Ans.* 0.634 in., say $\frac{11}{16}$ in.)

11-24. Determine the SMAW fillet weld size required for the connection shown in the accompanying illustration. E70. A36. AISC Specification.

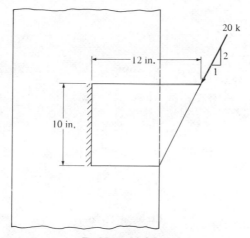

Problem 11-24

11-25. Determine the SMAW fillet weld size required for the connection shown in the accompanying illustration. E70. A36. AISC Specification. What angle thickness should be used? (*Ans.* $\frac{3}{16}$ in. weld, $3 \times 3 \times \frac{1}{4}$Ls)

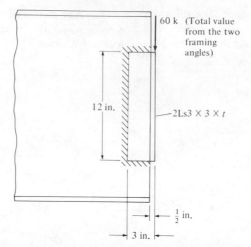

Problem 11-25

11-26. Assuming the SMAW process is to be used, determine the fillet weld size required for the connection shown in the accompanying illustration. Use A36 steel, the AISC Specification, and E70 electrodes.

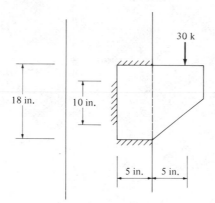

Problem 11-26

11-27. Determine the fillet weld size required for the connection shown in the accompanying illustration. A36. E70. AISC specification. The SMAW process is to be used. (*Ans.* 0.705 in., say $\frac{3}{4}$ in.)

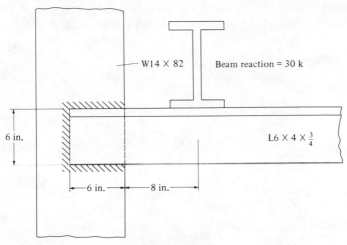

Problem 11-27

11-28. Determine the allowable value of the load P that can be applied to the connection shown in the accompanying illustration if $\frac{3}{8}$-in. fillet welds are used. E70. A36. AISC Specification. SAW.

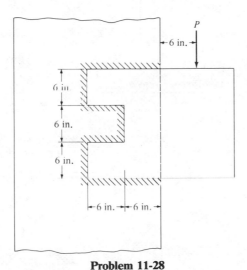

Problem 11-28

11-29. Determine the length of $\frac{1}{4}$-in. SMAW fillet welds 12 in. on center required to connect the cover plates for the section shown in the accompanying illustration at a point where the external shear is 120 k. A36. E70. AISC Specification. (*Ans*. 6.08 in., say 6-in. welds, 12-in. o.c.)

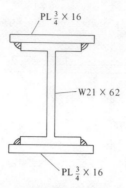

Problem 11-29

11-30. The welded plate girder shown in the accompanying illustration has an external shear of 480 k at a particular section. Determine the fillet weld size required to fasten the plates to the web if the SMAW process is used. AISC Specification. E70. A36.

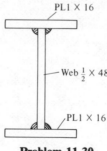

Problem 11-30

Chapter 12
Building
Connections

12-1. SELECTION OF TYPE OF FASTENER

This chapter is concerned with the actual beam-to-beam and beam-to-column connections commonly used in steel buildings. Under present-day steel specifications four types of fasteners are permitted for these connections. These are rivets, welds, unfinished bolts, and high-strength bolts.

Selection of the type of fastener or fasteners to be used for a particular structure usually involves many factors, including requirements of local building codes, relative economy, preference of designer, availability of good welders or riveters, loading conditions (as static or fatigue loadings), preference of fabricator, and equipment available. It is impossible to list a definite set of rules from which the best type of fastener can be selected for any given structure. One can only give a few general statements which may be helpful in making a decision. These are listed as follows.

1. Unfinished bolts are often economical for light structures subject to small static loads and for secondary members (such as purlins, girts, bracing, etc.) in larger structures.
2. Field bolting is very rapid and involves less skilled labor than welding or riveting. The purchase price of high-strength bolts, however, is rather high.
3. If a structure is later to be disassembled, riveting and welding are probably ruled out, leaving the job open to bolts.
4. For fatigue loadings friction-type high-strength bolts are excellent, while welds and bearing-type high-strength bolts are very good.
5. Welding requires the smallest amounts of steel, probably provides the most attractive-looking joints, and also has the widest range of application to different types of connections.
6. When continuous and rigid fully moment resisting joints are desired welding will probably be selected.
7. Welding is almost universally accepted as being satisfactory for shopwork. For fieldwork it is very popular in some areas of the

United States while in others it is stymied by the fear that field inspection is rather questionable.

8. Rivets that can be rapidly installed in the shop with heavy riveters are nevertheless very seldom used.

9. For field work, rivets are almost extinct except for some bridge-work.

A most interesting article entitled "Choosing the Best Structural Fastener" by Henry J. Stetina appeared in the November 1963 issue of *Civil Engineering*. It would be well worth the student's time to read this article.

A valuable reference on the relative economy of various types of beam to beam and beam to column connections is given in a handbook published by the United States Steel Corporation in October 1963 entitled *Building Design Data*.

12-2. TYPES OF BEAM CONNECTIONS

According to their rotational characteristics under load, connections can be classified as being simple, semirigid, and rigid. A connection which does not rotate at all or which has complete moment resistance is said to be a rigid connection while a connection which is completely flexible and free to rotate and has no moment resistance is said to be a simple connection. A semirigid connection is one whose resistance falls somewhere between the simple and rigid types.

From a practical standpoint, since there are no connections that are completely rigid or completely flexible, it is a common practice to classify them on a percentage of moment developed to complete rigidity or complete moment resistance. (A measure of the rotational characteristics of a particular connection cannot be practically obtained by a theoretical method and it is necessary to run tests and plot curves of the relationships between moments and rotations for each type of connection.) A rough rule is simple connections 0–20%, semirigid 20–90%, and rigid above 90%.

Each of these three general types of connections is briefly discussed in this section with little mention of the specific types of connectors used. The remainder of the chapter is concerned with detailed designs of these connections using specific types of fasteners. *In this discussion the author probably overemphasizes the semirigid and rigid type connections, because most of the building designs with which the average designer works will be assumed to have simple connections.* A few descriptive comments are given in the following paragraphs concerning each of these three types of connections.

Simple connections are quite flexible and are assumed to allow the beam ends to be substantially free to rotate downward under load as true simple beams should. Although simple connections do have some moment

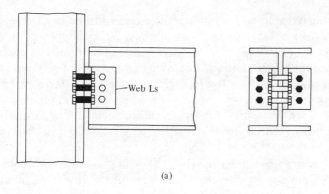

(a)

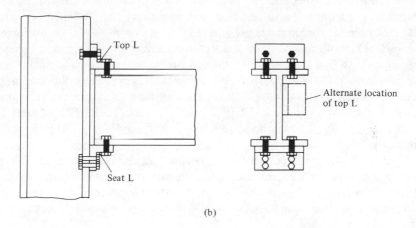

(b)

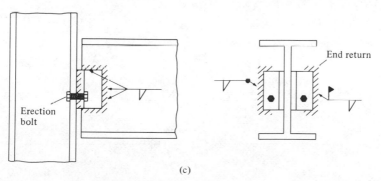

(c)

Figure 12-1   (a) Framed simple connection (bolted or riveted). (b) Seated simple connection. (c) Framed simple connection.

resistance (or resistance to end rotation) it is assumed to be negligible and they are assumed to be able to resist shear only. Several types of simple connections are shown in Fig. 12-1. More detailed descriptions of each of these connections and their assumed behavior under load are given in later sections of this chapter. In this figure each connection is shown as being made entirely with the same type of fastener while in actual practice two types of fasteners are often used for the same connection. For example, a very common practice is to shop-weld the web angles to the beam web and field-bolt them to the column or girder.

Semirigid connections are those which have appreciable resistance to end rotation thus developing appreciable end moments. In design practice it is quite common for the designer to assume all of his connections are simple or rigid with no consideration given to those situations in between thereby simplifying the analysis. Should he make such an assumption for a true semirigid connection he may miss an opportunity for appreciable moment reductions. To understand this possibility the student is referred to the moment diagrams shown in Fig. 12-2 for a group of uniformly loaded beams supported with connections having different percentages of rigidity. This figure shows that the maximum moments in a beam vary greatly with different types of end connections. For example, the maximum moment in the semirigid connection of part (d) of the figure is only 50% of the maximum moment in the simply supported beam of part (a) and only 75% of the maximum moment in the rigidly supported beam of part (b).

Actual semirigid connections are used fairly often but usually no advantage is taken of their moment reducing possibilities in the calculations. Perhaps one factor that keeps the design profession from taking advantage of them more often is the AISC Specification (Section 1.2)

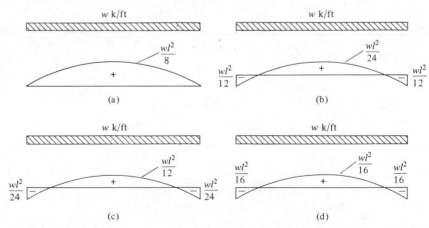

Figure 12-2 (a) Simple connections (0%). (b) Rigid connections (100%). (c) Semirigid connections (50%). (d) Semirigid connections (75%).

A semirigid beam-to-column connection, Ainsley Building, Miami, Fla. (Courtesy of The Lincoln Electric Company.)

which states that consideration of a connection as being semirigid is permitted only upon presentation of evidence that it is capable of resisting a certain percentage of the moment resistance provided by a complete rigid connection.

A second deterring factor is the need for a method of analysis which falls in between analysis for simple beams and analysis for a statically indeterminate structure with completely rigid joints. The student can see that analysis of a building by moment distribution could be drastically affected if the end connections were assumed to have varying percentages of moment restraint. This subject is discussed in detail by Bruce Johnston and Edward H. Mount in a paper entitled "Analysis of Building Frames with Semi-Rigid Connection."[1] The student is also referred to Chapter 8 of *Advanced Design in Structural Steel* by John E. Lothers (Prentice-Hall, 1960) for an excellent discussion of this subject. Several types of semirigid connections are shown in Fig. 12-3.

Rigid connections are those which theoretically allow no rotation at the beam ends and thus transfer 100% of the moment of a fixed end. Connections of this type may be used for tall buildings in which wind resistance is

[1] *Trans. ASCE* **107** (1942), p. 993.

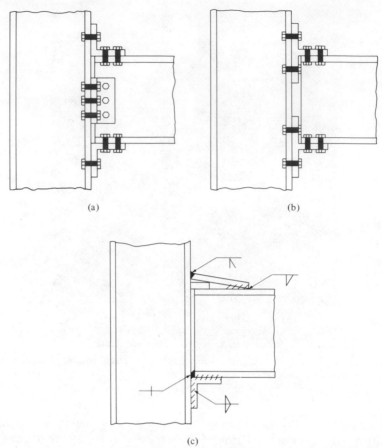

(a)

(b)

(c)

Figure 12-3 (a) Web Ls with seat L and top L. (b) Structural tee connection. (c) Welded semirigid connection.

developed by providing continuity between the members of the building frame. Connections that provide almost 100% restraint are shown in Fig. 12-4. In this figure the student might compare the heavy, awkward riveted or bolted connections of parts (a) and (b) of the figure with the welded types shown in parts (c) and (d). From the standpoint of appearance alone she might see why the welded moment connections are more popular than the others. The type of connection shown in part (b) is becoming obsolete with only occasional use today.

In part (c) a welded connection is shown on one end of a beam where the beam can be placed right up against the column. Although it may be possible to butt one end of a steel beam against a supporting column or girder, practical field dimensions in steel erection are not usually precise enough to permit such close fitting on the other end. There it is necessary to use some type of connection which permits a little variation in fitting dimensions. Part (d) of Fig. 12-4 presents such a connection.

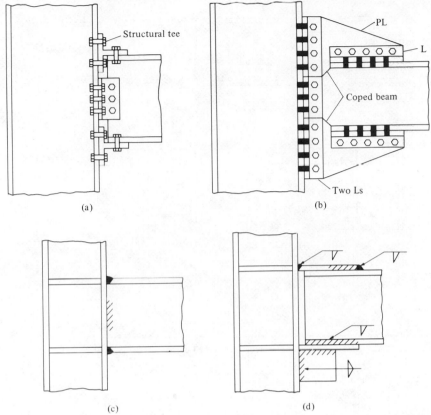

Figure 12-4 (a) Structural tee connection. (b) Bracket connection. (c) Welded moment-resisting connection. (d) Welded moment-resisting connection.

12-3.   STANDARD RIVETED OR BOLTED BEAM CONNECTIONS

Several types of standard riveted or bolted connections are shown in Fig. 12-5. These connections are usually designed to resist shear only, as testing has proved this practice to be quite satisfactory. Part (a) of the figure shows a connection between beams with the so-called *framed connection*. This type of connection consists of a pair of flexible web angles probably shop connected to the web of the supported beam web and field connected to the supporting beam or column. When two beams are being connected it is usually necessary to keep their top flanges at the same elevation, with the result that the top flange of one will have to be cut back (called *coping*) as shown in part (b) of the figure.

Simple connections of beams to columns can be either framed or seated as shown in Fig. 12-5. In part (c) of the figure a framed connection is shown in which two web angles are connected to the beam web in the shop, after which bolts are placed through the angles and column in the

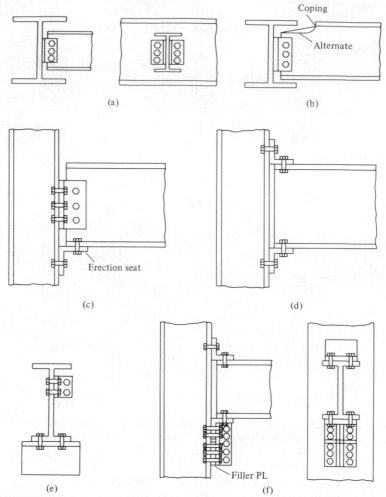

Figure 12-5 (a) Framed connection. (b) Framed connection. (c) Framed connection. (d) Seated connection. (e) Seated connection. (f) Seated connection with stiffener angles.

field. It is often convenient to have an angle, called an *erection seat*, to support the beam during erection. Such an angle is shown in the figure.

The seated connection has an angle under the beam similar to the erection seat just mentioned, which is shop-connected to the column. In addition, there is another angle probably on top of the beam which is field-connected to the beam and column. A seated connection of this type is shown in part (d) of the figure. Should space limitation prove a problem above the beam, the top angle may be placed in the optional location shown in part (e) of the figure. The top angle, at either of the locations mentioned, is very helpful in keeping the top flange of the beam from being accidentally twisted out of place during construction.

The amount of load which can be supported by the types of connections shown in parts (c), (d), and (e) of Fig. 12-5 is severely limited by the flexibility or bending strength of the horizontal legs of the seat angles. For heavier loads it is necessary to use stiffened seats such as the one shown in part (f) of the figure.

The majority of these connections are selected by referring to standard tables. The AISC Manual has excellent tables for selecting bolted or welded beam connections of the types shown in Fig. 12-5. After a rolled-beam section has been selected it is quite convenient to refer to these tables and select one of the standard connections, which will be suitable for the vast majority of cases. No numerical examples are given here because several examples together with a very good explanation are given in the AISC Manual. It is strongly suggested that the student study these examples very carefully and see if she can duplicate the shearing and bearing strengths given in the tables for the various connections.

In order to make these standard connections have as little moment resistance as possible the angles used in making up the connections are usually light and flexible. To qualify as simple end supports the ends of the beams should be free to rotate downwards. Figure 12-6 shows the manner in which framed and seated end connections will theoretically deform as

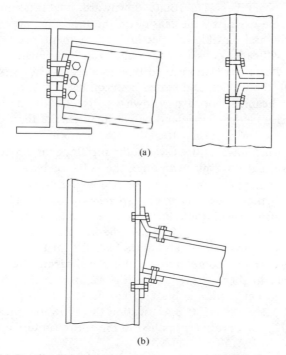

(a)

(b)

Figure 12-6 (a) Bending of framed-beam connection. (b) Bending of seated-beam connection.

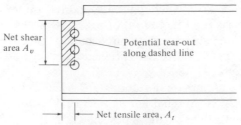

Figure 12-7 Block shear.

the ends of the beams rotate downward. The designer does not want to do anything that will hamper these deformations.

For the rotations shown in Fig. 12-6 to occur there must be some deformation of the angles. As a matter of fact, if end slopes of the magnitudes which are computed for simple ends are to occur, the angles will actually bend enough to be stressed beyond their yield points. If this situation occurs, they will be permanently bent and the connections will quite closely approach true simple ends. The student should now see why it is desirable to use rather thin angles and large gages for the bolt spacing if flexible simple end connections are the goal of the designer.

These connections do have some resistance to moment. When the ends of the beam begin to rotate downward, the rotation is certainly resisted to some extent by the tension in the top bolts, even if the angles are quite thin and flexible. Neglecting the moment resistance of these connections will cause conservative beam sizes. If moments of any significance are to be resisted, more rigid-type joints need to be provided than available with the framed and seated connection.

When a beam is coped as shown in Fig. 12-5(b) the part cut out usually includes the top flange and only a small part of the web. The result is that the calculated shearing strength of the beam is reduced very little. It has been shown,[2] however, that where the top flange is coped failure may occur due to shear on a plane through the bolts, or by a combination of shear on that plane plus tension on a perpendicular plane as shown in Fig. 12-7. This is called *block shear* or *web tear-out shear*. The situation is particularly serious when there are only a few bolts in the beam web and when they are not uniformly placed over the full depth of the beam.

Tests[2] have shown that the failure load for such a situation can be estimated very well by combining the ultimate shearing strength of the net section subject to shear stress (A_v in Fig. 12-7) with the ultimate tensile strength of the net section subject to tensile stress (A_t in Fig. 12-7). In Section 1.5.1.2.2 of the AISC Specification the allowable shear stress F_v is given as $0.30 F_u$ conservatively applied to the entire rupture area ($A_v + A_t$).

[2] P. C. Birkemoe and M. I. Gilmor, "Behavior of Bearing Critical Double-Angle Beam Connections," *Engineering Journal*, AISC, **15**, no. 4 (4th Quarter, 1978), pp. 109–115.

The AISC Commentary (1.5.1.2) says that, as an alternative, the tension and shear stresses may be treated separately and the total allowable capacity allowed to equal $0.30A_v F_u + 0.50A_t F_u$. Using these latter values the tables on pages 4-10 and 4-11 of the AISC Manual (8th ed.) provide allowable block shears for coped sections.

12-4.   SEMIRIGID AND RIGID RIVETED OR BOLTED CONNECTIONS

The standard beam connections discussed in Section 12-3 are usually designed to transfer shear only. Although they do transfer moments, the amounts are very small due to the flexible angles used. When a frame is to be continuous or when resistance to wind or other lateral loads is to be provided by the joints, connections will be used which have much greater moment resistance.

Several types of semirigid and rigid connections were shown in Figs. 12-3 and 12-4. Perhaps the simplest semirigid connection type is the one shown in part (a) of Fig. 12-3, where a beam is connected to a column with a pair of standard web angles together with top and bottom clip angles. Usually the web angles are designed to resist the shears and are picked from the AISC Manual while the clip angles are designed to resist the moments. Another very satisfactory type of semirigid connection is the one shown in part (b) of Fig. 12-3. In this case the connection is made with a pair of structural tees and it is called a structural tee connection or a split beam connection.

In the bolted semirigid connection of Fig. 12-9 the end moment is resisted by an equal and opposite couple produced by the tensile force in the bolts in the top clip angles and a compressive force in the lower flange bolts. The bolts passing through the clip angle and the beam flange are placed in single shear by a force equal to the pull at the top of the connection. From this information the number of bolts can be easily determined.

To consider a method of determining the thickness of the clip angles in a connection of this type, Fig. 12-8 is presented showing the *assumed* bending condition of the top angle. The true bending condition of this

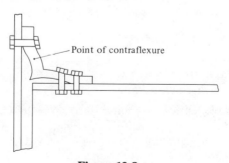

Point of contraflexure

Figure 12-8

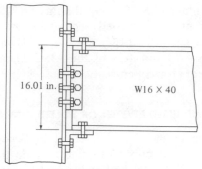

Figure 12-9

angle is extremely complicated. An assumption is made here that no clamping force is provided by the bolt. It is further assumed that the angle bends in such a manner as to have a point of contraflexure midway between the top of the horizontal leg of the angle and the center of the bolt.

If this bending condition were correct, the bending moment in the angle could be estimated as being the force being transferred times the distance from the top of the horizontal leg or from the center of the bolt to the point of contraflexure. The angle thickness can be determined from the flexure formula using the moment estimated in the vertical angle leg. It is fairly common design practice to make some rough assumption as to the thickness of these clip angles without bothering to estimate the moment developed. The frequent result is angles which do not have sufficient thickness.

Example 12-1

Design a connection of the type shown in Fig. 12-9 to resist a moment of 50 ft-k and a shear of 25 k. Use $\frac{7}{8}$-in. A325 bolts in a bearing-type connection threads excluded from shear plane, A36 steel, and the AISC Specification.

SOLUTION

Design of web Ls to resist full shear: From AISC Manual, "Framed Beam Connections" table:

Use 2 Ls$4 \times 4 \times \frac{1}{2} \times 8\frac{1}{2}$ in. with 3 rows of bolts

Selection of clip L bolts:

    (a) Bolts through column flange:

        allowable tension per bolt $= (0.6)(44) = 26.4$ k

$$\text{pull on bolts} = \frac{(12)(50)}{16.01} = 37.5 \text{ k}$$

$$\text{no. of bolts reqd.} = \frac{37.5}{26.4} = 1.42 \quad \left(\underline{\text{say } 2}\right)$$

(b) Bolts through beam flange: bolts in single shear and bearing on 0.505 in. (t_f of W16×40)

$$\text{single shear} = (0.6)(30.0) = 18\ \text{k} \leftarrow$$

$$\text{bearing} = \left(\tfrac{7}{8}\right)(0.505)(1.5 \times 58) = 38.44\ \text{k}$$

$$\text{no. reqd.} = \frac{37.5}{18} = 2.08 \quad (\ \text{say } 3\ )$$

Selection of clip Ls: Assume Ls7×4×1×0 ft 8 in. for bolts selected. Distance from center of tension bolts to top of horizontal leg of $L = 2\tfrac{1}{2} - 1 = 1.50$ in. Moment in vertical leg of clip $L = (37.5)(1.50/2) = 28.12$ in.-k

$$S = \frac{1}{6}bt^2 = \frac{M}{F_b}$$

$$t = \sqrt{\frac{6M}{bF_b}}$$

$$= \sqrt{\frac{(6)(28.12)}{(8)(27)}}$$

$$= 0.88 \text{ in.} < 1.0 \text{ in.} \qquad\qquad \text{OK}$$

Use 7×4×1×0 ft 8 in. clip L

A split beam connection of the type shown in Fig. 12-3(b) is designed in Example 12-2. This type of connection is the most rigid of the semirigid types. To understand its increased rigidity over the clip angle type the reader should compare the deformation diagrams for the tension sides of these two types of connections shown in Figs. 12-8 and 12-10. In Fig. 12-8 the tension part of the couple is acting eccentrically on the clip angle, while in Fig. 12-10 the structural tee is loaded axially and has less deformation.

With respect to Fig. 12-10, the pull is divided into $T/2$ times the distance x. The other dimensions of the structural tee can be obtained

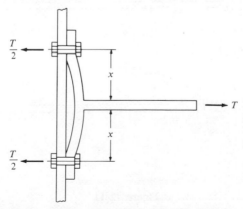

Figure 12-10 Tension side of structural tee construction.

from the space required to place the bolts. Another variation from the clip-angle problem is that no web angles are used to resist the shear, with the result that the bolts from structural tees to the columns are placed in a combination of shear and tension. In this discussion clamping force has again been very conservatively neglected. Truthfully, there are present rather large clamping forces. Another factor of importance is that on the compression side of the connection the bolts are pressed against the column and a large percentage of the total shear on the column is carried on that side by friction.

Example 12-2

A split beam connection is to be designed for the W18×55 shown in Fig. 12-11 to resist a shear of 50 k and a moment of 80 ft-k. Design the connection if A36 steel, 1-in. A325 bearing-type bolts in a friction-type connection, and the AISC Specification are used.

SOLUTION
Selection of bolts:

$$\text{pull on tee} = \frac{(12)(80)}{18.11} = 53.01 \text{ k}$$

Bolts through beam flange in single shear and bearing on 0.630:

$$\text{shear} = (0.785)(17.5) = 13.74 \text{ k} \leftarrow$$

$$\text{bear} = (1)(0.630)(1.5 \times 58) = 54.81 \text{ k}$$

$$\text{no. reqd.} = \frac{53.01}{13.74} = 3.86 \qquad (\underline{\text{say 4}})$$

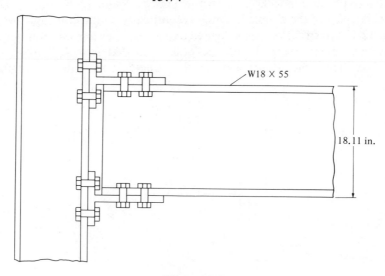

Figure 12-11

Bolts through column flange in tension and shear:

Assume four bolts in each tee.

$$f_v = \frac{50}{(8)(0.785)} = 7.96 \text{ ksi} < 17.5 \text{ ksi}$$

$$f_t = \frac{53.01}{(4)(0.785)} = 16.88 \text{ ksi}$$

$$\text{allowable } F_v = \left(1 - \frac{16.88 \times 0.785}{51}\right) 17.5$$

$$= 12.95 \text{ ksi} > 7.96 \text{ ksi} \qquad\qquad \text{OK}$$

Selection of structural tee: Assume gage of flange bolts $= 5\frac{1}{2}$ in. After running through the following steps a few times on scratch paper, a WT16.5 × 100.5 ($b_f = 15.745$, $t_f = 1.15$, $t_w = 0.715$, $d = 16.84$) is tried.

$$x = \left(\tfrac{1}{2}\right)\left(5\tfrac{1}{2} - 0.715\right) = 2.39 \text{ in.}$$

$$M = \frac{53.01}{2}(2.39) = 63.35 \text{ in.-k}$$

Assuming length of WT = 10 in.,

$$t = \sqrt{\frac{6M}{bF_b}} = \sqrt{\frac{(6)(63.35)}{(10)(27)}} = 1.186 \text{ in.} > 1.15 \text{ in.} \qquad \text{But OK}$$

Use WT16.5 × 100.5 × 0 ft 10 in.

Space is not taken here to present the design of a rigid bolted connection because it is felt that all of the principles necessary to work such a problem are presented in this chapter and in Chapters 9 and 10. Particular reference should be made to Example 12-6 of this chapter.

12-5. TYPES OF WELDED BEAM CONNECTIONS

The sections to follow present several methods of making welded connections between beams and girders and between beams and columns. These include web angles (Fig. 12-12), beam seats (Fig. 12-15), stiffened beam seats (Fig. 12-20), and moment-resistant connections (Fig. 12-23). The first three of these types are designed as simple connections and as such are expected to transfer end reactions while having no appreciable moment restraint. They are often referred to simply as *shear connections*. The moment-resistant connections provide resistance to both reactions and moments and are probably referred to as *moment connections*.

To design welded beam connections properly the designer must understand the stress conditions in the structure and how those stresses can be transmitted through welded connections. Two illustrations are as follows.

1. Most of the bending forces in beams occur in the flanges and if welds are to be designed to transfer these forces they should be primarily located at the beam flanges.
2. Similarly, most of the shearing forces in a beam occur in its web and welds designed to transmit these forces desirably need to be primarily located on the webs.

Members may be welded directly to each other without the use of connecting plates and angles; however, such connections may have serious drawbacks. An illustration of this fact can be seen where a beam end is welded directly to another beam or to a column. If the designer uses a high vertical weld along the web, he will have a connection that can resist an appreciable moment whether that was his intention or not. The resulting fairly large moments will produce fairly large bending stresses. A previous discussion described the difficulties of getting members to fit together perfectly in the field as would be required for this type of connection.

12-6. WELDED WEB ANGLES

Beams for which simple end supports are desired are often connected with web angles as shown in Fig. 12-12. Web-angle connections are normally designed to transmit shear and as little moment as possible by minimizing rotation resistance. These angles are probably shop-welded to the beam and may be field-welded to the columns or girders. Perhaps a more common practice today is to shop-weld the angles to the beam and field connect them to the columns or girders with high-strength bolts. The angles extend out from the beam web by approximately $\frac{1}{2}$ in., as shown in the figure. This distance is often referred to as the setback.

Erection bolts are usually necessary for erecting these beams. They are probably placed near the bottom of the angles so they will not appreciably reduce the flexibility of those angles. In some situations they may be

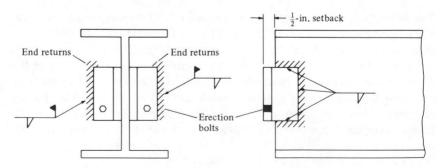

Figure 12-12

necessary at the top of the angles to stabilize the joint during erection. Such bolts can be removed at a later date if it is felt that they provide too much restraint against rotation.

As previously described for bolted beam connections it is desirable to use the thinnest possible angles. Thin angles will deflect easily and let the beam ends rotate enough to approach the desired simple end conditions. As the load is applied to the beam, the beam end tends to rotate downward and the flexibility of the angles allows the top of the beam to move away from the column or girder. Three- to four-inch legs are usually sufficient for connections to the beam web and the same or perhaps slightly longer legs are used for connections to the column or girder. The depth of the angle is limited by the beam depth and the space required for welding. The desired maximum weld depth equals the beam depth minus its width, but may be a little deeper. The thickness of the web angles will be $\frac{1}{16}$ to $\frac{1}{8}$ in. thicker than the weld size.

A good many designers assume that web angles are subject only to a vertical shear (equal to the end reaction) and no moment. Without question, however, the end reaction is eccentric with respect to both the shop and field welds and causes moment. For the field welds applied to the column or girder the load is transferred a distance e away, as shown in Fig. 12-13; and the moment in each weld equals $(R/2)(e)$. (Notice these vertical welds are returned about $\frac{1}{2}$ in. at the top of the angles because tests have shown such returns greatly strengthen the connection.)

These field welds are subject to a rotation effect which causes the web angles to be forced together against the beam web at the top and pushed apart at the bottom, tending to shear horizontally the fillet weld. The usual practice is to consider that the neutral axis dividing the tension from the compression is located one-tenth of the way down from the top of the angles. The horizontal shear is assumed to vary from zero at the one-tenth point to a maximum at the bottom of the angles.

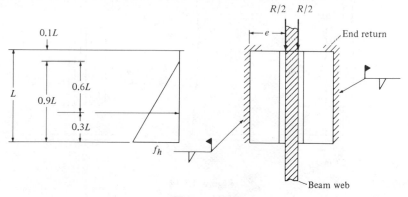

Figure 12-13

The top pressure is assumed to be concentrated at the one-tenth point and as is seen in the figure the horizontal shear will be concentrated at the center of gravity of the triangle $0.3L$ from the bottom of the web angles. Since the couple produced by these forces must be equal and opposite to the external moment, the value of f_h can be determined as follows.

$$\left(\frac{1}{2}\right)(0.9L)(f_h)(0.6L)=\frac{R}{2}e$$

$$f_h=\frac{Re}{0.54L^2}$$

The vertical shear (f_s) equals $R/2$ divided by the height of the weld and can be combined with f_h to obtain the maximum stress ($f_r=\sqrt{(f_h)^2+(f_s)^2}$) after which the weld size can be determined. Example 12-3 illustrates the calculations involved in the design of this type problem. The AISC Manual again has tables from which these values can be picked directly.

Example 12-3

Design shop and field welds for the W21×68 beam shown in Fig. 12-14 using E70 electrodes, the SMAW process, and the AISC Specification. Assume web angles are $3\times3\times\frac{3}{8}\times0$ ft 10 in. Beam reaction is assumed to be 44 k.

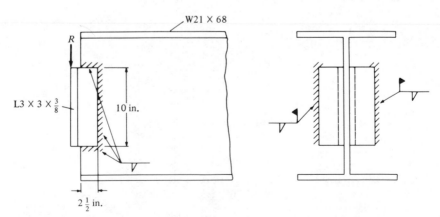

Figure 12-14

SOLUTION
Design of shop welds to beam web:

$$A = (2)(2.5) + 10 = 15 \text{ in.}^2$$

$$\bar{x} = \frac{(5)(1.25)}{15} = 0.42 \text{ in.}$$

$$I_x = \left(\tfrac{1}{12}\right)(1)(10)^3 + (2)(2.5)(5)^2 = 208.3 \text{ in.}^4$$

$$I_y = (2)\left(\tfrac{1}{3}\right)(1)(2.08^3 + 0.42^3) + (10)(0.42)^2 = 7.8 \text{ in.}^4$$

$$J = 208.3 + 7.8 = 216.1 \text{ in.}^4$$

$$f_h = \frac{(22)(2.58)(5)}{216.1} = 1.31 \text{ k/in.}$$

$$f_v = \frac{(22)(2.58)(2.08)}{216.1} = 0.546 \text{ k/in.}$$

$$f_s = \frac{22}{15} = 1.47 \text{ k/in.}$$

$$f_r = \sqrt{(0.546 + 1.47)^2 + (1.31)^2} = 2.40 \text{ k/in.}$$

$$\text{weld size reqd.} = \frac{2.4}{(0.707)(1.0)(0.30 \times 70)} = 0.162 \text{ in.} \quad \left(\text{say } \tfrac{3}{16} \text{ in.}\right)$$

Design of shop weld to column:

$$f_h = \frac{Re}{0.54L^2} = \frac{(44)(3)}{(0.54)(10)^2} = 2.44 \text{ k/in.}$$

$$f_s = \frac{44}{(2)(10)} = 2.2 \text{ k/in.}$$

$$f_r = \sqrt{(2.44)^2 + (2.2)^2} = 3.29 \text{ k/in.}$$

$$\text{weld size reqd.} = \frac{3.28}{(0.707)(1.0)(0.30 \times 70)} = 0.222 \text{ in.} \quad \left(\text{say } \tfrac{1}{4} \text{ in.}\right)$$

12-7. DESIGN OF WELDED SEATED BEAM CONNECTIONS

Another type of fairly flexible beam connection can be obtained by using a beam seat such as the one shown in Fig. 12-15. Beam seats are obviously of advantage to the workers performing the erection. The seat is probably shop-welded to the column and field welded to the beams. Seat angles, which are also called shelf angles, may be punched for temporary erection bolts as shown in the figure. These holes can be slotted to permit easy alignment of the members.

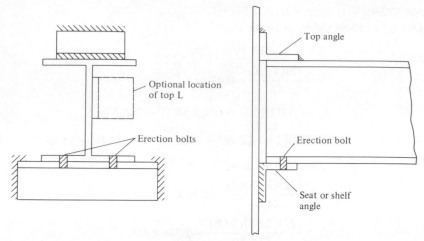

Figure 12-15

In addition to the seat angle, a top angle is used which furnishes lateral support. As it is not usually assumed to resist any of the load, its size is probably selected by judgment. Fairly flexible angles are used which will bend away from the column or girder to which it is connected when the beam tends to rotate downward when under load [see Fig. 12-6(b)]. A common size selected is the $4 \times 4 \times \frac{1}{4}$-in. angle.

Unstiffened seated beam connections of practical sizes can support light loads up to only 50 or 60 k (222 to 267 kN) when A36 steel is used. For these light loads two vertical end welds on the seat are sufficient. The top angle is welded only on its toes so that as the beam tends to rotate this flexible angle will be free to pull away from the column.

The minimum length of bearing required by the AISC for the beam on the seat angle can be determined from the usual web-crippling formula which was discussed in Chapter 7. A length of from 3 to 4 in. (75 to 100 mm) is usually satisfactory. The beam reaction causes a bending moment in the horizontal leg of the seat angle. The critical section for bending (represented by section 1-1 in Fig. 12-16) is usually assumed to be at the toe of the fillet which is approximately $t + \frac{3}{8}$ in. from the back of the

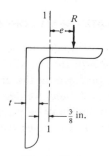

Figure 12-16

vertical leg for most angles and R (the beam reaction) is assumed to be concentrated at the center of the theoretical required bearing length.

Some designers, to obtain more flexible seats, use a higher allowable bending stress than normally permitted (an increase of approximately 20% being quite common). The resulting higher stress on the seat will cause it to rotate more and approach more nearly true simple end conditions. The length of the vertical leg of the seat angle can be determined from the weld size required. A depth can be assumed and the weld size required for that depth determined. After one trial a depth can probably be selected which will require a reasonable weld size.

There are many opinions concerning the stress variation in these vertical welds. Some designers assume that the neutral axis occurs at middepth, while others assume that the bottom third of the welds is in compression and the top two thirds in tension. The former method is used in the solution of Example 12-4. In this problem the author assumes there is a setback of $\frac{1}{2}$ in. while the tables for seated beam connections in the AISC Manual are based on setbacks of $\frac{3}{4}$ in. to take care of the fact that beams may "underrun" a little. This causes a little difference in answers.

Example 12-4

Design an unstiffened beam seat for the simple beam shown in Fig. 12-17, using A36 steel, SMAW fillet welds, E70 electrodes, and the AISC Specification.

SOLUTION
Length of bearing required:

$$\frac{R}{t(N+k)} = 0.75F_y$$

$$\frac{20{,}360}{0.295\left(N+1\frac{1}{8}\right)} = 27{,}000$$

$$N = 1.43 \text{ in.}$$

Use a 4-in. horizontal leg

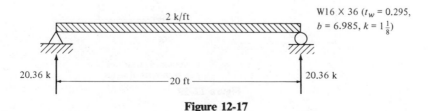

Figure 12-17

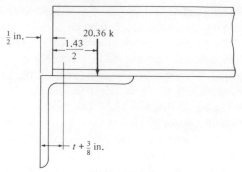

Figure 12-18

Selection of seat angle:

width of beam $=6.99$ in.
assume width of seat $L=8.00$ in.
assume thickness of $L=\frac{1}{2}$ in.

$$e=\frac{1.43}{2}+0.50-\frac{1}{2}-\frac{3}{8}=0.34 \text{ in.} \qquad \text{(see Fig. 12-18)}$$

$$M=(20.36)(0.34)=6922 \text{ in.-k}$$

Using allowable $f=0.75 \ F_y =27{,}000$ psi:

$$f=\frac{Mc}{I}$$

$$27{,}000=\frac{(6922)(t/2)}{\left(\frac{1}{12}\right)(8)(t^3)}$$

$$t=0.44 \text{ in.} \qquad \left(\text{use } \tfrac{1}{2} \text{ in.}\right)$$

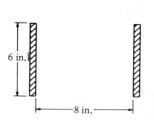
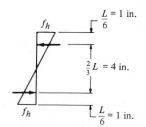

Assumed stress variation
in vertical welds

Figure 12-19

Design of vertical welds (see Fig. 12-19): Assume welds are 6-in. deep.:

$$\text{moment on each weld} = \left(\frac{R}{2}\right)(e)$$

$$e = \tfrac{1}{2} + 1.75 = 2.25 \text{ in.}$$

$$M = (2.25)\left(\frac{20.36}{2}\right) = 22.9 \text{ in.-k}$$

Resisting couple has a lever arm of $\frac{2}{3} L$ or 4 in.:

$$\left(f_h \times 3 \times \tfrac{1}{2}\right)(4) = 22.9$$

$$f_h = 3.82 \text{ k/in.}$$

$$f_v = \frac{20.36}{12} = 1.70 \text{ k/in.}$$

$$f_r = \sqrt{3.82^2 + 1.70^2} = 4.18 \text{ k/in.}$$

$$\text{weld size reqd.} = \frac{4.18}{(1)(0.707)(0.3 \times 70)} = 0.282 \text{ in.} \qquad \left(\text{use } \tfrac{5}{16} \text{ in.}\right)$$

Use $6 \times 4 \times \tfrac{1}{2}$-in. seat L $\times$ 0 ft 8 in. with $4 \times 4 \times \tfrac{1}{4}$-in. top L

12-8. WELDED STIFFENED BEAM SEAT CONNECTIONS

When beam reactions become fairly large (above 40 to 60 k depending on steel used) the thickness required for the seat angles becomes excessive and it becomes necessary to use stiffened seats. Stiffened beam seats probably consist of a structural tee or of a couple of plates welded into the shape of a tee and their design is relatively simple. Figure 12-20 shows one type of stiffened beam seat.

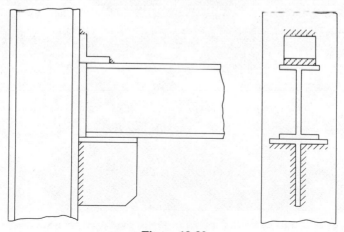

Figure 12-20

A stiffened beam seat must first provide sufficient bearing length for the beam from a web-crippling standpoint. After this length is determined it is necessary to estimate the position of the center of gravity of the beam reaction. When the beam is loaded and as its end begins to rotate, the center of gravity of the reaction tends to move toward the outer edge of the seat. The result is a larger eccentricity and a larger eccentric moment than would occur in a comparable unstiffened seat. Some designers assume that the center of gravity of the reaction is located at the ·center of the theoretical bearing length, while others assume it is located two-thirds of the actual bearing length toward the outer edge. For this discussion the author assumes the reaction is centered a distance from the outer edge of the seat equal to one-half of the theoretical bearing length required for web crippling.

The stems of beam seats usually are in little danger of buckling and they are not designed by a high-level theoretical method. In fact the usual practice followed is to make the stem thickness at least equal to the web thickness of the beam. Some more conservative designers limit the stem to certain maximum l/r ratios, and check it for axial load and bending. The depth of the stem is determined by the length of the weld required, and the width of the flange should be a little wider than the beam flange to permit reasonably simple field welds. On each side it is desirable for the seat to be at least twice the weld size wider than the beam flange.

Top flanges of beams connected by strap plates over a girder while lower flanges are butt-welded to the girder web, Ainsley Building, Miami, Fla. (Courtesy of The Lincoln Electric Company.)

Finally, the welds for the seat must be of sufficient size to adequately resist the shear and bending stresses applied to them. The vertical welds are close together and, therefore, have little resistance to transverse flexure. For this reason it is usually considered desirable to weld horizontally along the bottom of the top flange a distance equal to roughly one-fourth to one-half of the vertical web depth. These horizontal welds greatly increase the resistance of the connections to twisting. The horizontal weld can be made on top of the flange of the tee, but there is a slight possibility that such a weld may get in the way of the beam.

In the example to follow (Example 12-5) the stem is not considered to resist any of the moment as the weld is assumed to take it all. The maximum stress occurs at the bottom of the vertical welds and is determined by a combination of the shearing and flexure stresses. The weld dimensions are roughly estimated, and the size required for these dimensions is determined to see if the original dimensions are reasonable.

Example 12-5

Design a structural tee stiffened seat for the reactions of the beam shown in Fig. 12-21. The design is to be made with A36 steel, SMAW fillet welds, E70 electrodes, and the AISC Specification.

SOLUTION
Length required for bearing:

$$\frac{R}{t(N+k)}=0.75F_y$$

$$\frac{48{,}480}{(0.415)(N+1.375)}=27{,}000$$

$$N=2.95+0.5 \text{ (set back)}=3.45 \text{ in.} \qquad \left(\text{use } 3.5 \text{ in.}\right)$$

$$e=3.50-\frac{2.95}{2}=2.025 \text{ in.}$$

$$M=(48{,}480)(2.025)=98{,}172 \text{ in.-lb}$$

Estimate weld length by dividing reaction by strength of a $\frac{3}{16}$-in. fillet weld:

$$\text{estimated length}=\frac{48.480}{\left(\frac{3}{16}\right)(0.707)(0.30\times70)}=17.41 \text{ in.} \qquad \left(\text{say } 18 \text{ in.}\right)$$

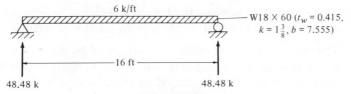

Figure 12-21

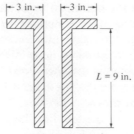

Figure 12-22

Assume the welds shown in Fig. 12-22.

$$A = (2)(9+3) = 24 \text{ in.}$$

$$y = \frac{(18)(4.5)}{24} = 3.37 \text{ in.}$$

$$I_x = \left(\tfrac{1}{12}\right)(2)(9)^3 + (18)(1.13)^2 + (6)(3.37)^2 = 212.7 \text{ in.}^4$$

$$f_v = \frac{48.48}{18} = 2.69 \text{ k/in.} \quad \text{(neglecting shear carried by horizontal segments)}$$

$$f_s = \frac{(98.172)(5.63)}{212.7} = 2.60 \text{ k/in.}$$

$$f_r = \sqrt{(2.69)^2 + (2.60)^2} = 3.74 \text{ k/in.}$$

weld size reqd. $= \dfrac{3.74}{(1)(0.707)(0.30 \times 70)} = 0.252 \text{ in.}$ $\left(\text{use } \tfrac{5}{16} \text{ in.}\right)$

Selection of structural tee: The structural tee must have a stem thickness at least equal to that of the beam web (0.415 in.) and must be thick enough to support a $\tfrac{5}{16}$-in. weld. It is common to use a minimum stiffener plate thickness at least equal to two times the required weld size when using A36 steel and E70 electrodes $(2 \times \tfrac{5}{16} = \tfrac{10}{16} \text{ in.})$. Its flange should be wider than that of the W18×60 by approximately $4 \times \tfrac{5}{16}$ in. to permit easy field welding. Its depth must also be at least 10 in.

Use WT 10.5×73.5 stiffened beam seat with a top L4×4×$\tfrac{1}{4}$ in.

12-9. WELDED MOMENT-RESISTANT CONNECTIONS

Perhaps the greatest efficiency in structural welding is reached in fully continuous structures. The two major reasons for this efficiency are

1. The negative moments which are produced in the ends of continuous beams cause appreciable reductions in the positive moments out in the spans. These reduced moments permit the use of smaller members.

2. When overloads occur, they are more easily redistributed in fully continuous structures. The student will remember that the AISC recognizing this plastic redistribution permits the design of fully continuous structures for only $\frac{9}{10}$ of the maximum negative moments caused by gravity loads. This reduction is permissible if the positive moments are increased by $\frac{1}{10}$ of the average negative moments. (See Section 6-8.)

The welded beam connections discussed prior to this section were designed with the object of eliminating most of the moment resistance. For fully continuous structures the connections are designed to resist the full calculated moments. Figure 12-23(a) shows a common type of moment-resisting connection. In the connection shown the tensile force at the top of the beam is transferred by fillet welds to the top plate and by groove welds from the plate to the column. For easier welding the top plate is probably tapered as shown in part (b) of the figure. The student may have noticed such tapered plates used for facilitating welding in other situations.

If the design is to be made for gravity loads alone the only forces (other than shear) which need to be considered are the total tension T in the top flange and the total compression C in the bottom flange. Should wind or other lateral forces be involved, it may become necessary to design all welds to resist tension as well as compression.

A common practice in moment-resisting welded connections is to butt weld the beam flanges flush with the column on one end and connect the beam on the other end with the type of connection just described. This practice is illustrated in Fig. 12-24, in Example 12-6.

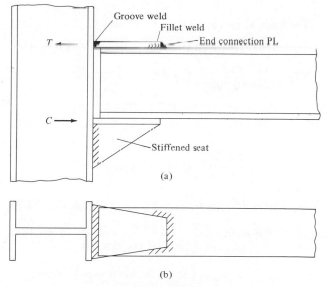

(a)

(b)

Figure 12-23

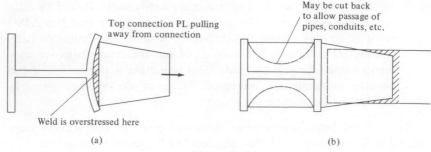

Figure 12-24

Example 12-6

Design moment-resistant connections of the type shown in Fig. 12-24 for the ends of a W18×46. The beam, which consists of A36 steel, has end reactions of 42 k and end moments of 165 ft-k. The AISC Specification, SMAW welds, and E70 electrodes are to be used. It is assumed that a 6×4×$\frac{3}{4}$-in. seat L has previously been selected.

SOLUTION

Shear connection: Try $\frac{1}{4}$-in. fillet welds on seat L:

$$\text{depth reqd.} = \frac{42}{(2)(0.707)\left(\frac{1}{4}\right)(0.30\times70)} = 5.66 \text{ in.} \qquad \left(\underline{\text{use 6 in. each side}}\right)$$

Moment connections on flush end: Assuming a bevel butt weld for full flange width:

$$T = \text{force to be carried}$$
$$= \text{moment divided by c.-to-c. distance of flanges}$$
$$= \frac{12\times165}{18.06-0.605} = 113.4 \text{ k}$$

Strength of butt weld in tension for full width of flange = (6.060)(0.605)(24) = 88.0 k. Tension to be resisted by auxiliary PL = 113.4 − 88.0 = 25.4 k. Assuming a $\frac{3}{8}$-in. thick PL, its width will equal:

$$\frac{25.4}{\frac{3}{8}\times24} = 2.82 \text{ in.} \qquad \left(\text{say 3 in.}\right)$$

Assume a $\frac{3}{16}$-in. fillet weld on auxiliary PL as shown in Fig. 12-25:

$$\text{length of fillet weld} = \frac{25.4}{\left(\frac{3}{16}\right)(0.707)(0.30\times70)} - 3 = 6.12 \text{ in.}$$
$$\times \left(\underline{\text{say 4 in. each side}}\right)$$

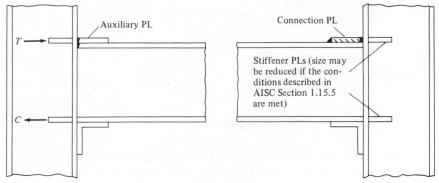

Figure 12-25

Allow 1 in. for bevel groove weld.

Use auxiliary $PL\frac{3}{8} \times 3 \times 4$

Note: Should wind moment be involved, a connection similar to this one would be designed for the lower flange; however, allowable stresses may be increased by one-third to resist wind forces, says AISC.

Moment connection at nonflush end: Assuming T and C a distance apart equal to the beam depth:

$$T = \frac{12 \times 165}{18.06} = 109.6 \text{ k}$$

Assuming top connection has a width something less than the width of the beam flange (say 5 in.), its thickness can be found as follows.:

$$t = \frac{109,600}{5.0 \times 24,000} = 0.913 \text{ in.} \qquad \left(\text{say 1.0 in.} \right)$$

Assume $\frac{3}{8}$-in. fillet welds on PL

reqd. length of weld $= \dfrac{109.6}{\left(\frac{3}{8}\right)(0.707)(0.30 \times 70)} = 19.69 \text{ in.}$ (see Fig. 12-26)

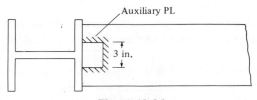

Figure 12-26

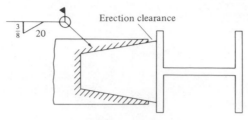

Figure 12-27

If the column to which a beam is being connected bends appreciably at the connection the moment resistance of the connection is going to be reduced no matter how good the connection may be. Furthermore, if the top connection plate in pulling away from the column tends to bend the column flange as shown in part (a) of Fig. 12-24, the middle part of the weld may be greatly overstressed. For these reasons it is a desirable practice to reinforce the column with plates opposite the column flange as shown in part (b) of the same figure. This practice is somewhat objectionable to architects, for they like to run pipes and conduits up the columns, but the plates can be cut back somewhat as shown in the figure.

The web stiffeners shown for the column of Figure 12-25 may or may not be required depending on the proportions of the member. Should the expression to follow (AISC Formula 1.15-1) yield a positive value for A_{st} (the combined area of a pair of column web stiffeners) stiffeners will be required.

$$A_{st} = \frac{P_{bf} - F_{yc}t(t_b + 5k)}{F_{yst}}$$

in which P_{bf} equals the computed force (kips) caused by the flange or moment connection plate. This value is to be multiplied by $\frac{5}{3}$ (to agree with the usual 1.67 safety factor) when the force is computed for dead and live loads only. Should the force be due to dead and live loads in conjunction with wind or earthquake forces it should be multiplied by $\frac{4}{3}$.

F_{yc} and F_{yst} = the yield stresses (ksi) of the column and stiffeners, respectively

t = thickness (in.) of the column web

t_b = thickness (in.) of the flanges or the moment connection plate delivering the concentrated force

k = the distance (in.) from the outer part of the column flange to the web toe of the fillet for a rolled section and an equivalent distance for a welded built-up section.

It is also necessary for the slenderness ratio of an unstiffened web of the supporting member to be limited to avoid the possibility of buckling. For this purpose AISC Formula 1.15-2 is used which states that a stiffener or a pair of stiffeners must be used opposite the compression flange if the column depth between the web toes of the fillets exceeds the following

$$\frac{4100t^3\sqrt{F_{yc}}}{P_{bf}}$$

By AISC Formula 1.15-3 a pair of stiffeners must be provided opposite the tension flange if the thickness of the column flange (t_f) is less than the following

$$0.4\sqrt{\frac{P_{bf}}{F_{yc}}}$$

When stiffeners are needed by the preceding expressions they must meet several requirements listed in AISC Section 1.15.5.4. For instance the width of each stiffener plus $\frac{1}{2}$ of the column web thickness must not be less than $\frac{1}{3}$ times the width of the flanges or moment connection plate delivering the concentrated force. In addition the thickness of the stiffener must not be less than $t_b/2$. If the force is applied to only one column flange the length of the stiffener need not be greater than $\frac{1}{2}$ the column depth. See Figure 12-26.

To facilitate the application of these formulas the AISC Manual in its column section provides numerical values for each section above for which web stiffeners are required. If the beam force P_{bf} (which has been factored by multiplying it by the $\frac{5}{3}$ or $\frac{4}{3}$ value as required) is equal to or less than the values of P_{wb}, P_{fb}, P_{wi}, and P_{wo} it is not necessary to use web stiffeners. These latter values are defined as follows

P_{wb} = maximum column web resisting force (kips) at beam compression flange

$$= \frac{4100t^3\sqrt{F_{yc}}}{d_c}$$

P_{fb} = maximum column web resisting force (kips) at beam tension flange

$$= \frac{F_{yc}t_f^2}{0.16}$$

$P_{wi} = F_{yc}t$ (kips/in.)

$P_{wo} = 5F_{yc}tk$ (kips)

Space is not taken in this chapter to present a numerical illustration since a very clear design example is provided on page 3-13 of the AISC Manual (8th ed.).

Problems

12-1. Determine the maximum end reaction that can be transferred through the web angle connection shown in the accompanying illustration if the moment produced by the connection is neglected. The steel is A36, the bolts are $\frac{7}{8}$ in. A325 friction type, and the specification is AISC.

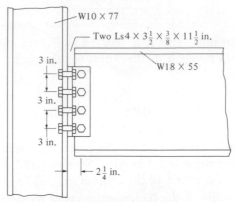

Problem 12-1

12-2. Using the AISC Manual, select a pair of standard web angles for a W33×241 with a maximum reaction of 180 k. The bolts are to be 1-in. A325 bearing-type threads excluded from shear plane and the beam consists of A441 steel.

12-3. Repeat Prob. 12-2 if the maximum reaction is 250 k.

12-4. Select a standard framed beam connection from the AISC Manual for a W16×40 with a maximum reaction of 50 k. The bolts are to be $\frac{7}{8}$-in. A325 friction-type and the beam is made from A36 steel.

12-5. Select a pair of framed beam angles from the AISC Manual to connect a W21×68 beam (A36 steel) with a 44-k reaction to a column. The connection is to be made with $\frac{3}{4}$-in. A325 high-strength bolts, threads excluded from shear plane, and a friction-type of connection is to be used.

12-6. Determine the resisting moment of the connection shown in the accompanying illustration if A36 steel, $\frac{7}{8}$-in. A325 high-strength bolts in a bearing-type connection with threads excluded from shear plane, and the AISC Specification are to be used.

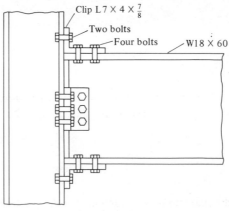

Problem 12-6

12-7. A W16×36 beam with an end shear of 20 k and an end moment of 35 ft-k is connected to a W12×58 column. Design an angle connection of the type used in Prob. 12-6 using $\frac{3}{4}$-in. A325 bearing type high-strength bolts, threads excluded from shear plane, and the AISC Specification. A36.

12-8. Rework Prob. 12-7 if a split beam connection of the type shown in Fig. 12-10 is to be used.

12-9. Design a structural tee connection of the type shown in Fig. 12-11 to develop one-third of the resisting moment of a W21×68 beam. Use A36 steel, $\frac{7}{8}$-in. A325 bearing-type high-strength bolts, threads excluded from shear plane, and the AISC Specification.

12-10. Rework Prob. 12-2 using SMAW welded shop and field connections and E70 electrodes.

12-11. Rework Prob. 12-4 using SMAW welded shop and field connections and E70 electrodes.

12-12. Select a framed beam connection from the AISC Manual for a W33×130 (A36) beam with a 120-k reaction. The web angles are to be SMAW shop-welded to the beam using E70 electrodes and are to be field-connected to the girder with $\frac{7}{8}$-in. A325 bearing-type high-strength bolts.

12-13. Rework prob. 12-12 using an SMAW welded shop and field connections.

12-14. Using the AISC Manual, select a seated beam connection using $\frac{3}{4}$-in. A325 high-strength bolts in a bearing-type connection threads excluded from shear plane to support a 30-k reaction from a W16×40 (A36) beam. Assume col. gage=$5\frac{1}{2}$ in. in col. web.

12-15. Rework Prob. 12-14 using SMAW welded seated beam connection and E70 electrodes.

12-16. Using the AISC Manual, select a stiffened beam seat for a W24×84 (A36) beam for a reaction of 65 k using $\frac{3}{4}$-in. A325 bearing-type high-strength bolts.

12-17. Rework Prob. 12-16 using a stiffened welded seat with E70 electrodes.

12-18. Design SMAW welded end connections for the ends of a W21×68 beam. The end moments produced are due to wind and are assumed to be equal to $\frac{2}{7}$ of the moment resistance of the beam. Use E70 electrodes, the AISC Specification, and assume the column flanges to be 14 in. wide.

12-19. Design SMAW welded end connections for the ends of W24×76 beam to resist gravity load moments equal to 50 percent of the moment resistance of the section. Use E70 electrodes, A36 steel, and the AISC Specification. Assume the column flange to be 16 in. wide.

12-20. The beam shown in the accompanying illustration is assumed to be attached at its ends with completely rigid connections. Select the beam assuming full lateral support using A36 steel, and design SMAW welded connections using E70 electrodes and the AISC Specification.

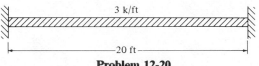

3 k/ft

20 ft

Problem 12-20

12-21. The two W16×40 beams shown in the accompanying illustrations are to be made continuous across the supporting W24×84 girder. If the end reactions of the beams are 20 k and their end moments are 40 ft-k, design the connection using E70 electrodes, SMAW welds, and the AISC Specification.

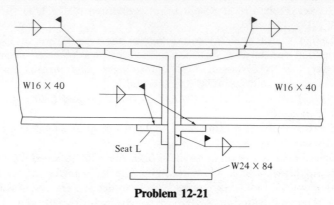

W16 × 40

W16 × 40

Seat L

W24 × 84

Problem 12-21

Chapter 13
Design
of Steel
Buildings

13-1. INTRODUCTION

The material in this chapter generally pertains to the design of steel buildings from one to several stories in height, while the design of multistory buildings is presented in Chapter 20. The present discussion applies to apartment houses, office buildings, warehouses, schools, and institutional buildings which are not very tall with respect to their least lateral dimensions. The separating factor between the buildings discussed here and those mentioned in the multistory chapter is the matter of wind forces and not the actual height of the building in question. The usual rule of thumb is that if the height of the building is not greater than twice its least lateral dimension, provision for wind forces is unnecessary. For buildings of these dimensions the walls and partitions probably provide sufficient resistance to wind forces except in unusual circumstances.

13-2. TYPES OF STEEL FRAMES USED FOR BUILDINGS

Steel buildings are usually classified as being in one of four groups according to their type of construction. These types are: bearing-wall construction, skeleton construction, long-span construction, and combination steel and concrete framing. More than one of these construction types can be used in the same building. Each of these types is briefly discussed in the following paragraphs.

Bearing-Wall Construction

Bearing-wall construction is the most common type of single-story light commercial construction. The ends of beams or joists or light trusses are supported by the walls that transfer the loads to the foundation. For taller buildings the walls by necessity must be thicker to provide sufficient strength and lateral stability. These increasing wall thicknesses usually establish an economical upper limit of roughly two or three stories for

Coliseum in Spokane, Wash. (Courtesy of Bethlehem Steel Company.)

bearing-wall construction; although the method may occasionally be advantageous for parts of taller buildings. (A great deal of research is being done today around the world concerning the use of brick bearing-wall construction for much taller buildings and many such structures have been successfully erected. Of particular note is the work done in this area by the department of civil engineering at the University of Edinburgh in Scotland.)

The average engineer is not very well versed in the subject of wall-bearing construction with the result that she may often require complete steel or reinforced concrete frames where wall-bearing construction might have been just as satisfactory and at the same time more economical. Wall-bearing construction is not very resistant to seismic loadings and does have an erection disadvantage for buildings of more than one story. For such cases it is necessary to place the steel members floor by floor as the masons complete their work below thus requiring alternation of the masons and ironworkers.

Bearing plates are probably necessary under the ends of the beams or light trusses which are supported by the masonry walls because of the relatively low bearing strength of the masonry. Although theoretically the beam flanges may on many occasions provide sufficient bearing without bearing plates, the plates are almost always used—particularly where the members are so large and heavy that they must be set by a steel erector.

The plates are probably shipped loose and set in the walls by the masons. Setting them in their correct positions and at the correct elevation is a very critical part of the construction. Should they not be properly set there will be some delay in correcting their positions. If a steel erector is used he will probably have to make an extra trip to the job.

When the ends of a beam are enclosed in a masonry wall, some type of wall anchor is desirable to prevent the beam from moving longitudinally with respect to the wall. The usual anchors consist of bent steel bars passing through beam webs. These are called *government anchors* and details of their sizes are given in the AISC Manual. Occasionally clip angles attached to the web are used instead of government anchors. Should longitudinal loads of considerable magnitude be anticipated, regular vertical anchor bolts may be used at the beam ends.

For small commercial and industrial buildings bearing-wall construction is quite economical when the clear spans are not greater than roughly 35 or 40 ft. If the clear spans are much greater it becomes necessary to thicken the walls and use pilasters to ensure stability. For these cases it may often be more economical to use intermediate columns if permissible.

Skeleton Construction

In skeleton construction the loads are transmitted to the foundations by a framework of steel beams and columns. The floor slabs, partitions, exterior walls, etc. are all supported by the frame. This type of framing, which can be erected to tremendous heights, is often referred to as beam-and-column construction.

In beam-and-column construction the frame usually consists of columns spaced 20, 25, or 30 ft apart with beams and girders framed into

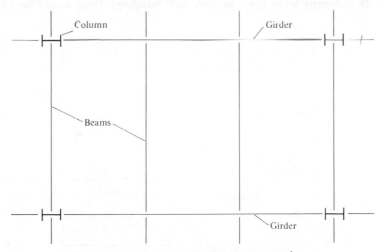

Figure 13-1 Beam and column construction.

them from both directions at each floor level. One very common method of arranging the members is shown in Fig. 13-1. The girders are placed in the longer direction between the columns, while the beams are framed between the girders in the short direction. With various types of floor construction other arrangements of beams and girders may be used.

For skeleton framing the walls are supported by the steel frame and are generally referred to as *nonbearing* or *curtain walls*. The beams supporting the exterior walls are called *spandrel beams*. These beams, which are illustrated in Fig. 13-2, can usually be placed so that they will serve as the lintels for the windows.

Long-Span Steel Structures

When it becomes necessary to use very large spans between columns as for field houses, auditoriums, theaters, hangars, or hotel ballrooms, the usual skeleton construction may not be sufficient. Should the ordinary rolled W sections be insufficient it may be necessary to use coverplated beams, plate girders, box girders, large trusses, arches, rigid frames, and the like. When depth is limited coverplated beams, plate girders, or box girders may be called upon to do the job. Should depth not be so critical trusses may be satisfactory. For very large spans arches and rigid frames are often used. These various types of structures are referred to as long-span structures and most of them are discussed at length in the chapters to follow. Figure 13-3 shows a few of these types of structures.

Combination Steel and Concrete Framing

A tremendous percentage of the buildings erected today make use of a combination of reinforced concrete and structural steel. If reinforced-concrete columns were used in very tall buildings they would be rather

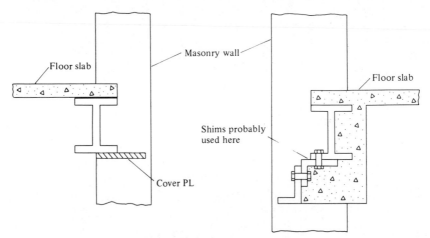

Figure 13-2 Spandrel beams.

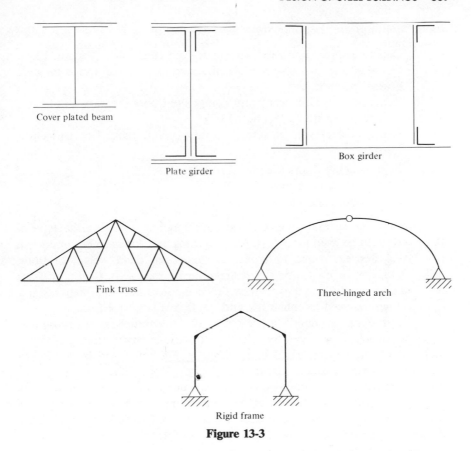

Cover plated beam

Plate girder

Box girder

Fink truss

Three-hinged arch

Rigid frame

Figure 13-3

large on the lower floors and take up considerable space. Steel column shapes surrounded by and bonded to reinforced concrete are commonly used and are referred to as combination columns. Chapter 15 is devoted to the design of composite floors where steel beams and reinforced concrete slabs are bonded together in such a manner that they act compositely in resisting loads. Other systems will be mentioned in later chapters.

13-3.   COMMON TYPES OF FLOOR CONSTRUCTION

Concrete floor slabs of one type or another are used almost universally for steel-frame buildings. Concrete floor slabs are strong, have excellent fire ratings, and good acoustic ratings. On the other hand, appreciable time and expense is required for providing the formwork necessary for most slabs. Concrete floors are heavy, they need some type of reinforcing bars or mesh included, and there may be a problem involved in making them watertight. Among the many types of concrete floors used today for steel frame buildings are the following.

1. Concrete slabs supported with open-web steel joists (Section 13-4).
2. One-way and two-way reinforced concrete slabs supported on steel beams (Section 13-5).
3. Concrete slab and steel beam composite floors (Section 13-6).
4. Concrete-pan floors (Section 13-7).
5. Structural clay tile, gypsum tile, and concrete block floors (Section 13-8).
6. Steel decking floors (Section 13-9).
7. Flat slab floors (Section 13-10).
8. Precast concrete slab floors (Section 13-11).

Among the several factors to be considered in selecting the type of floor system to be used for a particular building are: loads to be supported, fire rating desired, sound and heat transmission, dead weight of floor, ceiling situation below (to be flat or have beams exposed), facility of floor for locating conduits, pipes, wiring, etc.; appearance, maintenance required, time required to construct, and depth available for floor.

The student can obtain a great deal of information about these and other construction practices by referring to various engineering magazines and catalogues, particularly *Sweet's Catalog File* published by McGraw-Hill Information Systems Company of New York. The author cannot make too strong a recommendation for the student to examine these books to see the tremendous amount of data available therein which may be of use to him in his engineering practice. The sections to follow present brief descriptions of the floor systems mentioned in this section along with some discussions of their advantages and chief uses.

13-4. CONCRETE SLABS ON OPEN-WEB STEEL JOISTS

Perhaps the most common type of floor slab in use for small steel-frame buildings is the slab supported by open-web steel joists. The joists are really small parallel chord trusses whose members are often made from bars (from whence the common name *bar joist*) or small angles or other rolled shapes. A special paper reinforced with welded wire fabric may be attached to the top of the closely spaced joists and a thin concrete slab placed on top of the paper. This is probably the lightest type of concrete floor and comes very close to being the most economical one. A sketch of an open-web joist floor is shown in Fig. 13-4.

Open-web joists are particularly well suited to building floors with relatively light loads and for structures where there is not too much vibration. They have been used a good deal for fairly tall buildings but generally speaking they are better suited for the shorter buildings. They are very satisfactory for supporting floor and roof slabs for schools, apartment houses, hotels, office buildings, restaurant buildings, and other similar low-level buildings.

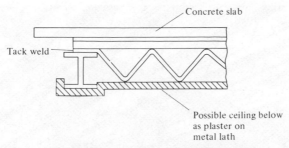

Figure 13-4 Open-web joists.

Maximum open-web joist spacings center to center are about 24 in. for concrete floor slabs and 30 in. for concrete roof slabs. For floors and roofs with steel decking (to be described in Section 13-9) they can possibly be spaced as high as 7 ft on centers. Open-web joists can be obtained in standard depths from 8 to 24 in. in 2-in. increments. Each of the joists has a maximum span of about 48 ft for the deeper sections.

Open-web joists must be braced laterally to keep them from twisting or buckling and also to keep the floors from being too springy. Lateral support is provided by *bridging*, which consists of continuous horizontal rods fastened to the top and bottom chords of the joists or of diagonal cross bracing. Bridging is desirably used at spaces not exceeding 7 ft on centers.

Open-web joists are easy to handle and are quickly erected. If desired, a ceiling can be attached to the bottom of the joists or suspended therefrom. The open spaces in the webs are admirably suited for placing conduits, ducts, wiring, piping, etc. The joists should be either welded to supporting steel beams or well anchored in the masonry walls. When concrete slabs are placed on top of the joists they are usually from 2 to $2\frac{1}{2}$ in. thick. Nearly all of the many concrete slabs on the market today can be used successfully on top of open-web joists.

Open-web joists are selected from manufacturers' catalogues. The student can learn a great deal about open-web joists and other construction materials by studying various manufacturers' literature. An exceptionally useful reference of this type is the previously mentioned *Sweet's Catalog File*.

The joists are designated as "J-Series" when the material used has a minimum yield point of 36 ksi (248 MPa) and as "H-Series" when the chord material consists of steel with a minimum yield point of 50 ksi (345 MPa). There are also long-span joists "L-J Series" and the deep long-span joists (DLJ Series with $F_y = 36$ ksi) intended for roof construction. The DLH Series are deep longspan joists with minimum yield points of 50 ksi. In the tables the number preceding the joist number is the nominal depth of the joist in inches, the letter or letters (as J or DLJ) indicates the joist series, and the number to the right designates the chord section.

Short-span joists in the Univac Center of Sperry-Rand Corporation, Blue Bell, Pa. (Courtesy of Bethlehem Steel Company.)

The tables provided in the manufacturers' catalogues give the safe uniform loads in pounds per foot which joists can support. In addition, the maximum end reaction in pounds which a joist can support from the standpoint of shear and the maximum uniform live load in pounds per foot which it can support without having a deflection larger than $\frac{1}{360}$ of the span are provided.

13-5. ONE-WAY AND TWO-WAY REINFORCED CONCRETE SLABS

One-Way Slabs

A very large number of concrete floor slabs in old office and industrial buildings consisted of one-way slabs about 4 in. thick supported by steel beams 6 to 8 ft on centers. These floors were often referred to as concrete arch floors because at one time brick or tile floors were constructed in approximately the same shape, that is in the shape of arches with flat tops.

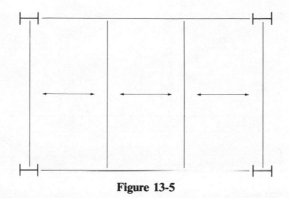

Figure 13-5

A one-way slab is shown in Fig. 13-5. The slab spans in the short direction shown by the arrows in the figure. One-way slabs are usually used when the long direction is two or more times the short direction. In such cases the short span is so much stiffer than the long span that almost all of the load is carried by the short span. The short direction is the main direction of bending and will be the direction of the main reinforcing bars in the concrete, but temperature and shrinkage steel is needed in the other direction.

A typical cross section of a one-way slab floor with supporting steel beams is shown in Fig. 13-6. When steel beams or joists are used to support reinforced concrete floors it may be necessary to encase them in concrete or other materials to provide the required fire rating. Such a situation is shown in the figure.

It may be necessary to leave steel lath protruding from the bottom flanges or soffits of the beam for the purposes of attaching plastered ceilings. Should such ceilings be required to cover the beam stems, this floor system will lose a great deal of its economy.

One-way slabs have an advantage when it comes to formwork in that the forms can be supported entirely by the steel beams with no vertical shoring needed. They have a disadvantage in that they are much heavier than most of the newer lightweight floor systems. The result is that they are not used as often as formerly for lightly loaded floors; but when a floor to support heavy loads is desired or a rigid floor or a very durable floor, the one-way slab may be a very desirable selection.

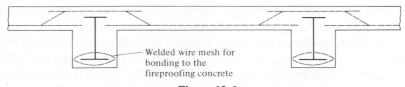

Welded wire mesh for
bonding to the
fireproofing concrete

Figure 13-6

The 104-ft steel joists for the Bethlehem Catholic High School, Bethlehem, Pa. (Courtesy of Bethlehem Steel Company.)

Two-Way Concrete Slabs

The two-way concrete slab is used when the slabs are square or nearly so and supporting beams are planned under all four edges. The main reinforcing runs in both directions. Other characteristics are similar to those of the one-way slab.

13-6. COMPOSITE FLOORS

Composite floors are those where the steel beams (rolled sections, coverplated beams, or built-up members) are bonded together with the concrete slabs in such a manner that the two act as a unit in resisting the total loads which the beam sections would otherwise have to resist alone. There can be a saving in sizes of steel beams when composite floors are used because the slab acts as part of the beam.

A particular advantage of composite floors is that they utilize concrete's high compressive strength by keeping all or nearly all of the concrete in compression, and at the same time stress a larger percentage of the steel in tension than is normally the case in steel-frame structures. The result is less steel tonnage in the structure. A further advantage of composite floors is that they can permit an appreciable reduction in total floor thickness which is particularly important in taller buildings.

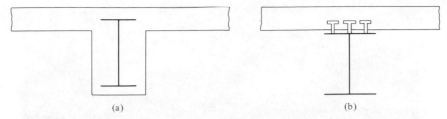

(a)   (b)

Figure 13-7 Composite floors. (a) Steel beam encased in concrete. (b) Steel beam bonded to concrete slab with shear connectors.

Two types of composite floor systems are shown in Fig. 13-7. The steel beam can be completely encased in the concrete and the horizontal shear transferred by friction and bond (plus some shear reinforcement if necessary). This type of composite floor is shown in part (a) of the figure. A second possibility is shown in part (b), where the steel beam is bonded to the concrete slab with some type of shear connectors. Various types of shear connectors have been used during the past few decades including spiral bars, channels, angles, studs, etc., but economic considerations usually lead to the use of round studs welded to the top flanges of the beams in place of the other types mentioned. Typical studs are $\frac{1}{2}$ to $\frac{3}{4}$ in. in diameter and 2 to 4 in. in length.

Cover plates may be welded to the bottom flanges of rolled steel sections. The student can see that with the slab acting as a part of the beam there is quite a large area available on the compressive side of the beam. By adding plates to the tensile flange a little better balance is obtained. Chapter 15 is devoted entirely to the design of composite floors for buildings and bridges.

One-piece metal dome pans. (Courtesy of Gateway Erectors, Inc.)

Temple Plaza parking facility, Salt Lake City, Utah. (Courtesy of Ceco Steel Products Corporation.)

13-7. CONCRETE-PAN FLOORS

There are several types of pan floors which are constructed by placing concrete on removable pan molds. (Some special light corrugated pans are also available which can be left in place.) Rows of the pans are arranged on wooden floor forms and the concrete is placed over the top of them producing a floor cross section similar to the one shown in Fig. 13-8. Joists are formed between the pans, giving a tee-beam type floor.

These floors, which are suitable for fairly heavy loads, are appreciably lighter than the one-way and two-way concrete slab floors. Although fairly light, they require a good deal of formwork including appreciable shoring underneath the stems. Labor is thus higher than for many floors but saving

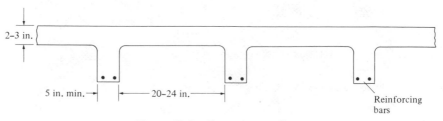

Figure 13-8   Concrete-pan floor.

due to weight reduction and reuse of standard-size pans may make them competitive. When suspended ceilings are required pan floors will have a decided economic disadvantage.

Two-way construction is available—that is, with ribs or stems running in both directions. Pans with closed ends are used and the result is a waffle type floor. This latter type floor is usually used when the floor panels are square or nearly so. Two-way construction can be obtained for reasonably economical prices and makes a very attractive ceiling below and one which has fairly good acoustical properties.

13-8. STRUCTURAL CLAY TILE, GYPSUM TILE, AND CONCRETE-BLOCK FLOORS

Similar to the concrete-pan floors are the clay tile, gypsum tile, and concrete-block floors. These floors have approximately the same formwork and roughly the same effect as the metal-pan floors. They are just about obsolete today, as pan floors are more competitive costwise and provide a more attractive undersurface. Tile floors were at one time economical for spans up to 25 or 30 ft. A typical cross section of this type of floor is shown in Fig. 13-9.

This system does have an advantage over the metal-pan floors in that it gives a level ceiling underneath to which plaster can be applied directly without the necessity of hanging a ceiling. The lower tile surfaces are very satisfactory for bonding the plaster and no metal lath is needed except for the main steel beams at edges of the floor if they are to be fireproofed. When tile is used it is desirable to use tile of the same material under the concrete stems (as shown in Fig. 13-9); otherwise it is difficult to get an even color of plaster on the different material soffits.

A concrete slab roughly $2\frac{1}{2}$ to 3 in. thick is cast monolithically with the supporting joists or stems spaced approximately 20 to 36 in. on centers. Among the standard tile sizes used are 12×12, 12×16, and 16×16 in plan with depths varying from 3 to 12 in. for the 12×12s and from 3 to 10 in. for the 12×16s and 16×16s. Tile floors have the advantage that conduits, wiring, piping, etc. can be conveniently placed through their cells. Should the panels be approximately square, joists can be used in both directions.

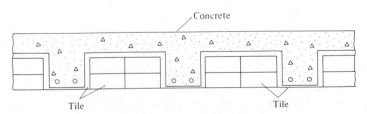

Figure 13-9 Tile or block floors.

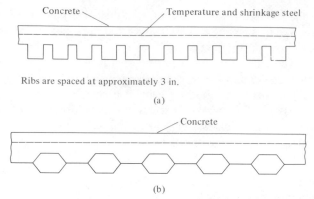

Ribs are spaced at approximately 3 in.

(a)

(b)

Figure 13-10   Steel-decking floors.

13-9.   STEEL-DECKING FLOORS

Typical cross sections of steel-decking floors are shown in Fig. 13-10. Only two types are shown in the figure but several other variations are available. In recent years steel-decking floors have become quite popular for some applications, particularly for office buildings. They are also popular for hotels and apartment houses and other buildings where the loads are not very large.

A particular advantage of steel-decking floors is that as soon as the decking is placed a working platform is available for the workers. The light steel sheets are quite strong and can span up to 20 ft or more. Due to the considerable strength of the decking the concrete does not have to be particularly strong. This fact permits the use of lightweight concrete as thin as 2 or $2\frac{1}{2}$ in.

The cells in the decking can be conveniently used for placing conduits, pipes, and wiring. The steel is probably galvanized and if exposed underneath can be left as it comes from the manufacturer or painted as desired. Should fire resistance be necessary, a suspended ceiling with metal lath and plaster will be used. The same is necessary if a flat ceiling is required below for the types of decking shown in Fig. 13-10, but decking is available with a flat soffit.

13-10.   FLAT SLABS

Formerly flat slab floors were limited to reinforced concrete buildings but today it is possible to use them in steel-frame buildings. A flat slab is a slab that is reinforced in two or more directions and transfers its loads to the supporting columns without the use of beams and girders protruding below. The supporting concrete beams and girders are made so wide that they are the same depth as the slab.

Flat slabs are of great value when the panels are approximately square, when more headroom is desired than is provided with the normal beam and girder floors, when heavy loads are anticipated, and when it is desired to place the windows as near to the tops of the walls as possible. Another advantage is the flat ceiling produced for the floor below. Although the large amounts of reinforcing steel required cause increasing costs, the simple formwork cuts expenses decidedly. The significance of simple formwork will be understood when it is realized that over one-half of the cost of the average poured concrete floor slab is in the formwork.

For some reinforced-concrete frame buildings with flat slab floors, it is necessary to flare out the tops of the columns forming column capitals and perhaps thicken the slab around the column with the so-called drop panels. These items, which are shown in Fig. 13-11, may be necessary to prevent shear failures in the slab around the column.

It is possible today in steel-frame buildings to use short steel cantilever beams connected to the steel columns and embedded in the slabs. These beams serve the purposes of the flared columns and drop panels in ordinary flat-slab construction. This latter arrangement is often called a *steel grillage* or *column head*. The flat slab is not a very satisfactory type of floor system for the usual tall building where lateral forces (wind or earthquake) are appreciable, because protruding beams and girders are desirable to serve as part of the lateral bracing system.

13-11. PRECAST CONCRETE FLOORS

Precast concrete sections are more commonly associated with roofs than they are with floor slabs but their use for floors is increasing. They are quickly erected and reduce the need for formwork. Lightweight aggregates are generally used in the concrete making the sections light and easy to handle. Some of the aggregates used make the slabs nailable and easily cut and fitted on the job. For floor slabs with their fairly heavy loads, the

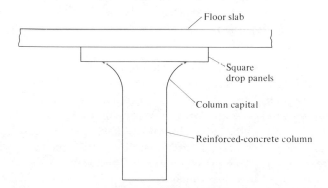

Figure 13-11 A flat slab floor for a reinforced-concrete building.

Interns' living quarters, Good Samaritan Hospital, Dayton, Ohio. (Courtesy of The Flexicore Company, Inc.)

aggregates should be of a quality that will not greatly reduce the strength of the resulting concrete.

The reader is again referred to *Sweet's Catalog File* at this point. In these catalogs a great amount of information is available on the various types of precast floor slabs on the market today. A few of the common types of precast floor slabs available are listed after this paragraph and a cross section of each type mentioned is presented in Fig. 13-12. Due to slight variations in the upper surfaces of precast sections it is necessary to use a mortar topping of 1 to 2 in. before asphalt tile or other floor coverings can be installed.

1. *Precast concrete planks* are roughly 2 to 3 in. thick, 12 to 24 in. wide, and are placed on joists up to 5 or 6 ft on center. They probably have tongue and groove edges.
2. *Hollow-cored slabs* are sections of roughly 6 to 8-in. depth and 12 to 18-in. widths, which have hollow perhaps circular cores formed in them in a longitudinal direction, thus reducing their weights by approximately 50% (as compared to solid slabs of the same dimensions). These sections, which can be used for spans of from roughly 10 to 25 ft, may be prestressed.

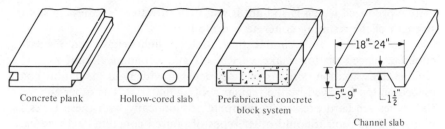

Concrete plank   Hollow-cored slab   Prefabricated concrete
block system

18"-24"

5"-9"   1½"

Channel slab

Figure 13-12 Precast concrete roof and floor slabs.

3. *Prefabricated concrete block systems* are slabs made by tying together precast concrete blocks with steel rods (may also be prestressed). They are generally 16 in. wide, 4 to 8 in. deep, and can be used for spans of roughly 8 to 32 ft.
4. *Channel slabs* are used for spans of approximately 8 to 24 ft. Rough dimensions of these slabs are given in Fig. 13-12.

13-12. TYPES OF ROOF CONSTRUCTION

The types of roof construction commonly used for steel-frame buildings include concrete slabs on open-web joists, steel-decking roofs, and various types of precast concrete slabs. For industrial buildings the predominant roof system in use today is one which involves the use of a cold rolled steel deck. In addition, the other types mentioned in the preceding sections on floor types are occasionally used but they usually are unable to compete economically. Among the factors to be considered in selecting the specific types of roof construction are strength, weight, span, insulation, acoustics, appearance below, and type of roof covering to be used.

The major differences between floor-slab selection and roof-slab selection probably occur in the considerations of strength and insulation. The loads applied to roofs are generally much smaller than those applied to floor slabs, thus permitting the use of many types of lightweight aggregate concretes which may be appreciably weaker. Roof slabs should have good insulation properties or they will have to have insulation materials placed on them and covered by the roofing. Among the many types of lightweight aggregates used are wood fibers, zonolite, foams, sawdust, gypsum, expanded shale, etc. Although some of these materials decidedly reduce concrete strengths, they provide very light roof decks with excellent insulating properties.

Precast slabs made with these aggregates are light, quickly erected, have good insulating properties, and usually can be sawed and nailed. For poured concrete slabs on open-web joists several lightweight aggregates work very well (zonolite, foams, gypsum, etc.), and in some cases the resulting concrete can easily be pumped up to the roofs, thereby facilitating construction. By replacing the aggregates with certain foams, concrete can be produced which is so light it will float in water. Needless to say the strength of the resulting concrete is quite low.

Steel decking with similar thin slabs of lightweight and insulating concrete placed on top make very good and economical roof decks. A competitive variation consists of steel decking with rigid insulation board placed on top followed by the regular roofing material. The other concrete-slab types are hard pressed to compete economically with these types for lightly loaded roofs. Should other types of poured concrete decks be used, the labor will be much higher. Furthermore, if an insulating type of lightweight aggregate is not used it will be necessary to place some type of insulating board on top of the slab before the roofing is applied.

13-13. EXTERIOR WALLS AND INTERIOR PARTITIONS

Exterior Walls

The purposes of exterior walls are to provide resistance to atmospheric conditions including insulation against heat and cold, satisfactory sound-absorption and light-refraction characteristics, sufficient strength, and fire ratings. They should have satisfactory appearance and be reasonably economical.

For many years exterior walls were constructed of some type of masonry, glass, or corrugated sheeting. In recent decades, however, the number of satisfactory materials for exterior walls has increased tremendously. Available and commonly used today are precast concrete panels, insulated metal sheeting, and many other prefabricated units. Of increasing popularity are the light prefabricated sandwich panels which consist of three layers. The exterior surface is made from aluminum, stainless steel, ceramic or plastic materials, and others. The center of the panel consists of some type of insulating material such as fiberglass or fiberboard, while the interior surface is probably made from metal, plaster, masonry, or some other attractive material.

Interior Partitions

The main purpose of interior partitions is to divide the inside space of a building into rooms. Their selection is based on appearance, fire rating, weight, or acoustical properties. Partitions are referred to as being bearing or nonbearing. These two types are described in the following two paragraphs.

1. *Bearing partitions* are those that support gravity loads in addition to their own weights and are thus permanently fixed in position. They can be constructed from wood or steel studs or from masonry units and faced with plywood, plaster, wallboard, or other material.

Table 13-1   Typical Building-Design Live Loads

Type of building	LL (psf)
Apartment houses	
Corridors	80
Apartments	40
Public Rooms	100
Office buildings	
Offices	50
Lobbies	100
Restaurants	100
Schools	
Classrooms	40
Corridors	80
Storage warehouses	
Light	125
Heavy	250

2. *Nonbearing partitions* are those that do not support any loads in addition to their own weights and thus may be fixed or movable. Selection of the type of material to be used for a partition is based on the answers to the following questions: Is the partition to be fixed or movable? Is it to be transparent or opaque? Is it to extend all the way to the ceiling? Is it to be used to conceal piping and electrical conduits? Are there fire-rating and acoustical requirements? Perhaps the more common types are made of metal, masonry, or concrete. For design of the floor slabs in a building that has movable partitions, some allowance should be given to the fact that the partitions may be moved. Probably the usual practice is to increase the floor-design live load by 15 or 20 psf.

13-14   DESIGN LIVE LOADS (VERTICAL)

Floor Loads

The designer is usually fairly well controlled in the design live loads by the building code requirements in the particular area. The values given in these various codes unfortunately vary from city to city and the designer must be sure that her designs meet the requirements of that locality. A few of the typical values for floor loadings are listed in Table 13-1. These values were adopted from the ANSI Code.[1] In the absence of a governing code this one seems to be an excellent one to follow.

---

[1]*American National Standard Building Code Requirements for Minimum Design Loads in Buildings and Other Structures*, ANSI A58.1-1972 (New York: American Standards Institute).

Roof Loads

The usual roof slab is designed for a minimum gravity load of 20 psf of horizontal projection of the roof whether the roof is sloping, curved, or flat. This live load does not include any wind or seismic effects which may also have to be considered. In some areas due to the shape of the roof (as it pertains to its snow-catching ability) and due to the particular climate, the designer may feel that 20 psf is not a sufficient load and she may increase it somewhat. Should the roof be used for other purposes than for supporting roofing, such as roof gardens or terraces, the live load should be increased to values ranging from 60 to 100 psf, or larger depending on the judgment of the designer.

One of the best sources for estimating snow loads is given by the ANSI Code.[1] In this Code information is given concerning the largest "on ground" snow depths throughout the United States for 50-year periods. Included are coefficients to be multiplied times the ground snow depths to give estimated depths on roofs depending on roof slopes, wind exposure, whether multilevel roofs are involved, etc. In general, for flat roofs, values from 20 psf in the southern states to 30 or 40 psf in the northern states are obtained.

Reduction in Design Live Loads

Each of the members of a building frame must be designed for the full dead loads that it supports, but it may be possible to design some members for lesser loads than their full theoretical live-load values. For example, it seems unlikely in a building frame of several stories for the absolutely maximum-design live load to occur on every floor at the same time. The lower columns in a building frame are designed for all of the dead loads above but probably for a percentage of live load appreciably less than 100%. Some specifications require that the beams supporting floor slabs be designed for full dead and live loads but permit the main girders to be designed under certain conditions for reduced live loads. (This reduction is based on the thought that it seems rather improbable that a very large area of a floor would be loaded to its full theoretical live-load value at any one time.) ANSI presents some very commonly used reduction expressions This information is listed as follows.

1. *Roofs and garages.* No live load reduction is permitted.
2. *For live loads of 100 psf or less on floor slabs.* For live loads of 100 psf or less for any member supporting an area of 150 ft^2 or more, the live load can be reduced by 0.08%/ft^2 of area supported except no reduction is permitted for places of public assembly. Furthermore the reduction may not exceed 60% nor the value given by the expression to follow for R (the percent reduction). *DL* and

LL are the dead and live loads, respectively, in pounds per square foot.

$$R = 23\left(1 + \frac{D}{L}\right)$$

3. *For live loads greater than 100 psf on floor slabs.* No reduction is permitted except for a 20% reduction in live loads used for column design.

13-15.   WIND AND EARTHQUAKE LOADINGS

Wind Forces

A great deal of research has been conducted in recent years on the subject of wind forces. Nevertheless a great deal more work needs to be done as the estimation of these forces can by no means be classified as an exact science. The magnitudes of wind loads vary with geographical locations, heights above ground, types of terrain surrounding the buildings including other nearby structures, and other factors.

Wind pressures are assumed to be uniformly applied to the windward surfaces of buildings and are assumed to be capable of coming from any direction. These assumptions are not quite accurate because wind pressures are not very even over large areas, the pressures near the corners of buildings being probably greater than elsewhere due to wind rushing around the corners, etc. From a practical standpoint, therefore, all of the possible variations cannot be considered in design although today's specifications are becoming more and more detailed in their requirements.

When the engineer working with large stationary buildings makes poor wind estimates the results are probably not too serious but this is not the case when tall slender buildings (or long flexible bridges) are being considered. The usual practice for buildings is to ignore wind forces unless their heights are at least twice their least lateral dimensions. For such cases as these it is felt that the floors and walls give the frame sufficient lateral stiffness to eliminate the need for a definite wind bracing system. Should the buildings have their walls and floors constructed of very modern light materials or should they be located in areas of unusual wind conditions the 2 to 1 ratio will probably not be followed. Building codes do not usually provide for estimated forces during tornadoes. The forces created directly in the paths of these storms are so violent that it is not considered economically feasible to design buildings to resist them.

Wind forces act as pressures on vertical windward surfaces, pressures or suction on sloping windward surfaces (depending on the slope) and suction on flat surfaces and on leeward vertical and sloping surfaces (due to the creation of negative pressures or vacuums). The student may have noticed this definite suction effect where shingles or other roof coverings have been lifted from the leeward roof surfaces of buildings. Suction or uplift can easily be demonstrated by holding a piece of paper at two of its corners and blowing above it. For some common structures uplift may be as large as 20 to 30 psf (958 to 1 436 N/m^2) or even more.

The average building code in the United States makes no reference to wind velocities in the area or to shapes of the building or to other factors. It just probably requires the use of some specified wind pressure in design such as 20 psf on projected area in elevation up to 300 ft with an increase of 2.5 psf for each additional 100-ft increase in height. The values given in these codes are thought to be rather inaccurate for modern engineering design.

For a period of several years the ASCE Task Committee on Wind Forces made a detailed study of existing information concerning wind forces. A splendid report of this information entitled *Wind Forces on Structures*[2] was presented by the committee in 1961. As stated in the report, its purpose was to provide a compact source of information which could be practically used by the civil engineering profession.

In the report there is much information presented concerning wind-pressure coefficients for various types of structures, information concerning maximum wind velocities for particular geographical areas, and much additional data which can be of great value in realistically estimating wind forces.

The wind pressure on a building can be estimated with the expression to follow in which p is the pressure in pounds per square foot acting on vertical surfaces, C_s is a shape coefficient, and V is the basic wind velocity in miles per hour estimated from weather bureau records.

$$p = 0.002558 C_s V^2$$

The coefficient C_s depends on the shape of the structure, primarily the roof. For box type structures C_s is 1.3 of which 0.8 is for the pressure on the windward side and 0.5 is for suction on the leeward side. For such a building the total pressure on the two surfaces equals 20 psf for a wind velocity of 77.8 mph.

The maximum wind and earthquake pressures for which design is made occur at large intervals of time and then last for only relatively short periods of time. It, therefore, may be reasonable to use higher allowable stresses (such as the one-third AISC increase) for lateral forces than for the relatively long-term gravity live loads.

Earthquake Forces

Many areas of the world, including the western part of the United States, fall in earthquake territory and in these areas it is necessary to consider

---

[2]*Trans. ASCE* **126**, part II (1961), pp. 1124–1198.

seismic forces in design for tall or short buildings. During an earthquake there is an acceleration of the ground surface. This acceleration can be broken down into vertical and horizontal components. Usually the vertical component of the acceleration is assumed to be negligible but the horizontal component can be severe.

Most buildings can be designed with little extra expense to withstand the forces caused during an earthquake of fairly severe intensity. On the other hand, carthquakes during recent years have clearly shown that the average building which is not designed for earthquake forces can be destroyed by earthquakes that are not particularly severe. The usual practice is to design buildings for additional lateral loads (representing the estimate of earthquake forces) which are equal to some percentage (5 to 10%) of the weight of the building and its contents. An excellent reference on the subject of seismic forces is a publication by the Structural Engineers Association of California entitled *Recommended Lateral Force Requirements and Commentary*, Seismology Committee, Structural Engineers Association of California (San Francisco, 1975).

It should be noted that many persons look upon the seismic loads to be used in designs as being merely percentage increases of the wind loads. This thought is not really correct, however, as seismic loads are different in their action and as they are not proportional to the exposed area but to the building weight above the level in question.

13-16. FIREPROOFING OF STRUCTURAL STEEL

Although structural steel members are incombustible their strength is tremendously reduced at temperatures normally reached in fires when the other materials of a building burn. Many disastrous fires have occurred in empty buildings where the only fuel for the fires were the buildings themselves. It is true that steel is an excellent heat conductor and nonfire-proofed steel members may transmit enough heat from one burning compartment of a building to ignite materials with which they are in contact in adjoining sections of the building.

Steels (particularly those with rather high carbon contents) may actually increase a little in strength as they are heated from room temperature up to approximately 600°F. As temperatures are raised into the 800° to 1000°F range steel strengths are drastically reduced and at 1200°F they have little strength left.

The fire resistance of structural steel members can be greatly increased by applying to them fire protective covers such as concrete, gypsum, mineral fiber sprays, special paints, and others. The thickness and kind of fireproofing used depends on the type of structure, the degree of fire hazard, and economics.

In the past, concrete was commonly used for fire protection. Although concrete is not a particularly good insulation material it is very satisfactory when applied in thicknesses of $1\frac{1}{2}$ to 2 or more in. due to its mass. Furthermore the water in concrete (16 to 20% when fully hydrated) improves its fireproofing qualities appreciably. This is due to the fact that the boiling off of the water from the concrete requires a great deal of heat. It is true, however, that in very intense fires the boiling off of the water may cause severe cracking and spalling of the concrete.

Though concrete is an everyday construction material and though in mass it is quite a satisfactory fireproofing material its installation cost is extremely high and its weight is large. As a result, for steel high-rise construction, sprayed on fireproofing materials have almost completely replaced concrete.

The spray-on materials usually consist of either mineral fibers or cementious fireproofing materials. The mineral fibers formerly used were comprised of asbestos. Today, however, due to the health hazards associated with this material, its use has been discontinued and other fibers are now used by manufacturers. The cementious fireproofing materials are probably composed of gypsum, perlite, vermiculite, and others. Sometimes, when plastered ceilings are required in a building, it is possible to hang the ceilings and light gauge furring channels by wires from the floor systems above and use the plaster as the fireproofing.

The cost of fireproofing structural steel buildings is high and hurts steel in its economic competition with other materials. As a result, a great deal of research is being conducted by the steel industry on imaginative new fireproofing methods. Among these ideas is the coating of steel members with expansive and insulative paints. When heated to certain temperatures these paints will char, foam, and expand, forming an insulative shield around the members.

Other techniques involve the isolation of some steel members outside of the building where they will not be subject to damaging fire exposure; the circulation of liquid coolants inside box or tube shaped members in the building. It is probable that major advances will be made with these and other fireproofing ideas in the near future.[3]

[3] W. A. Rains, "A New Era in Fire Protective Coatings for Steel," *Civil Engineering*, September, 1976, *ASCE*: New York, pp. 80–83.

Chapter 14
Introduction
to Steel Bridges

14-1. GENERAL

The first bridge has been estimated to have been built as early as 15,000 B.C. In fact it has been said that human beings knew how to build bridges before they knew how to build houses. The first bridges probably consisted of stream crossings made with tree trunks or large flat stones laid across the streams. These latter bridges have been called *clapper bridges* because of the sound the loose stones made when being crossed.

Another type of bridge built before the dawn of history was the vine type of suspension bridge. A person first seeing the magnificent Golden Gate or Mackinac suspension bridges would be amazed to learn that the suspension bridge idea probably originated with the use of vines across gorges in the Himalayan Mountains many centuries ago. In the same vein, early human beings in the Southwest must have been amazed at the sight of the great natural stone arches there. It is useless to speculate on how much time elapsed before they got the idea of putting two large stones together in the shape of an inverted V to ford a small stream with an arch.[1]

The exact dates of construction of the first bridges are impossible to obtain. There is no question, however, that beam, arch, and suspension bridges have been built in some form for several thousand years. It is said that the oldest bridge still in use in the world is the Caravan Bridge over the river Meles at Smyrna, Turkey. Supposedly Homer and St. Paul both walked across this 40-ft (12.2-m) stone-slab bridge many centuries apart.

Bridges built before 1840 in the United States were made of timber. In 1840 an all-iron bridge was built across the Erie Canal at Frankfurt, New York. The first bridge constructed with structural carbon steel was Ead's Bridge across the Mississippi River at St. Louis, begun in 1867 and completed in 1874. This bridge consists of three arches with a center span of 520 ft and end spans of 502 ft. It is well known for the tragic loss of life suffered during construction. Of 600 men who were employed in sinking

[1] H. S. Smith, *The World's Great Bridges* (New York: Harper & Row, 1953), p. 1.

the caissons there were 119 serious cases of the "bends" and 14 deaths.[2] This beautiful bridge was the essential link in the construction of the cross-continental railroad system and had a tremendous effect on the growth of the city of St. Louis. It has two levels, the upper one for highway traffic and the lower one for trains.

During the last part of the nineteenth century and the early part of the twentieth the majority of bridges built were for railway traffic, but this is no longer the case. Due to various problems in the railroad industry today very few new railroad bridges are being built. On the other hand, the increase in highway traffic during the past few decades has caused a rise in highway bridge construction almost beyond belief. Today more than 6000 bridges of all types are built each year in the United States.

14-2. THROUGH, DECK, AND HALF-THROUGH BRIDGES

The student has often seen highway bridges in which the trusses were on the sides. As he rode across the bridge he could see overhead lateral bracing between the trusses. This type of bridge is said to be a *through bridge*. The floor system is supported by floor beams that run under the roadway and between the bottom chord joints of the trusses.

In the *deck bridge* the roadway is placed on top of the trusses or beams. Deck construction has every advantage over through construction except for underclearance. There is unlimited overhead horizontal and vertical clearance, and future expansion is more feasible. Another very important advantage is that supporting trusses or beams can be moved closer together, reducing lateral moments in the floor system. Other advantages of the deck truss are simplified floor systems and possible reduction in the sizes of piers and abutments due to reductions in their heights. Finally, the very pleasing appearance of deck structures is another reason for their popularity.

Should the roadway lie in between the top and bottom chords and should there be no room for overhead bracing, the bridge is said to be a *half-through* or a *pony bridge*. One major problem with the pony truss is the difficulty of providing adequate lateral bracing for the top chord compression members. This type of bridge is rarely practical today.

Today's bridge designer tries to prevent any sense of confinement for the users of his bridges. He will in trying to achieve this goal attempt to eliminate any overhead bracing or truss members which will protrude above the roadway level. The result is that the deck structure is again desirable unless underclearance requirements prevent their use or the spans are so large as to make them impractical.

[2] H. S. Smith, op. cit., p. 84.

14-3. ERECTION METHODS FOR BRIDGES

Before the various types of bridges are discussed it is felt that a few general comments should be made about the erection methods used for bridges of varying span lengths. A point that cannot be overemphasized is the necessity for carefully preparing an erection schedule. This schedule should actually be outlined before the final design is completed. Stresses may be caused in some members during erection in excess of those produced by the working loads; perhaps even more serious, the character of the stresses may be changed, as from tension to compression. Very large stresses may be caused by such things as heavy cranes moving on the trusses or by the forcing of parts to make them fit, as in getting opposite ends of a bridge to come together at the middle.

The erection schedule should be developed showing the order of erecting various truss members. After this plan is made the joints can be designed showing which connections are to be field-made and which ones are to be shop-made. The gusset plates should be shop-connected to the correct members so they will be available in the proper order of erection.

In the paragraphs to follow a few comments are made about the common erection procedures used for short-, medium-, and long-span bridges.

Swinging a plate girder into place on Goat Island Bridge across the Niagara River, Niagara Falls, N.Y. (Courtesy of Bethlehem Steel Company.)

Short-Span Bridges (Up to Approximately 125 ft)

For these short spans it is usually possible to assemble the spans in the shop or at the job site and put them in place in one piece with crawler or truck cranes working from below or with a stiffleg derrick traveler. A traffic problem may be involved below (land or water) and it is important to limit the time of the interference to the absolute minimum.

Medium-Span Bridges (Approximately 125 to 400 ft)

The usual procedure for medium spans is to use falsework, the temporary material used to support a structure until it is self-supporting. When falsework is used, erection stresses are kept at a minimum. In some cases deep fast-flowing water or deep gorges below may prevent the use of falsework and the cantilever erection procedure will probably be used. This method, which is particularly applicable to cantilever and continuous bridge trusses, is illustrated in Fig. 14-1. As shown in the figure, the end span has been erected (probably on falsework) and the bridge is extended beyond the pier or tower, member by member with the use of small crane travelers. The truss is designed to be self-supporting and the members are designed for the erection stress reversal that will occur. The same erection procedure will be followed from the opposite shore.

It should be realized that cantilever erection has the disadvantages of being more expensive and dangerous. Other erection methods which have been used for medium spans include floating the bridge section into place when water is being spanned, and the so-called *protrusion method* where the structure is rolled into place (illustrated in Fig. 14-2 for a girder).

Long Spans (Approximately 400 ft and Above)

Falsework is usually not feasible for long-span bridges and the cantilever erection procedure will probably be used. Suspension bridges are erected from the cable and there is no interference with navigation. A few more comments about their erection is given in Section 14-11.

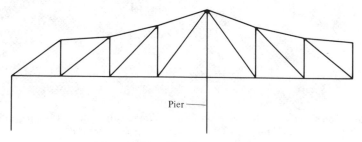

Pier

Figure 14-1 Cantilever erection.

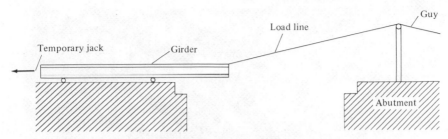

Figure 14-2   Protrusion method of erection.

14-4.   THE BEAM BRIDGE

The beam bridge, in which a concrete roadway slab is supported by W section beams, is very popular because of its simple design and construction. This type of span is economical for railroad spans up to roughly 50 ft and for highway spans of up to roughly 80 ft. It will be remembered from Chapter 7 that the AREA and the AASHTO specify certain minimum depth-span ratios ($\frac{1}{15}$ and $\frac{1}{25}$, respectively). Shallower beams can be used, provided the deflections are limited to what they would have been if the design had been made using the recommended minimum ratios.

The W36 × 300, which is the deepest standard W section, has a total depth of 36.74 in. For the $\frac{1}{15}$ AREA ratio the maximum span for which this section could be used is 45.9 ft and for the $\frac{1}{25}$ AASHTO ratio the maximum span is 76.5 ft. These values can be extended somewhat by using cover plates and/or limiting the deflections as required for smaller depth-span ratios. As previously described in Chapter 13, steel beams and concrete slabs may be constructed so they act together as composite sections. Simple-span composite highway spans have been economically used for spans as high as 120 ft.

14-5.   THE PLATE GIRDER BRIDGE

For many years bridge trusses were quite commonly used for spans of from roughly 70 to 150 ft as well as for spans greater than 150 ft. Today, however, beam and plate girder bridges have almost completely captured the market for spans up to about 150 ft as well as for many larger spans. They are actually common for 200-ft spans and have been used for spans of well over 400 ft. The main span of the continuous Bonn-Beuel plate-girder bridge over the Rhine in Germany is 643 ft.

It is very difficult for trusses to compete with beam and plate-girder structures as long as the beams or girders can be shipped in one piece. Plate girders have been shipped in one piece for spans above 150 ft. The problems of fabrication, erection, and transportation usually limit built-up sections to maximum depths of roughly 10 or 12 ft. When deeper sections than these are required, trusses may prove to be competitive. Generally

New York State Thruway crossing a railroad near Suffern, N.Y. (Courtesy of The Lincoln Electric Company.)

speaking, plate girders are very economical for railroad bridges from about 50 to 130 ft and for highway bridges from about 80 to 150 ft. They are frequently very competitive for much higher spans than these, particularly where continuous spans are used. Actually simple spans greater than 100 ft are rarely economical. Chapter 16 is devoted to the design of plate girders.

14-6.   TRUSS BRIDGES FOR MEDIUM SPANS

As spans become longer and loads heavier, trusses will begin to be competitive. For spans greater than those economical for plate girders, truss bridges will probably prove to be economical. Although many bridge trusses are still in existence along the older highways, one-span, simply supported trusses are built quite infrequently today. A few types of bridge trusses which can be used for medium spans are shown in Fig. 14-3.

The Pratt truss, which is reasonably economical for spans of 150 to 200 ft, has one particular advantage in that the diagonals are all in tension under dead loads. The end posts, however, are always in compression and the movement of live loads back and forth across the span may cause stress reversals in some of the diagonals.

The Warren truss is thought by many to have a little more attractive appearance than the Pratt and may be a little more popular for the same spans. It is probably used more as a deck truss than as a through truss because it is particularly economical when so used.

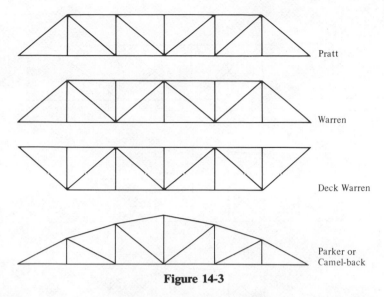

Figure 14-3

The deeper a truss is for the same size chord members the greater is its resisting moment. For longer spans it is economical to increase truss depths where the moments are larger. These varying depth trusses are definitely lighter than corresponding parallel chord trusses, but their fabrication costs are higher. Above approximately 180 to 200 ft, the weight saving will more than cancel the extra fabrication costs and the so-called curved-chord trusses become economical. The Parker truss (also called a camel-back truss) is quite satisfactory for spans from roughly 180 to 360 ft. For long continuous and cantilever span trusses the depths are almost always varied with the moments. The greatest depths in these trusses will occur at the supports where the moments are largest.

14-7. SUBDIVIDED TRUSS BRIDGES

Various economic studies for truss bridges have shown that the diagonals should be kept at angles of approximately 45° with the horizontal and the depth to span ratios should vary from about $\frac{1}{5}$ to $\frac{1}{8}$ with the smaller ratios used for the larger spans. Should these recommendations be followed for spans much greater than 300 ft, the panel lengths will become excessive.

When panels are very long the compression member sizes get out of hand because of their excessive unsupported lengths. Furthermore, the floor system between panel points becomes unreasonably heavy and expensive for large panels. In fact floor beam spacings greater than 25 ft are not recommended.

The subdivided trusses shown in Figs. 14-4(a) and (b) can be used to keep the panel lengths within the desired values for longer trusses while at the same time following economic requirements previously mentioned

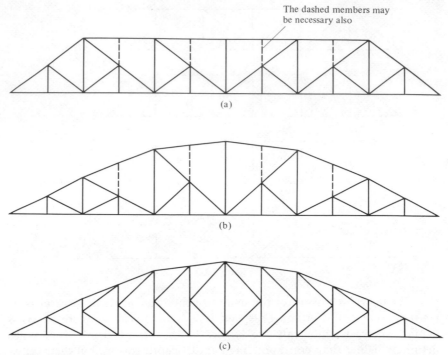

The dashed members may
be necessary also

(a)

(b)

(c)

Figure 14-4 (a) Baltimore or Subdivided Pratt. (b) Pennsylvania or Petit. (c) K truss.

(diagonal slope and depth-to-span ratio). In part (a) of the figure the parallel chord Baltimore or subdivided Pratt truss is shown. Again it is usually more economical to use trusses that vary in depth with the moments and a curved chord Baltimore is shown in part (b) of the figure. This truss is usually called a Pennsylvania or Petit truss.

Subdivision introduces its own disadvantages. The trusses are usually not too attractive with their large number of members, their pound price is high due to the large number of connections, and secondary stresses are often rather high. Another truss that accomplishes the same ends achieved with subdivision and without their disadvantages is the K truss shown in part (c) of Fig. 14-4. Simple-span bridges probably have an upper economical limit of roughly 600 ft although longer spans have been built.

14-8. CONTINUOUS BRIDGE TRUSSES

Continuous bridge trusses are normally economical for highway bridges for spans of from roughly 250 to 700 or 800 ft and for railroad bridges for spans of from roughly 450 to 700 or 800 ft. A very popular truss for continuous spans is the Warren, an example of which is shown in Fig. 14-5. For economic reasons continuous trusses may be subdivided as were

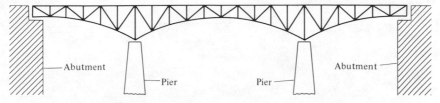

Figure 14-5 Continuous Warren deck bridge.

the simple spans. Among the several advantages possessed by continuous bridge trusses as compared to simple-span trusses are the following.

1. There can be an appreciable saving of steel weight perhaps running as high as 10 or 20% with the higher values being possible for highway bridges. For railroad bridges the highest weight saving potential is probably in the range of 10% because of the more frequent stress reversals.
2. Fewer supports are required as longer spans are possible.
3. Continuous trusses are ideally suited for the cantilever type of erection.
4. Continuous bridges have been clearly shown to support higher ultimate loads than simple spans.

On the other hand, continuous trusses have some shortcomings. These include the following.

1. They are statically indeterminate to outer forces and are desirably used only where good foundation conditions are present.
2. There are more stress reversals in a continuous truss, with the result that part of the weight saving advantage is canceled.
3. A statically indeterminate structure is more difficult to analyze and design than is a statically determinate one.

14-9. CANTILEVER BRIDGES

Cantilever construction consists essentially of two simple spans each with an overhanging or cantilevered end (as shown in Fig. 14-6) and with another simple span in between, called the suspended span, supported by the cantilevered ends (as shown in Fig. 14-7).

From Figs. 14-6 and 14-7 the student can see why these structures are sometimes referred to as see-saw construction because the bridge seems to

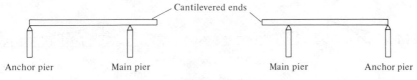

Figure 14-6

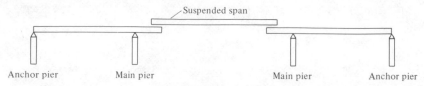

Figure 14-7

see-saw over the main piers. In cantilever construction unsupported hinges or points of zero moment are introduced at each end of the suspended span. Figure 14-8 shows a sketch of a cantilever bridge truss showing one method of introducing the unsupported hinges.

The method of forming the hinges should be carefully noted in this figure. The student might claim that she has ridden over some cantilever bridges and has not noticed any missing truss members. Should the hinges be formed as shown in this figure, the missing members would probably be installed but would not be connected to resist stress. The purposes of their installation are for better appearance and for keeping the users of the bridge from being worried.

It is possible to locate the hinges in a cantilever bridge at such positions so as to obtain the same moments (and this roughly equivalent economy) as would occur in a continuous truss of the same spans so far as dead loads are concerned. This could be done by locating the hinges at points of zero moment or points of inflection in the center span of the continuous structure. Unfortunately such a situation does not apply to live loads. For long spans with their heavy dead loads, cantilever bridge trusses compare quite favorably with continuous construction, but in shorter spans continuous construction will probably prove to be more economical.

The cantilever bridge is said to have come into general use for long railroad bridges several decades ago because the suspension bridges of that time were not considered to be sufficiently rigid to withstand the heavy impact, nosing of locomotives, and other types of train loadings.

The usual economical cantilever spans range from approximately 500 to 1800 ft for highway bridges and from 600 to 1800 ft for railroad bridges. The longest cantilever bridge in the world is the Quebec Bridge which has an 1800-ft center span. Perhaps the most famous cantilever bridge is the one which spans the Firth of Forth in Scotland. This bridge, which was completed in 1890, was the first long railroad bridge built with steel. It has two 1710-ft spans and is the second longest cantilever bridge in the world.

Figure 14-8

Among the several advantages possessed by cantilever bridges are the following.

1. Cantilever bridges can be used where foundation conditions are poor because they are statically determinate and pier settlement is not a serious matter as it affects stresses.
2. As they are statically determinate, their analysis and design is simpler.
3. They are well adapted to cantilever erection.

The cantilever bridge does have several disadvantages including the following.

1. They are less rigid than continuous bridge trusses.
2. The construction of the hinges is expensive.
3. The reactions, which have to be supplied at the supports, can be disadvantageous. For example, there may be uplifts at the anchor piers while tremendously large reactions may have to be provided at the interior supports or main piers.

As to erection, the anchor spans are probably built on falsework and the interior spans by the cantilever erection method (previously illustrated in Fig. 14-1). The portion of the bridge truss between the interior hinges may often be floated to the site and lifted into position. It does not take much imagination to understand the importance of having an absolutely perfect check and recheck on dimensions for the final fitting.

During the construction of the Firth of Forth Bridge, calculations showed that a minimum temperature of 60°F was required to bring the bolt holes in position so as to get the lower chords to join for the final fitting. Construction was delayed due to various reasons until October and it appeared that the required temperature could not be obtained that year. To make the holes come in line, wood shavings and oily waste soaked in naphtha were laid along the chord and set on fire for 60 ft on either side of the connection, with the result that the required expansion was obtained and the connection was successfully completed. A similar but opposite situation was encountered during the construction of Ead's Bridge at St. Louis (an arch bridge previously mentioned in Section 14-1), when it was necessary to pack the chords in ice to reduce their lengths.[3]

14-10. STEEL ARCHES

The arch is certainly one of the more attractive types of bridges. It is often defined by the structural engineer as a structure which under gravity loads tends to produce converging horizontal reactions. If the student will examine the curved arch of Fig. 14-9, she will see that under the vertical

---

[3] H. S. Smith, op. cit., p. 126.

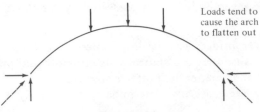

Figure 14-9

loads shown the arch will tend to spread or flatten out. The result is that large inward or converging reactions must be provided at the supports. Because of this fact there is a well-known expression about arches to the effect that "an arch never sleeps" because the horizontal thrust is always there, even due to the weight of the arch alone.

The horizontal reactions or thrusts tend to cause moments in an arch which in turn tend to cancel the moments caused by the vertical reaction components. In other words, at a support the resultant of the vertical and horizontal reaction components (shown dotted in Fig. 14-9) tends to cause primarily axial compression in the arch. It is theoretically possible to construct an arch with a parabolic shape, which when loaded with a uniform load, will have no bending moment.

Arches are classified as being three-hinged, two-hinged, one-hinged (rare), or hingeless (fixed). They are further classified as being solid ribbed, tied, braced rib, or spandrel braced. These types are shown in Fig. 14-10.

Henry Hudson Bridge over Harlem River, New York City. (Courtesy of American Bridge Division, U.S. Steel Corporation.)

West End North Side Bridge, Ohio River, Pittsburgh, Pa. (Courtesy of American Bridge Division, U.S. Steel Corporation.)

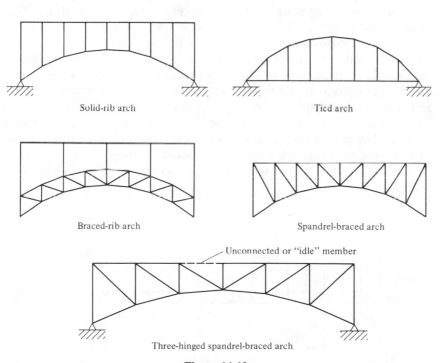

Solid-rib arch

Tied arch

Braced-rib arch

Spandrel-braced arch

Unconnected or "idle" member

Three-hinged spandrel-braced arch

Figure 14-10

The solid rib steel arch is used more for highway bridges than railroad bridges. Its upper economical limit is roughly 500 to 600 ft. The braced-rib arch, which is mainly used for highway bridges, is ideally suited for cantilever erection and is used for long spans with an upper economical limit of roughly 1600 to 1800 ft. The spandrel-braced arch is used primarily for railroad bridges. It has an upper economical limit of about 1000 ft because it is heavier than the braced-rib arch and is not as well adapted to cantilever erection.

The 1700-ft New River Gorge Bridge in Fayette County, West Virginia is the longest steel arch bridge in the world. It is 876 ft above the river. The Kill van Kull highway bridge (a two-hinged, braced rib arch) from Bayonne, New Jersey to Staten Island has a 1652-ft span. The Sydney Harbour, Australia, steel railway arch has a span of 1650 ft and is also a two-hinged, braced-rib arch.

14-11. SUSPENSION BRIDGES

Probably the suspension bridge with its gracefully curving cables is the most beautiful of all bridge types. A suspension bridge is shown in Fig. 14-11. The bridge is supported by the cables that pass over the towers and are anchored at the bridge ends. The stiffening truss stiffens the cable against vibration caused by the live loads and holds it in its regular shape.

The suspension bridge provides an excellent method of reducing moments in long-span structures. The usual practice is to erect the structure with its weight being entirely carried by the cables. This is done by waiting to connect the stiffening trusses until all of the dead weight is carried by the hangers to the cables. The result is that these trusses carry none of the dead load and only a part of the live load. A large part of the live loads is transferred by the hangers to the cable, with the result that the amount of bending moment to be resisted by the trusses is tremendously reduced. The majority of the load on a suspension bridge is carried by the cable in tension, a very efficient and economical method.

The late Dr. D. B. Steinman said that suspension bridges were potentially economical for highway bridges for clear spans from 400 ft to

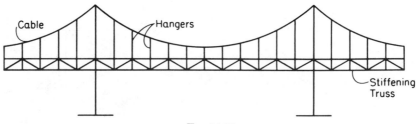

Fɪɢ. 14-11

San Francisco-Oakland Bay Bridge. (Courtesy of American Bridge Division, U.S. Steel Corporation.)

10,000 ft.[4] For many years the longest suspension bridge in the world was the Golden Gate Bridge with its 4200-ft suspended span. That title, however, was captured by the Verrazano-Narrows Bridge with its 4260-ft suspended span across the entrance to New York's harbor from Brooklyn to Staten Island and then by the Humber Estuary Bridge in England with its 4626-ft suspended span.

14-12. MISCELLANEOUS BRIDGE TYPES

There are several other types of steel bridges which are used, among them being rigid-frame bridges, orthotropic systems, Vierendeel trusses, and prestressed steel bridges. A brief discussion is made of each of these types in the following paragraphs.

Steel Rigid-Frame Bridges

A type of bridge which is quite often used for single spans is the steel rigid-frame bridge. This type of bridge which is very satisfactory for overpasses is economical for spans of from roughly 30 to 150 ft. The legs of the rigid frame serve as the abutments and provide continuity. The design

[4] D. B. Steinman, *A Practical Treatise on Suspension Bridges* (New York: Wiley, 1929), p. 83.

of this type of structure can be handled in a manner quite similar to the method described for rigid frame buildings in Chapter 19.

Orthotropic Systems

In the usual types of bridges each of the parts (slab, stringers, floor beams, trusses, etc.) is assumed to perform a certain individual task without assistance by the other parts. In the orthotropic bridges the stringers and floor beams and main girders are rigidly connected together and a continuous deck plate is placed over them. The beams are welded to the underside of the plate to make the entire bridge a single unit in resisting loads. The various parts do not have clear-cut single functions—they all support the load together.

A paving surface (probably asphalt because of its light weight) is placed on the steel plate. Orthotropic bridges have been used primarily in Europe, the labor situation being a major factor. A great deal of labor is involved in constructing the bridges although there is a large saving in steel.

Orthotropic bridges have been primarily used for spans of 200 ft or more, but they seem to have the potential for competing economically for spans of 100 ft or even less. One orthotropic plate girder bridge has a span of 856 ft. It is the Save River Bridge in Belgrade, Yugoslavia.[5] An excellent reference on the design of orthotropic bridges is the *Design Manual for Orthotropic Steel Plate Deck Bridges*, published in 1963 by the AISC.

Vierendeel Trusses

A type of truss which is uncommon in the United States but which has had frequent application in Europe for bridges is the Vierendeel truss, developed by M. Vierendeel of Belgium in 1896. This type of truss, illustrated in Fig. 14-12, is actually not a truss by the usual definition and requires the use of moment-resisting joints. Loads are supported by means of the bending resistance of its short heavy members. Although analysis and design are quite difficult, it is a fairly efficient structure.

Figure 14-12

[5] Jerry C. L. Chang, "Orthotropic-Plate Construction for Short Span Bridges," *Civil Engineering*, December 1961, p. 53.

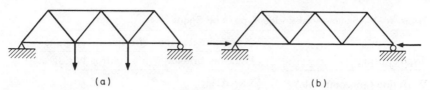

(a)   (b)

Figure 14-13

Prestressed Steel Bridges

Another type of bridge construction which may be used a great deal in the future is the prestressed steel bridge. In a prestressed structure, stresses are introduced which are opposite in character to those that will be produced by the working loads. Another way of defining this process is to say that where the working loads cause positive moments, prestressing is introduced to produce negative moments, and vice versa.

Prestressing tends to counteract the effect of the working loads thus enabling the same structure to support increased loads or to be used for larger spans with the same loads. As an illustration of the prestressing idea the student is referred to Fig. 14-13. In part (a) of the figure the gravity loads shown tend to cause the truss to bend downward, causing tension in the bottom chords and compression in the top ones (i.e., positive moments). In part (b) the bottom chord is compressed and the truss tends to bend upward (i.e., negative moments). One method of producing the prestressing force in a truss is by tensioning a group of high-strength cables to very large initial stresses—perhaps as high as 175 ksi—and attaching them to the bottom chord of the truss.

Shop fabrication of a Vierendeel truss. (Courtesy of Bethlehem Steel Company.)

Table 14-1 Summary of Bridge Types and Spans

	Approximate range of economical spans	
Type of bridge	Highway bridges	Railroad bridges
W section (noncomposite)	0–80	0–50
W section (composite)	50–120	—
PL girder $\left(\begin{array}{l}\text{composite and}\\\text{noncomposite}\end{array}\right)$	80–150	50–130
Simple truss	150–600	130–600
Continuous truss	250–800	450–800
Cantilever truss	500–1800	600–1800
Arch	100–1800	100–1000
Suspension	400–10,000	1500–10,000

14-13. SUMMARY OF BRIDGE TYPES

A brief summary of the general types of steel bridges is given in Table 14-1, together with approximate economical spans. There are many variables involved in each of the values given. In this regard, notice that the table indicates that simple truss bridges may be economical for spans up to approximately 600 ft. This does not mean that any type of simple truss can be economical for this size of span. Obviously the subdivided and K trusses are the only ones which would be economical for such large spans. A similar argument can be made for arches. A few of the values in the table are only roughly estimated. For example, the 10,000 ft estimated economical span for suspension bridges has not even been approached.

14-14. COMPARISON OF HIGH-LEVEL BRIDGES AND LOW-LEVEL MOVABLE BRIDGES

Because our U.S. highways are subject to tremendous traffic volumes and to high-speed traffic, any kind of movable bridge on their route is objectionable. The result is that the use of fixed high-level bridges furnishing the clearances required by the Corps of Engineers (which has jurisdiction over all structures crossing navigable streams) are usually desirable.

Nevertheless a bridge-design organization may be faced with the problem of spanning a navigable stream with a high bridge above the naval traffic or a low-level bridge with a movable span. (The discussion of tunnels, another feasible solution, is not considered in this text.)

Several types of low-level movable bridges are shown in Figs. 14-14 and 14-15. Low-level bridges have smaller initial costs, take up less of the surrounding land areas at their ends, permit quicker crossing by vehicles (when a ship is not passing) and the operating costs of the crossing vehicles

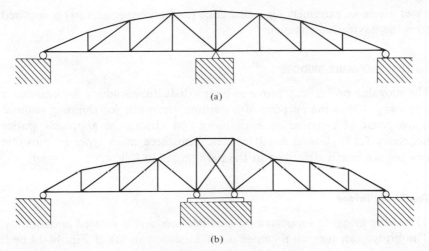

(a)

(b)

Figure 14-14   Swing bridges

are reduced. This latter fact is of appreciable importance when railway traffic is involved. Low-level bridges may have some appreciable advantage when a high but narrow navigation channel is to be spanned.

On the other hand, low-level bridges with movable sections are always a nuisance to traffic above and below and are a real hazard for land traffic in cases of emergency use by fire engines, ambulances, etc. They require additional costs for the operators (24 hours per day seven days per week) for opening and closing of the bridges as well as the cost of the machinery and power for the opening and closing.

High-level bridges have much higher initial costs than low-level bridges, their long approaches damage or take up much of the surrounding area on the shores, they have steeper grades, and they may block off streets and through traffic for parts of urban areas.

The high-level bridge is probably selected for highway traffic or for a crossing in a rural area or for a location where the navigation channel to be spanned is wide. On the other hand, low-level movable bridges may be given serious consideration for railway traffic or for urban areas or for

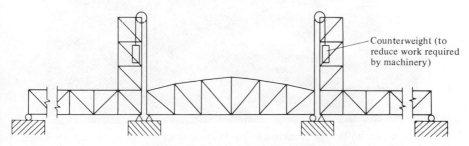

Counterweight (to reduce work required by machinery)

Figure 14-15   Vertical-lift bridge.

cases where an extremely high clearance (and probably narrow) is required over the navigation channel.

14-15. MOVABLE BRIDGES

The movable bridge may prove to be a satisfactory solution for crossing a waterway. It has the purpose of providing clearance for shipping without the expense of constructing high piers and raising the approach grades necessary for high-level fixed bridges. The three main types of movable bridges are briefly described in the paragraphs to follow.

The Swing Bridge

The swing bridge is supported on a center pier and is rotated horizontally. The bridge can turn on a center pivot as shown in (a) of Fig. 14-14 or it can turn on a turntable as shown in (b) of the same figure.

With the swing bridge there is no problem with vertical clearance as it is unlimited but the center pier will serve as an obstruction to ships.

The Vertical Lift Bridge

As the name implies, the movable span is lifted vertically above the navigational clearance area as shown in Fig. 14-15. This type of movable bridge is used where the horizontal clearance required is greater than the vertical clearance required.

George P. Coleman Bridge, Yorktown, Va. (Courtesy of American Bridge Division, U.S. Steel Corporation.)

Arthur Kill Bridge, Staten Island, N.Y. (Courtesy of American Bridge Division, U.S. Steel Corporation.)

Manasquan River highway bridge from Brielle to Point Pleasant, N.J. (Courtesy of American Bridge Division, U.S. Steel Corporation).

Bascule Bridge

The bascule bridge is one in which the span is turned up vertically at one end, usually by some type of counterweight system. It may prove to be a satisfactory solution when a narrow but high clearance is required.

14-16. SELECTION OF BRIDGE TYPE

Among the many factors that affect the choice of the type of bridge to be used for a given site are the following.

1. Span required.
2. Foundation conditions. (Poor foundation conditions may rule out statically indeterminate structures.)
3. Clearance required. (This factor can have a great effect on the erection method used and hence the type of structure.)
4. Probable erection methods required by conditions at the site.
5. Live loads to be supported.
6. First cost.
7. Operation and maintenance cost.
8. Harmony of bridge with surrounding terrain.

For a particular situation the designer could list all of the bridge types that might be economical for the span length involved. His next step might be to study the preceding factors 2 through 8 to determine if any of the possibilities were impractical. Should more than one possibility remain, preliminary designs and cost estimates for each may have to be made before a decision can be reached.

14-17. SPACING OF PIERS AND ABUTMENTS

This section is devoted to a discussion of the spacings of the piers and abutments for bridges. A well-known rule in bridge engineering on the subject of costs states that for best economy the cost of the superstructure should equal the cost of the substructure. Simply stated, the rule means that when foundation conditions are poor and thus substructure costs high per pier, it is desirable to use longer spans in order to reduce the number of piers.

For this rule to be completely applicable the pier depths, foundation materials, and other construction conditions must be the same at each pier. (It should be noted here that the cost of the superstructure varies roughly as the square of the span, whereas the cost of the piers varies roughly as the square of the depth below water level.) For uniform conditions the rule can be easily proved, as shown in several texts,[6] but space is not taken here

---

[6]E. H. Gaylord, Jr., and C. N. Gaylord, *Design of Steel Structures* (New York: McGraw-Hill, 1957), pp. 405–407.

for that purpose. Since the rule is perfect only for uniform conditions it should probably be taken in general as being an approximate rule of the greatest importance.

14-18. LIVE LOADS FOR HIGHWAY BRIDGES

Although highway bridges must support several different types of vehicles, the heaviest possible loads are caused by a series of trucks. The AASHTO specifies that highway bridges shall be designed for lines of motor trucks occupying 10-ft wide lanes. Only one truck is placed in each span for each lane. The truck loads specified are designated with an H prefix followed by a number indicating the total weight of the truck in tons. The weight is followed by another number indicating the year of the specifications. For example, an H20-44 loading indicates the 20-ton truck and the 1944 specifications. A sketch of the truck and the distances between axle centers, wheel centers, and other dimensions is given in Fig. 14-16.

The selection of the particular truck loading to be used in design depends upon the bridge location, anticipated traffic, and related factors. These loadings may be broken down into three groups as follows.

Two-Axle Trucks, H20-44, H15-44, and H10-44. The weight of an H truck is assumed to be distributed two-tenths to the front axle (e.g., 4 tons, or 8 k, for an H20-44 loading) and eight-tenths to the rear axle. The axles are spaced 14 ft–0 in. on centers, while the center-to-center lateral spacing

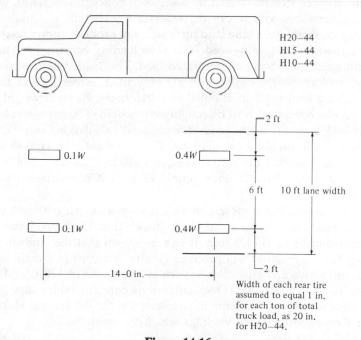

Figure 14-16

of the wheels is 6 ft–0 in. Should a truck loading varying in weight from these be desired, one that has axle loads in direct proportion to the standard ones listed here may be used. A loading as small as the H10-44 may be used only for bridges supporting the lightest traffic.

Two-Axle Trucks Plus One-Axle Semitrailer; HS20-44 and HS15-44. For today's highway bridges carrying a great amount of truck traffic, the two-axle truck loading with a one-axle semitrailer weighing 80% of the main truck load is commonly specified for design. The HS20-44 truck has 4 tons on the front axle, 16 tons on the rear axle, and 16 tons on the trailer axle. The distance from the rear truck axle to the semitrailer axle is varied from 14 to 30 ft, depending on which spacing will cause the most critical conditions.

Bridges for the new interstate highway system are designed in accordance with the AASHTO Specifications using the HS20-44 loading. An alternative load can be used in place of the HS20-44 truck. This alternate system, which consists of a pair of 24-k axle loads spaced 4 ft on center, is critical for short spans only. This loading will produce maximum moments for simple spans from 11.5 to 37 ft and maximum shears for spans from 6 to 22 ft. For other spans the HS20-44 loading or its equivalent lane loading will be critical.

Uniform Lane Loadings

Computation of stresses caused by a series of concentrated loads, whether they represent two-axle trucks or two-axle trucks with semitrailers, is a tedious job; therefore, a lane loading which will produce approximately the same stresses is frequently used. The lane loading consists of a uniform load plus a single moving concentrated load. This load system represents a line of medium-weight traffic with a heavy truck somewhere in the line. The uniform load per foot is equal to 0.016 times the total weight of the truck to which the load is to be roughly equivalent. The concentrated load equals 0.45 times the truck weight for moment calculations and 0.65 times the truck weight for shear calculations. These values for an H20-44 loading would be as follows: 0.016×20 tons equals 640 lb/ft of lane; concentrated load for moment 0.45×20 tons equals 18 k; and concentrated load for shear 0.65×20 tons equals 26 k.

The lane loading is more convenient to handle but may not be used unless it produces stresses equal to or greater than those produced by the corresponding H or HS loading. It can be shown that the equivalent lane loading for the HS20-44 will produce greater moments in simple spans of 145 ft and above and greater shears for simple spans of 128 ft and above. Appendix A of the AASHTO specifications contains tables that give the maximum shears and moments in simple spans for the various H loadings or for their equivalent lane loadings whichever controls.

The possibility of having a continuous series of heavily loaded trucks in every lane of a bridge having more than two lanes does not seem as great as that for a bridge having only two lanes. The AASHTO, therefore, permits the stresses caused by full loads in every lane to be reduced if the bridge has more than two lanes (10% reduction for three lanes and 25% for four or more lanes).

Impact

The truck and train loads on highway and railroad bridges are not applied gently and gradually but rather violently, causing stress increases. Additional loads, called impact loads, must be considered; these are taken into account by increasing the live-load stresses by some percentage, the percentage being obtained from purely empirical expressions. Numerous formulas have been presented for estimating impact. One example is the following 1949 AASHTO formula for highway bridges, in which I is the percent of impact and L is the length of the span, in feet, over which the live load is placed to obtain a maximum stress. The AASHTO says that it is unnecessary to use an impact percentage greater than 30%, regardless of the value given by the formula. Notice that the longer the span length becomes the smaller becomes the impact.

$$I = \frac{50}{L + 125}$$

Wind loads

Wind can produce very large forces on bridges. On several occasions wind forces have been of sufficient magnitude to cause collapse. Accurately estimating the magnitude of these loads is very difficult because there are so many variables involved such as terrain, wind velocity, and sizes and shapes of members.

The present bridge specifications have been influenced appreciably by considerable testing of bridge models in wind tunnels.[7, 8] The AASHTO says that in applying wind loads the exposed area shall equal the sum of the areas of all floors, railing, members, etc., as seen in elevation. The loads they specify are estimated for a 100-mph wind but the values may be increased or decreased provided the designer can show good evidence for such changes. The magnitudes of the forces for different velocities are to be determined from the specified values in direct proportion to the squared

[7]John M. Biggs, Saul Namyet, and Jiro Adachi, "Wind Loads on Girder Bridges," *Proc. ASCE*, Sep. No. 587 (January 1955).
[8]John M. Biggs, "Wind Loads on Truss Bridges," *Proc. ASCE*, Sep. No. 201 (July 1953).

values of the different velocities. A moving wind load is to be applied as follows.

> For trusses and arches—75 psf
> For girders and beams—50 psf

The total load so calculated may not be less than 300 lb per linear foot in the plane of the loaded chord nor 150 lb per foot in the plane of the unloaded chord for truss spans nor less than 300 lb per linear foot for girder spans.

When wind loads are combined with other loads the AASHTO permits some increase in allowable stresses. For the dead load plus full wind load combination, the allowable stresses can be increased by 25%. When wind loads are considered to be applied at the same time as dead load plus all of the other live loads, only 30% of the wind load is considered applied and the allowable stresses can be increased by 25%; however, a wind load must be included in this combination and applied to the live load 6 ft above the deck and equalling 100 lb per linear foot.

14-19. LIVE LOADS FOR RAILWAY BRIDGES

Railway bridges are commonly analyzed for a series of loads devised by Theodore Cooper. His loads, referred to as E loadings, represent two locomotives followed by a line of freight cars. A series of concentrated loads is used for the locomotives, while a uniform load represents the freight cars. Cooper introduced his loading system in 1894; it was the so-called E-40 load, pictured in Fig. 14-17. The train is assumed to have a 40-k load on the driving axle of the engine. Since his system was introduced, the weights of trains have been increased considerably, until at the present time bridges are designed on the basis of loads in the vicinity of an E-72 loading; and the use of E-80 and E-90 loadings is not uncommon.

Various tables that are obtainable present detailed information pertaining to Cooper's loadings such as axle loads, moments, and shears. If information is available for one E loading, the information for any other E loading can be obtained by direct proportion. The axle loads of an E-75

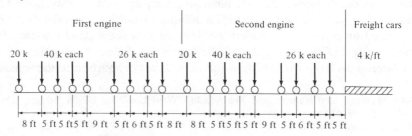

Figure 14-17 E-40 Cooper loading.

are 75/40 those for an E-40; those for an E-60 are 60/72 of those for an
E-72.

Cooper's loadings do not accurately picture today's trains, but they
are still in general use despite the availability of several more modern and
more realistic loadings such as Dr. D. B. Steinman's M-60 loading.[9]

Impact

Some AREA impact formulas are given at the end of this paragraph. The
values given, which are to be applied vertically at the top of rails, are for
open-deck bridges. For ballasted-deck bridges 90% of these values are to
be used. In these expressions S is the distance between centers of longitudinal beams, girders or trusses, or the length, in feet, between the supports of
floor beams or transverse girders; L is the length, in feet, center to center
of supports for stringers, transverse floor beams without stringers, longitudinal girders and trusses or the length of the longer adjacent supported
stringers, longitudinal beam, girder or truss for impact in floor beams,
floor beam hangers, subdiagonals of trusses, transverse girders, etc. For
rolling equipment without hammer blow (diesels, electric locomotives,
tenders alone, etc.)

$$\text{for } L \text{ less than 80 ft,} \qquad \frac{100}{S} + 40 - \frac{3L^2}{1600}$$

$$\text{for } L \text{ 80 ft or more,} \qquad \frac{100}{S} + 16 + \frac{600}{L-30}$$

The AREA also specifies impact percentages for steam locomotives
with hammer blow and for members receiving load from more than one
track.

Wind Loads

The wind loads specified by the 1969 AREA Specifications are handled
quite similarly to those required by the AASHTO. A moving wind force of
30 psf is to be applied to $1\frac{1}{2}$ times the vertical projections of leeward
trusses which are not shielded by the floor system. The total values so
determined may not be less than 200 lb per linear foot for the loaded
chord or flange nor 150 lb per linear foot on the unloaded chord or flange.
In addition a wind load is to be applied to the train equal to 300 lb per
linear foot on the one track and applied 8 ft above the rail.

The AREA makes one further requirement as it pertains to load
combinations. Should stresses computed for a 50-psf load applied as
described in the preceding paragraph plus dead load be in excess of those
caused by the wind loads also described in the last paragraph plus all other
loads (dead, live, impact, and centrifugal), the members will be designed
for the larger stresses.

[9] *Trans. ASCE* **86** (1923), pp. 606–636.

Chapter 15
Composite Design

15-1. COMPOSITE CONSTRUCTION

When a concrete slab is supported by steel beams and there is no provision for shear transfer between the two the result is a noncomposite section. There is no doubt in noncomposite construction but that the load causes the slab to deflect along with the beam, with the result that some of the load is carried by the slab. Unless, however, there is a great deal of bond between the two (as would be the case where the steel beam is completely encased in the concrete, or where a system of mechanical shear connectors is provided) the load carried by the slab is small and may be neglected.

For many years steel beams and reinforced-concrete slabs were used together with no consideration being made for any composite effect. In recent years, however, it has been shown that a great strengthening effect can be obtained by tying the two together to act as a unit in resisting loads. Steel beams and concrete slabs joined together compositely can often support a one-third or even greater increase in load than could the steel beams alone in noncomposite action.

Composite construction for highway bridges was given the green light by the adoption of the 1944 AASHTO Specifications which approved the method. Since about 1950 the use of composite bridge floors has rapidly increased until today they are commonplace all over the United States. In these bridges the longitudinal shears are transferred from the stringers to the reinforced concrete slab or deck with shear connectors (to be described in Section 15-6) causing the slab or deck to assist in carrying the bending moments. This type of section is shown in part (a) of Fig. 15-1.

The first approval for composite building floors was given by the 1952 AISC Specification and today they are very common. These floors may either be encased in concrete as shown in part (b) of Fig. 15-1 or be nonencased with shear connectors as shown in part (c) of the figure. The larger percentage of composite building floors being built today are of the nonencased type. If the steel sections are encased in concrete the shear transfer is made by bond and friction between the beam and the concrete

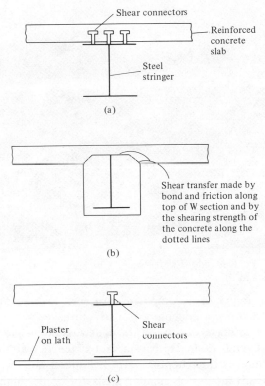

Figure 15-1   (a) Composite bridge floor with shear connectors. (b) Encased section for building floors. (c) Building floors with shear connectors.

and by the shearing strength of the concrete along the dotted lines shown in part (b) of Fig. 15-1. If more shearing strength is required, some type of steel reinforcing is provided along the sections indicated with the dotted lines.

15-2. ADVANTAGES OF COMPOSITE CONSTRUCTION

The floor slab in composite construction acts not only as a slab for resisting the live loads but also as an integral part of the beam. It actually serves as a large cover plate for the upper flange of the steel beam, appreciably increasing the beam's strength.

A particular advantage of composite floors is that they take advantage of concrete's high compressive strength by putting all or almost all of the slab in compression. At the same time a larger percentage of the steel is kept in tension (also advantageous) than is normally the case in steel-frame structures. The result is less steel tonnage required for the same loads and spans (or longer spans for the same sections). Composite sections have greater stiffnesses than noncomposite sections and they have smaller deflections—perhaps only 20 to 30% as large. Furthermore, tests have

Bridge on John F. Kennedy Expressway, Chicago, Ill. (Courtesy of Robert McCullough, Chicago, Ill.)

shown that the ability of a composite structure to take overload is decidedly greater than for a noncomposite structure.

A further advantage of composite construction is the possibility of having smaller overall floor depths—a fact of particular importance for tall buildings. Smaller floor depths permit reduced building heights with the consequent advantages of smaller costs for walls, plumbing, wiring, ducts, elevators, and foundations. Another important advantage available with reduced beam depths is a saving in fireproofing costs for the beams. A disadvantage for composite construction is the cost of furnishing and installing the shear connectors. This extra cost will usually exceed the cost reductions mentioned when spans are short and lightly loaded.

15-3. DISCUSSION OF SHORING

After the steel beams are erected the concrete slab is placed on them. The formwork, wet concrete, and other construction loads must, therefore, be supported by the beams or by temporary shoring. Should no shoring be used, the steel beams must support all of these loads as well as their own weights. Most specifications say that after the concrete has gained 75% of its 28-day strength the section has become composite and all loads applied thereafter may be considered to be supported by the composite section. When shoring is used it supports the wet concrete and the other construction loads. It does not really support the weight of the steel beams unless they are given an initial upward deflection (probably impractical). When the shoring is removed (after the concrete gains at least 75% of its 28-day strength) the weight of the slab is transferred to the composite section, not just to the steel beams. The student can see that if shoring is used it will be possible to use lighter and thus cheaper steel beams. The question then arises, "Will the saving in steel cost be greater than the extra cost of shoring?" Probably the answer is "No." The usual decision is to use heavier steel beams and do without shoring for several reasons. These include the following.

1. Apart from reasons of economy, the use of shoring is a tricky operation, particularly where settlement of the shoring is possible, as is often the case in bridge construction.

2. Both theory and load tests show that the ultimate strengths of composite sections of the same sizes are the same whether shoring is used or not. If lighter steel beams are selected for a particular span because shoring is used, the result is a smaller ultimate strength.

3. Another disadvantage of shoring is that after the concrete hardens and the shoring is removed the slab will participate in composite action in supporting the dead loads. The slab will be placed in compression by these long-term loads and will have substantial creep and shrinkage parallel to the beams. The result will be a great decrease in the stress in the slab with a corresponding increase in the steel stresses. The probable consequence is that most of the dead load will be supported by the steel beams anyway and composite action will really apply only to the live loads as though shoring had not been used. A common practice when shoring is used is to reduce the calculated effective area of the concrete (as described in Section 15-5).

15-4. EFFECTIVE FLANGE WIDTHS

The portion of the slab or flange which can be considered to participate in the beam action is controlled by the specifications. For buildings the AISC permits the same method of computing effective flange widths as is permitted by the ACI for reinforced concrete tee beams. Reference is made here to Fig. 15-2.

The maximum effective flange width permitted by the AISC can be determined by computing the following values and will equal the least value of b so obtained.

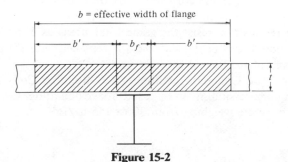

Figure 15-2

1. b not greater than one-fourth of the span of the beam.
2. b' not greater than one-half of the clear distance to the adjacent beam.
3. b' not greater than eight times the slab thickness.

Should the slab exist only on one side of the beam, the following requirements control.

1. b' not greater than one-twelfth of beam span.
2. b' not greater than one-half of the clear distance to the adjacent beam.
3. b' not greater than six times the slab thickness.

The AASHTO requirements for maximum permissible flange widths are slightly different. The maximum flange width may not exceed one-fourth of the beam span, twelve times the least thickness of the slab, nor the distance center to center of the beam. Should the slab exist on only one side of the beam its effective width may not exceed one-twelfth of the beam span, six times the slab thickness, nor one-half of the center to center distance to the next beam.

15-5. STRESS CALCULATIONS FOR COMPOSITE SECTIONS

For stress calculations the properties of a composite section are computed with the transformed area method. In this method the cross-sectional area of one of the two materials has to be replaced or transformed into an equivalent area of the other. For composite design, it is customary to replace the concrete with an equivalent area of steel, whereas the reverse procedure is used in the working stress design method for reinforced concrete design.

In the transformed area procedure the concrete and steel are assumed to be bonded tightly together so that their strains will be the same at equal distances from the neutral axis. The unit stress in either material can then be said to equal its strain times its modulus of elasticity (ϵE_c for the concrete or ϵE_s for the steel). The unit stress in the steel is then $\epsilon E_s/\epsilon E_c = E_s/E_c$ times as great as the corresponding unit stress in the concrete. The E_s/E_c ratio is referred to as the modular ratio n; therefore, n in.^{2} of concrete are required to resist the same total stress as 1 in.^{2} of steel; and the cross-sectional area of the slab (A_c) is replaced with a transformed or equivalent area of steel equal to A_c/n.

The American Concrete Institute (ACI) Building Code suggests that the following expression may be used for calculating the modulus of elasticity of concrete weighing from 90 to 155 lb/ft^3.

$$E_c = w^{1.5} 33 \sqrt{f_c'}$$

In this expression w is the weight of the concrete in pounds per cubic foot and f'_c is the 28-day compressive strength in pounds per square inch.

Should the neutral axis fall in the slab the concrete below may not be considered to contribute to the moment of inertia (except for deflection calculations) as this concrete is assumed to be cracked. In the example problems of this chapter, M_D is the dead-load moment which must be supported by the steel beam alone, M_L is the live-load moment applied to the composite section, and M_d is the dead-load moment applied to the composite section.

When temporary shoring is not used, the actual stresses in the concrete and steel under full load may be determined with the flexure expressions to follow in which S_S is the section modulus of the steel beam alone and S_{TR} is the section modulus of the transformed composite section. Both values are referred to the flange where the stress is being computed.

$$f_s = \pm \frac{M_D}{S_S} \pm \frac{M_d + M_L}{S_{TR}}$$

$$f_c = \frac{M_d + M_L}{n S_{TR}}$$

As indicated in the discussion of shoring, both theory and tests show that the ultimate strengths of composite sections are the same whether shoring is used or not. On this basis the AISC Specification in its Section 1.11.2.2 states that all loads are to be considered as being applied to the composite section regardless of the use or nonuse of shoring.

For composite sections with no temporary shoring the AISC limits the total service load bending stress in the steel for unshored construction to 1.35 times the allowable stress for the steel beam alone (equal to about $0.81F_y$ for noncompact sections and $0.89F_y$ for compact sections). These allowable values are for the situations where all of the moment (M_D, M_d, and M_L) are assumed to be applied to the steel beam alone. This allowable value is checked by means of the expression given at the end of the next paragraph for S_{TR}, the section modulus of the transformed composite section referred to the tension flange. When temporary shoring is not used, the value of S_{TR} may not exceed the value determined by substitution in this expression.

When sections are subjected to positive bending moment, stresses must be computed in the tensile flanges. At sections subjected to negative moments it is necessary to calculate stresses in both tension and compression flanges of the steel beams. In applying the formula it is to be remembered that the loads supported by the steel beam before the concrete hardens may not cause the allowable stresses in the beam to be exceeded. Notice that M_L in this expression is the moment caused by all loads (*dead or live*) which are applied after the concrete has reached 75%

of its 28-day strength.

$$S_{tr} = \left(1.35 + 0.35\frac{M_L}{M_D}\right)S_S$$

Stresses are calculated in Example 15-1 for a composite section using the AISC Specification and in Example 15-2 for another section in accordance with the AASHTO requirements. In this latter example it is necessary to take into account the effect of creep in the concrete caused by the long-term loads. The AASHTO says that for short-term loads applied to the composite section, the regular value of n should be used to compute the transformed area of the concrete. For long-term loads, however, there will be an appreciable creep or plastic flow in the concrete. As a result the AASHTO Specifications require that $3n$ be used to determine the transformed area of the concrete. Thus for long-term loads applied to composite sections, the concrete becomes much less effective and more of the moment is thrown onto the steel section. In summary there will be stress in the tension flange of the steel beam due to M_D acting on the steel section alone, plus stress due to M_d acting on the composite section ($n=30$), plus stress due to M_L acting on the composite section ($n=10$). For the concrete no stress is caused by M_D, so the stress is caused by M_d acting on the composite section ($n=30$), plus the effect of M_L acting on the composite section ($n=10$).

Example 15-1

Determine the stresses for the composite beam section shown in Fig. 15-3 according to the AISC Specification. Adequate shear connectors are assumed to be present and no shoring is used. Assume simple spans and the following data.

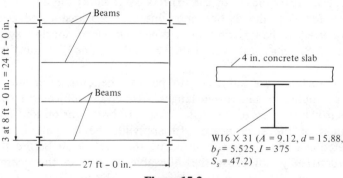

Figure 15-3

$LL = 100$ psf
Partition weight $= 18$ psf
4-in. concrete slab weighs 50 psf
$f_c' = 3000$ psi
$f_c = 1350$ psi
$n = 9$
A36 steel

SOLUTION

Calculation of Moments: Loads applied before concrete hardens (75% of 28-day strength):

$$\text{slab} = (8)(50) = 400 \text{ lb/ft}$$

$$\text{beam} = 31 \text{ lb/ft}$$

$$\text{total} = \overline{431} \text{ lb/ft}$$

$$M_D = \frac{(0.431)(27)^2}{8} = 39.3 \text{ ft-k}$$

Loads applied after concrete hardens:

$$\text{partitions} = (8)(18) = 144 \text{ lb/ft}$$

$$LL = (8)(100) = 800 \text{ lb/ft}$$

$$\text{total} = \overline{944} \text{ lb/ft}$$

$$M_L = \frac{(0.944)(27)^2}{8} = 86.0 \text{ ft-k}$$

$$M_T = 39.3 + 86.0 = 125.3 \text{ ft-k}$$

Effective Width of Flange:

(1) $b = (\frac{1}{4})(27 \times 12) = 81$ in.

(2) $b' = (\frac{1}{2})(8 \times 12 - 5.525) = 45.24$ in.

$b = (2)(45.24) + 5.525 = 96$ in.

(3) $b' = (8)(4) = 32$ in.

$b = (2)(32) + 5.525 = 69.525$ in.   (controls)

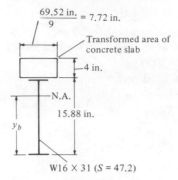

W16 × 31 (S = 47.2)

Figure 15-4

Properties of Composite Section: (Reference made here to Fig. 15-4):

$$A = 9.12 + (4)(7.72) = 40.00 \text{ in.}^2$$

$$y_b = \frac{(9.12)(7.94) + (30.88)(17.88)}{40.00} = 15.61 \text{ in.}$$

$$I = 375 + (9.12)(7.67)^2 + \left(\tfrac{1}{12}\right)(7.72)(4)^3 + (30.88)(2.27)^2$$

$$= 1112 \text{ in.}^4$$

Review of Stresses: Before concrete hardens:

$$f_s = \frac{(12)(39.3)(7.94)}{375} = 9.99 \text{ ksi} < 24 \text{ ksi} \qquad\qquad \text{OK}$$

Actual stresses after concrete hardens:

$$f_c = -\frac{(12)(86.0)(4.27)}{(9)(1112)} = -0.440 \text{ ksi} < 1.35 \text{ ksi} \qquad\qquad \text{OK}$$

$$f_s = +9.99 + \frac{(12)(86.0)(15.61)}{1112} = +24.48 \text{ ksi} > 24 \text{ ksi} \qquad \text{NG}$$

But Section 1.11.2.2 of the AISC Specification permits the calculation of the steel stress for the total service load moment using S_{TR} provided it does not exceed the maximum permissible value.

$$\text{actual } S_{tr} = \frac{1112}{15.61} = 71.24 \text{ in.}^3$$

$$\text{max permissible } S_{TR} = \left(1.35 + 0.35 \times \frac{86.0}{39.3}\right) 47.2 = 99.87 > 71.24 \text{ in.}^3 \text{OK}$$

Thus stresses under full load may be calculated as follows by AISC:

$$f_c = \frac{(12)(125.3)(4.27)}{(9)(1112)} = 0.642 \text{ ksi} < 1.35 \text{ ksi} \qquad \text{OK}$$

$$f_s = \frac{(12)(125.3)}{71.24} = 21.11 \text{ ksi} < 24 \text{ ksi} \qquad \text{OK}$$

Example 15-2

Calculate the stresses in the composite section shown in Fig. 15-5 using the AASHTO Specifications. Adequate shear connectors are assumed to be present and no shoring is used. Assume simple spans and the following data.

$$M_D = 300 \text{ ft-k}$$
$$M_d = 100 \text{ ft-k} \qquad \text{(due to curbs, railing, surface treatment, etc.)}$$
$$M_L = 450 \text{ ft-k}$$
$$M_T = 850 \text{ ft-k}$$
$$f_s = 18 \text{ ksi}$$
$$f_c = 1.2 \text{ ksi}$$
$$n = 10$$

SOLUTION

Properties of Composite Section: For $n = 10$:

$$A = \left(\tfrac{66}{10}\right)(6) + 41.6 = 81.2 \text{ in.}^2$$

$$y_b = \frac{(41.6)(16.65) + (39.6)(36.30)}{81.2} = 26.23 \text{ in.}$$

$$I = 7450 + (41.6)(9.58)^2 + \left(\tfrac{1}{12}\right)(6.6)(6)^3 + (39.6)(10.07)^2$$

$$= 15,402 \text{ in.}^4$$

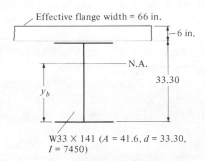

Effective flange width = 66 in.

6 in.

N.A.

33.30

y_b

W33 × 141 ($A = 41.6, d = 33.30,$ $I = 7450$)

Figure 15-5

For $n = 30$:

$$A = \left(\tfrac{66}{30}\right)(6) + 41.60 = 54.8 \text{ in.}^2$$

$$y_b = \frac{(41.6)(16.65) + (13.2)(36.30)}{54.8} = 21.38 \text{ in.}$$

$$I = 7450 + (41.6)(4.73)^2 + \left(\tfrac{1}{12}\right)(2.2)(6)^3 + (13.2)(14.92)^2$$

$$= 11,359 \text{ in.}^4$$

Final Stresses:

$$f_c = \frac{(12)(450)(13.07)}{(15,402)(10)} + \frac{(12)(100)(17.92)}{(11,359)(30)} = 0.521 \text{ ksi} < 1.2 \text{ ksi} \quad \text{OK}$$

$$f_s = \frac{(12)(300)(16.65)}{7450} + \frac{(12)(450)(26.23)}{15,402} + \frac{(12)(100)(21.38)}{11,359}$$

$$= 19.50 \text{ ksi} > 18.00 \text{ ksi} \quad\quad\quad\quad\quad\quad\quad\quad\quad \text{NG}$$

15-6.   SHEAR TRANSFER

The concrete slabs may rest directly on top of the steel beams or the beams may be completely encased in concrete for fireproofing purposes. The longitudinal shear can be transferred between the two by bond and shear and possibly some type of shear reinforcing if needed when the beams are encased. When not encased, mechanical connectors must transfer the load. Fireproofing is not necessary for bridges and the slab is placed on top of the steel beams. Bridges are subject to heavy impactive loads and the bond between the beams and the deck, which is easily broken, is considered negligible. For this reason shear connectors are designed to resist all of the shear between bridge slabs and beams.

Various types of shear connectors have been tried including spiral bars, channels, zees, angles, and studs. Several of these types of connectors are shown in Fig. 15-6. Economic considerations have usually led to the use of round studs welded to the top flange of the beams. These studs are available in diameters from $\frac{1}{2}$ to 1 in. and in lengths from 2 to 8 in. but the most commonly used sizes are $\frac{3}{4}$ or $\frac{7}{8}$ in. and 2 to 4 in. in length. They actually consist of rounded steel bars welded on one end to the steel beams. The other end is upset to prevent vertical separation of the slab from the beam. These studs can be quickly attached to the steel beams with stud-welding guns by semiskilled workers.[1]

Shop installation of shear connectors is initially more economical but there is a growing tendency to use field installation. There are two major reasons for this trend: the connectors may easily be damaged during

[1]Ivan M. Viest, R. S. Fountain, and R. C. Singleton, *Composite Construction in Steel and Concrete* (New York: McGraw-Hill, 1958), pp. 50–51.

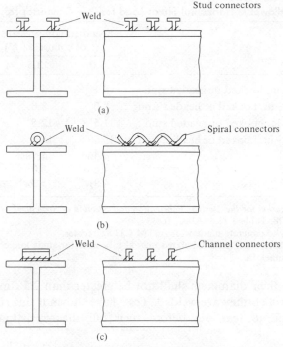

Figure 15-6 Shear connectors.

transportation and setting of the beams, and they serve as a hindrance to the workers walking along the top flanges during the early phases of construction.

AISC Requirements

For composite sections it is permissible to use normal weight stone concrete (made with aggregates conforming to ASTM C33) or light-weight concrete weighing not less than 90 lb/ft^3 (made with rotary kiln-produced aggregates conforming to ASTM C330).

The shear values permitted by the AISC for buildings for various types of connectors are presented in Table 15-1 for normal weight concrete. This table is a reproduction of Table 1.11.4 of the AISC Specification. The values given for q are the ultimate strengths of the connectors reduced by a factor of safety of 2.5 to obtain working load values.

When light-weight concretes are used it is necessary to reduce the allowable connector shear loads. This is done by multiplying the allowable shear loads given in Table 15-1 by the appropriate coefficient from Table 15-2. This latter table is a reproduction of Table 1.11.4A from the AISC Specification.

The AISC requires that shear connectors have at least 1 in. of concrete cover in all directions. In addition, unless they are located directly

Table 15-1 Allowable Horizontal Shear Load for One Connector (q), Kips[a]

Connector[b]	Specified compressive strength of concrete (f_c'), ksi		
	3.0	3.5	≥4.0
$\frac{1}{2}$ in. diam.×2 in. hooked or headed stud	5.1	5.5	5.9
$\frac{5}{8}$ in. diam.×$2\frac{1}{2}$ in. hooked or headed stud	8.0	8.6	9.2
$\frac{3}{4}$ in. diam.×3 in. hooked or headed stud	11.5	12.5	13.3
$\frac{7}{8}$ in. diam.×$3\frac{1}{2}$ in. hooked or headed stud	15.6	16.8	18.0
Channel C3×4.1	$4.3w^c$	$4.7w^c$	$5.0w^c$
Channel C4×5.4	$4.6w^c$	$5.0w^c$	$5.3w^c$
Channel C5×6.7	$4.9w^c$	$5.3w^c$	$5.6w^c$

SOURCE: *Specification for the Design, Fabrication and Erection of Structural Steel Buildings*, November 1, 1978, Table 1.11.4. AISC, Inc.
[a]Applicable only to concrete made with ASTM C33 aggregates.
[b]The allowable horizontal loads tabulated may also be used for studs longer than shown.
[c]w-length of channel, in.

over the web their diameters shall not be greater than 2.5 times the flange thickness to which they are welded. Tests have shown if this rule is not met they will tend to tear out before their full shear-resisting capacity is reached.

Shear connectors must be capable of resisting both horizontal and vertical movement because there is a tendency for the slab and beam to separate vertically as well as to slip horizontally. Most designers feel that because of the tendency to slip vertically the longitudinal spacing of connectors should not exceed approximately 2 ft. The upset heads of the common studs help to prevent vertical separation.

Careful attention should be given to the AISC method of determining the horizontal shear to be taken by the connectors. Rather than basing their designs on the shear computed by the VQ/I formula, shear is estimated at ultimate load conditions. When a composite beam is being tested, failure will probably occur with a crushing of the concrete. At that time it seems reasonable to assume that the concrete and steel have both reached a plastic stress condition.

For this discussion reference is made to Fig. 15-7. Should the neutral axis fall in the slab the maximum horizontal shear (or horizontal force on

Table 15-2 Coefficients for Use with Concrete Made with C330 Aggregates

Specified compressive strength of concrete (f_c')	Air dry unit weight of concrete, pcf						
	90	95	100	105	110	115	120
≤ 4.0 ksi	0.73	0.76	0.78	0.81	0.83	0.86	0.88
≥ 5.0 ksi	0.82	0.85	0.87	0.91	0.93	0.96	0.99s

SOURCE: *Specification for the Design, Fabrication, and Erection of Structural Steel Buildings*, November 1, 1978, Table 1.11.4A. AISC, Inc.

Channel-section shear connectors, Grand Rapids, Mich. (Courtesy of The Lincoln Electric Company.)

the plane between the concrete and the steel) is said to be $A_s F_y$; and if the neutral axis is in the steel section, the maximum horizontal shear is considered to equal $0.85 f_c' A_c$. (For the student unfamiliar with the strength design theory for reinforced concrete, the average stress at failure on the compression side of a concrete beam is usually assumed to be $0.85 f_c'$.)

From this information, expressions for V_h (the shear to be taken by the connectors) are written below. These values are divided by 2 to

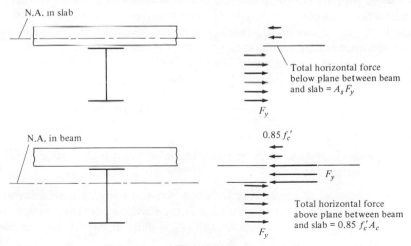

Figure 15-7

estimate conditions at working loads. The AISC says that the two expressions for V_h are to be solved and the smaller value used. V_h is the total shear to be resisted between the point of maximum positive moment in a simple span and the end of the beam. (For continuous beams, the wording would be "between the point of maximum positive moment and the point of contraflexure.") The number of shear connectors required on each side of the point of maximum positive moment can be determined by dividing V_h by q the strength of one connector of the size and type being used. As previously described, the AISC permits a uniform spacing of connectors.

$$V_h = \frac{0.85 f'_c A_c}{2}$$

$$V_h = \frac{A_s F_y}{2}$$

$$\text{no. of connectors} = \frac{V_h}{q}$$

When longitudinal reinforcing steel (with area A'_s and specified minimum yield stress F_{yr}) is located within the effective width of the concrete flange and when it has been included in the calculated properties of the composite section, the first of the V_h expressions should be written as follows.

$$V_h = \frac{0.85 f'_c A_c}{2} + \frac{A'_s F_{yr}}{2}$$

Shear connectors are relatively flexible with the result that there is some slippage between the steel girders and the concrete slabs. Since slip does occur there is a redistribution of load between the connectors similar to that discussed for bolted connections in Section 9-10. Because of this redistribution of unequal connector loads the AISC permits, for the usual building with its design based on uniform loads, a uniform spacing of connectors between the points of maximum positive moment and the points of zero moment. However, for concentrated loads the AISC says that the number of connectors between the concentrated load and the nearest point of zero moment shall not be less than determined by the following formula.

$$N_2 = \frac{N_1 (M\beta / M_{\max} - 1)}{\beta - 1} \qquad \text{(AISC Formula 1.11-7)}$$

where

$M =$ moment (less than the maximum moment) at a concentrated load point

$N_1 =$ number of connectors required between point of maximum moment and point of zero moment, determined by the relationship V_h / q or V'_h / q, as applicable (V'_h is defined later in this section)

$\beta = \dfrac{S_{tr}}{S_s}$ or $\dfrac{S_{eff}}{S_s}$ as applicable

For continuous beams connectors may be spaced uniformly between the points of maximum negative moment and the points of zero moment says the AISC.

The resisting moment of a steel beam and concrete slab increases directly with the number of shear connectors used up to a maximum value when the required number of connectors for full composite action is used. At times it may not be feasible to use shear connectors, at other times it may be feasible or necessary to use only a part of the total number required, thus providing only partial composite action. For this reason the AISC permits the use of full composite action and of partial composite action.[2]

For cases where the number of shear connectors required for full composite action is not used the AISC permits an effective section modulus to be determined by the following expression.

$$S_{eff} = S_S + \frac{V_h'}{V_h}(S_{TR} - S_S)$$

In this expression V_h is as previously defined and V_h' is the total shear developed between the steel and the concrete on each side of the point of maximum moment and equals q times the number of connectors furnished between the point of maximum moment and the nearest zero moment point.

If a steel beam is selected which has a section modulus (S_{tr}) larger than needed (even by a small amount) the number of shear connectors required will be substantially reduced if this partial composite action expression is applied. The AISC Manual (p $2-104$ of 8th ed) states that if S_{tr} is 3 to 10% higher than required the number of shear connectors will be reduced from 10 to 30%. If fewer shear connectors are used there will be a slight increase in the live load deflection due to the smaller effective moment of inertia of the section.

AASHTO Requirements

The AASHTO requirements for the design of shear connectors are more conservative than those permitted for buildings by the AISC due to the nature of the highly impactive bridge loadings. The AASHTO says that there shall be a minimum of 2 in. of concrete over the tops of shear connectors and the connectors must penetrate a minimum of 2 in. up from the bottom of the slab.

A maximum pitch of 24 in. is permitted except over the interior supports of continuous beams where it may be desirable to avoid placing them at points where tensile stresses are high in the flanges of the steel

[2] R. G. Slutter and G. C. Driscoll, "Flexural Strength of Steel-Concrete Composite Beams" *Proc. ASCE* **91**, no. ST2 (April 1965), pp. 71–99.

girders. The required number of connectors is still the same even if the pitch is modified in the negative moment region.

The AASHTO says that shear connectors shall be designed for (a) fatigue, and checked for (b) ultimate strength, and they may be spaced at regular or variable intervals.

FATIGUE

The range of horizontal shear is to be calculated with the following expression in which V_r is the range of shear due to live load plus impact. The range of shear at any section is to be considered to equal the difference in the maximum and minimum shear envelope values. Q is the statical moment about the neutral axis of the transformed compressive concrete area (or at sections of negative moment equals the statical moment of the reinforcement embedded in the concrete). I is the moment of inertia of the transformed composite girder in positive moment regions (or in negative moment regions, is the moment of inertia of the steel girder and may include the I of any reinforcing embedded in the concrete).

$$S_r = \frac{V_r Q}{I}$$

The AASHTO allowable shear (pounds) for connectors, Z_r, is determined from the expressions to follow.

1. Channels:

$$Z_r = Bw$$

where

$w = $ length of channel shear connector measured in inches transverse to the steel girder

$B = 4000$ for 100,000 cycles of loading, 3000 for 500,000 cycles, 2400 for 2,000,000 cycles, 2100 for over 2,000,000 cycles

2. Welded studs when H/d is equal to or greater than 4:

$$Z_r = \alpha d^2$$

where

$H = $ height of stud in inches

$d = $ diameter of stud in inches

$\alpha = 13,000$ for 100,000 cycles, 10,600 for 500,000 cycles, 7850 for 2,000,000 cycles, 5500 for over 2,000,000 cycles.

Section 1.7.2 of the AASHTO Specifications specifies the number of cycles of maximum stress to be considered for design depending on the types of highways, spans, etc.

The required pitch of shear connectors at any one section is to be determined by dividing the allowable range of horizontal shear of all connectors at one transverse girder cross section (ΣZ_r) by the horizontal range of shear S_r at the section in question.

$$\text{spacing} = \frac{\Sigma Z_r}{S_r}$$

ULTIMATE STRENGTH

After the connectors are designed for fatigue the AASHTO says their number, N, used between the points of maximum moment and the end points or dead load points of contraflexure or between points of maximum negative moment and dead load points of contraflexure must be checked for ultimate strength by the expression to follow.

$$N_1 = \frac{P}{\phi S_u}$$

where

S_u = the ultimate strength of 1 shear connector in pounds

ϕ = a reduction factor = 0.85

P = the force in the slab defined as the smaller of P_1 or P_2 in the following.

For maximum positive moment points

$$P_1 = A_s F_y$$
$$P_2 = 0.85 f_c' bc$$

in which b is the effective flange width and c is the slab thickness.

The number of connectors N_2 required between points of maximum positive moment and adjacent points of maximum negative moment shall not be less than the following.

$$N_2 = \frac{P + P_3}{\phi S_u}$$

At points of maximum negative moment, the force in the slab P_3 is to be taken as

$$P_3 = A_s{}' F_y{}'$$

where

$A_s{}'$ = total area of longitudinal reinforcing bars at interior support within effective flange width of slab

$F_y{}'$ = yield stress of reinforcing steel

The AASHTO ultimate strength value for a shear connector (S_u) is to be determined from the expressions to follow where t is the thickness of the web of the channel shear connector in inches and W is the length of channel shear connector in inches.

For channels

$$S_u = 550\left(h + \frac{t}{2}\right) W \sqrt{f_c'}$$

For welded studs ($H/d \geqslant 4$),

$$S_u = 0.4 d^2 \sqrt{f_c' E_c}$$

Should the longitudinal reinforcing bars in negative moment regions not be used in computing section properties, the AASHTO presents another expression for computing the number of connectors to be used at the points of contraflexure.

Design of shear connectors in accordance with the AISC Specification is illustrated in Example 15-3 while AASHTO spacing requirements are illustrated in Example 15-7.

15-7. PROPORTIONING COMPOSITE SECTIONS

Composite construction is of particular advantage economically when loads are heavy, spans are long, and beams are spaced at fairly large intervals. For steel-building frames, composite construction is economical for spans varying roughly from 25 to 50 ft, with particular advantage in the longer spans. For bridges, simple spans have been economically constructed up to approximately 120 ft and continuous spans 50 or 60 ft longer. Composite bridges are generally economical for simple spans greater than about 40 ft and for continuous spans greater than about 60 ft.

Sometimes cover plates are welded to the bottom flanges of steel beams with improved economy. The student can see that with the slab acting as part of the beam there is a very large compressive area available and by adding cover plates to the tensile flange a little better balance is obtained.

For best economy, thicker slabs and cover-plated beams are usually of advantage. An important point to realize is that the major cost of cover plates regardless of their size is in fabrication. The result is that heavy cover plates do not cost proportionately more than thin ones; it is logical, therefore, to use larger cover plates or none at all from an economical standpoint. Cover plates are not usually economical when light-weight concrete is used because the high modular ratios of such concrete greatly reduce the effective concrete areas.

In tall buildings where headroom is a problem it is desirable to use the minimum overall floor thicknesses possible. For buildings, minimum depth -span ratios of approximately $\frac{1}{24}$ are recommended if the loads are fairly static and $\frac{1}{20}$ if the loads are of such a nature as to cause appreciable vibration. The thicknesses of the floor slabs are known (from the concrete design) and the depths of the steel beams can be fairly well estimated from these ratios. For bridges the AASHTO gives a desirable minimum ratio of $\frac{1}{25}$ for total depth-to-span and $\frac{1}{30}$ for steel-beam depth alone to span. Should shallower sections be used than these there is the usual requirement that the resulting deflections may not exceed those developed had the recommended ratios been followed.

Perhaps the greatest difficulty in composite design has been the selection of economical beam sizes as a lengthy trial-and-error process has been involved. In the past few years, however, a great deal of data has been published which appreciably abbreviates the problem. For example, the AISC Manual includes a set of tables for selecting sizes. These tables developed for building floors are given for 3000 psi concrete, 36 ksi steel, and for 4-, $4\frac{1}{2}$-, $5\frac{1}{2}$- and 6 in.- slabs.

An excellent method of estimating steel beam sizes for building or bridge composite floors is presented on page 104 of *Advanced Design In Structural Steel* by John E. Lothers (Prentice-Hall, Inc., 1960). Lothers suggests that the steel beams be designed as though they alone have to support all of the loads but using a higher allowable stress to account for the composite action. He presents a table of estimated allowable steel stress increases for varying values of n. The increases are as follows: 20% for $n = 15$, 24% for $n = 12$, 28% for $n = 10$, 33% for $n = 8$, and 37% for $n = 6$. Increases for n values in between the ones given are roughly proportional. In applying these values it must be remembered that the usual allowable steel stresses may not be exceeded by M_D alone. This information provides a very good initial trial section but another trial or two may still be necessary before a final selection is made.

Other information is available from several sources. The previously mentioned book by Viest, Fountain, and Singleton gives a great deal of information on the subject including design procedures, curves, and tables of data. Additional data is contained in a book entitled *Properties of Composite Sections for Bridges and Buildings* published by the Bethlehem Steel Company (Bethlehem, Pa., 1962). Another excellent reference is Chapters 3 and 11 and the accompanying nomograph of *USS Highway Structures Design Handbook* published by the United States Steel Corporation (Pittsburgh, Pa.).

Example 15-3 illustrates the design of a composite section with a cover plate on the bottom flange of the steel beam. As previously mentioned using bottom cover plates with composite sections involves such high labor costs that they are seldom used. Nevertheless the author used a W section with a cover plate in this example to illustrate the calculations involved. The trial section was taken from the 7th edition of the AISC Manual since the very abbreviated composite section tables of the 8th edition do not include any cover plated sections.

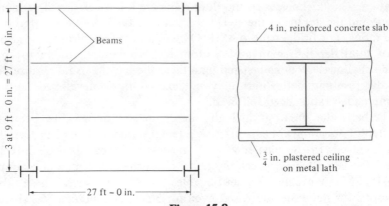

Figure 15-8

Example 15-3

Design a composite section using A36 steel and the AISC Specification for the situation shown in Fig. 15-8. No shoring is to be used and simple spans are assumed. The following data are supplied.

$$LL = 120 \text{ psf}$$

$$\text{ceiling weight} = 10 \text{ psf}$$

$$\text{partition weight} = 15 \text{ psf}$$

$$\text{4-in. concrete slab} = 50 \text{ psf}$$

$$f_c' = 3000 \text{ psi}$$

$$f_c = 1350 \text{ psi}$$

$$n = 9$$

$$\underline{\text{A36 steel}}$$

SOLUTION

Calculation of Moments: Loads applied before concrete hardens:

$$\text{slab} = (9)(50) = 450 \text{ lb/ft}$$

$$\text{assumed beam weight} = \quad 31 \text{ lb/ft}$$

$$\text{total} = \overline{481} \text{ lb/ft}$$

$$M_D = \frac{(0.481)(27)^2}{8} = 43.8 \text{ ft-k}$$

Loads applied after concrete hardens:

$$\text{ceiling} = (9)(10) = \quad 90 \text{ lb/ft}$$

$$\text{partitions} = (9)(15) = \quad 135 \text{ lb/ft}$$

$$LL = (9)(120) = 1080 \text{ lb/ft}$$

$$\text{total} = \overline{1305} \text{ lb/ft}$$

$$M_L = \frac{(1.305)(27)^2}{8} = 119 \text{ ft-k}$$

$$M_T = M_D + M_L = 162.8 \text{ ft-k}$$

$$\text{reqd. } S_b = \frac{(12)(162.8)}{24} = 81.4 \text{ in.}^3$$

Selection of Trial Section from the 7th ed. of AISC Manual: Try W14×22 with 1×4 cover PL (several other available sections.) Properties of steel section with cover PL.

$$A = 6.49 + 4.00 = 10.49 \text{ in.}^2$$

$$y_b = \frac{(6.49)(7.87) + (4.00)(0.5)}{10.49} = 5.06 \text{ in.}$$

$$y \text{ to top of steel} = 14.74 - 5.06 = 9.68 \text{ in.}$$

$$I = 199 + (6.49)(2.81)^2 + (4.00)(4.56)^2 = 333.4 \text{ in.}^4$$

$$S_s \text{ for tension flange} = \frac{333.4}{5.06} = 65.89 \text{ in.}^3$$

Properties of Composite Section from 7th ed. of AISC Manual (these values can be computed by usual method):

$$y_b = 13.75 \text{ in.}$$

$$b = \text{effective width of flange} = 69.00 \text{ in.}$$

$$I_{tr} = 1440 \text{ in.}^4$$

$$S_{tr} = 105 \text{ in.}^3 \text{ for tension flange}$$

$$S_{tr} = 289 \text{ in.}^3 \text{ for concrete flange}$$

Review of Section: Before concrete hardens:

$$f_s = \frac{(12)(43.8)(9.68)}{333.4} = 15.26 \text{ ksi} < 24 \text{ ksi} \qquad \text{OK}$$

Maximum permissible S_{tr} by AISC is

$$\left(1.35 + 0.35 \times \frac{119}{43.8}\right) 65.89 = 151.6 \text{ in.}^3 > 105 \text{ in.}^3 \qquad \text{OK}$$

After concrete hardens:

$$f_s = \frac{(12)(162.8)}{105} = 18.61 \text{ ksi} < 24 \text{ ksi} \qquad \text{OK}$$

$$f_c = \frac{(12)(162.8)}{(9)(289)} = 0.751 \text{ ksi} < 1.35 \text{ ksi} \qquad \text{OK}$$

Design of Shear Connectors:

$$V_h = \frac{0.85 f'_c A_c}{2} = \frac{(0.85)(3.0)(4 \times 69.00)}{2} = 352 \text{ k}$$

$$V_h = \frac{A_s F_y}{2} = \frac{(10.49)(36)}{2} = 188.8 \text{ k} \qquad \text{(controls)}$$

Assuming the $\frac{3}{4} \times 3$-in. headed stud which can resist 11.5 k (see Table 15-1):

$$\text{no. reqd.} = \frac{188.8}{11.5} = 16.4 \text{ each side of } \text{Ⱡ}$$

Use 17 connectors on each side of Ⱡ evenly spaced

The cover plate on the bottom flange of the steel beam is needed only for the sections of the span where the moment is largest. When the moment decreases to a value equal to the resisting moment of the slab and steel shape above, the cover plate can theoretically be cut off. The AISC and most other specifications state, however, that a cover plate should be extended some distance beyond its theoretical point of cutoff. The distance is that required for the connectors (rivets, welds, or bolts) to develop or transfer the portion of the flexural stresses taken by the plate at the theoretical point of cutoff. Example 15-4 illustrates the calculations required for determining cover-plate lengths.

Particular attention should be given to the design of the intermittent welds connecting the cover plate to the W shape. The student should be able to follow this design by reading the paragraph numbers referred to in the AISC Specification.

Example 15-4

Determine the required length of the cover plate for the section designed in Example 15-3 if $\frac{5}{16}$-in. SMAW fillet welds, E70 electrodes, and the AISC Specification are used.

SOLUTION
From Example 15-3:

$$y_b = 13.75 \text{ in.}$$

$$I_{tr} = 1440 \text{ in.}^4$$

S_{TR} for tension flange for section with cover PL $= 105$ in.^{3}

S_{tr} for section without cover PL $= 46.4$ in.^{3} (from AISC Manual)

Resisting moments:

$$M \text{ for section with cover PL} = \frac{(24)(105)}{12} = 210 \text{ ft-k}$$

$$M \text{ for section without cover PL} = \frac{(24)(46.4)}{12} = 92.8 \text{ ft-k}$$

Theoretical point of cutoff (see Fig. 15-9):

For a uniformly loaded section with its parabolic moment diagram, the following expression can be written.

$$\frac{x^2}{(L/2)^2} = \frac{210 - 92.8}{210}$$

$$x = \sqrt{\frac{(13.5)^2(117.2)}{210}}$$

$$x = 10.09 \text{ ft}$$

Total stress in plate at theoretical point of cutoff:

$$\text{total stress} = \text{average stress} \times \text{PL area}$$

$$\text{average stress} = \frac{(12)(92.8)(13.25)}{1440} = 10.25 \text{ ksi}$$

$$\text{total stress} = (10.25)(4.0) = 41.0 \text{ k}$$

Length of weld required:

$$\text{strength of } \tfrac{5}{16} \text{ fillet weld} = \left(\tfrac{5}{16}\right)(0.707)(0.30 \times 70) = 4.64 \text{ k/in.}$$

$$L \text{ reqd.} = \frac{41.0}{4.64} = 8.84 \text{ in.}$$

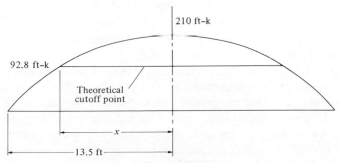

Figure 15-9

Use $\frac{5}{16}$-in. fillet welds on end of PL and for 6 in. on each side
as required by AISC Section 1.10.4.

total cover PL length $= (2)(10.09) + (2)(6 \text{ in.}) = 21.18$ ft (say 22 ft–0 in.)

Intermediate cover-plate welds:

min length of intermittent
welds by Section 1.17.5 AISC $= 1\frac{1}{2}$ in.

V at theoretical point of cutoff $= (13.5 - 3.41)(1.305 + 0.481) = 18.02$ k

$$v = \frac{VQ}{I} = \frac{(18.02)(4.00 \times 13.25)}{1440} = 0.663 \text{ k/in.}$$

spacing of $1\frac{1}{2}$ in. long welds $= \dfrac{(1.5)(2)(4.64)}{0.663} = 21$ in. c.-to-c.

max spacing permitted
(Section 1.18.3.1 AISC) $= 24 t_f$ or 12 in.

$$= (24)(0.335) = 8.04 \text{ in.}$$

Use 8-in. spacing

15-8. DESIGN OF ENCASED SECTIONS

For fireproofing purposes the steel beams in building floors may be completely encased in concrete. Under certain conditions the horizontal shears between the slabs and beams can be considered to be transferred by natural bond and friction between the two. The AISC says that for this transfer to be permissible the encasing concrete must be placed integrally with the slab concrete and must cover the steel by at least 2 in. on the sides and bottom (or soffit). It is further required that the top of the steel section be at least $1\frac{1}{2}$ in. below the top of the slab and 2 in. above the bottom of the slab. Finally, the encasing concrete must have adequate mesh or other reinforcing for its full depth and across the soffit of the beam. The exact amount, which is not specified by the AISC, can be very nominal in size (from "chicken wire size" on up).

The AISC Specification states that steel sections which are to be encased in concrete and are not to be shored must be able to resist M_D and, in conjunction with the slab, must be able to resist $M_D + M_d + M_L$ without a steel stress exceeding $0.66 F_y$. An alternative design method is permitted which says that the steel beam alone must be able to resist all moments without a steel stress exceeding $0.76 F_y$. This value might be used where a reduction in calculations is desired. The higher allowable stress recognizes the fact that composite action permits the section to support more load than the steel beam could alone. Should this latter method be used it is unnecessary to compute the properties of the composite section.

The critical sections for resistance to longitudinal shear in an encased beam are shown in Fig. 15-10. The total resistance to the longitudinal

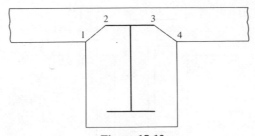

Figure 15-10

shear can be estimated to equal the bond along the top of the steel shape (line 2–3 in the figure) plus the shearing resistance of the concrete along lines 1–2 and 3–4.

A typical value used for the allowable bond between the steel and concrete is $0.03f_c'$ while a common allowable shear in the concrete as along sections 1–2 and 3–4 is $0.12f_c'$. (A much smaller shear allowable is used in Example 15-5.) Should the longitudinal shear be greater than the sum of the shearing and bond resistance along the sections mentioned, some type of reinforcing will be necessary. The usual types of shear connectors on top of the beam flange will probably not be of much value because relatively large deformations will probably have to occur before much load can be applied to the connectors. By the time that much deformation occurs, the natural bond between the steel and concrete will probably have broken. Some type of shear reinforcing placed along lines 1–2 and 3–4 will probably be the most effective method of increasing the shearing resistance. The longitudinal shear will have to be extremely large to require this shear reinforcing.

Example 15-5 illustrates the calculations involved in reviewing the design of an encased section. Notice in this problem that since no shoring is used the only longitudinal shear to be resisted is that produced by the loads applied after the concrete hardens. The ultimate-strength theory is today considered not to apply to these encased sections due to the lack of shear connectors, and longitudinal shears are computed by the familiar VQ/I expression.

Example 15-5

Using the AISC Specification review the encased beam section shown in Fig. 15-11 for longitudinal shear. No shoring is used and the following data are assumed.

$$\text{allowable bond} = 90 \text{ psi}$$
$$\text{allowable shear} = 180 \text{ psi}$$
$$LL \text{ maximum external shear} = 22 \text{ k}$$
$$\text{effective flange width} = 60 \text{ in.}$$
$$n = 9$$

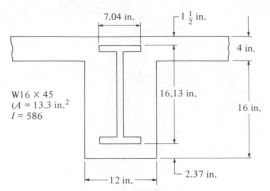

Figure 15-11

SOLUTION

Calculated Properties of Composite Section: Neglecting concrete area below flange,

$$A = 13.3 + \frac{(4)(60)}{9} = 39.96 \text{ in.}^2$$

$$y_b = \frac{(13.3)(10.44) + (26.66)(18)}{39.96} = 15.50 \text{ in.}$$

$$I = 586 + (13.3)(5.06)^2 + \left(\tfrac{1}{12}\right)\left(\tfrac{60}{9}\right)(4)^3 + (26.66)(2.5)^2$$

$$= 1129 \text{ in.}^4$$

Q for area above section 1–2–3–4 in Fig. 15-12,

$$Q = \tfrac{1}{9}\left[(4)(60)(2.5) - (7.04 \times 2.50)(1.75) - (2)\left(\tfrac{1}{2} \times 2.48 \times 2.5\right)(1.33)\right]$$

$$= 62.8 \text{ in.}^3$$

Checking Longitudinal Shear

$$v = \frac{VQ}{I} = \frac{(22,000)(62.8)}{1129} = 1224 \text{ lb/in.}$$

allowable bond $= (7.04)(90) = $ 634 lb/in.

allowable shear $= (2)(3.52)(180) = $ 1270 lb/in.

total shear resistance $= $ 1904 lb/in. $> $ 1224 lb/in.   OK

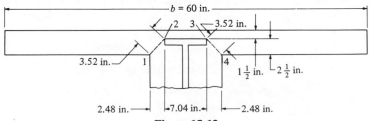

Figure 15-12

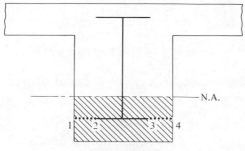

Figure 15-13

For buildings continuous composite construction with encased sections is permissible. For continuous construction the positive moments are handled exactly as has been illustrated by the preceding examples. For negative moments, however, the transformed section is taken as shown in Fig. 15-13. The cross-hatched area represents the concrete in compression and all concrete on the tensile side of the neutral axis (that is above the axis) is neglected. Notice also the fact that the resistance to longitudinal shear is provided along line 1–2–3–4 in the figure. Example 15-6 illustrates the review of a composite section for negative moment.

Example 15-6

For the composite section shown in Fig. 15-14 and previously considered in Example 15-5 determine the bending stresses for negative moments of $M_D = 50$ ft-k and $M_L = 65$ ft-k assuming no shoring is used. If the maximum live load shear is 32 k, determine if extra shear reinforcing is needed in the negative moment region of the beam. Other data are the same as for Example 15-5.

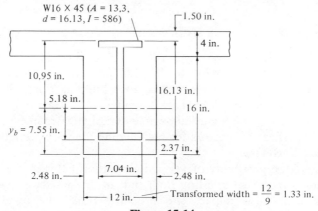

Figure 15-14

SOLUTION

Properties of Section:

$$(1.33y_b)\left(\frac{y_b}{2}\right) = (13.3)(10.44 - y_b)$$

where 10.44 is the distance from soffit to c.g. of W section.

$$y_b = 7.55 \text{ in.}$$

$$I_{tr} = 586 + (13.3)(2.89)^2 + \left(\tfrac{1}{3}\right)(1.33)(7.55)^3 = 888 \text{ in.}^4$$

Review of Stresses: Before concrete hardens:

$$f_s = \frac{(12)(50)(8.07)}{586} = 8.26 \text{ ksi} < 24.0 \text{ ksi} \qquad\qquad \text{OK}$$

After concrete hardens:

$$f_c = \frac{(12)(115)(7.55)}{(9)(888)} = 1.30 \text{ ksi} < 1.35 \text{ ksi} \qquad\qquad \text{OK}$$

$$f_s = \frac{(12)(115)(10.95)}{888} = 17.0 \text{ ksi} < 24.0 \text{ ksi} \qquad\qquad \text{OK}$$

Check Shear

$$V = 32 \text{ k}$$

$$Q = (1.33)(2.37)\left(7.55 - \frac{2.37}{2}\right) = 20.1$$

$$v = \frac{VQ}{I} = \frac{(32,000)(20.1)}{888} = 724 \text{ lb/in.}$$

$$\text{allowable bond} = (7.04)(90) = 634 \text{ lb/in.}$$

$$\text{allowable shear} = (2)(2.48)(180) = \underline{893 \text{ lb/in.}}$$

$$\text{total shear resistance} = 1527 \text{ lb/in.} > 724 \text{ lb/in.} \qquad \text{OK}$$

15-9.   DESIGN OF COMPOSITE SECTIONS USING AASHTO SPECIFICATIONS

The AISC Specification for composite sections does not include cases where there are heavy moving loads with large vibrations and impact. Should loads of this nature be applied to a building floor, the designer might decide to use the AASHTO Specifications. He might, however, reasonably decide to use higher allowable design stresses in both the steel beams and the shear connectors for his building than are permitted by the conservative AASHTO.

There is little possibility of vertical separation between the slabs and steel beams in the usual building floor, but in bridges the heavy moving loads with their large impact will probably break the bond if shear connectors are not used. For this reason the only shear resistance per-

Continuous-welded plate girders in Henry-Jefferson Co., Iowa. (Courtesy of The Lincoln Electric Company.)

mitted by the AASHTO is to be provided by specifically designed shear connectors regardless of the arrangement of the beams and slabs.

Frequently a designer will not consider composite action for continuous bridge stringers but will in all probability use shear connectors in the negative-moment range because they give greater stiffness and prevent vertical separation and horizontal slipping between the slabs and stringers. Actually the AASHTO says that for continuous spans the concrete on the tension side should be considered cracked, and that if the reinforcing steel there is not used in computing the section, it will be unnecessary to use shear connectors.

Example 15-7 illustrates the design of a composite section using the AASHTO Specifications. In bridge construction the top flange of the steel beam is usually encased in a concrete haunch as shown in Fig. 15-15. For

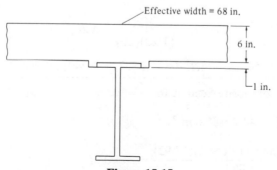

Figure 15-15

simplicity, the effect of the haunch is not usually included in computing the properties of the composite sections. In this example fictitious values for maximum external shears applied to the composite section are assumed at three sections. These values are used to illustrate the design of shear connectors in accordance with AASHTO requirements.

Example 15-7

Using the AASHTO Specifications and A36 steel design the simple composite section shown in Fig. 15-15 for the data given. No shoring is used and the following data are assumed.

$$\text{span}=40 \text{ ft}$$
$$M_D=380 \text{ ft-k}$$
$$M_d=80 \text{ ft-k}$$
$$M_L=500 \text{ ft-k}$$
$$M_T=\overline{960} \text{ ft-k}$$
$$f_s=20.0 \text{ ksi}$$
$$f_c'=3.0 \text{ ksi}$$
$$E_c=3.14\times10^6 \text{ psi}$$
$$f_c=1.2 \text{ ksi}$$
$$n=10$$

Assumed maximum external shears due to live load and impact applied to composite section are 70 k at end, 52 k at 10 ft, and 36 k at 20 ft. An estimated 500,000 cycles of maximum stress is assumed to be applied to the shear connectors.

SOLUTION
Trial Section (see Fig. 15-16): Using Lothers' increased allowable stress method (28% increase for $n=10$):

$$S \text{ reqd.}=\frac{(12)(960)}{(1.28)(20)}=450 \text{ in.}^3$$

Try W36×150 ($A=44.2$, $d=35.85$, $t_f=0.940$, $I=9040$)

Properties of Composite Section for $n=10$:

$$A=\left(\frac{68}{10}\right)(6)+44.2=85.0 \text{ in.}^2$$
$$y_b=\frac{(40.8)(38.91)+(44.2)(17.92)}{85.0}=28.00 \text{ in.}$$
$$I=9040+(44.2)(10.08)^2+\left(\tfrac{1}{12}\right)(6.8)(6)^3+(40.8)(10.91)^2=18,510 \text{ in.}^4$$

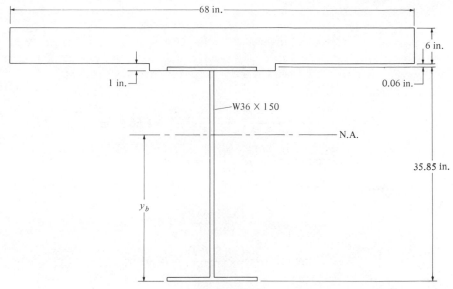

Figure 15-16

For $n=30$:

$$A = \left(\tfrac{68}{30}\right)(6) + 44.2 = 57.8 \text{ in.}^2$$

$$y_b = \frac{(13.6)(38.91) + (44.2)(17.92)}{57.8} = 22.86 \text{ in.}$$

$$I = 9040 + (44.2)(4.94)^2 + \left(\tfrac{1}{12}\right)\left(\frac{68}{30}\right)(6)^3 + (13.6)(16.05)^2 = 13{,}663 \text{ in.}^4$$

Review of Stresses: Before concrete hardens:

$$f_s = \frac{(12)(380)(17.92)}{9.040} = 9.04 \text{ ksi} < 20.0 \text{ ksi} \qquad \text{OK}$$

After concrete hardens:

$$f_s = 9.04 + \frac{(12)(80)(22.86)}{13{,}663} + \frac{(12)(500)(28.00)}{18{,}510} = 19.72 \text{ ksi} < 20.0 \text{ ksi}$$

OK

(overstressed)

$$f_c = \frac{(12)(80)(19.05)}{(30)(13{,}663)} + \frac{(12)(500)(13.91)}{(10)(18{,}510)} = 0.496 \text{ ksi} < 1.2 \text{ ksi} \qquad \text{OK}$$

(a) Design of shear connectors for fatigue

Assuming three $\tfrac{3}{4} \times 3$ headed studs ($H/d = 4.0$) at each section:

$$\alpha = 10{,}600 \text{ for } 500{,}000 \text{ cycles of maximum stress}$$

$$Z_r = \alpha d^2 = (10{,}600)\left(\tfrac{3}{4}\right)^2 = 5962 \text{ lb/stud}$$

$$Q = (6.8)(6)(10.91) = 445 \text{ in.}^3$$

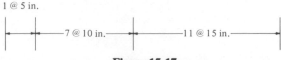

Figure 15-17

S_r at end

$$S_r = \frac{V_r Q}{I} = \frac{(70)(445)}{18,510} = 1.68 \text{ k/in.}$$

$$\text{spacing reqd.} = \frac{(3)(5.962)}{1.68} = 10.65 \text{ in.}$$

S_r at 10 ft

$$S_r = \frac{(52)(445)}{18,510} = 1.25 \text{ k/in.}$$

$$\text{spacing reqd.} = \frac{(3)(5.962)}{1.25} = 14.31 \text{ in.}$$

S_r at 20 ft

$$S_r = \frac{(36)(445)}{18,510} = 0.87 \text{ k/in.}$$

$$\text{spacing reqd.} = \frac{(3)(5.962)}{0.87} = 20.56 \text{ in.}$$

Assuming the following spacing diagram (Fig. 15-17):

(b) Checking number of connectors for ultimate strength

$$P_1 = A_s F_y = (44.2)(36) = 1591.2 \text{ k}$$

$$P_2 = 0.85 f_c' bc = (0.85)(3)(68)(6) = 1040.4 \text{ k}$$

$$P = 1040.4 \text{ k}$$

$$S_u = (0.4)\left(\tfrac{3}{4}\right)^2 \sqrt{(3000)(3.14 \times 10)^6} = 21,838 \text{ lb} = 21.838 \text{ k}$$

$$N_1 = \frac{P}{\phi S_u} = \frac{1040.4}{(0.85)(21.838)} = 56.05$$

Total number of shear connectors $= 3 \times 19 = 57$ each side of beam $\ell > 56.05$ required by ultimate design. Use 56 connectors each side of ℓ with one placed at beam ℓ. These may be spaced uniformly according to the AASHTO Specifications.

15-10. MISCELLANEOUS

Continuous Spans

The AISC (Section 1.11.4) and the AASHO (Article 1.7.48) permit the use of continuous composite design. If shear connectors are used as required by the applicable specification in the negative moment regions the reinforcing bars which are parallel to the steel girders and within the calculated effective width of the slab can be used in calculating the properties of the composite section.

The composite section properties are calculated by the elastic theory and any concrete in tension is neglected. The total cross-sectional area of a section in the negative moment region equals the area of the steel girder plus the area of the reinforcing bars plus the transformed area of any concrete which is in compression as in encased construction. The centroid of the section and its moment of inertia can be calculated by the usual methods.

Extra Reinforcing

For building design calculations the spans are often considered to be simply supported, but the steel beams do not generally have perfectly simple ends. The result of this situation is that some negative moment may occur at the beam ends with possible cracking of the slab above. To prevent or minimize cracking, some extra steel can be placed in the top of the slab extending 2 or 3 ft out into the slab. The amount of steel added is often roughly equal to the temperature and shrinkage requirements and is used in addition to those requirements.

Deflections

Deflections for composite beams may be calculated by the same methods used for other types of beams. The student must be careful to compute deflections for the various types of loads separately. For example, there are dead loads applied to the steel section alone (if no shoring is used), dead loads applied to the composite section, and live loads applied to the composite section. The deflections of the steel beam can be determined for the noncomposite loads and added to the deflections for the dead and live loads applied to the composite section.

To obtain reasonable results from deflection calculations, it is absolutely essential to use different moduli of elasticity (or different modular ratios) for long-term dead and live loads and short-term live loads applied to the slab. The common practice is to divide the modulus by 3 (or use $3n$ for the modular ratio in calculating the section properties) for the long-term loads. It is said that members designed by the AISC Specification with structural carbon steel will deflect from 10 to 20% more than the theoretically calculated values. Generally speaking, shear deflections are neglected although on occasion they can be quite large.[3] The steel beams can be cambered for all or only some portion of the deflection. It may be feasible

---

[3] L. S. Beedle et al., *Structural Steel Design* (New York: Ronald Press, 1964), p. 452.

in some situations to make a floor slab a little thicker in the middle than on the edges to compensate for deflection.

Steel Formed Decks

In recent years formed metal decks have become commonly used in composite floor and roof systems. The composite systems consist of steel beams and concrete slabs with shear connectors welded through the metal decks to the steel beams. Usually the steel deck corrugations are run perpendicular to the beams although sometimes they are placed in the other direction.

Section 1.11.5 of the AISC Specification provides detailed requirements for composite beams or girders with steel formed decks. Included in the specification are maximum rib depths, minimum widths of concrete ribs, minimum lengths of shear connectors, minimum slab thickness, maximum spacing of shear connectors, allowable shear loads on shear connectors, and other items. Space is not taken herein to present a numerical example but such examples are available in several texts as well as in the 8th edition of AISC Manual.[4]

Problems

15-1. Using the AISC Specification, calculate the bending stresses for the section shown in the accompanying illustration. The section is used for a simple span of 24 ft and is to have a dead uniform load of 40 psf applied after composite action develops and a live uniform load of 150 psf. Assume no shoring and $n=9$. (*Ans.* $f_s = 19.98$ ksi, $f_c = -0.63$ ksi)

15-2. Rework Prob. 15-1 if a W8×18 with a $\frac{1}{2}$×8-in. cover plate on the tension flange is used instead of the W14×30.

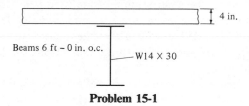

Beams 6 ft – 0 in. o.c.

4 in.

W14 × 30

Problem 15-1

15-3. Rework Prob. 15-1 using the AASHTO Specifications and $n=10$. (*Ans.* $f_s = 23.2$ ksi, $f_c = -0.552$ ksi)

15-4. Rework Prob. 15-2 using the AASHTO Specifications and $n=10$.

15-5. Using the AISC Specification review the stresses in the encased section shown in the accompanying illustration if no shoring is used. The section is assumed to be used for a simple span of 30 ft and to have a dead uniform load of 30 psf applied after composite action develops and a live uniform load of 125 psf. Assume $n=9$. (*Ans.* $f_s = 24.3$ ksi, $f_c = -1.06$ ksi)

[4]S. H. Marcus, *Basics of Structural Steel Design* (Reston, Va.: Reston Publishing Co., 1977), pp. 330–337.

15-6. Using the AISC Manual, design a nonencased composite section for the simple span beams shown in the accompanying illustration assuming a 5-in. concrete slab and using A36 steel and 3000-lb concrete. Use $n = 9$. Load applied after composite action = 150 psf. Assume no shoring. Use $\frac{3}{4} \times 3$-in. headed studs, assume full composite action, and do not use cover plates on beam section. Maximum dead load deflection = 1 in. and maximum live load deflection = $\frac{1}{360}$ span.

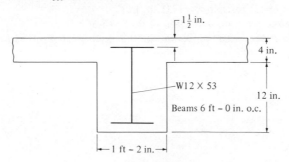

Problem 15-5

15-7. For the floor plan shown for Prob. 15-6 design a nonencased section for the girders, assuming the floor plan is symmetrical on both sides of the girder. (*Ans.* W24 × 94 with 87 $\frac{3}{4}$ × 1 studs)

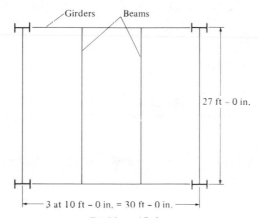

Problem 15-6

15-8. Repeat Prob. 15-6 using an encased section.

15-9. Repeat Prob. 15-7 using an encased section.

15-10. Using the AASHTO Specifications select a composite section for the following information: 6-in. concrete slab, $f_s = 20$ ksi, $f'_c = 3000$ psi, $n = 10$, no shoring, beams spaced 6 ft–0 in. on centers, and each beam subjected to a live load caused by the movement of two 20-k loads 10 ft–0 in. on centers. Assume a simple span of 50 ft.

15-11. Rework Prob. 15-10 if shoring is used.

15-12. Design shear connectors for Prob. 15-1. Use $3\frac{1}{2} \times \frac{7}{8}$-in. headed studs. $f'_c = 3000$ psi.

15-13. Design shear connectors for Prob. 15-1. Use $3 \times \frac{3}{4}$-in. headed studs and assume full composite action. $f_c' = 3000$ psi. A36 steel. (*Ans.* 28 connectors)

15-14. Design shear connectors for Prob. 15-10. Use $3\frac{1}{2} \times \frac{7}{8}$-in. headed studs. $f_c' = 3000$ psi.

15-15. Check to see if extra shear reinforcing is needed in the encased section of Prob. 15-5. $f_c' = 3000$ psi. Assume allowable bond stress = 90 psi and allowable shear stress for concrete = 180 psi. (*Ans.* Calculated shear = 975 lb/in. < shear resistance = 1731 lb/in.)

15-16. Determine the total deflection in Prob. 15-3.

15-17. Calculate the total deflection in the section of Prob. 15-5 if $n = 9$. (*Ans.* 1.53 in.)

Chapter 16
Built-up Beams
and Plate Girders

16-1. COVER-PLATED BEAMS

Should the largest available W section be insufficient to support the loads anticipated for a certain span, several possible alternatives may be taken. These include the use of the following: (1) two or more regular W sections side by side (an expensive solution), (2) a cover-plated beam, (3) a built-up section or plate girder, or (4) a steel truss. This section is concerned with the cover-plated beam alternative, and the remainder of the chapter is devoted to plate girders.

In addition to being practical for cases where the moments to be resisted are slightly in excess of those which can be supported by the deepest W sections, there are other useful applications for cover-plated beams. On some occasions the total depth may be so limited that the resisting moments of W sections of the specified depth are too small. For instance, the architect may show a certain maximum depth for beams in his drawings for a building. In a bridge, beam depths may be limited by clearance requirements. Cover-plated beams will nearly always prove to be the best solution for situations of these types. Furthermore, there can be economical uses for cover-plated beams where the depth is not limited and where there are standard W sections available to support the loads. A smaller W section than required by the maximum moment can be selected and have cover plates attached to its flanges. These plates can be cut off where the moments are smaller with resulting saving of steel. Applications of this type are quite common for continuous beams.

Should the depth be fixed and a cover-plated beam seem to be a feasible solution, the usual procedure will be to select the strongest standard section which has a depth leaving room for top and bottom cover plates, then select the coverplate sizes. A cover-plated beam is shown in Fig. 16-1. If the section has been selected, the moment of inertia of the entire section equals the moment of inertia of the W section (I_s) plus the moment of inertia of the unknown plate sizes. The moment of inertia of these plates about the x axis (neglecting the minute values about their own axes) equals $A(d/2)^2$ for each plate.

Figure 16-1

If d represents the distance between the centers of gravities of the flange cover plates, the moment of inertia of the entire section can be expressed as

$$I \text{ reqd.} = I_s + 2A\left(\frac{d}{2}\right)^2$$

It is usually more convenient to work with section modulus values rather than with moments of inertia. The section modulus required can be written *approximately* as follows (noting that $d/2$ is not exactly the correct distance to the outermost fiber of the plates and that the two section moduli values are not correctly added together since their c distances to their extreme fibers are not equal).

$$S \text{ reqd.} = S_s + \frac{2A(d/2)^2}{d/2}$$

$$= S_s + Ad$$

From this last expression it is possible to estimate very closely the cover plate area required. The value actually obtained will be very slightly on the small side due to the slightly incorrect distance between the outermost fibers used in the derivation. A cover-plated beam design is illustrated in Example 16-1. The connections between the cover plates and the W section of this example are not designed, as this matter will be discussed in the plate-girder design material to follow. Actually a bolted connection of this type was illustrated previously in Example 9-4.

Example 16-1

Select a beam limited to a maximum depth of 28.50 in. for the load and span of Fig. 16-2. The beam is assumed to have full lateral support for its compression flange and to have an allowable bending stress of 20 ksi.

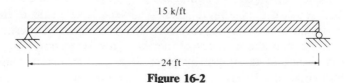

15 k/ft

24 ft

Figure 16-2

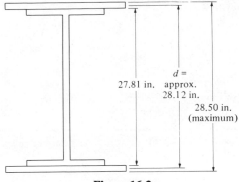

Figure 16-3

SOLUTION
Assume beam weight = 225 lb/ft:

$$M = \frac{(15.225)(24)^2}{8} = 1096.2 \text{ ft-k}$$

$$S \text{ reqd.} = \frac{(12)(1096.2)}{20} = 657.7 \text{ in.}^3$$

Try W27×178 ($d=27.81$, $b=14.085$, $I=6990$, $S=502$) (see Fig. 16-3)

$$S \text{ reqd.} = S_s + Ad$$

$$657.7 = 502 + (A)(28.12)$$

$$A = 5.54 \text{ in.}^2$$

Try one PL$\frac{5}{16}$ ×20 each flange ($A = 6.25$ in.^{2})

Computing actual S furnished:

$$S = \frac{6990 + (2)(6.25)(14.06)^2}{14.22} = 665.3 \text{ in.}^3 > 657.7 \text{ in.}^3 \qquad \text{OK}$$

Use W27×178 with one PL$\frac{5}{16}$ ×20 each flange

16-2. INTRODUCTION TO PLATE GIRDERS

Plate girders are large I-shaped sections built up from plates and rolled sections. They have resisting moments somewhere between those of rolled beams and steel trusses. Several possible plate girder arrangements are shown in Fig. 16-4. Possible riveted or bolted girders are shown in parts (a) and (b) of the figure, while several welded types are shown in parts (c) through (f). Nearly all plate girders constructed today are welded, although they may frequently have bolted field splices.

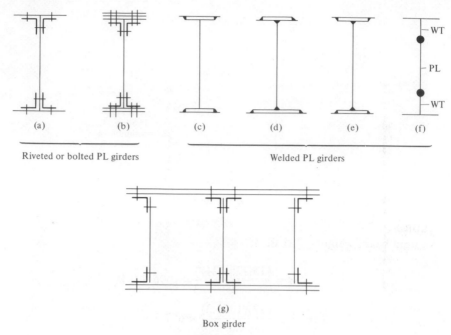

(a)   (b)   (c)   (d)   (e)   (f)

Riveted or bolted PL girders   Welded PL girders

(g)

Box girder

Figure 16-4

The welded girder of part (d) of Fig. 16-4 is arranged to reduce overhead welding as compared to the girder of part (c), but in so doing may be creating a little worse corrosion situation if the girder is exposed to the weather. A box girder, illustrated in part (g), is occasionally used where moments are large and depths are quite limited. Box girders also have great resistance to torsion and lateral buckling.

Plates and shapes can be arranged to form plate girders of almost any reasonable proportions. This fact may seem to give them a great advantage for all situations, but for the smaller sizes the advantage is usually canceled by the higher fabrication costs. For example, it is possible to replace a W36 with a plate girder roughly twice as deep which will require considerably less steel and will have much smaller deflections; however, the higher fabrication costs will almost always rule out such a possibility.

Most steel highway bridges built today for spans less than about 80 ft are steel-beam bridges. For longer spans the plate girder begins to compete very well economically. Where loads are extremely large as for railroad bridges plate girders are competitive for spans as low as 45 or 50 ft.

The upper economical limits of plate girder spans depend on several factors, including whether the bridge is simple or continuous, whether a highway or railroad bridge is involved, the largest section which can be shipped in one piece, etc. Generally speaking plate girders are very economical for railroad bridges for spans from 50 to 130 ft (15 to 40 m) and for highway bridges for spans from 80 to 150 ft (24 to 46 m). However,

Buffalo Bayou Bridge, Houston Tex.—a 270-ft span. (Courtesy of The Lincoln Electric Company.)

they are often very competitive for much longer spans, particularly when continuous. In fact they are actually common for 200-ft (61 m) spans and have been used for many spans in excess of 400 ft (122 m).

Plate girders are not only used for bridges. They are also fairly common in various types of buildings where they are called upon to support heavy concentrated loads. Frequently a large ballroom or dining room with no interfering columns is desired on a lower floor of a multi-story building. Such a situation is shown in Fig. 16-5. The plate girder shown must support some tremendous column loads for many stories

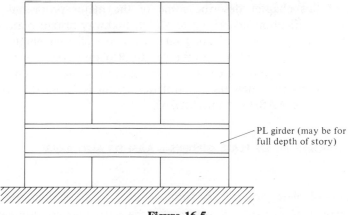

PL girder (may be for full depth of story)

Figure 16-5

above. The usual building plate girder of this type is simple to analyze because it probably does not have moving loads, although some building girders may be called upon to support traveling cranes.

The usual practical alternative to plate girders in the spans for which they are economical is the truss. In general plate girders have the following advantages with particular comparison made to trusses.

1. The pound price for fabrication is lower than for trusses but higher than for rolled beam sections.
2. Erection is cheaper and faster than for trusses.
3. Due to their compactness, vibration and impact are not serious problems.
4. Plate girders require smaller vertical clearance than trusses.
5. The plate girder has fewer critical points for stresses than do trusses.
6. A bad connection here or there is not as serious as in a truss where such a situation could spell disaster.
7. There is less danger of injury to plate girders in an accident as compared to trusses. Should a truck run into a bridge plate girder it would probably just bend it a little, but a similar accident with a bridge truss member could cause a broken member and perhaps failure.
8. A plate girder is more easily painted than a truss.

On the other hand, plate girders are usually heavier than trusses for the same spans and loads, and they have a further disadvantage in the large number of connections required between webs and flanges.

16-3.   COMMENTS ON SPECIFICATIONS

There is an appreciable difference between the design requirements for plate girders in the AISC and AASHTO specifications. The next few sections of this chapter describe some of the major provisions of the AASHTO Specification according to which highway-bridge plate girders are designed today. The concluding sections of the chapter emphasize the AISC Specification as it applies to the design of plate girders for buildings. Actually after the heading of each of the remaining sections of this chapter the specifications to which the general discussion is applicable are indicated (as AISC or AASHTO and AREA).

16-4.   PROPORTIONS OF PLATE GIRDERS—AASHTO AND AREA

Depth

The depths of plate girders vary from about $\frac{1}{6}$ to $\frac{1}{15}$ of their spans with average values of $\frac{1}{10}$ to $\frac{1}{12}$, depending on the particular conditions of the

Indiana Harbor Works, East Chicago, Ind. (Courtesy of Inryco, Inc.)

job. One condition that may limit the proportions of the girder is the largest size that can be fabricated in the shop and shipped to the job. There may be a transportation problem such as clearance requirements that limit maximum depths to 10 or 12 ft along the shipping route.

The shallower girders will probably be used when the loads are light, and the deeper ones when very large concentrated loads need to be supported as from the columns in a tall building. If there are no depth restrictions for a particular girder it will probably pay the designer to make rough designs and corresponding cost estimates to arrive at a depth decision.

Web Size

After the total girder depth is estimated, the general proportions of the girder can be established from the maximum shear and the maximum moment. As previously described for I-shaped sections in Section 7-1, the web of a beam carries nearly all of the shearing stress; this shearing stress is assumed by the AASHTO to be uniformly distributed throughout the web. The web depth can be closely estimated by taking the total girder depth and subtracting a reasonable value for the depths of the flanges (roughly 2 in. each). The web depths are selected to the nearest even inch because these plates are not stocked in fractional dimensions.

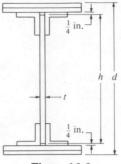

Figure 16-6

For this discussion reference is made to Fig. 16-6 in which d is the total girder depth, h is the web depth, and t is the web thickness. In this figure it will be noted that the flange angles are placed so they extend about $\frac{1}{4}$ in. beyond the top and bottom of the web. If these clearances were not used, the small irregularities in depth of the web plate as it comes from the rolling mill might cause it to protrude slightly, resulting in expensive chipping or other leveling processes.

The web thickness can be determined from the expression to follow in which V is the maximum total shear and v is the allowable shearing stress given by the specifications.

$$t = \frac{V}{hv}$$

Shear is not the only factor that must be considered in selecting the web thickness. Where a large reaction or concentrated load is acting on the plate girder there is danger of buckling of the high thin web. To prevent such buckling the web can be stiffened with plates or angles at various intervals as described in Section 16-7.

Stress analysis shows that high, thin plate-girder webs may buckle due to a combination of bending and shearing stresses unless stiffeners are used at certain intervals. Tests have shown that for carbon steels there is little danger of buckling if the web thickness is at least $\frac{1}{60}$ of the unsupported depth of the web. Should a web be selected which is thinner than the value given by this ratio it will be necessary to provide stiffeners spaced no further apart than the clear web depth. It is usually more economical for bridge plate girders to use thinner webs with stiffeners than to use the thicker ones without stiffeners.

The AREA and AASHTO specifications give minimum permissible ratios of web thickness to unsupported depths of 1/165 for structural carbon steel regardless of the number and placing of stiffeners. Design by these specifications is, therefore, based on the so-called buckling strength of the girders. They do, however, recognize the fact that plate girders have considerable postbuckling strengths by using small safety factors.

From a corrosion standpoint the usual practice is to use some absolute minimum web thickness. For bridge girders, $\frac{3}{8}$ in. is a common minimum, while $\frac{1}{4}$ or $\frac{5}{16}$ in. are probably the minimum values for the more sheltered building girders.

Flange

After the web dimensions are selected the next step is to select the area of the flange. The goal, of course, is to select a flange of sufficient area which will not be overstressed in bending. For this discussion reference is made to Fig. 16-6. The total bending strength of the plate girder equals the bending strength of the web plus the bending strength of the flanges.

The moment of inertia of the entire plate girder equals the moment of inertia of the web about its own centroidal axis plus the area of each flange times the distance from its center of gravity to the centroid of the section squared (thus neglecting the very small moment of inertia of the flange about its own centroidal axis). To simplify the expression, the centers of gravity of the flanges are assumed to fall exactly at the top and bottom of the web (a fairly good assumption).

The approximate gross moment of inertia of the plate girder can now be written as follows, where A_f is the area of one flange.

$$I = \frac{th^3}{12} + 2A_f\left(\frac{h}{2}\right)^2$$

It is usually more convenient to work with section modulus values in design as in ordinary beams. The gross section modulus of the girder can be approximately written as

$$S = \frac{th^3/12}{h/2} + \frac{2A_f(h/2)^2}{h/2}$$

$$= \frac{th^2}{6} + A_f h$$

The required section modulus M/f is equated to S and the resulting expression is solved for A_f as follows.

$$\frac{M}{f} = \frac{th^2}{6} + A_f h$$

$$A_f = \frac{M}{fh} - \frac{th}{6}$$

The last term in this expression $th/6$ is referred to as the *web equivalent*. The AREA says that the web equivalent should be multiplied by 75% for bolted and riveted girders to estimate the effect of the holes in the web. There are undoubtedly a large number of holes in the webs of these girders where the stiffeners are connected. Should 1-in. holes be placed 4 in. on centers, one-fourth of the web will be theoretically cut out.

After the flange area formula is applied it is necessary to proportion the parts of the flange. For best economy it is desirable to place as large a proportion of the flange area in cover plates as possible. The plates are the greatest distance from the centroidal axis of the girder, and the larger their areas the greater the moment of inertia developed for the same total steel area. Cover plates can be conveniently cut off where moments are smaller, so another reason for putting more area in the cover plates is that by so doing more steel area can be cut off where moments are small. It has been found to be good economy to place from 40 to 60% of the total flange area in cover plates.

The flange angles selected will probably be unequal leg angles with the long legs horizontal to increase the moment of inertia and thus the resisting moment. Should a large number of rivets or bolts be needed to connect the angles to the web, equal leg angles may be necessary. Some small plate girders have only angles used for the flanges, but these are not too common. In fact, most specifications require at least one cover plate to run for the full girder length. A cover plate ties the flange angles together into a more compact unit and prevents water from collecting in the $\frac{1}{4}$-in. groove between the angles on top of the web and encouraging corrosion.

Suggestions as to proportioning the flange can usually be found in the specifications being used. For instance some specifications limit the maximum cover plate area to 70% of the flange area. It is often economical to use several cover plates because they can be cut off at the points where they are no longer required by moment (with the probable exception of the inside plate on each flange). It is probably reasonable to select plates that are all of the same widths and thicknesses. The specifications usually state that the plates may not be thicker than the flange angles and if of different thicknesses should decrease in thickness, going away from the centroid of the girder. As previously indicated, from a corrosion standpoint the cover plates should probably not decrease in width away from the centroid. (One exception to this statement was shown in Fig. 16-4 where the plates were so arranged as to reduce the amount of overhead welding.)

After the flanges are selected it is necessary to review the stresses to determine if the proportions selected are satisfactory. The method of computing the moment of inertia depends on the specifications being used. For instance, the AISC says the gross moment of inertia shall be used unless the holes in either flange exceed 15% of the flange area, in which case only the excess area is deducted. The AREA and AASHTO require that the gross moment of inertia be used for calculating compressive stresses and the net moment of inertia be used for calculating tensile stresses. Both moments of inertia are to be calculated about the centroidal axis of the gross section. In either case when holes are being subtracted they are to be subtracted from both flanges—not just the tensile flange.

Arc-welded plate girders, Louisiana State Fair Grounds, Shreveport, La. (Courtesy of The Lincoln Electric Company.)

Example 16-2 illustrates the selection of the proportions of a plate girder and the review of stresses in the resulting section. The critical shears and moments for this highway bridge were assumed by the author. In Chapter 18 the calculations of critical shears and moments using the HS20-44 truck loadings are demonstrated.

Example 16-2

Select proportions and review stresses for a bolted deck plate girder given the following data. Specifications: 1977 AASHTO. Steel: structural carbon (A36).

$$\text{span} = 100 \text{ ft}$$
$$\text{maximum total moment} = 8{,}200 \text{ ft-k}$$
$$\text{maximum total shear} = 330 \text{ k}$$
$$\text{allowable shear in web} = 12{,}000 \text{ psi}$$
$$\text{allowable compression due to bending} = 20{,}000 \text{ psi (assuming full lateral}$$
$$\text{support for the compression flange)}$$

Bolts: $\frac{7}{8}$-in. diameter A325 friction type.

SOLUTION

Depth of Girder:

$$\text{assume } d = \tfrac{1}{12} \text{ span} = \left(\tfrac{1}{12}\right)(100 \text{ ft}) = 100 \text{ in}$$

Size of Web: Web depth:

$$h = d - 4 \text{ in.} = 96 \text{ in.}$$

Web thickness:

(1) For shear $t = \dfrac{330}{(12)(96)} = 0.286$ in.

(2) For corrosion minimum $t = \tfrac{5}{16}$ in.

(3) For diagonal compression $t = \tfrac{1}{170}$ of clear web depth
$= \text{approx.} \left(\tfrac{1}{165}\right)(84) = 0.509$ in.

Try $\tfrac{1}{2} \times 96$-in. web   (See also Article 1.7.43C)

Flange Area: Allowable bending stress in outermost fibers $= 20$ ksi. Approximate stress at top of web (or c.g. of flange) $= \left(\tfrac{48}{50}\right)(20) = 19.2$ ksi.

$$A_f = \frac{M}{fh} - \frac{th}{6}$$

$$= \frac{(12)(8200)}{(19.2)(96)} - \frac{(0.50)(96)}{6} = 45.4 \text{ in.}^2$$

plus estimated area of bolt holes $= 7.0$ in.^{2}

total flange area $= 52.4$ in.^{2}

Flange Proportions: Placing approximately 40% of area in angles, try

$$2 \text{ Ls} 8 \times 6 \times \tfrac{3}{4} \text{ in.} = 19.88 \text{ in.}^2$$

$$3 \text{ cover PLs } \tfrac{5}{8} \times 18 \text{ in.} = 33.75 \text{ in.}^2$$

$$\text{total area} = \overline{53.63} \text{ in.}^2$$

Moments of Inertia (see Fig. 16-7):

Part	Area	Gross I
Web	48.00	36,864
Angles	19.88	86,675
Inside PL	11.25	53,057
Middle PL	11.25	54,442
Outside PL	11.25	55,823
	$I_{\text{gross}} = 286,861$ in.^{4}	

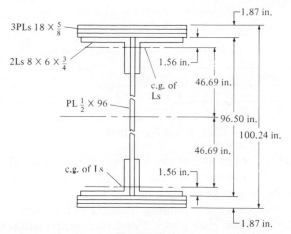

Figure 16-7

Assuming holes for $\frac{7}{8}$-in. bolts spaced 4 in. on centers in web plus one line through vertical L legs and two lines through plates.

I_{holes}:

$$web = 2\left[(1)(\tfrac{1}{2})(4^2 + 8^2 + 12^2 + 16^2 + 20^2\right.$$

$$\left. + 24^2 + 28^2 + 32^2 + 36^2 + 40^2)\right] = 6,160$$

$$\text{vertical L legs} = 2\left[(1)(2)(44.75)^2\right] = 8,010$$

$$\text{cover PLs} = 2\left[(2)(1)(2.62)(48.81)^2\right] = \underline{25,015}$$

$$\Sigma I_{holes} \qquad\qquad\qquad\qquad\qquad = 39,185 \text{ in.}^4$$

$$I_{net} = 286,861 - 39,185 = 247,676 \text{ in.}^4$$

Review of Stresses:

$$f_c = \frac{(12)(8200)(50.12)}{286,861} = 1719 \text{ psi} < 20,000 \text{ psi} \qquad \text{OK}$$

$$f_t = \frac{(12)(8200)(50.12)}{247,676} = 1991 \text{ psi} < 20,000 \text{ psi} \qquad \text{OK}$$

16-5. CUTTING OFF COVER PLATES—AASHTO, AREA, AND AISC

For those parts of the span where the bending moment has decreased sufficiently some cover plates may be cut off. Theoretically the resisting moment of the section with one plate cut off from each flange can be calculated and for the parts of the beam where the moment is that small or smaller those plates can be eliminated. The resisting moment of the section

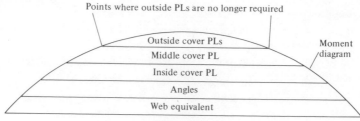

Figure 16-8

can be calculated when two plates are cut off from each flange, etc. The usual practice, however, is to use a simpler method such as the one described in the following paragraphs.

Figure 16-8 shows the bending moment curve (here assumed to be parabolic) for a certain plate girder. For a plate girder subjected to moving loads, this curve should be a curve of the maximum possible moments throughout the beam. In this figure the moment is assumed to be resisted by the web equivalent, angles, and cover plates. Usually there is a little more area than theoretically required but this excess is conservatively neglected. Each square inch is assumed to resist an equal amount of the moment.

The points where the outside cover plates are no longer required, where the middle plates are no longer required, etc., can be accurately determined by scaling from a carefully plotted diagram. Perhaps a more common practice is to assume that the bending-moment curve is parabolic and write expressions for the cutoff points. Reference is made here to Fig. 16-9.

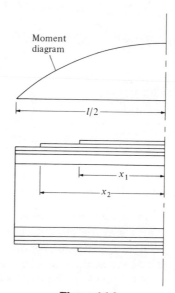

Figure 16-9

In the expressions to be developed A_1 is the area of the outside plate, A_2 is the area of the middle plate, A_{FL} is the total area of the flange plus the web equivalent, and x_1 and x_2 are the distances from the girder centerline to the theoretical points of cutoff of the outside plates and the middle plates, respectively.

$$\frac{A_1}{A_{FL}} = \frac{{x_1}^2}{(l/2)^2}$$

$$x_1 = \frac{l}{2}\sqrt{\frac{A_1}{A_{FL}}}$$

Similarly,

$$x_2 = \frac{l}{2}\sqrt{\frac{A_1 + A_2}{A_{FL}}}$$

As previously described, the inside cover plate on each flange will probably have to extend for the full girder length. Should the moment diagram (neglecting the beam weight) consist of straight lines due to heavy stationary concentrated loads (as from columns in a multistory building), the theoretical points of cutoff can be found even more simply.

Highway bypass bridge, Stroudsburg, Pa. (Courtesy of Bethlehem Steel Company.)

Most specifications require that cover plates be extended for some little distance beyond their theoretical points of cutoff. The plate is supposed to be carrying stress right up to its theoretical end, but it cannot do so unless there is some extra plate beyond the point connected to the girder for stress transfer. These extra distances depend on the individual designer and upon the specification being used. The AASHTO Specifications say that if bolted cover plates are used the plate shall extend for a sufficient distance beyond its theoretical cutoff point to develop the capacity of the plate or it shall continue to a point where the computed stress in the remainder of the girder flange equals the allowable fatigue stress, whichever extension is greater. In other words, there must be enough bolts through the plate outside of the theoretical cutoff point to have a shearing strength equal to the allowable tension or compression in the plate.

16-6. COVER-PLATE AND FLANGE CONNECTORS—AASHTO, AREA, AND AISC

Cover-Plate Connectors

The rivets or bolts passing through the cover plates and flange angles must be of sufficient strength to resist the transfer of longitudinal shear between these parts. The cover-plate connectors shown in Fig. 16-10 are in single shear and bearing on the angle thickness (or the sum of the cover plate thicknesses at the section, whichever is smaller). In the figure, p represents the pitch of the connectors, given two connectors every p in. Each of the connectors must be able to resist half of the longitudinal shear for a distance of p in.

In the expression to be developed, R is the allowable load on one connector under that condition. The allowable strength of the connector is equated to the load which it must carry and the theoretical pitch or spacing of the connectors can be determined as follows.

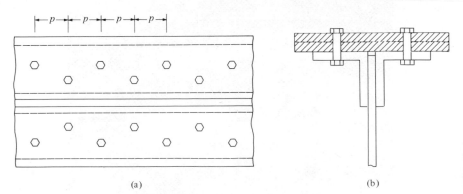

(a)   (b)

Figure 16-10 Cover-plate connectors.

Longitudinal shear in pounds per inch of girder which must be resisted by the connectors equals

$$v = \frac{VQ}{I}$$

Shear to be resisted per connector equals

$$\left(\frac{1}{2}\right)(v)(p) = \frac{pVQ}{2I}$$

$$R = \frac{pVQ}{2I}$$

$$p = \frac{2RI}{VQ}$$

In this expression for pitch of cover-plate connectors I is the gross moment of inertia of the girder at the section being considered, V is the external shear at the section, and Q is the statical moment of the plates about the neutral axis.

Flange Connectors

The flange connectors are the rivets or bolts that pass through the angles and web. These horizontal connectors are in double shear and in bearing on the web thickness. They must have sufficient strength to resist the usual longitudinal shearing stress and they may also be required to resist a vertical shear due to loads applied to the top of the girder. The longitudinal shearing stress expression VQ/I is used in which Q is the statical moment of the flange about the neutral axis of the girder. The area involved is crosshatched in Fig. 16-11(a). If a welded girder of the type shown in part (b) of the figure is used, the weld will be subjected to longitudinal shear only.

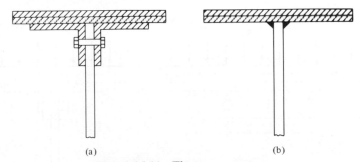

(a) (b)

Figure 16-11 Flange connectors.

The strength of the connector R is equated to the horizontal shearing force to be resisted and the resulting expression is solved for p as follows.

$$R = \frac{VQ}{I} p$$

$$p = \frac{RI}{VQ}$$

Plate girders used in buildings are usually subjected to heavy concentrated loads. Stiffeners are placed under these loads and keep the flange connectors from being subjected to a direct vertical shear. In bridges, however, the wheel loads of trucks or trains move back and forth across the girders. It is not feasible to use continuous stiffeners all across the span, and the flange connectors are subjected to vertical shears due to the concentrated wheel loads. In Fig. 16-12 a uniform load of w lb/in. is assumed to be applied to the top of the girder and the vertical shear applied to each connector will equal wp. (For the moving concentrated loads the designer probably will make some assumed distribution of the concentrated wheel load to obtain the value of w described here. The AREA says that when the ties rest on the flanges a wheel load (including 80% impact) shall be spread over 3 ft. Sometimes the load may be assumed to be spread out at 45° in each direction through the roadway slab until it reaches the top of the girders.)

These connectors discussed are subjected to a combination of longitudinal and vertical shears and their allowable strength must be at least equal to the resultant of the two shears.

$$R = \sqrt{(wp)^2 + \left(p\frac{VQ}{I}\right)^2}$$

$$= p\sqrt{w^2 + \left(\frac{VQ}{I}\right)^2}$$

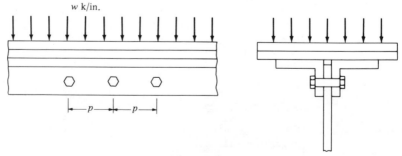

Figure 16-12

Solving this expression for the required pitch of the connector, the following expression is obtained.

$$p = \frac{R}{\sqrt{w^2 + (VQ/I)^2}}$$

16-7.   STIFFENERS—AASHTO

As previously described it is usually necessary to stiffen the high thin webs of plate girders to keep them from buckling. For riveted or bolted girders, angles are connected to the webs while for welded girders plates can be welded to the webs. Figures 16-13 and 16-14 show these types of stiffeners. Stiffeners are divided into two groups: the *bearing stiffeners* which transfer heavy reactions or concentrated loads to the full depth of the web, and the *nonbearing stiffeners* which are placed at various intervals along the web to prevent buckling due to diagonal compression. An additional purpose of bearing stiffeners is to transfer heavy loads to the web without putting all the load on the flange connectors.

Bearing Stiffeners

Bearing stiffeners should fit tightly against the flanges being loaded and should extend out toward the edges of the flange plates and angles as far as possible. In order to obtain a snug fit or good bearing between the flange and stiffeners it is desirable either to weld the stiffeners to the flanges or to mill their outstanding legs.

The usual specifications only allow the part of the stiffener legs outside of the fillets of the flange angles to be counted in supporting the

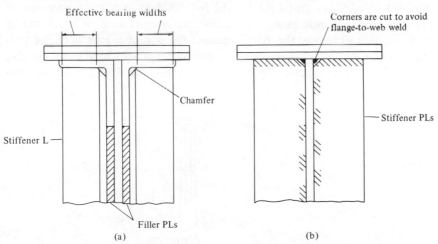

Figure 16-13   (a) Angle bearing stiffeners. (b) Welded girder stiffeners.

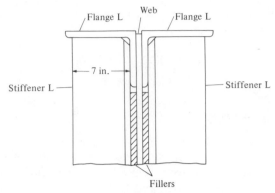

Figure 16-14

load in pure bearing [see Fig. 16-13(a)]. The distance to the outside of the fillet can be found in the angle tables of the AISC Manual. The reaction or load is divided by the allowable bearing stress given by the specifications, and that much area must be provided outside of the flange-angle fillets. A riveted or bolted bearing stiffener is shown in Fig. 16-13(a) and a welded one in part (b) of the same figure. The stiffeners shown in this figure are cut off so they will fit in the corners of the fillet. Notice in part (a) of the figure how filler plates are needed between the web and the stiffeners. Riveted or bolted bearing stiffeners cannot be crimped as they can for nonbearing stiffeners as shown in Fig. 16-15.

Bearing stiffeners are really a special type of column which is difficult to analyze because it supports the load in conjunction with the web. The amount of support supplied by each is difficult to estimate. The specifications usually say that the load divided by the effective bearing area may not exceed the allowable bearing as previously described. Some specifications say in addition that P/A for the full stiffener area and some part of the web area may not exceed the allowable compression in a column of that size. For example, the AISC permits the use of a length of the web equal to 12 times its thickness plus the full stiffener area to act as a column for end-bearing stiffeners. The effective length to be used for the column

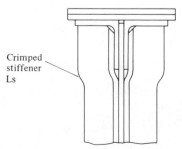

Figure 16-15 Intermediate stiffener.

(again difficult to estimate) is $\frac{3}{4}$ of the stiffener height, says the AISC. For interior bearing stiffeners they permit a length of web equal to 25 times the web thickness to be included in the effective column area.

Since bearing stiffeners cannot be crimped, they require the use of filler plates of thickness equal to that of the flange angle. Should the filler plates be held in place only by connectors that pass through the stiffeners, filler plates, and the web they are said to be *loose fillers*. *Tight fillers* are those that are connected by rivets or bolts passing through the stiffeners, filler plates, and web plus extra connectors passing through the filler plates (which are extended outside the stiffeners) and web, provided the number of extra connectors is equal to at least 50% of the number of connectors required by the stiffeners.

The usual procedure for determining the number of connectors required in bearing stiffeners is to divide the load or reaction by the connector strength. If loose fillers are used, the connectors are considered to be in double shear and bearing on the web thickness. Should tight fillers be used, the connectors are considered to be in double shear and bearing on a thickness equal to the web thickness plus that of the fillers or a thickness equal to that of the two stiffeners, whichever is the least. (A fairly common practice if loose fillers are used is to increase the number of connectors by 50% because it is felt that the connectors are subject to moment along with bearing and shear.)

Nonbearing Stiffeners

Nonbearing stiffeners are also called intermediate stiffeners or stability stiffeners or transverse intermediate stiffeners. The AASHTO Specification (1.7.43D) states that these stiffeners may be omitted if (1) the web thickness is not less than $D/150$ where D is the unsupported height of the web and if (2) the average calculated shearing stress F_v in the gross section of the web is less than the value given by the equation to follow.

$$ F_v = \frac{5.625 \times 10^7}{(D/t_w)^2} \ll F_y/3 $$

The size of intermediate stiffeners is usually obtained by applying some empirical rule, or it may be definitely controlled by the specifications being used. The AASHTO says that the length of the outstanding leg of a stiffener angle or a plate stiffener may not exceed 16 times its thickness nor be less than 2 in. $+ \frac{1}{30}$ of the total girder depth and preferably shall not be less than $\frac{1}{4}$ the full width of the girder flange. An outstanding leg will probably be selected with a length equal to approximately $\frac{1}{30}$ of the girder depth plus 2 in. and with a thickness equal to approximately $\frac{1}{16}$ of the width but probably not less than $\frac{3}{8}$ in. The other leg of a stiffener angle need only be large enough to accommodate the connectors.

The spacing of intermediate stiffeners is governed by the specifications. The AASHTO states that the first intermediate stiffener at the simple support end of a plate girder shall be such that the shearing stress in the end panel shall not exceed the value obtained from the equation to follow.

$$F_v = \frac{7 \times 10^7 \left[1 + (D/d_o)^2 \right]}{(D/t_w)^2} \leqslant \frac{F_y}{3}$$

The AASHTO Specifications state that these stiffeners may not have a spacing greater than $D/2$ nor may the spacing be so large that the average shearing stress in the gross section of the web is greater than the following.

$$F_v = \frac{F_y}{3} \left[C + \frac{0.87(1 - C)}{\sqrt{1 - (d_o/D)^2}} \right]$$

where

$$d_o = \text{the spacing of the intermediate stiffeners}$$

$$C = \frac{2.2 \times 10^8 \left[1 + (D/d_o)^2 \right]}{F_y (D/t_w)^2} \leqslant 1$$

Article 1.7.43D of the AASHTO specifies a minimum I and other items for these stiffeners.

16-8. LONGITUDINAL STIFFENERS—AASHTO

Although longitudinal stiffeners are not as effective as transverse stiffeners they are frequently used in highway girders because of their attractive appearance. The best location for longitudinal stiffeners can be theoretically proved to lie approximately one-fifth of the distance from the compression flange to the tension flange. On this basis the AASHTO says that the gage lines of longitudinal stiffeners are to be placed $\frac{1}{5}$ of D (where D is the clear distance between flanges) from the toe of the compression flange. The required moment of inertia of a longitudinal stiffener about the edge of the stiffener in contact with the web is required by these specifications to equal

$$I = D t_w^{3} \left(2.4 \frac{d_o^{2}}{D^2} - 0.13 \right)$$

In this expression D is the clear distance between flanges, d_o is the actual distance between transverse stiffeners, and t_w is the web-plate thickness. A one-sided stiffener (that is on one side of the web only) designed by the AASHTO Specifications is normally more economical than a two-sided one. The connection problem for a one-sided stiffener is only one-half what it is for a two-sided stiffener and the area required to furnish I_E is only about $\frac{5}{8}$ of the area required for a two-sided plate. The student can verify this figure by substituting for a given condition the required moment of inertia into the equation. Longitudinal stiffeners do not have to be continuous and may be cut at their intersection with transverse stiffeners. Their thickness may not be less than the following.

$$\frac{b'\sqrt{f_b}}{2250}$$

where b' is the width of the stiffeners and f_b is the calculated bending compressive stress in the flange.

16-9. WEB SPLICES—AASHTO, AREA, AND AISC

Splicing of plate-girder webs is not too desirable, if avoidable, but is frequently necessary for one or more of the following reasons.

1. The length of the girders may be so great that plates of the required length are not available from the steel mills.
2. Even if available, extremely large plates are difficult to handle without their twisting out of shape. (A 100-ft web plate of the size selected in Example 16-2 would weigh 16,300 lb. Imagine the difficulties of handling a PL$\frac{1}{2}$×96 in., 100 ft long weighing over 8 tons without twisting it out of shape before the stiffeners are installed!)
3. The sizes of girders that can be transported over the highway and railroad routes to the job may be severely limited by clearances and other factors.
4. The equipment available to handle and erect the steel may also limit the sizes of girder sections.

Many engineers would not place a splice at the centerline of a simple span because of the large centerline moment even if only one splice is needed. They would probably use a pair of splices located at the stiffeners nearest the third points of the span. This practice is probably unnecessary because a single splice at the centerline is just as satisfactory from a strength standpoint, and is certainly cheaper than two separate splices. Also the moment at the third points is not substantially reduced from its value at the centerline anyway.

For welded girders it is unnecessary to use splice plates because the webs can be joined by butt welding. Several possible splices are shown in Fig. 16-16 for riveted or bolted plate girders. The simplest and cheapest type of splice is the one shown in part (a) of the figure in which a single

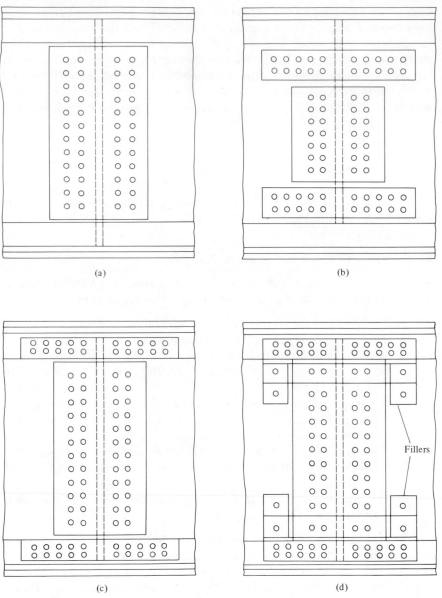

Figure 16-16 (a) Single-plate or shear splice. (b) Triple-plate or moment splice. (c) Triple-plate or moment splice. (d) Triple-plate splice with fillers.

splice plate is used on each side of the web. This type of splice, called a *single-plate splice* or *shear splice*, is fairly satisfactory but is definitely not as good as the other types shown in the figure. In part (b) a much stronger type of splice called the *triple-plate splice* or *moment splice* is shown.

Each of these first two types of splices have a disadvantage in that the plates do not run for the full depth of the web because the flange angles cover part of the web. In other words, the largest stress that occurs in the web in its outermost fiber is several inches away from the nearest part of the splice. It seems probable, therefore, that an appreciable portion of this maximum stress will be transferred over into the flange (at the break in the web), increasing the stress in the flange rather than going completely to the splice as desired.

The splices of parts (c) and (d) of Fig. 16-16 are probably the most desirable types from a theoretical standpoint. They consist of plates between the flanges (called shear plates) plus narrow plates over the vertical legs of the flange angles (called moment plates). In this type of splice the plates over the angle legs serve to splice the part of the web under the angles. Should the moment plate be deeper than the angle legs as shown in part (d), several filler plates will be needed. These plates are rather objectionable.

The common practice in designing a web splice is to select one that will replace the full moment resistance of the web and the maximum actual external shear at the section. Some engineers say that for statically determinate structures the splices should be designed only for the calculated maximum external shear, but for statically indeterminate structures they should be designed for the total shearing resistances of the webs. Their reasoning is that should plastic action occur in a statically indeterminate girder and the loads be redistributed, the splice would have a strength equal to that of the web.

The triple-plate splice has three plates on each side of the web, actually giving it a total of six plates. The top and bottom plates are referred to as the moment plates and are proportioned to have a resisting moment equal to that of the web. The resisting moment of the web equals fI/C or $\frac{1}{6}fth^2$ and can be said to equal one-sixth of the web area concentrated at the top and bottom of the web. An equivalent procedure was followed in developing the flange-area formulas in Section 16-4. The resisting moment of the splice plates is assumed to equal the area of one plate times its average stress, times the distance between the centers of gravities of the plates. The stress in a splice plate is in proportion to the stress in the outermost fiber of the web as are their respective distances from the neutral axis. For this discussion reference is made to Fig. 16-17.

$$\text{total force in moment plate} = \left(\frac{d}{h}f\right)(A_p)$$

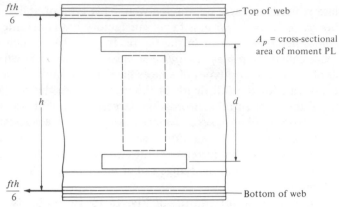

Figure 16-17

Equating resisting moment of web to that of moment plates and solving for A_p.

$$\left(\frac{fth}{6}\right)(h) = \left(\frac{d}{h}\right)(f)(A_p)(d)$$

$$A_p = \frac{th^3}{6d^2}$$

To use this expression it will be noted that d is not known. An excellent estimate of its value can be made, however, by assuming two or three horizontal rows of connectors in the moment plates. The widths of these plates necessary to include the connectors can be assumed; and the value of d, the distance center-to-center of the moment plates calculated. The area required for the moment plates can then be calculated. Since their width has already been assumed their thickness can be determined.

Finally, the length of the plates is established by the total number of connectors required. The connectors in the moment splice are designed for the total force to be carried by the plates fA_p (remembering that the stress in the moment plates is d/h times the stress in the outermost fiber of the web and using A_p as the theoretically required plate area). The number of connectors required equals the total force in the plates divided by the strength of one connector. The connectors are in double shear and bearing on the web thickness or the sum of the two-splice plate thickness. The force that is developed by the connectors is also assumed to equal $(d/h)R$.

$$\text{Number of connectors} = \frac{(d/h)fA_p}{R}$$

The space available between the moment plates determines the depth of the shear plates using a clearance of about $\frac{1}{4}$ in. between the plates. The two shear plates (one on each side of the web) are selected to give an area sufficient to resist the total shear at the splice. The number of connectors used on each side of the web separation equals the external shear divided by the strength of one connector. The design of a triple-plate splice is illustrated by Example 16-3.

Example 16-3

Design a triple-plate splice for the plate girder shown in Fig. 16-18 if the maximum calculated external shear is 175 k. Assume an allowable bending stress of 20 ksi and $\frac{7}{8}$-in. A325 friction-type bolts with an allowable shear of 13.5 ksi and an allowable bearing of 44.0 ksi. Allowable shear in web $= 12.0$ ksi.

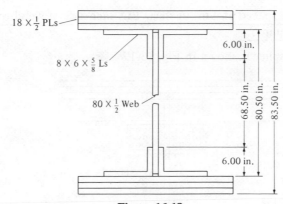

Figure 16-18

SOLUTION

Design of Moment Plates: Assuming three horizontal rows of $\frac{7}{8}$-in. bolts spaced 3 in. on centers and using $1\frac{1}{2}$-in. edge distance, depth of PLs $= (2)(3) + (2)(1\frac{1}{2}) - 9$ in.:

$$d = 68.50 - (2)\left(\tfrac{1}{4}\right) - 9 = 59.00 \text{ in.}$$

$$A_p = \frac{th^3}{6d^2} = \frac{\left(\tfrac{1}{2}\right)(80)^3}{(6)(59)^2} = 12.25 \text{ in.}^2 \text{ for 2PLs}$$

$$\text{thickness of PLs} = \frac{12.25}{(2)(9)} = 0.681 \text{ in.} \qquad \left(\text{say } \tfrac{11}{16} \text{ in.}\right)$$

R for bolts in double shear and bearing on $\frac{1}{2}$ in. $= (2)(0.6)(13.5)(\frac{59}{80}) = 11.95$ k each:

$$\text{no. of bolts} = \frac{(d/h)fA_p}{R} = \frac{\left(\tfrac{59}{80}\right)(20)(12.25)}{11.95} = 15.1$$

Use 15 bolts each side of web break

Design of Shear Plates: Depth of shear $PLs = 68.50 - (2)(\frac{1}{4}) - (2)(9) - (2)(\frac{1}{4})$
$= 49.50$ in.

$$A \text{ shear } PLs = \frac{175}{12} = 14.6 \text{ in.}^2 \text{ for 2PLs}$$

$$t = \frac{14.6}{(2)(49.50)} = 0.147 \text{ in.} \qquad \left(\text{say } \tfrac{1}{4} \text{ in. minimum } t\right)$$

Bolts in double shear and bearing on $\frac{1}{2}$ in., $R = 16.2$ k.

$$\text{no. of bolts required} = \frac{175}{16.2} = 10.8$$

Use bolts spaced at 4 in. in shear PLs each side of web break giving more than required.

Detail of the splice is shown in Fig. 16-19.

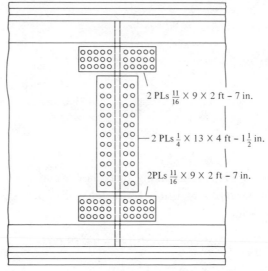

2 PLs $\frac{11}{16}$ × 9 × 2 ft – 7 in.

2 PLs $\frac{1}{4}$ × 13 × 4 ft – 1$\frac{1}{2}$ in.

2PLs $\frac{11}{16}$ × 9 × 2 ft – 7 in.

Figure 16-19 Detail of three-plate splice of Example 16-3.

16-10. PROPORTIONS OF WEBS—AISC

Buckling Considerations

The depths of plate girders have been previously discussed in Section 16-4 and were said to have average values varying from $\frac{1}{10}$ to $\frac{1}{12}$ of spans. After the total depth is assumed the web depth can be estimated to be from 2 to 4 in. less than the total depth and selected to the nearest inch. A plate-girder web must have sufficient thickness to prevent *vertical buckling of the compression flange*. As the compression flange of a simply supported beam deflects or curves downward, it pushes against the web subjecting it to vertical compression. The amount of this downward force can be estimated to equal the total bending stress in the compression flange times the sine of the angle made by the curved flange with the horizontal.

The total load that can be applied to the web in this manner before the web buckles can be estimated by the Euler formula. Expressing this another way, the Euler formula can be used to determine a limiting height-to-thickness ratio to prevent buckling. Based on such a derivation (including an assumption for residual stress variation) the AISC (Section 1.10.2) says that the clear depth between flanges may not exceed the value computed by the expression given below. Substitution into this expression yields a value of $322t$ (where t is the web thickness) for A36 steel.

$$\frac{14,000}{\sqrt{F_y(F_y + 16.50)}}t$$

When transverse stiffeners are used (spaced no more than $1\frac{1}{2}$ times the girder depth), the limiting ratio is $2000/\sqrt{F_y}$ where F_y is the yield stress of the compression flange.

The possibility of lateral buckling of the web may require a reduction in the allowable bending stress in the compression flange. It has been found that if the depth-to-thickness ratio of the web does not exceed $760/\sqrt{F_b}$ where F_b is the applicable allowable bending stress, no reduction is required in the allowable bending stress in the compression flange. This ratio equals 162 for A36 steel ($F_b = 22$ ksi).

For a trial girder section using A36 steel, a ratio of web depth to thickness somewhere between 162 and 322 will probably be selected, noting that the allowable bending stress in the flange will have to be reduced (to be described).

Web Shear

The AISC Specification for plate girders permits their design on the basis of postbuckling strength. Designs on this basis give better economy and provide a more realistic idea of the actual strength of a girder. Should a girder be loaded until initial buckling occurs, it will not collapse because of a phenomenon known as *tension field action*.

After initial buckling a plate girder acts much like a truss. The web acts as a truss with tension diagonals and is able to resist additional shear. A diagonal strip of the web acts similarly to the diagonal of a parallel chorded truss (see Fig. 16-20). The stiffeners keep the flanges from coming together and the flanges keep the stiffeners from coming together. The intermediate stiffeners, which before initial buckling were assumed to resist no load, will after buckling resist compressive loads (or will serve as the compression verticals of a truss) due to diagonal tension. The result is that

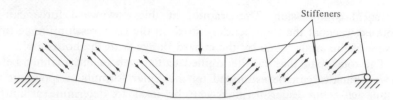

Figure 16-20 Tension field action in plate-girder web.

a plate-girder web can probably resist loads equal to two or three times those present at initial buckling before complete collapse will occur.

Until the web buckles initially, deflections are relatively small but after initial buckling the girder's stiffness decreases considerably and deflections may increase to several times the values estimated by the usual deflection theory.

The estimated total or ultimate shear that a panel (a part of the girder between a pair of stiffeners) can withstand equals the shear initially causing web buckling plus the shear which can be resisted by tension field action. The amount of tension field action is dependent on the proportions of the panels.

Based on a consideration of the ultimate shearing strength of a girder, AISC Formulas 1.10-1 and 1.10-2 were derived to estimate the maximum permissible shearing stresses in the webs. The average shearing stress f_v in the web of a girder can be computed by $V/t_w h_w$, where t_w and h_w refer to the thickness and height of the web, respectively. The AISC says the values of the average shear (f_v) so computed may not exceed the values given by the following expression.

$$F_v = \frac{F_y}{2.89}(C_v) \leqslant 0.4F_y \qquad \text{(AISC Formula 1.10-1)}$$

Alternately for girders, other than hybrid girders, with intermediate stiffeners designed as required by the AISC, and if $C_v \leqslant 1.0$, the allowable shear of Formula 1.10-2 may be used instead of that given by Formula 1.10-1.

$$F_v = \frac{F_y}{2.89}\left[C_v + \frac{1-C_v}{1.15\sqrt{1+(a/h)^2}} \right] \leqslant 0.4F_y \qquad \text{(AISC Formula 1.10-2)}$$

In these expressions a is the clear stiffener spacing in inches. The definitions of the other terms are given in AISC Section 1.10.5.2. The allowable values of the shear stress computed from these formulas are given in Tables 10-50, and 11-50 in Appendix A of the AISC Manual for each of the steels.

Before an allowable shearing stress can be obtained for a particular girder it is necessary to try a stiffener spacing. Should the computed shearing stress not exceed the value given by these equations, no more intermediate stiffeners are required.

Using the AISC Specification, overall economy can usually be obtained when unstiffened webs, or those which are sufficiently thick to carry shear without buckling, are used. Even though plate girders with unstiffened webs are heavier than those with intermediate stiffeners they usually cost less due to smaller fabrication costs. (In addition design calculations are reduced by about 50%). The girder of Example 16-4 is proportioned with intermediate stiffeners to show how they are designed. An unstiffened girder would probably have been more economical.

Combined Bending and Shear

Another factor pertaining to the web design which needs to be investigated is the combined action of the shearing stress and the tension due to the moment in the plane of the web. The AISC says that the bending tensile stress in a plate girder web which depends upon tension field action subject to a combination of shear and tension cannot exceed $0.6F_y$, nor the value given by the following expression in which f_v is the computed average shearing stress and F_v is the allowable shearing stress from AISC Formula 1.10-2.

$$\left(0.825 - 0.375\frac{f_v}{F_v}\right)F_y \qquad \text{(AISC Formula 1.10-7)}$$

The Commentary on AISC Specification (1.10.7) says that plate-girder webs can be proportioned on the basis of bending stress alone if the shearing stress does not exceed 0.6 of its permissible value; or can be proportioned on the basis of shear alone when the bending stress does not exceed $\frac{3}{4}$ of its allowable value. Otherwise Formula 1.10-7 is to be applied.

16-11.   PROPORTIONS OF FLANGE—AISC

The following expression for the area of the flange to resist a certain bending moment was developed in Section 16-4.

$$A_f = \frac{M}{fh} - \frac{th}{6}$$

It was assumed that there was an internal couple producing a resisting moment. Each force in the couple is equal to its flange area and web equivalent times a stress slightly less than the allowable stress in the extreme fiber. The lever arm is the distance between the centers of gravities of the forces.

Substitution into this expression gives the estimated flange area required from which the actual proportions of the flange can be decided. It will be remembered that the AISC (Section 1.9.1.2) to prevent local buckling of the compression flange says that the extended length of the flange divided by its thickness may not exceed $95.0/\sqrt{F_y}$. For A36 steel the protruding length may not be greater than $16t$, or it may be said that the maximum flange width is $32t$.

The usual beam theory says that the flexure stress in a plate girder varies proportionately with the distance from the neutral axis. Tests, however, have shown that the stress in the web on the compression side of the girder is less than this theory would indicate because the web deflects a small amount laterally. These lateral deflections cause the flange compression stresses to be in excess of those anticipated. The result is that nearly all of the compression stress is carried by the compression flange. On some occasions it is necessary to reduce the allowable bending stress in that flange. It has been found that if the web depth to thickness ratio exceeds $760/\sqrt{F_b}$ there is a slight buckling of the web. AISC Formula 1.10-5 is to be used when the ratio is exceeded and the AISC says that the allowable flange stress F_b' may not be greater than the following.

$$F_b' \leqslant F_b \left[1.0 - 0.0005 \frac{A_w}{A_f} \left(\frac{h}{t} - \frac{760}{\sqrt{F_b}} \right) \right] \qquad \text{(AISC Formula 1.10-5)}$$

In girders with sufficient lateral support F_b in the preceding expression is $0.60F_y$. Should sufficient lateral support not be available it is necessary to use the appropriate formulas of AISC Section 1.5.1.4.5 to obtain F_b.

16-12.   DESIGN OF STIFFENERS—AISC

Bearing Stiffeners

The purpose and arrangement of bearing stiffeners was previously discussed in Section 16-7 and examples were shown in Figs. 16-13 and 16-14. The AISC says that the area of the stiffeners outside of the welds between the web and the flange [see Fig. 16-13(b)] must be sufficiently large to keep the bearing stress from exceeding $0.90F_y$ (Section 1.5.1.5.1).

The AISC (1.10.5.1) says that bearing stiffeners should be reviewed as columns. For an end bearing stiffener the "column" is assumed to consist of the parts of the stiffener outside of the web welds plus a length of the web equal to 12 times its thickness ($25t$ for intermediate bearing stiffeners). The stresses so obtained should not exceed the allowable column stress using an effective length equal to $\frac{3}{4}$ of the actual stiffener lengths. If the plate girder is securely fastened to columns at its ends by plates and/or angles, end bearing stiffeners are usually unnecessary.

Intermediate Stiffeners

As previously described the post buckling behavior of a plate girder involving tension field action causes the stiffeners to act as compression members. AISC Formula 1.10-3 provides an area of intermediate stiffeners which is supposedly sufficient to withstand the compression so produced.

$$A_{st} = \frac{1-C_v}{2}\left[\frac{a}{h} - \frac{(a/h)^2}{\sqrt{1+(a/h)^2}}\right]YDht \qquad \text{(AISC Formula 1.10-3)}$$

The various terms included in this expression are given along with the formula in Section 1.10.5.4 of the AISC Specification. Fortunately values of this rather complicated expression are recorded in Tables 11-36 and 11-50 of Appendix A of the AISC Specification in the AISC Manual. The values given are expressed as percentages of the web area.

If a single plate or angle stiffener is used (instead of one angle or plate on both sides of the web), the AISC says the area obtained from Formula 1.10-3 should be multiplied by 1.8 if single-angle stiffeners are used and 2.4 if single-plate stiffeners are used. One-sided stiffeners are subject to moment in addition to axial load and it is necessary to increase their cross-sectional area to take into account their less efficient operation.

In Section 1.10.5.4 an expression is given to estimate the shear in pounds per inch which must be transferred between the intermediate stiffeners and web during shear transfer due to tension field action. This expression follows.

$$f_{vs} = h\sqrt{\left(\frac{F_y}{340}\right)^3} \qquad \text{(AISC Formula 1.10-4)}$$

If the actual web shear ($f_v = V/ht_w$) is less than the allowable shear at the point (Formula 1.10-2), the shear to be transferred can be reduced in direct proportion.

16-13.   EXAMPLE PLATE GIRDER DESIGN—AISC

Example 16-4 illustrates the design of a welded plate girder supporting column loads at its one-third points using the AISC Specification. Since the girder is assumed to have full lateral support the allowable bending stress is $0.6F_y$ or the value determined from Formula 1.10-5, which is the formula used when there is lateral buckling of the web on the compression side of the girder. For cases where full lateral support is not provided it is necessary to apply the appropriate formulas of AISC Section 1.5.1.4.5 in addition to Formula 1.10-5.

The cover-plate size can be reduced in those areas where the moments are smaller, the smaller plates being butt-welded to the larger plate. Space is not taken in this example to show such calculations. If the plates are changed in size, it is to be remembered that the smaller plates as well as the larger ones must meet the $95.0/\sqrt{F_y}$ requirement for their width-thickness ratios.

Example 16-4

A welded plate girder is to be designed for a 54-ft simple span and is to support 150-k column loads at its one-third points together with a uniform load of 2 k/ft. The design is to be made with A36 steel, SMAW fillet welds, E70 electrodes, and the AISC Specification, assuming that full lateral support is provided for the compression flange. For the convenience of the student, AISC section and formula numbers are shown throughout the example.

SOLUTION

(a) Selection of web:

$$\text{assume girder depth} = \tfrac{1}{10}l = 65 \text{ in.}$$

$$\text{assume web depth} = 62 \text{ in.}$$

max h/t ratio for no stress reduction in allowable flange stress:

$$= 760/\sqrt{22} = 162 \text{ (Section 1.10.6)}$$

Therefore, minimum t for no stress reduction $= 62/162 = 0.383$ in.

Absolute maximum clear distance between flanges $= 322t$. Therefore, absolute minimum $t = 62/322 = 0.193$ in.

Try web PL$\tfrac{5}{16} \times 62$

(b) Shear and moment diagrams:

$$\text{area of web} = A_w = (62)(\tfrac{5}{16}) = 19.4 \text{ in.}^2$$

$$\text{weight of web} = (19.4/144)(490) = 66 \text{ lb/ft}$$

estimated weight of flanges and stiffeners (after some scratch-paper work)

$$= 200 \text{ lb/ft}$$

Total estimated girder weight $= \overline{266}$ lb/ft

Moment and shear diagrams are shown in Fig. 16-21.

(c) Flange selection:

$$\text{assume average flange stress} = (31.00/32.5)(22) = 20.98 \text{ ksi}$$

$$A_f = \frac{M}{fh} - \frac{1}{6}th = \frac{(12)(3522)}{(20.98)(62.0)} - \left(\frac{1}{6}\right)\left(\frac{5}{16}\right)(62) = 29.26 \text{ in.}^2$$

Try PL $1\tfrac{1}{2} \times 20$ each flange $\left(A_f = 30.0 \text{ in.}^2\right)$

Checking local buckling (AISC 1.9.1.2)

$$\frac{b_f}{2t_f} = \frac{10}{1.5} = 6.67 < \frac{95}{\sqrt{36}} = 15.8$$

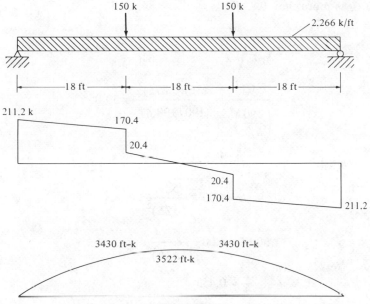

Figure 16-21

Review of flange stresses:

$$I = \left(\tfrac{1}{12}\right)\left(\tfrac{5}{16}\right)(62)^3 + (2)(30.0)(31.75)^2 = 66,700 \text{ in.}^4$$

$$f_b = \frac{(12)(3522)(32.5)}{66,700} = 20.6 \text{ ksi}$$

Since $h/t = 62/\tfrac{5}{16} = 198 > 162$, allowable flange stress must be reduced from 22.0 ksi using AISC Formula 1.10-5.

$$F_b' = 22.0\left[1.0 - 0.0005\left(\frac{19.4}{30.0}\right)\left(198 - \frac{760}{\sqrt{22}}\right)\right] \qquad \text{OK}$$

$$= 21.7 \text{ ksi} > 20.6 \text{ ksi}$$

(d) Location of first intermediate stiffener (AISC 1.10.5.2 and 1.10.5.3):

$$V = 211.2 \text{ k at end}$$

$$f_v = \frac{211.2}{\tfrac{5}{16} \times 62} = 10.90 \text{ ksi} < 0.40F_y \qquad \text{OK}$$

$$F_v = \frac{F_y}{2.89}(C_v) \leqslant 0.40F_y \qquad \text{(AISC Formula 1.10-1)}$$

Assume $F_v = 10.90$ ksi.

$$C_v = \frac{2.89F_v}{F_y} = \frac{(2.89)(10.90)}{36} = 0.875$$

When C_v is more than 0.8,

$$C_v = \frac{190}{h/t}\sqrt{\frac{k}{F_y}} \quad \text{and} \quad \frac{h}{t} = \frac{62}{\frac{5}{16}} = 198.4$$

$$k = \frac{F_y C_v^{\ 2}}{\left(\dfrac{190}{h/t}\right)^2} = \frac{(36)(0.875)^2}{(190/198.4)^2} = 30.05$$

When $a/h < 1.0$,

$$k = 4.00 + \frac{5.34}{(a/h)^2}$$

$$30.05 = 4.00 + \frac{5.34}{(a/h)^2}$$

$$\frac{a}{h} = 0.453$$

$$\text{max } a = (0.453)(62) = 28.09 \text{ in.}$$

Tentatively place first intermediate stiffener 2 ft–0 in. from end.

$$\frac{a}{h} = \frac{24}{62} = 0.387$$

$$\frac{h}{t} = 198.4$$

From Table 11-36 AISC Appendix A or from AISC Formulas 1.10-1 and 1.10-2,

$$F_v = \text{approx. } 11.9 \text{ ksi} > 10.90 \text{ ksi} \qquad\qquad \text{OK}$$

(e) Spacing of other intermediate stiffeners.

V at 2 ft–0 in. from end of girder $= 211.2 - (2.0)(2.266) = 206.7$ k

$$f_v = \frac{206.7}{\frac{5}{16} \times 62} = 10.67 \text{ ksi}$$

Intermediate stiffeners (AISC 1.10.5.3) not required if $h/t < 260$ and if $f_v < F_v$ permitted by Formula 1.10-1.

$$\frac{h}{t} = 198.4 < 260$$

A bearing stiffener will be located in the girder under each column load (AISC 1.10.5.1 and 1.10.10.2). This stiffener will be 16 ft or 192 in.

from the first intermediate stiffener.

$$\frac{a}{h} = \frac{192}{62} = 3.10$$

$$k = 5.34 + \frac{4.00}{(3.10)^2} = 5.76$$

NG

$$C_v = \frac{(45,000)(5.76)}{(36)(198.4)^2} = 0.183$$

F_v = about 2.1 ksi from AISC Table 11-36 < 10.67 ksi

Therefore, intermediate stiffeners required for the 192 in.

$$\text{max value of } \frac{a}{h} = \left(\frac{260}{h/t}\right)^2 \qquad \text{(AISC Section 1.10.5.3)}$$

$$\text{max } a = (62)\left(\frac{260}{198.4}\right)^2 = 106.5 \text{ in.}$$

Try intermediate stiffener at $\frac{1}{2}$ point (96 in.).

$$\frac{a}{h} = \frac{96}{62} = 1.55, \qquad \frac{h}{t} = 198.4$$

F_v from Table 11-36 = about 7.3 ksi < 10.67 ksi   NG

Try intermediate stiffeners at $\frac{1}{6}$ points (at 32 in.).

$$\frac{a}{h} = \frac{32}{62} = 0.516, \qquad \frac{h}{t} = 198.4$$

F_v from Table 11-36 = about 11.8 ksi > 10.67 ksi   OK

Use one intermediate stiffener at 2 ft–0 in. and six at 2 ft–8 in.

Combined shear and bending at column loads:

$$f_v = \frac{170.4}{62 \times \frac{5}{16}} = 8.79 \text{ ksi}$$

$$F_b = 0.60F_y = 22 \text{ ksi} \quad \text{or}$$

$$F_b = \left(0.825 - 0.375\frac{8.79}{11.8}\right)36 = 19.64 \text{ ksi}$$

(AISC Section 1.10.7)

I of plate girder $= \left(\frac{1}{12}\right)\left(\frac{5}{16}\right)(62)^3$

$$+ (2)(1.5)(20)(31.75)^2 = 66,700 \text{ in.}^4$$

$$f_b = \frac{(12)(3430)(31.25)}{66,700} = 19.28 \text{ ksi} < 19.64 \text{ ksi} \quad \text{OK}$$

Intermediate stiffeners between column loads

$$f_v = \frac{20.4}{\frac{5}{16} \times 62} = 1.05 \text{ ksi}$$

If no stiffeners are used, $a = (12)(18) = 216$ in.

$$\frac{a}{h} = \frac{216}{62} = 3.48, \qquad \frac{h}{t} = 198.4$$

F_v from Table 11-36 $= 2.1$ ksi > 1.05 ksi

However Section 1.10.5.3 of AISC Specification states that a/h not $>$ 3.0. Therefore use an intermediate stiffener at girder $\mathbb{L}$.

Check web at $\mathbb{L}$ of center span to see if stiffener needed as per Section 1.10.10.2 and Formula 1.10-11. Assume construction prevents rotation of compression flange.

As described in Section 1.10.10.2 the calculated compression stress $=$ $2.266/(12 \times \frac{5}{16}) = 0.604$ ksi. Allowable compression stress is

$$\left[5.5 + \frac{4}{(216/62)^2} \right] \frac{10,000}{(198)^2} = 1.49 \text{ ksi} > 0.604 \text{ ksi} \qquad \text{OK}$$

Place stiffeners as shown in Fig. 16-24

(f) Design of interior bearing stiffeners (Section 1.10.5 and 1.10.10): Try two 8-in. PLs

$$\max \frac{w}{t} \text{ ratio} = \frac{95.0}{\sqrt{36}} \cong 16$$

$$\min t = \frac{8}{16} = 0.500 \text{ in.}$$

Checking bearing stress: Allowable bearing $F_p = 0.90 F_y = 32.4$ ksi (Section 1.5.1.5.1 and AISC Table 11-36). Assuming $\frac{3}{8}$-in. welds between web and flange, $7\frac{5}{8}$ in. of PLs left to support load. Therefore,

$$f_p = \frac{150}{(2)(7\frac{5}{8})(\frac{1}{2})} = 19.7 \text{ ksi} < 32.4 \text{ ksi} \qquad \text{OK}$$

Checking column action (Sec. 1.10.5): "Column" area is shown in Fig. 16-22.

$$A_v = (7.8)(\tfrac{5}{16}) = 2.44 \text{ in.}^2$$
$$A \text{ PLs} = (2)(7\tfrac{5}{8})(\tfrac{1}{2}) = 7.62 \text{ in.}^2$$
$$A_{\text{column}} = \overline{10.06} \text{ in.}^2$$

$$I = \left(\tfrac{1}{12}\right)(7.8)\left(\tfrac{5}{16}\right)^3 + (2)\left(\tfrac{1}{12}\right)\left(\tfrac{1}{2}\right)\left(7\tfrac{5}{8}\right)^3$$
$$+ (2)(3.81)(4.34)^2 = 180.5 \text{ in.}^4$$

$$r = \sqrt{\frac{180.5}{10.06}} = 4.24 \text{ in.}$$

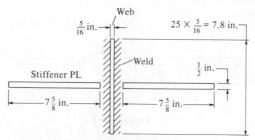

Figure 16-22

effective length $= Kl = \frac{3}{4} \times 62 = 46.5$ in.   (Section 1.10.5)

$$\frac{Kl}{r} = \frac{46.5}{4.24} = 10.97$$

allowable $F_a = 21.1$ ksi from Table 3-36

actual $f_a = \dfrac{150}{10.06} = 14.9$ ksi < 21.1 ksi   OK

Use 2PLs $\frac{1}{2} \times 8$ under column loads

(g) Design of end bearing stiffeners:
Try two 8-in. PLs

$$\min t = \tfrac{8}{16} = \tfrac{1}{2} \text{ in.}$$

Check bearing stress: Allowable bearing $F_p = 0.90 F_y = 32.4$ ksi. Assuming $\frac{3}{8}$-in. welds, $7\frac{5}{8}$ in. of PL left to support load. Therefore,

$$f_p = \frac{211.2}{(2)\left(7\frac{5}{8}\right)\left(\frac{1}{2}\right)} = 27.7 \text{ ksi} < 32.4 \text{ ksi} \qquad \text{OK}$$

Checking column action: "Column" area is shown in Fig. 16-23.

$$A = (3.75)\left(\tfrac{5}{16}\right) + (2)\left(7\tfrac{5}{8}\right)\left(\tfrac{1}{2}\right) = 8.80 \text{ } in.^2$$

$I = 180.5$ in.^{4} same as for column bearing stiffeners

$$r = \sqrt{\frac{180.5}{8.80}} = 4.53$$

effective length $= Kl = \left(\tfrac{3}{4}\right)(62) = 46.5$ in.

$$\frac{Kl}{r} = \frac{46.5}{4.53} = 10.26$$

allowable $F_a = 21.10$ ksi

actual $f_a = \dfrac{211.2}{8.80} = 24.0$ ksi   NG

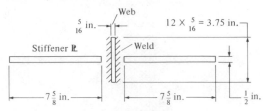

Figure 16-23

Subsequent calculations show $\frac{9}{16}$ stiffener satisfactory.

Use 2PLs $\frac{9}{16} \times 8$ in. at girder ends

(h) Design of intermediate stiffeners: Minimum required gross area of interior stiffener can be determined from AISC Formula 1.10-3, but values are recorded in AISC Table 11-36 for A36 steel.

For $h/t = 198.4$ and $a/h = 32/62 = 0.516$ minimum area = approx. 2.9% web area $= (.029)(62)(\frac{5}{16}) = 0.56$ in.²

Try 2PLs $\frac{1}{4} \times 4$

$$\text{approximate } I \text{ about web} = \left(\frac{1}{3}\right)\left(\frac{1}{4}\right)(4)^3(2) = 10.67 \text{ in.}^4$$

$$\text{min } I \text{ permissible (Section 1.10.5.4)} = \left(\frac{h}{50}\right)^4 = \left(\frac{62}{50}\right)^4$$

$$= 2.36 < 10.67 \qquad \text{OK}$$

$$\frac{w}{t} = \frac{4}{\frac{1}{4}} = 16 \qquad \text{OK}$$

Intermediate stiffeners may be stopped a distance equal to four times the web thickness from the tension flange (Section 1.10.5.4):

$$L = 62.0 - (4)\left(\frac{5}{16}\right) = 60\frac{3}{4}$$

Use 2PLs $\frac{1}{4} \times 4 \times 60\frac{3}{4}$ in. bearing on compression flange only

(i) Design of welds between web and intermediate stiffeners: Total shear transferred $= f_{vs}$ (Section 1.10.5.4):

$$f_{vs} = h\sqrt{\left(\frac{F_y}{340}\right)^3}$$

$$= 62\sqrt{\left(\frac{36}{340}\right)^3} = 2.136 \text{ k/in. on two stiffener welds}$$

Assuming $\frac{3}{16}$ in. welds 3 in. long $= (\frac{3}{16})(0.707)(0.30 \times 70) \times 2 = 5.57$ k/in.,

$$\text{spacing} = \frac{(3)(5.57)(2)}{l} = 2.136$$

$$l = 15.65 \text{ in.}$$

Use $\frac{3}{16}\times 3$ welds 15 in. on centers

(j) Design of flange to web welds:

$$v = \text{horizontal shear between flange and web} = \frac{VQ}{I}$$

$$Q = (30)(31.75) = 953 \text{ in.}^3$$

$$f_v \text{ at girder end} = \frac{(211.2)(953)}{66,700} = 3.02 \text{ k/in.}$$

min weld size $= \frac{5}{16}$ in. for $1\frac{1}{2}$PL   (AISC Table 1.17.2A)

Assuming $\frac{5}{16}$ welds (E70 electrodes),

$$\text{allowable shear} = (2)\left(\frac{5}{16}\right)(0.707)(0.30\times 70) = 9.28 \text{ k/in.}$$

allowable shear on $\frac{5}{16}$ web $= \left(\frac{5}{16}\right)(14.5) = 4.53 \text{ k/in.}$

min length of fillet welds $= (4)\left(\frac{5}{16}\right) = 1\frac{1}{4}$ in. (Section 1.17.4)

Try $\frac{5}{16}\times 4$ welds located $(4)(4.53)/3.02 = 6$ in.

Maximum spacing of compression flange to web welds (Section 1.18.3.1) is

$$24\times \frac{5}{16} = 7.5 \text{ in.} > 6 \text{ in.} \qquad\qquad \textbf{OK}$$

Use $\frac{5}{16}\times 4$-in. welds 6 in. on centers}

(k) Design of welds between bearing stiffeners and web: Use continuous welds both sides since bearing stiffeners are major load carrying members:

$$\text{min weld size} = \tfrac{1}{4} \qquad \left(\text{or } \tfrac{1}{4}\times 0.707\times 0.30\times 70 = 3.71 \text{ k/in.}\right)$$

$$\text{total strength furnished} = (2)(2)(62)(3.71) = 920 \text{ k} > 211.2 \text{ k} \qquad\qquad \textbf{OK}$$

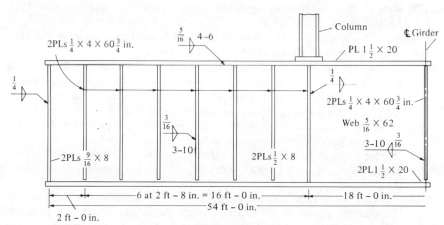

Figure 16-24

Use $\frac{1}{4}$ continuous fillet welds for all bearing stiffeners

(l) Design sketch of final girder design shown in Fig. 16-24.

Problems

16-1. A 50-ft plate girder consists of a $\frac{1}{2} \times 54$-in. web, $6 \times 6 \times \frac{1}{2}$-in. flange angles, and two $\frac{1}{2} \times 16$-in. plates on each flange. Determine the maximum tension and compression stresses for a bending moment of 2,400 ft-k. Assume 1-in. holes are placed 4 in. on centers in the web and assume there are two cover plate bolts and one flange bolt at a section. AASHTO Specifications. (*Ans.* $f_c = 17.6$ ksi, $f_t = 20.8$ ksi)

16-2. Repeat Prob. 16-1 assuming the girder to be welded and the angles omitted.

16-3. Using the AASHTO Specifications and structural carbon steel (A36), compute the maximum allowable bending moment for a bolted plate girder consisting of a $\frac{5}{8} \times 80$-in. web, $8 \times 6 \times \frac{3}{4}$-in. flange angles, and three $\frac{5}{8} \times 18$-in. cover plates on each flange. Assume holes of the size and distribution mentioned in Prob. 16-1. (*Ans.* 6868 ft-k)

16-4. Determine the theoretical cutoff points of the outside and middle cover plates of the girder of Prob. 16-1 if it is assumed to be uniformly loaded.

16-5. Repeat Prob. 16-3 assuming the girder to be welded leaving off flange angles and assuming $F_b = 20$ ksi. (*Ans.* 5564 ft-k)

16-6. Using the AASHTO Specifications design end stiffeners for the girder of Prob. 16-3 if the maximum end reaction is 260 k.

16-7. Design a single-plate web splice for the girder of Prob. 16-3 if the external shear is assumed to be 150 k at the splice. Use $\frac{7}{8}$ in A325 high-strength friction bolts. (*Ans.* 2 PLs $\frac{9}{16} \times 25 \times 5$ ft–8 in. with four rows of bolts)

16-8. Design a triple-plate splice for the conditions described in Prob. 16-7.

16-9. Preliminary designs are to be made for a plate girder with A36 steel, the AASHTO Specifications and $\frac{7}{8}$-in. bolts for a maximum bending moment of 3,000 ft-k and a maximum shear of 250 k. Proportion the girder with total depths of 48, 64, and 80 in. and compare the resulting cross-sectional areas. (*Ans.* 121.8, 96.5, and 104.5 in.^{2} respectively)

16.10. Rework Prob. 16-9 using a welded girder, A242 steel, and the AASHTO Specifications.

16-11. Rework Prob. 16-9 using a welded girder, A36 steel, and the AISC Specification. (*Ans.* 80.0, 69.0, and 61.3 in.^{2} respectively)

16-12. If the external shear is 220 k at a section in the girder of Prob. 16-3, determine the theoretical spacing required for $\frac{7}{8}$-in. flange and cover-plate A325 high-strength friction bolts using the AASHTO Specifications and A36 steel. It is assumed that a 30-k concentrated load can be applied at the section. Assume this load is distributed over 12 in. by the time it reaches the girder and that bolts are arranged as in Fig. 16-10.

16-13. Repeat Prob. 16-12 if $\frac{7}{8}$-in. A325 high-strength bearing bolts and the AISC Specification are used. (*Ans.* Theoretical values: flange bolts 7.67 in., cover plate-plate bolts 17.35 in.)

16-14. Design a bolted plate girder with A36 steel using the AISC Specification for the situation shown in the accompanying illustration. The girder is to be connected with $\frac{7}{8}$-in. A325 high-strength bearing bolts. Full lateral support is assumed for the compression flange.

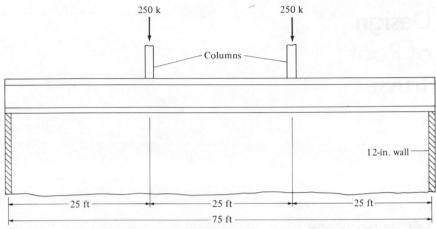

250 k   250 k

Columns

12-in. wall

|← 25 ft →|← 25 ft →|← 25 ft →|

|← 75 ft →|

Problem 16-14

16-15. Rework Prob. 16-14 using a welded girder and SMAW E70 electrodes.

16-16. Repeat Prob. 16-15 if lateral support for the compression flange is provided at the ends and column points only.

Chapter 17
Design
of Roof
Trusses

17-1. INTRODUCTION

Trusses may be defined as large, deep beams with open webs. They are usually formed by members arranged in triangles or groups of triangles, and the number of possible types is almost endless. The purposes of roof trusses are to keep the elements out (rain, snow, wind) and to support the loads connected underneath (ducts, piping, ceiling). While performing these functions they must also support the roofs and their own weight.

The engineer is often concerned with the problem of selecting a truss or a beam to span a given opening. Should no other factors be present, the decision would probably be based on consideration of economy. The smallest amount of material will nearly always be used if a truss is selected for spanning a certain opening; however, the cost of fabrication and erection of trusses will probably be appreciably higher than required for beams. For shorter spans the overall cost of beams (material plus fabrication and erection) will definitely be less but as the spans become greater, the higher fabrication and erection costs of trusses will be more than canceled by their weight saving. A further advantage of trusses is that for the same amounts of material they have greater stiffnesses than do beams.

On the subject of truss depths, it should be realized that the deeper a truss is made for a given span and loading the smaller will be the chord members, but that with deeper trusses the lengths of the web members increase. This fact means that the slenderness ratios of the web members may become a factor and require the use of heavier members.

It is impossible to give a lower economical span for steel trusses. They may be used for spans as small as 30 or 40 ft and as large as 300 to 400 ft. Actually they are not often used today for spans of less than 60 ft. Beams may be economical for some applications for spans much greater than the lower limits mentioned for trusses. The AISC Specification permits the design of very thin webbed plate girders (see Chapter 16) which are quite competitive with roof trusses for many spans.

Owen Steel Company Fabrication Plant, Columbia, S.C. (Courtesy of Owen Steel Co., Inc.)

In the pages to follow the terms *pitch* and *slope* are often used. The pitch of a symmetrical truss is referred to as the rise of the top chord of the truss divided by the span. Should the truss be unsymmetrical, the numerical value of its pitch is not of much use. For such cases the slope of the truss on each side can be given. The slope is the rise of the top chord to its horizontal length often given as so many inches per horizontal foot. For symmetrical trusses the slope will equal twice the pitch.

17-2. TYPES OF ROOF TRUSSES

Roof trusses can be flat or peaked. In the past the peaked roof trusses have probably been used more for short-span buildings and the flatter trusses for the longer spans. The trend today for spans long or short, however, seems to be away from the peaked trusses and towards the flatter ones, the change being due to the appearance desired and perhaps more economical construction of roof decks.

Several of the commonly used types of roof trusses are shown in Fig. 17-1. A good many of these trusses have been named for the engineers or architects who first developed them. A remark or two is made about each of these trusses in the paragraphs to follow. The letter at the beginning of each of these paragraphs corresponds to the letter in Fig. 17-1 beneath the type of truss being considered.

(a) The Warren and Pratt trusses have probably been used more for the flatter roofs (slopes of from $\frac{3}{4}$ to $1\frac{1}{4}$ in./ft) where built-up roofing can be satisfactorily applied, than have the other types of trusses. These trusses can be economically used for flat roofs for spans of roughly 40 to 125 ft,

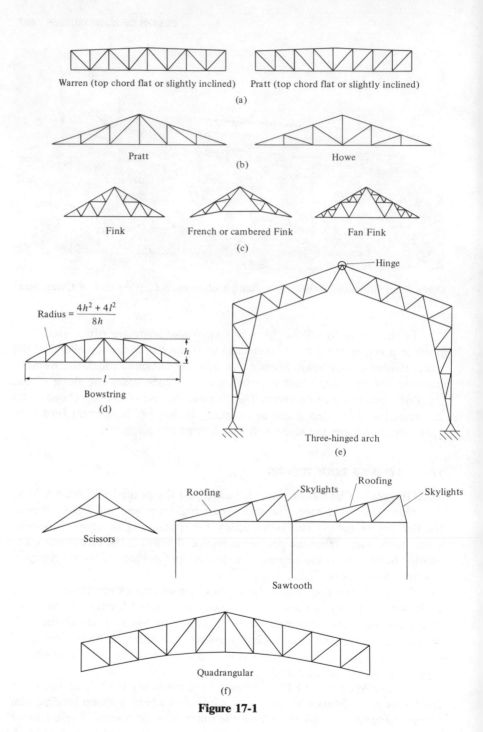

Warren (top chord flat or slightly inclined)   Pratt (top chord flat or slightly inclined)

(a)

Pratt   Howe

(b)

Fink   French or cambered Fink   Fan Fink

(c)

$$\text{Radius} = \frac{4h^2 + 4l^2}{8h}$$

h

l

Bowstring

(d)

Hinge

Three-hinged arch

(e)

Roofing   Skylights   Roofing   Skylights

Scissors

Sawtooth

(f)

Quadrangular

(f)

Figure 17-1

498

although they have been used for spans as great as 200 ft. The Warren is usually a little more satisfactory than the Pratt. The roofs may be completely flat for spans not exceeding 30 or 40 ft, but for longer spans the slopes mentioned are used for drainage purposes.

(b) The pitched Pratt and Howe trusses are probably the most common types of medium-rise trusses. The slopes intended here fall in between those given for (a) and (c). They have maximum economical spans of about 90 or 100 ft.

(c) For steep roofs (with slopes of 5 or 6 in./ft) the Fink truss is very popular. The Pratt and Howe trusses may also be used for steep slopes but they are usually not as economical. The Fink truss has been used for spans as great as 120 ft. A fact that makes it more economical is that most of the members are in tension while those that are in compression are fairly short. The panel layout of a truss may be controlled by the purlin spacings. As it is usually desirable to have purlins placed at panel points only, the main panels may be subdivided. Fink trusses can be divided up into a large number of panels to suit almost any span or purlin spacing. The Fan Fink shown illustrates subdivision very well.

(d) If a curved roof is acceptable, the Bowstring truss can be used economically for spans of up to 120 ft, although on occasions it has satisfactorily been used for much longer spans. When properly designed, this truss has the unusual feature of having very small stresses in the web members. Despite the fact that there is some expense in bending the top chord, the Bowstring has proved quite popular for warehouses, supermarkets, garages, and small industrial buildings. A recommended radius of curvature for the top chord is given in the figure.[1]

(e) When spans appreciably above 100 ft are planned consideration might be given to using steel arches as they may provide the most economical solutions. The three-hinged arch is the only one shown here. As compared to the two-hinged and hingeless arches it has the following advantages.

1. Analysis is easier as it is statically determinate.
2. Poor foundations are not such a serious matter as they may be for a statically indeterminate arch.
3. Erection is often simplified as the two halves of an arch can be erected separately and pinned together at the crown.

(f) In this part of the figure several miscellaneous types of trusses are shown. The scissors truss (which is so named because of its resemblance to a pair of scissors) may be satisfactory for supporting short-span churches and other buildings with steep roofs. Sawtooth trusses may be used when adequate natural lighting is desired from skylights in wide buildings. Their

[1] John E. Lothers, *Design in Structural Steel*, 3d ed. (Englewood Cliffs, N.J.: Prentice-Hall, 1972), p. 342.

steep faces support skylights and these faces usually face the north for more evenly diffused light. They are used when their numerous columns are not objectionable. A long-span roof truss, that has been used for spans well over 100 ft is the Quadrangular truss. Near the centerline of this truss the diagonals are reversed for the purpose of keeping as many of them in tension as possible.

17-3.   SELECTION OF TYPE OF ROOF TRUSS

The choice of the particular type of roof truss depends on a number of items including span, loading, preferred type of roof construction from an architectural viewpoint, climate, lighting, insulation, and ventilation. The following paragraphs present a discussion of some of the more important factors which may affect the selection.

Pitch

The desired pitch of a roof truss to a great extent controls the selection of the type of truss to be used because, as described in Section 17-2, different types of trusses are economical for different slopes of roofs. For instance, the Fink truss is quite satisfactory for the steeper roofs.

Roof Coverings

The type of roof covering used has a great deal to do with the selection of the roof slope. For instance, the built-up tar-and-gravel roofs are probably unsatisfactory for slopes greater than 1 in. per foot, for the tar will tend to run downhill during the warm summer months despite the availability of the so-called steep-roofing tar or pitch. When built-up asphalt-and-gravel roofs are used, somewhat steeper slopes are possible. The desirable slopes for slate roofs are about 7 in. per foot because water tends to work up under the slate on flatter slopes. Other desired slopes can be obtained from the manufacturers for the particular types of roof covering to be used.

Fabrication and Transportation Considerations

Fabrication and transportation difficulties need to be considered in selecting the type of truss to be used for a given situation. It is desirable economically to fabricate as much as possible of the truss in the shop—the whole truss if feasible—and to ship it to the job for erection. From a transportation standpoint the depth of the truss is often a controlling factor. As an example for a particular building it is assumed that the designer estimates that a 15-ft-deep truss will be the most economical, but the route along which the truss is to be shipped limits the maximum depths that can be transported to 12 ft. The truss may therefore be limited to a

Tightening high-strength bolts. (Courtesy of Bethlehem Steel Company.)

maximum depth of 12 ft, although it will be a little heavier, so that the whole truss can be assembled in the shop.

Architectural Effect Desired

The ideas of the architect as to the aesthetic effect desired can quite well be the controlling factor. For example, she may want a flat roof and that will pretty well narrow the choice down to one or two trusses.

Climate

The climate in the particular area may be particularly important as to drainage, or retention of snow and ice.

17-4. SPACING AND SUPPORT OF ROOF TRUSSES

The spacing of roof trusses depends upon the type of roof construction, truss spans, and foundation conditions. The usual center-to-center spacings vary from 12 to 30 ft, the lower spacings used for the shorter spans and the larger spacings for the longer spans. For the 50- and 60-ft spans, spacings of roughly 12 to 20 ft on centers are common, while values of from 15 to 24 ft are common for 90- to 100-ft spans. For very long-span roof trusses, say more than 140 or 150 ft, the center-to-center spacings may be as great as 50 or 60 ft. The purlins for such large spacings are probably trusses themselves framing into the sides of the main trusses. These trusses will provide a large part of the lateral bracing needed. When poor soil conditions are present causing expensive foundations, the designer may use larger center-to-center truss spacings than he would under ordinary conditions.

The usual truss is supported on brick, block, or concrete walls or on steel or reinforced concrete columns. Trusses are usually attached at their ends to these walls or columns with anchor bolts. To provide for temperature expansion and contraction it is usually necessary to have the anchor bolts at one end set in a slotted hole in the bearing plate to permit the required shortening or lengthening. The estimated change in length of the truss equals the coefficient of expansion times the estimated temperature change times the span length plus the diameter of the anchor bolt plus a little margin. Some consideration should be given to the temperature at the time of erection as it pertains to placing the anchor bolt or positioning the slot. Should the truss be erected in the middle of the summer, it seems logical that the anchor bolt should be near the outer end of the slot, as the truss will be expanded to a point near its greatest anticipated length. The dead and live loads supported by a truss also affect the length of the truss.

17-5. ESTIMATED WEIGHT OF ROOF TRUSSES

The usual loads to which roof trusses may be subjected were discussed in Chapter 13; therefore, the only discussion of loads in this chapter pertains to the estimated weights of the roof trusses themselves.

The weight of a roof truss can be estimated by the designer on the basis of his previous experience or by some reference to the various tables, curves, or formulas which have been developed for this purpose. An important fact to remember is that the designer certainly cannot estimate snow, ice or wind loads to the nearest 1%. Furthermore, he can only roughly estimate what the users of a building may hang on the trusses from below. These facts should show that it is unrealistic to expect him to estimate truss weights to the nearest 1%. In fact, estimates to the nearest 10% are probably quite reasonable.

One method of approximating the weight of a roof truss and its bracing is to estimate it to equal about 10% of the load it is to be required to support. For long spans the percentage probably should be increased a little. When the truss design is completed its weight should be roughly calculated and compared with the original weight estimate to see if that estimate was within reason.

Based on the previous experience of the design profession, the designer can estimate the weight of roof trusses as equaling so many pounds per square foot of roof surface. Dr. L. E. Grinter[2] recommends the following values, which vary somewhat with different spans and roof pitches.

1. For 40-ft spans and pitches varying from $\frac{1}{3}$ to $\frac{1}{4}$ estimate truss weight equal to between 2 and $3\frac{1}{2}$ lb for each square foot of roof surface.
2. For each 10 ft increase in span up to 80 ft the previous values should be increased by approximately 1 lb.
3. Increase the values by roughly $\frac{1}{2}$ to 1 lb/ft^2 of roof surface for flat roofs.
4. Decrease the values by roughly $\frac{1}{2}$ to 1 lb/ft^2 of roof surface for steeper roofs.

Through the years quite a few empirical formulas have been developed for estimating the weight of steel roof trusses. Nearly any of these expressions will give reasonable estimates if they are properly applied. One comment, however, should be made concerning expressions that are several decades old. Unless they take into account allowable stresses they will probably give estimated weights on the high side with today's steels, which have considerably higher allowable stresses than those used when the formulas were first presented.

One satisfactory expression for estimating the weight of steel roof trusses was presented in the *Engineering News Record* in 1919 in an article by Robins Fleming entitled "Weight of Roof Trusses by Empiric Formulas." This expression, which does include an allowable stress value, is as follows.

$$W = \sqrt{\frac{wa}{S}}\ (4l^2 + 60l)$$

where
  W = total roof truss weight
  w = total gravity load supported per horizontal square foot
  S = average allowable stress in pounds per square inch used in design
  a = center-to-center spacing of trusses
  l = truss span in feet

17-6. DISCUSSION OF ROOF TRUSS ANALYSIS

Before the members of a roof truss can be selected it is necessary for the truss to be analyzed for the different types of loading which can occur.

[2] L. E. Grinter, *The Design of Steel Structures* (New York: Macmillan, 1960), p. 284.

These loads include dead loads, snow and ice loads, and wind loads. In the more southern states a live load representing the roofers and their materials can be used in place of the snow loads.

A truss will probably be analyzed separately for each of the different types of loads because all of the different loads will probably not occur at the same time. As an illustration, it does not seem altogether logical to assume that a pitched roof will be covered with a full snow load when a 90-mph wind is blowing. Most engineers feel that the majority of the snow would blow off under such conditions, at least on the windward side. (Other engineers think that the snow may have crusted over and be substantially able to remain on the roof during the windstorm.)

There are several possible combinations of loads which may logically be applied at the same time to a roof truss. The three combinations listed at the end of this paragraph seem to cover the situation fairly well. The forces in each truss member are computed for each of the different loads and combined for each of the combinations given, and the maximum force for each member, regardless of the combination from which it came, is used for design. If a member is in tension for one combination and in compression for the other, it will have to be designed for both values.

1. Dead load + full snow load
2. Dead load + full wind load
3. Dead load + full wind load + $\frac{1}{2}$ snow load

Another combination often used in addition to the preceding ones is a lesser wind ($\frac{1}{3}$ to $\frac{1}{2}$) combined with dead load + full snow load assuming that such a wind would not blow off the snow.

In the following paragraphs are several further comments which may be of considerable importance for some cases because they pertain to the analysis of roof trusses.

1. The AISC Specification, which is used for an extremely large percentage of steel roof truss designs in the United States, says that allowable stresses can be increased by one-third for forces caused by wind or earthquake action alone or in combination with dead or live loads. This reduction is conveniently handled by multiplying the second and third load combinations (just given) for maximum stresses by $\frac{3}{4}$. This specification is followed in Example 17-1 which is presented in Section 17-7.

2. The procedure recommended today by the ASCE for estimating wind pressures results in suctions on the leeward sides of roof trusses and perhaps on the windward sides, depending on the slopes. A rather large percentage of designers neglect this suction in their designs.

3. If the AISC one-third reduction and the recommended ASCE wind forces are used for design, it will usually be found that the wind will not require any increases in the member sizes of roof trusses. Some engineers are not too happy about this situation as they feel the wind is another load and should cause greater stresses and thus require larger members. For this

reason the Duchemin formula[3] for estimating wind pressures is still occasionally used because it probably will cause stress increases.

4. Consideration may have to be given for the wind blowing from both directions. If a truss has a roller support on one end and a hinge support on the other, wind from the left may produce different forces in some members than wind from the right.

5. One further comment is made here concerning the computation of truss forces, and this pertains to the actual types of end supports as they affect member forces due to the wind loads. For fairly short trusses, probably no provision is made for expansion. This type of truss would actually be statically indeterminate, but the usual practice is to assume that the horizontal load splits equally between the supports. As trusses become longer, expansion supports are used on one end. The designer may assume all the horizontal reaction is provided at the other support or he may more realistically assume that some proportion of the horizontal reaction (as $\frac{1}{4}$ or $\frac{1}{3}$) is actually provided at the expansion end. From this discussion it is obvious that there are several possible variations in the member forces caused by the horizontal loads, depending on the assumptions made.

17-7. EXAMPLE ANALYSIS OF A ROOF TRUSS

Member forces are computed in this section for a roof truss with several possible loading conditions, and these various forces are combined as described in the previous section to obtain the most critical values. The average designer might not go to quite as much detail for a small roof truss as was used here by the author.

The actual force calculations are not shown in the figure, and only the answers are given. Forces can be obtained algebraically or graphically, depending on the preference of the individual designer. Perhaps there is little advantage of one method over the other for gravity loads, but the graphical procedure can certainly be a time saver when lateral loads and complicated trusses are involved.

In Example 17-1 forces are determined for a roof truss separately for dead load, full snow load, half snow load, and wind load. The full snow and half snow forces can be determined by ratio from the dead-load forces. The author has assumed that the half snow-load situation consists of half the snow load all the way across the roof. He feels this is a more feasible condition with full wind blowing than assuming a full snow load on one side and none on the other as is often the practice.

The wind can blow from the left as it can from the right; therefore, in the maximum force table of Example 17-1 the values of the wind force in a member are given for both directions. To obtain the design force shown in the final column of the table the most critical value of the wind force is used. Should the wind or other loads cause force reversal in a member,

[3]J. C. McCormac, *Structural Analysis*, 3d ed. (New York: Harper & Row, 1975), p. 7.

both values would be given in the design force column. In the example problem presented here the dead-load plus snow load condition controlled for every member, but this definitely may not be true for other trusses.

Purlins are placed only at the joints in Example 17-1 to simplify the problem (although the resulting spacing is rather large) and it is unnecessary to consider bending in the top chord members due to intermediate loads. Should the top-chord panel lengths become exceptionally long, it may be economical to place purlins in between the joints. Another case where intermediate purlins might be used arises with certain types of roofing. Should the spacing of purlins not be greater than 4 or 5 ft, certain types of corrugated steel, gypsum tile, or other roofing can be placed directly on the purlins. Average purlins usually weigh from 2 to 5 psf of roof surface.

Example 17-1

Determine the critical forces for design for each of the members of the left-hand side of the truss of Fig. 17-2. Assume the following conditions.

1. Friction to be neglected at the expansion support.
2. Truss weight = 5 psf of roof surface and four purlins (W10×33 s) have been selected for each side of the truss.
3. Snow load = 20 psf of horizontal roof surface projection.
4. Wind loads as follows: 4.2 psf suction on windward side and 9 psf suction on leeward side as recommended by Committee 31 of the ASCE.
5. Roofing = 15 psf.

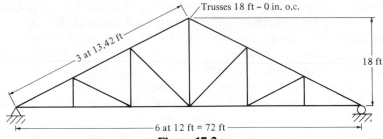

Figure 17-2

SOLUTION

(a) Dead load forces (Fig. 17-3):

$$\text{truss weight} = (5)(18)(13.42) = 1205 \text{ lb}$$

$$\text{roof covering} = (15)(18)(13.42) = 3615 \text{ lb}$$

$$\text{purlins} = (18)(33) = 594 \text{ lb}$$

$$\text{full panel load} = 5414 \text{ lb}$$

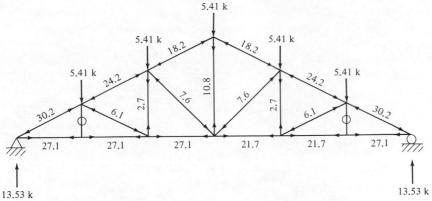

Figure 17-3   Member forces due to dead loads.

(b) Forces due to full snow load (Fig. 17-4):

$$\text{full panel load} = (20)(12)(18) = 4320 \text{ lb}$$

Forces determined by ratio from dead load forces.
   (c) Forces due to half snow load (Fig. 17-5):
   (d) Forces due to wind from left (Fig. 17-6) (wind from right not shown):

$$\text{full panel load, windward side} = (4.2)(13.42)(18) = 1.02 \text{ k}$$

$$\text{full panel load, leeward side} = (9)(13.42)(18) = 2.18 \text{ k}$$

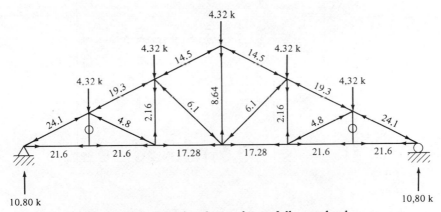

Figure 17-4   Member forces due to full-snow load.

Maximum Force Table

	Design forces due to					Force combinations			
Member	Dead load	Snow load	Half snow load	Wind from left	Wind from right	$DL+SL$	$(DL+WL)^{\frac{3}{4}}$	$(DL+1/2SL+WL)^{\frac{3}{4}}$	Design stresses
L_0L_1	+27.1	+21.6	+10.8	−4.73	−8.90	+48.7	+16.8	+24.9	+48.7
L_1L_2	+27.1	+21.6	+10.8	−4.73	−8.90	+48.7	+16.8	+24.9	+48.7
L_2L_3	+21.7	+17.28	+8.64	−3.57	−6.50	+39.0	+13.6	+20.1	+39.0
L_0U_1	−30.2	−24.1	−12.05	+7.25	+8.74	−54.3	−17.2	−26.2	−54.3
U_1U_2	−24.2	−19.3	−9.65	+6.46	+7.10	−43.5	−13.3	−20.5	−43.5
U_2U_3	−18.2	−14.5	−7.25	+5.76	+5.48	−32.7	−9.5	−15.0	−32.7
U_1L_1	0	0	0	0	0	0	0	0	0
U_1L_2	−6.1	−4.8	−2.40	+1.30	+2.73	−10.9	−3.6	−5.4	−10.9
U_2L_2	+2.7	+2.16	+1.08	−.58	−1.22	+4.86	+1.6	+2.4	+4.9
U_2L_3	−7.6	−6.1	−3.05	+1.66	+3.42	−13.7	−4.5	−6.8	−13.7
U_3L_3	+10.8	+8.64	+4.32	−3.59	−3.59	+19.44	+5.4	+8.7	+19.4

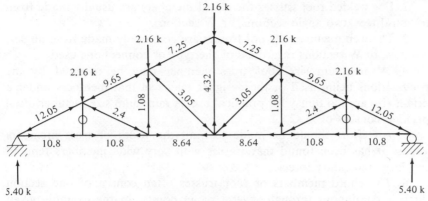

Figure 17-5 Member forces due to half-snow load.

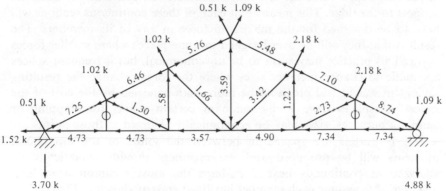

Figure 17-6 Member forces due to wind from left.

17-8. DESIGN OF A ROOF TRUSS

The members of the truss analyzed for various loading conditions in Example 17-1 are designed in this section, assuming welded connections are to be used. Before these designs are presented a few general remarks are made concerning the selection of the members of roof trusses. These are as follows.

1. For riveted and bolted trusses a pair of angles back-to-back is probably the most common type of member, but for short spans and lightly loaded trusses the single angle is sometimes used. It will be remembered that the single-angle tension member does have the disadvantage of eccentricity. Should the eccentric moments produced by the eccentricity be considered in the design of single-angle tension members, the resulting sections will probably be no more economical than if pairs of angles had been selected initially. For larger riveted or bolted roof trusses W, M, or S sections may be used for some of the members.

2. For welded roof trusses the chord members are usually made from structural tees, two angle sections, or W sections.

3. The web members of roof trusses are commonly made from angles, channels, or W sections regardless of the type of connections used.

4. A minimum size roof truss member may be required by the specifications being used or the designer may feel that members under a certain size are too flimsy for practical use. A minimum size member often specified consists of $2Ls2 \times 2 \times \frac{1}{4}$.

5. An effort should be made to limit the width of truss members because it has been found that trusses with very wide members tend to have large secondary forces.

6. The chord members of roof trusses often consist of one section which is continuous through several panel points. In the example given later in this section, the bottom chord is assumed to be continuous for half of the entire span while the top chord members run continuously from the support to the ridge. This means that each of these continuous sections will have to be designed for the maximum force in any of its members. The result is that they will be overdesigned in some places where smaller forces occur. This practice may seem to be uneconomical, but if frequent splices are made to change member sizes where the forces change the resulting savings in weight will probably be more than canceled by the cost of the splices. If splices have to be made at certain points for shipping or handling purposes, sizes may be economically changed at those points.

7. If purlins are spaced in between the joints of the top chord, moments will be produced and the members should theoretically be analyzed as continuous beams. Perhaps the most common procedure, however, is to assume each member has fixed ends as shown in Fig. 17-7.

Another possibility is to analyze the top chord as a continuous member by moment distribution. Whichever method is used to determine the moments, the member should be designed for the resulting moments plus its axial force as determined by the usual truss analysis. The design of members subject to bending and direct stress was discussed in Chapter 8.

8. In Example 17-2 the tension and compression members are selected and shown in tables. The practicing engineer with the aid of various tables and other design data would probably not show his work in such detail, but it is felt that this detail may be of benefit to the student. The tables used are thought to be self-explanatory. Perhaps a few comments will be helpful concerning the "minimum r" column in each of the tables. It will be remembered that the AISC permits maximum slenderness ratios of 240 and 200 respectively for main tension and compression members. The smallest permissible values of r are calculated for each of the members with their different lengths. This minimum value is of great assistance in preventing the designer from selecting members that are too slender, a fact he might not otherwise immediately discover. The minimum r values are of particular benefit in selecting members subjected to rather small forces.

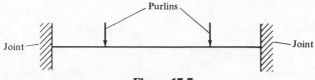

Figure 17-7

9. Should a member be subjected to force reversal, a situation that does not occur in this example, it would have to be designed to resist both tension and compression.

Example 17-2

Select member sizes for the truss analyzed in Example 17-1 assuming welded connections. Use A36 steel and the AISC Specification.

SOLUTION

Design of Tension Members

Member	Design force (k)	Area required (in.²)	Minimum permissible r (in.)	Section selected	Area furnished (in.²)	l/r
L_0L_1	+48.7	2.21	0.60	WT3×7.5	2.21	181
L_1L_2	+48.7	2.21	0.60	WT3×7.5	2.21	181
L_2L_3	+39.0	1.77	0.60	WT3×7.5	2.21	181
U_2L_2	+4.9	0.22	0.60	2Ls2×2×$\frac{1}{4}$	1.88	236
U_3L_3	+19.4	0.88	0.90	2Ls3×3×$\frac{1}{4}$	2.88	232
U_1L_1	0	0	0.30	2Ls2×2×$\frac{1}{4}$	1.88	118

Design of Compression Members

Member	Design force (k)	Minimum permissible r (in.)	Section selected	l/r
L_0U_1	−54.3	0.81	WT7×17	105
U_1U_2	−43.5	0.81	WT7×17	105
U_2U_3	−32.7	0.81	WT7×17	105
U_1L_2	−10.9	0.81	2Ls3×2×$\frac{5}{16}$ [a]	178
U_2L_3	−13.7	1.02	2Ls3$\frac{1}{2}$×2$\frac{1}{2}$×$\frac{5}{16}$ [a]	185

[a] Assuming $\frac{3}{8}$ in. spacing between Ls.

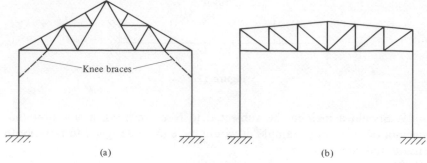

Figure 17-8

17-9.   TRUSSES FOR INDUSTRIAL BUILDINGS

For many industrial buildings the roof is supported by a steel truss rigidly connected to supporting columns. Examples of this type of construction are shown in Fig. 17-8. For the arrangement shown in part (a) of the figure there is very little lateral rigidity unless knee braces (shown with dotted lines) are used. Knee braces are generally placed at angles of approximately 45°. Another type of industrial building truss is shown in part (b). It will be noted that knee braces are unnecessary for this type.

For a good many years one-story industrial buildings were of the general type shown in Fig. 17-8. The name given to them was "mill buildings." These buildings, which provide large open areas, are quite economical to construct; but they are not especially attractive and lighting may be a problem. In recent years a large percentage of the market formerly taken by mill buildings has been taken over by rigid frame structures (considered in Chapter 19).

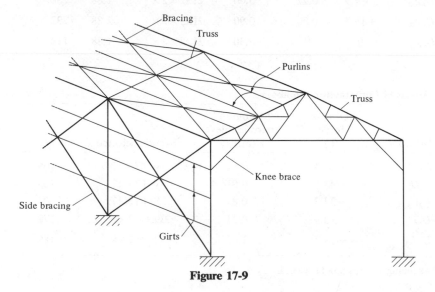

Figure 17-9

Mill buildings usually have exterior walls constructed from brick or block or perhaps some type of metal siding or covering. The sheathing, which supports the roofing material, consists of wood, precast concrete slabs, some type of corrugated metal, etc. The sheathing is supported by the purlins which transfer the roof loads to the trusses as concentrated loads.

The trusses are supported by the columns and the walls are usually nonbearing. Although wind forces do not usually have a great effect on member sizes for one-story mill buildings, the installation of a good bracing system perpendicular to the trusses is important. A typical arrangement of the members of a mill building frame is shown in Fig. 17-9. More information is presented in Section 17-10 concerning the bracing of these and other roof truss systems.

17-10. ROOF TRUSS BRACING

Properly designed roof trusses or mill-building bents can satisfactorily resist horizontal and vertical loads applied to them in the plane of the truss, as shown in Fig. 17-10. However, to make a series of roof trusses or mill-building bents stable normal to the plane of the trusses, a definite system of lateral bracing has to be installed between the trusses.

Before these systems of lateral bracing are discussed it is necessary to make a comment or two about the erection of individual trusses. The usual size roof truss has very little lateral strength and the result is that its handling and erection may present something of a problem. Many designers, to ensure reasonable lateral stiffness, will not permit the ratio of the width of the bottom chord to its total length to be less than a certain value ($\frac{1}{125}$ being common).

Small trusses are usually picked up at their peaks, causing the bottom chords to be put in compression. The result is that there is danger of lateral buckling of these members. A common practice among some steel erectors is to strap a timber (such as a 4×6) along the side of the truss to prevent it

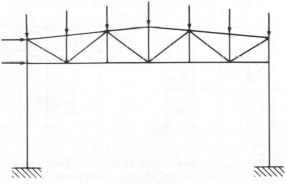

Figure 17-10

from bending laterally during erection. The steel erector probably sets the first truss and guys it off. After a few more trusses are erected the purlins can be installed and the structure begins to have appreciable lateral strength.

The bracing placed between roof trusses and mill-building bents serves the purpose of transferring the lateral loads to the building foundation. The amount of bracing used varies a great deal from engineer to engineer. In the paragraphs to follow a few general statements are made about the various types of lateral bracing.

Complete bracing systems are not often required in both the planes of the top and bottom chords; however, the presence of heavy moving loads and considerable vibration (as caused by moving cranes) may change the situation. Diagonal cross bracing should be used in the planes of the upper

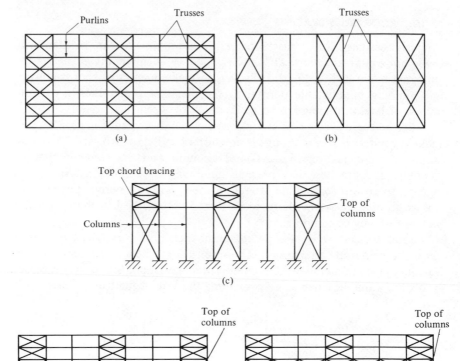

Figure 17-11 (a) Bracing in plane of top chord. (b) Bracing in plane of bottom chord. (c) Bracing in plane of columns. (d) Bracing in plane of columns. (e) Bracing in plane of columns.

chord members. These members tie the trusses together and with the assistance of the purlins (which act as struts) provide the necessary upper chord bracing. The usual practice is to place upper chord bracing only every three or four bays as shown in part (a) of Fig. 17-11.

Diagonal bracing in the plane of the bottom chords is often omitted in small structures unless heavy vibrating type loads are anticipated in the building. For larger structures bracing is usually put in the plane of the bottom chords every three or four bays. A typical arrangement of this type of bracing is shown in part (b) of Fig. 17-11.

Longitudinal bracing is also needed in the plane of the columns. The wind acting against the end of the building is transferred by this bracing to the column bases. Examples of this type of bracing are shown in parts (c) and (d) of Fig. 17-11. An alternate method of providing this longitudinal bracing is shown in part (e). This latter bracing has the advantage that it gives little interference to the placement of windows.

The members of the three bracing systems described in the preceding paragraphs rarely have forces of sufficient magnitude to affect their design and the usual practice is to select them on the basis of minimum sizes and controlling maximum slenderness ratios. To achieve tightly braced structures it is fairly common practice to detail the bracing members a little short so they will have to be slightly stretched to get them in place.

Problems

17-1. Using the load combinations suggested in Section 17-6 and the AISC Specification, determine the critical forces in the members of the left-hand side of the Pratt truss shown in the accompanying illustration. Use the following data.
1. Snow = 24 psf of horizontal projection.
2. Wind = 11.5 psf pressure on windward side only.
3. Purlins = 3 psf of roof surface.
4. Roofing = 18 psf of roof surface.
5. Truss weight to be estimated by the *Engineering News Record* expression given in Section 17-5, assuming A36 steel is to be used.

Trusses 18 ft – 0 in. o.c.

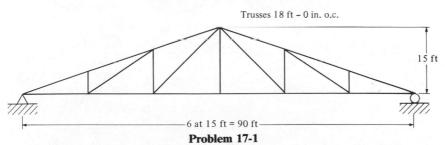

15 ft

6 at 15 ft = 90 ft

Problem 17-1

17-2. Design the members of the left-hand side of the truss of Prob. 17-1 assuming the connections are to be welded.

17-3. Determine the design forces for the members of one side of the Fink truss shown in the accompanying illustration, using the AISC Specification. Use the following data.

1. Snow = 20 psf of horizontal projection.
2. Wind = 9 psf suction on leeward surfaces and 4.1 psf suction on the windward side.
3. Purlins = 4 psf of roof surface.
4. Roofing = 20 psf of roof surface.
5. Truss weight to be estimated using Dr. Grinter's recommendations mentioned in Section 17-5.

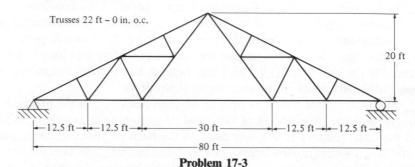

Problem 17-3

17-4. Design the members of the truss of Prob. 17-3 using A36 steel. The members are to be welded.

17-5. Determine the design forces in the members of the three-hinged arch shown in the accompanying illustration, using the AISC Specification and the following data.

1. Snow load = 20 psf of horizontal projection.
2. Wind = 15 psf pressure on sloping windward side, 7.5 psf suction on sloping leeward side, 20 psf pressure on vertical windward side, and 10 psf suction on vertical leeward side.
3. Purlins = 5 psf of roof surface.
4. Roofing = 15 psf of roof surface.
5. Truss weight to be estimated as in Prob. 17-1.

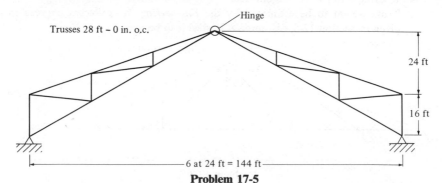

Problem 17-5

Chapter 18
Design
of Bridges

18-1. INTRODUCTION

The general procedure for analyzing and designing steel bridges is quite similar to that used for steel buildings. The major differences between the two are caused by the different loading conditions. The analysis of bridges is greatly affected by heavy moving loads and the impact which they cause. Impact is usually a minor problem in the analysis of buildings, although it has to be considered in a few cases where it may be caused by such things as elevators, moving cranes, and vehicles in off-ground garages.

Bridge specifications are much more conservative than the AISC Building Specification. Among the several reasons for this conservatism are

1. Live loads make up a larger percentage of the total load applied to bridges than to buildings and they are more violently applied.
2. Steel bridge frames are nearly always exposed to the weather with consequent corrosion problems, while the structural steel used in buildings is usually enclosed and protected from the elements.
3. Perhaps another factor that may explain the higher allowable stresses in building design is the fact that the more liberal architectural profession has had considerable influence on building specifications.

A discussion of the various types of bridge trusses in common use was presented in Chapter 14 and will not be repeated here. This chapter is devoted to the design of a simple-span truss bridge using the AASHTO Specifications.

Bridge specifications very clearly define the maximum loads to be applied to bridges for design purposes as well as spelling out the distribution of loads, values of impact, limiting depths, and various other specific design data. Example 18-1 illustrates the application of the AASHTO Specifications in selecting the proportions of a 144-ft simple-span, through Pratt truss. Admittedly a plate girder might win out economically over a

Fort Duquesne Bridge, Pittsburgh, Pa. (Courtesy of American Bridge Division, U.S. Steel Corporation.)

truss for this span, but the purpose of this example is to illustrate briefly the steps involved in laying out a bridge truss. Subsequent examples in this chapter will illustrate the design of the individual parts of this truss. The numbers in parentheses in this chapter refer to AASHTO article numbers and are included for the student's benefit.

Example 18-1

Select proportions for a two-lane, 144-ft through simple-span Pratt truss using the 1977 AASHTO Specifications. Also determine roadway width and stringer spacing.

SOLUTION
Layout of truss:

$$\text{min clearance} = 14 \text{ ft (Article 1.1.6)}$$

$$\text{min depth of sway bracing} = 5 \text{ ft (Article 1.7.44G)}$$

$$\text{estimated floor depth} = 5 \text{ ft}$$

$$\text{estimated min truss depth} = \overline{24 \text{ ft}}$$

$$\text{min preferable depth} = \tfrac{1}{10} \text{ span (Article 1.7.4)} = 14.4 \text{ ft} < 24 \text{ ft}$$

OK

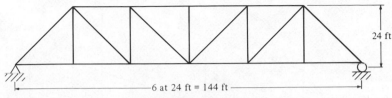

Figure 18-1

6 at 24 ft = 144 ft

24 ft

Assume diagonals at 45° as shown in Fig. 18-1. Truss spacing and stringer layout:

$$\text{width of roadway} = 26 \text{ ft}$$

$$\text{assume } 1\tfrac{1}{2}\text{-ft curbs each side} = 3 \text{ ft}$$

$$\underline{\text{assume truss widths} = 2 \text{ ft}}$$
$$\text{distance c.-to-c. trusses} = 31 \text{ ft}$$

Assume outside stringers 2 ft–6 in. from center of trusses, placing them 26 ft–0 in. on centers. Use stringers 6 ft–6 in. on centers as shown in Fig. 18-2.

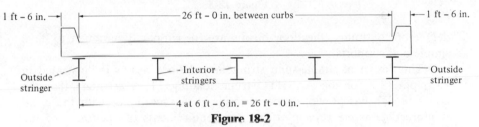

1 ft – 6 in.   26 ft – 0 in. between curbs   1 ft – 6 in.

Outside stringer   - Interior stringers   Outside stringer

4 at 6 ft – 6 in. = 26 ft – 0 in.

Figure 18-2

18-2.   BRIDGE FLOOR SYSTEM

To understand the manner of application of loads to a supporting girder or truss it is necessary for the arrangement of the members of the floor system to be carefully studied. The most common type of bridge floor is supported by a series of beams parallel to traffic and running the length of each panel. These beams, called *stringers*, frame into and are supported by transverse beams, called *floor beams*. The floor beams frame into the panel points of the supporting trusses. This arrangement is shown in Fig. 18-3.

The slab may be bonded to the stringers with shear connectors to form composite construction (as described in Chapter 15), permitting the use of smaller stringers.

Stringers are usually conservatively assumed to be simply supported but actually they have some continuity in their construction. Sometimes they are specially designed to be continuous with connecting plates lapping over the flanges of the stringers and floor beams. There is probably more continuity present in the floor beams than in the stringers. They are usually quite rigidly connected to the truss verticals, thus appreciably adding to the lateral rigidity of the bridge. Despite this probably high

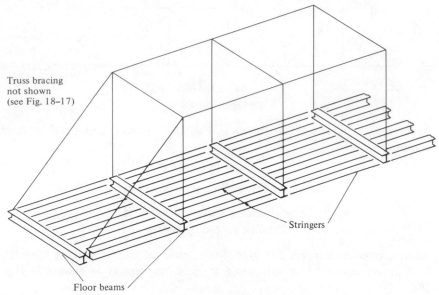

Truss bracing
not shown
(see Fig. 18–17)

Stringers

Floor beams

Figure 18-3

degree of continuity, the floor beams are for simplicity assumed to have simple end supports.

The design of simple-span stringers and floor beams is illustrated in Example 18-2 for the AASHTO truck loadings. It is assumed that the concrete slab has previously been designed in accordance with the usual reinforced concrete principles and the requirements of Section 5 of the AASHTO Specifications.

A few explanatory remarks are made here concerning the application of the wheel loads. The truck is assumed to be centered in the middle of a 10-ft lane; thus no wheel load can be closer to the curb than 2 ft (Article 1.2.5). Even though a wheel load is placed directly over an interior stringer it causes the adjacent stringers as well as the stringer in question to deflect due to the stiffness of the concrete slab. The obvious result is distribution of the load and the AASHTO Specifications (Article 1.3.1) attempt to estimate the effect of this distribution. They say the percentage of a wheel load to be taken is the center-to-center spacing of the stringers divided by a certain value which is determined from the type of floor, spacing of stringers, and number of traffic lanes. The outside stringers are assumed to support a portion of the wheel loads determined by assuming the flooring to act as a simple span between stringers.

It will be noted that the values for maximum live-load moments calculated in the example can be checked with tables in Appendix A of the AASHTO Specifications. These tables are for axle loads and do not include effects of load distribution and impact.

Example 18-2

Design the stringers and floor beams for the bridge of Example 18-1 using the HS20-44 loading of the AASHTO. Assume that an 8-in. concrete slab with a crown varying from 1 to 4 in. thick has previously been designed. The design is to be made with structural carbon steel (A36) and $\frac{7}{8}$-in. friction-type A325, high-strength bolts. Assume the concrete weighs 150 lb/ft^3.

SOLUTION

Design of Interior Stringers: Fraction of wheel load carried by each stringer (Article 1.3.1.B) is:

$$\frac{S}{5.5} = \frac{6.5}{5.5} = 1.18$$

slab weight $= \left(\frac{12}{12}\right)(150)(6.5) = 975$ lb/ft of stringer

(assuming 4-in. crown)

assume stringer weight $= 70$ lb/ft

total load on stringer $= 1045$ lb/ft

$$DL \text{ moment} = \frac{(1.045)(24)^2}{8} = 75.2 \text{ ft-k}$$

Place wheel loads as shown in Fig. 18-4 to cause maximum LL moment. LL moment $= (11.33)(8.5)(1.18) = 113.8$ ft-k (can be checked in Appendix A of AASHTO).

$$I = \frac{50}{l + 125} = \frac{50}{24 + 125} = 33.6\%$$

[use 30% maximum (Article 1.2.12.C)]

$$I \text{ moment} = (0.30)(113.8) = 34.1 \text{ ft-k}$$

total moment $= 223.1$ k ($F_b = 20$ ksi, Article 1.7.1)

$$S_{reqd.} = \frac{(12)(223.1)}{20} = 134 \text{ in.}^3$$

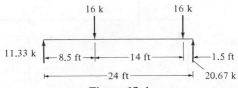

Figure 18-4

Use W21×68

Design of Outside Stringers:

$$\text{slab weight} = (6.5)(\tfrac{9}{12})(150)(\tfrac{1}{2}) = 366 \text{ lb/ft} \qquad \text{(assuming 1-in. crown)}$$
$$\text{assumed weight of curbs, railings, etc.} = 150 \text{ lb/ft}$$

(Article 1.3.1.B permits this load to be spread evenly to all stringers if placed after slab has cured.)

$$\text{assume stringer weight} = 50 \text{ lb/ft}$$
$$\text{total } DL = \overline{566} \text{ lb/ft}$$
$$DL \text{ moment} = \frac{(0.566)(24)^2}{8} = 40.8 \text{ ft-k}$$

Part of wheel load supported by exterior stringers is $4.5/6.5 = 0.692$ but not less than

$$\frac{6.5}{4.0 + \tfrac{1}{4} \times 6.5} = 1.16$$

(See Article 1.3.1.B and Fig. 18-5.)

Wheels arrangement for maximum moment same as for interior stringers.

$$LL \text{ moment} = (11.33)(8.5)(1.16) = 112.0 \text{ ft-k}$$
$$I \text{ moment} = (0.30)(112.0) = 33.6 \text{ ft-k}$$
$$\text{total moment} = 186.4 \text{ ft-k}$$
$$S_{\text{reqd.}} = \frac{(12)(186.4)}{20} = 112 \text{ in.}^3$$

Use W24×55

Bolts Required Between Stringers and Floor Beams: Interior stringers:

$$DL \text{ shear} = (12)(975 + 68) = 12.50 \text{ k}$$

Place wheel loads as shown in parts (a) and (b) of Fig. 18-6 for maximum

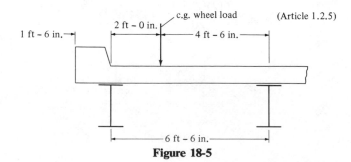

Figure 18-5

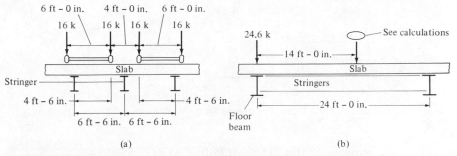

Figure 18-6 (a) Lateral placement of trucks. (b) Longitudinal placement of trucks.

LL shear. Load going to stringer from part (a) of figure $= (2)(16)(4.5/6.5)$
$= 22.2$ k. These loads placed as shown in part (b) of figure.

$$\text{max } LL \text{ shear} = 22.2 + \left(\frac{10}{24}\right)(16)(1.18) = 30.1 \text{ k}$$

(Notice no slab distribution is assumed to occur for beams placed right at floor beam but distribution is assumed for the other wheel loads.)

$$I \text{ shear} = (0.30)(30.1) = 9.0 \text{ k}$$

$$\text{max total shear} = 51.6$$

Bolts from stringer to floor beam:

$$\text{single shear value of bolts} = (0.6)(13.5) = 8.1 \text{ k}$$

$$\text{no. reqd.} = \frac{51.6}{8.1} = 6.37 \quad \text{(say 4 each side)}$$

Bolts from connection to web of stringer (web $t = 0.430$ in.):

$$\text{double shear value of bolts} = (2)(0.6)(13.5) = 16.2 \text{ k}$$

$$\text{bearing value} = \left(\frac{7}{8}\right)(0.430)(40) = 15.1 \text{ k}$$

$$\text{no. reqd.} = \frac{51.6}{15.1} = 3.42 \quad \text{(say 4)}$$

Outside stringers: Bolt calculations not shown since they would merely be a repetition of those made for interior stringers.

Design of Floor Beams:

dead-load reactions from interior stringers $= (24)(1.043) = 25.0$ k

dead-load reactions from outside stringers $= (24)(0.571) = 13.7$ k

These reactions are applied as concentrated loads to floor beams as shown in Fig. 18-7.

assume beam weight $= 230$ lb/ft

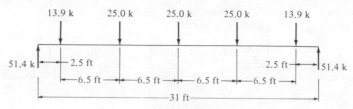

Figure 18-7

$$DL \text{ moment} = (51.4)(15.5) - (25.0)(6.5) - (13.9)(13) + \frac{(0.23)(31)^2}{8}$$

$$= 481.1 \text{ ft-k}$$

AASHTO does not permit transverse distribution of wheel loads in slab for design of floor beams (Article 1.3.1.C). Therefore, concentrated loads applied to floor beams from stringers are

$$16 + \left(\frac{10}{24}\right)(16) + \left(\frac{10}{24}\right)(4) = 24.34 \text{ k} \qquad \text{(see Fig. 18-8)}$$

Truck loads placed in traffic design lanes of 13-ft width and placed as shown in Fig. 18-9 to cause maximum *LL* moment.

$$LL \text{ moment} = (48.68)(15.5) - (24.34)(2+8) = 512 \text{ ft-k}$$

$$I \text{ moment} = (0.30)(512) = 153.6 \text{ ft-k}$$

$$\text{total moment} = 1146.7 \text{ ft-k}$$

$$S_{\text{reqd.}} = \frac{(12)(1146.7)}{20} = 688 \text{ in.}^3$$

Use W36×210

Connections can be designed as previously described.

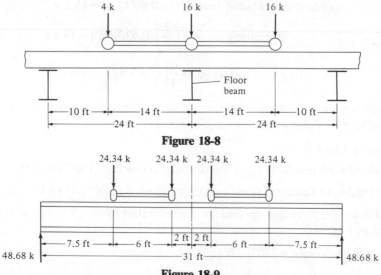

Figure 18-8

Figure 18-9

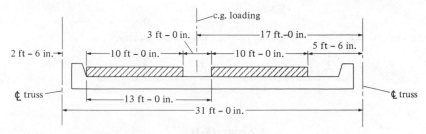

Figure 18-10

18-3. LIVE-LOAD TRUSS FORCES

The live-load forces are to be computed for the HS20-44 loading or the equivalent lane loadings, whichever causes the maximum values. It will be found that the equivalent lane loadings will control for moment when the loaded length exceeds 145 ft and for shear when the loaded length exceeds 128 ft. (These lengths are for the HS20-44 loading only.)* Forces can be computed with the aid of influence lines, by the panel-load method,[1] or by some other procedure.

In Example 18-3 influence lines are drawn for the members of the left-hand side of the truss under consideration in this chapter. From these influence lines it can be seen that the lane loadings control for all members except U_1L_1, U_1L_2, U_2L_2, and U_2L_3. The concentrated wheel loads are used for calculating the forces in these latter members while the lane loadings are used for all other members.

The AASHTO says the uniform lane loading is distributed across two 10-ft widths and is to be placed in design traffic lanes of 13-ft widths for a two lane roadway. As shown in Fig. 18-10, these two lane loadings are moved as close as possible to one of the trusses to cause maximum loads on that truss. From this figure the part of the live load going to the left truss can be seen to equal $\frac{17}{31}$ or 54.8%.

From the same AASHTO article (Article 1.2.6) the maximum percentage of the concentrated wheel loads can be determined as shown in Fig. 18-11. The maximum percentage in this case is the same as for the uniform lane loading case but can be slightly different for bridges of more than two lanes. The percentage is again $\frac{17}{31}$ or 54.8%.

The impact percentage in the various truss members will be determined from the usual AASHTO expression. This percentage is not constant for all members as it varies with the length over which the live load is placed to cause maximum force.

---

*For the example problem presented in this section the concentrated HS20-44 loads should have been used for all of the members but the author, to illustrate the application of the equivalent lane loadings, used them for the 144-ft lengths. The results are almost identical.
[1] J. C. McCormac, *Structural Analysis*, 3d ed. (New York: Harper & Row, 1975), pp. 210–212.

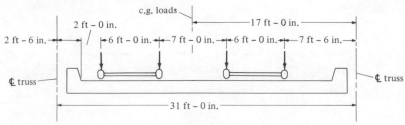

Figure 18-11

Example 18-3

Determine the maximum live-load forces in each of the members of the truss selected in Example 18-1.

SOLUTION
Drawing influence lines (Fig. 18-12).

Force Calculations for Uniform Lane Loadings:

uniform lane loading to one truss $= (2)(0.640)(0.548) = 0.703$ k/ft

concentrated load to one truss $= (2)(18)(0.548) = 19.75$ k

impact percentage for 144-ft loaded length $= \dfrac{50}{144 + 125} = 18.6\%$

$L_0 L_1$ and $L_1 L_2$:

LL force $= (0.703)(+59.8) + (19.75)(+.83) = +58.4$ k

I force $= (0.186)(+58.4) = +10.8$ k

$L_2 L_3$:

LL force $= (0.703)(+96) + (19.75)(+1.33) = +93.8$ k

I force $= (0.186)(+93.8) = +17.4$ k

$L_0 U_1$:

LL force $= (0.703)(-85) + (19.75)(-1.18) = -83.1$ k

I force $= (0.186)(-83.1) = -15.4$ k

$U_1 U_2$:

LL force $= (0.703)(-96) + (19.75)(-1.33) = -93.8$ k

I force $= (0.186)(-161.3) = -17.4$ k

$U_2 U_3$:

LL force $= (0.703)(-108) + (19.75)(-1.50) = -105.4$ k

I force $= (0.186)(-105.4) = -19.6$ k

Force Calculations for Axle Loads: Concentrated wheel loads:

$(2)(32)(0.548) = 35.1$ k

$(2)(8)(0.548) = 8.8$ k

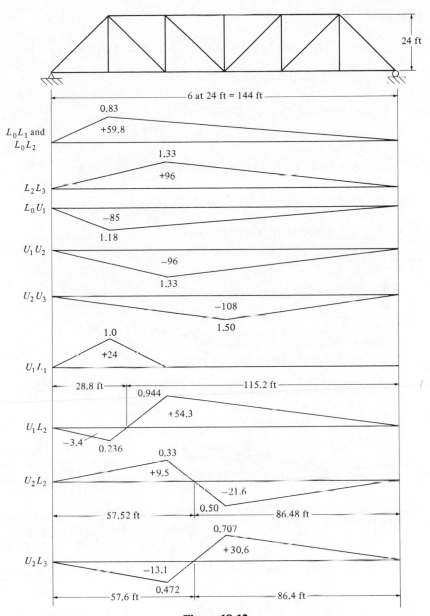

Figure 18-12

$U_1 L_1$:

$$LL \text{ force} = (35.1)\left[+1.0 + \left(\frac{10}{24}\right)(+1.0)\right] + (8.8)\left(\frac{10}{24}\right)(+1.0) = +53.3 \text{ k}$$

$$I = \frac{50}{48 + 125} = 28.9\%$$

$$I \text{ force} = (0.289)(+53.3) = +15.4 \text{ k}$$

U_1L_2:

$$\text{positive } LL \text{ force} = (35.1)\left[0.944 + \left(\frac{82}{96}\right)(0.944)\right] + (8.8)\left(\frac{4}{19.2}\right)$$

$$(0.944) = +63.1 \text{ k}$$

$$I = \frac{50}{115.2 + 125} = 20.8\%$$

$$I \text{ force} = (0.208)(+63.1) = +13.1 \text{ k}$$

$$\text{negative } LL \text{ force} = (35.1)\left[-0.236 + \left(\frac{10}{24}\right)(-0.236)\right] = -11.7 \text{ k}$$

$$I = \frac{50}{28.8 + 125} = 32.5\% \text{ (maximum value is 30\%)}$$

$$I \text{ force} = (0.30)(-11.7) = -3.5 \text{ k}$$

U_2L_2:

$$\text{positive } LL \text{ stress} = (35.1)\left[+0.33 + \left(\frac{34}{48}\right)(0.33)\right] = +19.8 \text{ k}$$

$$I = \frac{50}{57.52 + 125} = 27.4\%$$

$$I \text{ stress} = (0.274)(+19.8) = +5.4 \text{ k}$$

$$\text{negative } LL \text{ force} = (35.1)\left[-0.50 + \left(\frac{58}{72}\right)(-0.50)\right] + (8.8)\left(\frac{0.48}{14.48}\right)$$

$$(-0.50) = -31.8 \text{ k}$$

$$I = \frac{50}{86.48 + 125} = 23.6\%$$

$$I \text{ force} = (0.236)(-31.8) = -7.5 \text{ k}$$

U_2L_3:

$$\text{positive } LL \text{ force} = (35.1)\left[+0.707 + \left(\frac{58}{72}\right)(+0.707)\right] + (8.8)\left(\frac{0.4}{14.4}\right)$$

$$(+0.707) = +44.8 \text{ k}$$

$$I = \frac{50}{86.4 + 125} = 23.6\%$$

$$I \text{ force} = (0.236)(+44.8) = +10.6 \text{ k}$$

$$\text{negative } LL \text{ force} = 35.1\left[-0.472 + \left(\frac{34}{48}\right)(-0.472)\right] = -28.3 \text{ k}$$

$$I = \frac{50}{57.6 + 125} = 27.4\%$$

$$I \text{ force} = (0.274)(-28.3) = -7.7 \text{ k}$$

18-4. TRUSS DEAD LOADS

The dead load of a truss bridge consists of the weight of the floor system, truss, and bracing. The weight of the floor system, which comprises a large percentage of the total weight, can be closely estimated by making a preliminary design of the floor. From the weight of the floor system the load applied to the truss can be determined. Before dead-load analysis can be made of the main trusses the weights of the trusses must be estimated.

To expedite the design procedure it is essential to make a good estimate of the truss weight. The designer can make a rough estimate, analyze and design the structure, calculate its weight, and revise her estimate, until her estimate and the design weight are reasonably close. This procedure can be rather time-consuming, particularly for long-span bridges, and a good truss weight estimate can definitely speed up the process.

The weight of the truss can be estimated by increasing the other dead loads by some percentage or by using some approximate formula. There is a great deal of published data available concerning the estimated weights of short- and medium-span bridges. There are, however, so many variables in the weight of a bridge such as different specifications, different floor types (as concrete or steel grid floors), different steels, etc., that it is necessary to check the weight of the designed structure to be sure the estimate was reasonable in each case.

Perhaps the most complete data on the weight of bridge structures was presented by J. A. L. Waddell in his book *Bridge Engineering* (Wiley, 1916) and revised in the *Transactions of the ASCE* in 1936.

Charles W. Hudson is credited with developing a formula (called the Hudson formula) for estimating the weight of a truss. In his formula, given at the end of this paragraph, W is the total weight of either a railway or bridge truss including its bracing, S is the maximum total tensile force in the most stressed chord member, l is the length of the truss in feet, and F_t is the allowable tensile stress. Since S is the maximum total tensile force due to live load plus impact plus dead load, a value has to be assumed for the truss weight to apply the formula. If the largest tensile chord member is assumed to extend for the full length of the truss and its weight assumed to equal 20% of the weight of the entire truss and bracing, this formula can be derived.[2,3]

$$W = \frac{17Sl}{F_t}$$

Example 18-4 illustrates the application of the Hudson formula, the calculation of the dead-load forces, and a summary of the maximum total forces.

[2] John E. Lothers, *Design in Structural Steel*, 3d ed. (Englewood Cliffs, N.J.: Prentice-Hall, 1972), p. 420.
[3] Thomas C. Shedd, *Structural Design in Steel* (New York: Wiley, 1934).

Example 18-4

Estimate the weight of the truss considered in the previous examples of this chapter, compute the dead-load forces, and summarize the total forces for all members.

SOLUTION

Estimating truss weight by Hudson formula: Member $L_2 L_3$ is the largest stressed tensile chord member. *DL* floor beam reaction from Example 18-2 = 51.4 k. Adding 20% to this value to estimate truss weight = 61.7 k.

Force in $L_2 L_3$ by taking moments at U_2 when a 61.7-k load is placed at each joint:

$$L_2 L_3 = \frac{(48)(154.3)-(24)(61.7)}{24} = 246.9$$

estimated total force in $L_2 L_3 = +246.9 + 93.8 + 17.4 = +358.1\ k$

$$W = \frac{(17)(358.1)(144)}{20} = 43{,}831\ \text{lb or } 7.3\ \text{k at each joint}$$

Dead-load forces (Fig. 18-13):

$$\text{dead-panel loads} = 51.4 + 7.3 = 58.7\ k$$

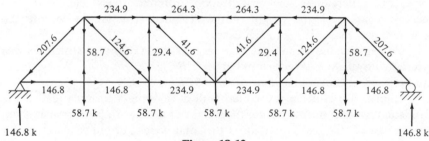

Figure 18-13

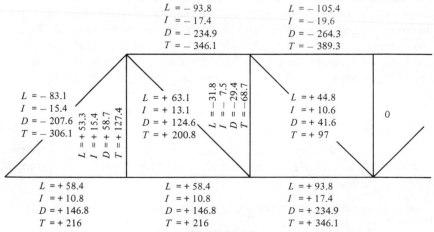

Figure 18-14

Part of the estimated truss weight should theoretically be placed at the top chord joints, but this refinement is considered unnecessary here.

Summary of maximum forces (Fig. 18-14):

Notice no force reversals occurred and the forces obtained in considering that possibility are not shown.

18-5. SELECTION OF TRUSS MEMBERS

For a small truss bridge of the type being considered, it may be more economical to use W sections for nearly all of the members. For larger trusses, built-up sections will be used to provide the necessary stiffnesses for the compression members. For the truss being designed in this chapter W sections are used for all of the members.

When built-up sections are used, the upper chord of the truss will probably consist of one of the two arrangements shown in Fig. 18-15. In part (a) of the figure is shown an upper chord member built up from two channels and a cover plate. The gusset plates are attached to the back of the channels and the verticals fit between the gussets. A similar arrangement is shown in part (b) of the figure where the chord member is built up from four angles and three plates. For this type of bolted truss it is common practice to design the verticals first because the spacing between the aforementioned gusset plates is not known until the vertical member sizes are known.

After the verticals are designed in a truss with built-up members the upper chord members are designed, starting with the one with the greatest force. The width of the cover plate for the compression chord is quite important because it appreciably affects the lateral stiffness of the compression flange and thus its allowable stress. In Article 1.7.44N of the AASHTO, maximum width/thickness ratios for cover plates are provided, the values varying with F_y, f_a, etc. After the upper chord members are designed it is probably desirable to design the remaining web members (probably the compression members followed by the tension ones). The

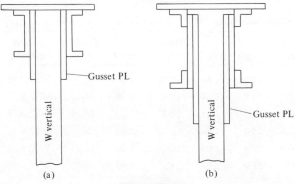

(a)   (b)

Figure 18-15

remaining tension members can be designed after which consideration is given to bracing, connections, supports, etc.

In Example 18-5 the members of the truss of Example 18-1 are designed using the maximum forces previously obtained in Example 18-3, with W sections used for all of the members. The depth of these sections vary from member to member and it may be necessary in places to use filler plates to equalize depths. In designing the tension chord members two holes for $\frac{7}{8}$-in. bolts are assumed to be present in each flange and web. The web deduction was considered unnecessary for the tension diagonals and only flange holes were deducted.

Article 1.7.5 of the AASHTO Specifications needs to be carefully studied before the truss members are selected. In this article the maximum permissible slenderness ratios are presented. For main tension members 200 is the maximum value and 240 is the maximum for bracing members. In computing kl/r values k is assumed to equal 1.0 for tension members. For compression members the comparable values permitted are 120 and 140, respectively. For main members subject to reversal of stress, the maximum permissible slenderness ratio is 140. Sometimes r for a member is larger if part of the member is neglected in the calculations. In this regard the AASHTO says any part of a member can be neglected in the calculations provided the strength of the remaining part (using the kl/r value with the part neglected) and the strength of the whole section (using the true kl/r value) are equal to or greater than the design stress.

Example 18-5

Select W sections for each of the members of the truss of Example 18-1 using the maximum forces obtained in Example 18-4. The members are to be designed with structural carbon A36 steel using the 1977 AASHTO Specifications. Allowable tension = 20,000 psi on the net section and allowable axial compression = $16,000 - 0.30 \, (l/r)^2$. Assume there are two holes for $\frac{7}{8}$-bolts in each flange and web and that $k = 1.0$ for each member.

SOLUTION

Design of Tension Members

Member	Design force	Length	Min. r	Net A reqd.	Est. hole A	Gross A reqd.	Section selected	Actual kl/r	Actual net A
L_0L_1	+216	24.0	1.44	10.80	3.30	14.10	W12×50	147	11.40
L_1L_2	+216	24.0	1.44	10.80	3.30	14.10	W12×50	147	11.40
L_2L_3	+346.1	24.0	1.44	17.31	3.54	20.85	W12×72	95	17.66
U_1L_1	+127.4	24.0	1.44	6.37	1.86	8.23	W12×30	189	6.51
U_1L_2	+200.8	33.9	2.03	10.04	2.30	12.34	W12×53	164	12.61
U_2L_3	+97	33.9	2.03	4.85	1.60	6.45	W12×53	164	12.61

Example design of $L_0 U_1$: Assume allowable $F_a = 10.75$ ksi.

$$A_{reqd.} = \frac{306.1}{10.75} = 28.47 \text{ in.}^2$$

Try W12×120 ($A = 35.3, r = 3.13$)

$$\text{allowable } F_a = 16,000 - (0.30)\left(\frac{12 \times 33.9}{3.13}\right)^2 = 10.93 \text{ ksi}$$

Design of Compression Members

Member	Design force	Length	Min. r	Section selected	Allowable load	Actual kl/r
$L_0 U_1$	−306.1	33.9	3.39	W12×120	−329	130[a]
$U_1 U_2$	−346.1	24.0	2.40	W12×106	−419	92.6
$U_2 U_3$	−389.3	24.0	2.40	W12×106	−419	92.6
$U_2 L_2$	−68.7	24.0	2.40	W12×40	−72.2	149[a]

[a] for $L_0 U_1$ and $U_2 L_2$ are greater than 120 but permissible under certain circumstances as described in AASHTO Article 1.7.5. See example design to follow.

$$\text{allowable } P = (35.3)(10.93) = 386 \text{ k} > 312.5 \text{ k}$$

$$\text{but } \frac{l}{r} = \frac{(12)(33.9)}{3.13} = 130 > 120$$

If l/r of 120 used, allowable $F_a = 16,000 - (0.30)(120)^2 = 11.68$ ksi.

$$A_{reqd.} = \frac{306.1}{11.68} = 26.21 \text{ in.}^2$$

$$\text{effective area} = (35.3)\left(\frac{3.13}{3.39}\right)^2 = 30.1 \text{ in.}^2 > 26.21 \text{ in.}^2 \qquad \text{OK}$$

$$\text{allowable } P = (30.1)(10.83) = 326 \text{ k} > 312.5 \text{ k} \qquad \text{OK}$$

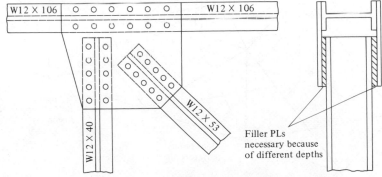

Figure 18-16 Joint U_2.

Design of $U_3 L_3$: Computed force in this member is zero (although it truthfully has a total DL force of 2 or 3 k due to the weight of the overhead members) and the smallest W12, a W12×14, is used.

Figure 18-16 shows a sketch of joint U_2 of the truss being designed in this example.

18-6.   LATERAL BRACING

A bridge truss must be braced laterally and longitudinally if it is to withstand lateral wind loads, vibrations from live loads, centrifugal forces on curves, nosing of locomotives, and the longitudinal forces produced by traffic loads when brakes are applied. On the subject of braking forces imagine the effect on a bridge when the driver of a trailer truck going 60 mph suddenly applies the brakes while crossing the bridge. As shown in Fig. 18-17, there are actually four distinct types of bracing systems in a through bridge. These include the following.

1. A lateral bracing system in the plane of the top chord.
2. A lateral bracing system in the plane of the bottom chord.
3. A portal frame at each end of the bridge.
4. Intermediate sway frames.

Top and Bottom Lateral Trusses

The wind forces are resisted by the top and bottom lateral bracing systems of a through bridge. These bracing systems usually consist of cross bracing desirably connected at their intersection points to reduce unsupported

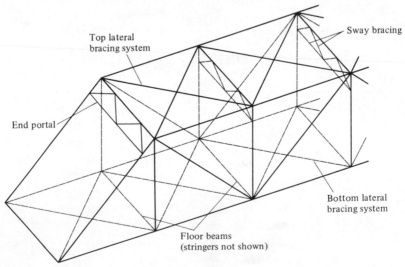

Figure 18-17

lengths. The truss bracing is actually statically indeterminate but is probably analyzed by some approximate procedure. One common method is to assume the shear divides equally between the diagonals and the members designed to take their respective forces (the compressive forces controlling sizes). Another procedure is to assume the diagonals, which would ordinarily be in tension for a shear of that sign, resist the entire shear, after which the member is designed and checked for the case in which the shear is assumed to divide equally.

Members of the bottom lateral truss of a through bridge probably consist of single angles or double angles connected to the bottom flanges of the chords by means of gusset plates. The AASHTO says the smallest angle that can be used for bracing is the $3 \times 2\frac{1}{2}$-in. angle and that no less than two bolts can be used for their end connections. The top lateral bracing is preferably (says the AASHTO) connected to both flanges of the chord. Four angles arranged in an I shape with a depth equal to that of the chord and connected to the top and bottom flanges of the chord is an example. In addition to serving as wind bracing the top lateral system also serves to reduce the unsupported lengths of the top chord compression members.

The chord members of the lateral trusses are also the chords of the main trusses. It is therefore necessary to consider the combination of wind forces in the chords as part of the lateral trusses with the forces caused by dead load plus live load plus impact as part of the main truss.

When wind loads are combined with other loads the AASHTO permits some increase in allowable stresses. For the dead load plus full wind load combination the allowable stresses can be increased by 25%. When wind loads are considered to be applied at the same time as dead load plus all the other live loads only 30% of the wind load is considered applied and the allowable stresses can be increased by 25%; however, a wind load applied to the live load 6 ft above the deck and equaling 100 lb per linear foot must be included in this combination.

Sway Bracing and End Portals

The top lateral truss of a bridge (previously discussed) serves the purpose of resisting the wind forces along the top chords of the trusses. In addition to this bracing, however, intermediate sway frames and end portals are considered necessary to provide lateral stiffness and increase the torsional strength of the bridge. These systems were shown in Fig. 18-17.

The intermediate sway frames are difficult to analyze and are usually not designed for calculated stresses but are proportioned on the basis of kl/r limitations. A different situation exists at the ends where the portals are located. The end posts of a through bridge are tied together with bracing and the result is called a *portal*. The purpose of this frame is to transfer the end reaction of the top lateral truss to the foundation. They

Fabrication of curved girders for I-90 highway bridge. (Courtesy of Inryco, Inc.)

are assumed to serve as the end supports for the top lateral trusses and as such must be able to resist half of the total top chord wind load. The deeper the bracing in the portals the smaller will be the bending in the end posts. Portals are actually statically indeterminate and are probably analyzed by some approximate method similar to the one used for mill buildings.[4]

18-7. BRIDGE-TRUSS DEFLECTIONS

The deflections of bridge trusses are a very important subject to consider. Where underclearance is close it is essential to compute the deflections carefully. During cantilever erection, deflection calculations are even more critical because the various members must come together at exactly the right points for fitting. Even if underclearance is not a problem and the cantilever erection process is not being used, the appearance of a bridge truss is an important matter. Trusses with horizontal bottom chords will sag down in the middle, injuring their appearance and perhaps worrying the users of the bridge. These trusses should desirably be cambered to avoid such sagging.

Several methods are available for determining truss deflections. Perhaps the most practical one is the Williot-Mohr diagram with which the

[4] J. C. McCormac, op. cit., pp. 277–281.

deflection of all truss joints can be obtained at the same time. Trusses can be cambered by one of the following methods.

1. The tension members can be made shorter and the compression members longer by the amounts they would theoretically lengthen or shorten, respectively. After the loads are applied the truss would presumably return to its original theoretical dimensions. This method is generally followed for trusses roughly 300 ft or longer, while the second method described is probably used for the shorter trusses.

2. The second method used is to merely shorten the fabricated dimensions of the bottom chords. A common practice is to shorten these members by $\frac{1}{8}$ in. for each 10 ft of length.

The next question is "For what loads is camber to be made?" If the camber is made for all loads, the truss will probably be bowed up a little most of the time (perhaps not a bad situation). The usual practice is to camber bridge trusses for dead load plus some part of the live load. The AASHTO merely says that camber shall be at least equal to the deflections produced by dead loads. A common practice for highway bridges, however, is to camber them for dead load plus live load but no impact. The AREA says that railroad bridges should be cambered for the deflections caused by dead load plus a uniform live load of 3 k/ft.

18-8. END BEARINGS FOR BRIDGES

This section is devoted to a discussion of the various types of bridge bearings which may be used by the bridge engineer. As for so many other subjects in this book the student can learn a great deal by taking time to stop at a few of the bridges in his locality and examine the bearings used. In general, bearings are classified as being of the expansion or fixed types. As the names imply, the expansion bearings are those that supposedly allow the bridge to freely expand or contract, and the fixed bearings are those that are fixed against longitudinal movement. The term fixed bearing is rather misleading because it does not necessarily mean the bearing is fixed against rotation as commonly intended in structural analysis. It instead means the position of bearing is fixed. These two types of end bearings are discussed briefly in the following paragraphs.

Expansion Bearings

Expansion bearings: (1) allow the bridge ends to freely move back and forth with temperature expansion and contraction, (2) allow the bridge to move freely at its ends with changes in the length of the bridge caused by the live loads, and (3) keep horizontal loads from being applied to some of the bridge supports where such forces may be undesirable.

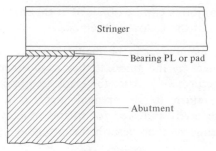

Figure 18-18

Expansion bearings may be of the sliding type or of the roller or rocker types, depending on the spans and loads. For short spans a simple bearing plate can be used such as the one shown in Fig. 18-18 for a steel stringer. In this arrangement the beam end is allowed to slide on a smooth metal plate. The expansion plates may be made from bronze with the sliding surfaces planed and polished, or from some copper alloy with smooth, true surfaces. Instead of following these requirements it is possible to use some type of bearing pad probably made from neoprene or rubber. These pads, which are generally referred to as elastomeric pads, are cheaper initially than the metal plates and require less maintenance.[5]

These kinds of bearing plates are not satisfactory for long spans where a large part of the reaction is caused by the live load. Downward deflections of the beam or truss cause the inside pressure for the plate to become excessive, with the result that the masonry support may be injured. It is, therefore, necessary to use some type of bearing that takes into account the deflections of longer span structures. The plate types of bearings also have considerable friction against temperature and stress-length changes. These items are more important for longer spans and the AASHTO says plates can be used for spans of less than 50 ft. For spans greater than these it is necessary to use some type of support involving sliding plates, rockers, or segmental rollers.

A sliding plate type of expansion bearing that can be used for spans of up to 100 ft is shown in Fig. 18-19. In the first plate type of expansion support the plate did not move as is the case for this sliding plate type of bearing.

For longer spans and heavier loads a more satisfactory type of bearing which can be used is the rocker type of bearing such as the one shown in Fig. 18-20. It can be seen in this figure that expansion and contraction is permitted by rolling on the curved surfaces of the rocker.

Theoretically speaking, there is only a line of contact between a rocker and its bearing plate but application of load spreads out this contact. From

[5]H. E. Fairbanks, "Elastomeric Pads as Bearings for Steel Beams," *Proc. ASCE* **87**, no. ST8 (December 1961).

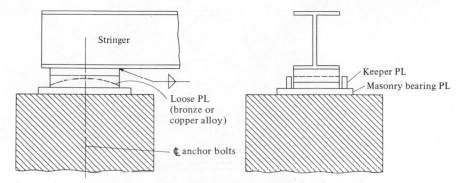

Figure 18-19 Sliding expansion plates.

various tests empirical formulas have been developed for rocker designs. These can be found in the specifications being used.

From this information it is obvious that for very large reactions the radius of the rocker required to provide the necessary bearing area and the length of the rocker will become impractical. To overcome this problem it may be necessary to use a series of segmental rollers of the types shown in Fig. 18-21. The segmental rollers are preferred to the round ones because for the same diameter they require less space. A very important feature of this type of bearing is its ability to be cleaned to keep the bearing from binding. For this reason dust guards are often used and the rollers may have hourglass shapes to make the cleaning operation simpler.

Fixed Bearings

The AASHTO permits the use of regular bearing plates connected to the abutment with anchor bolts for spans less than 50 ft. For spans greater than 50 ft it is considered necessary to use a bearing which takes into

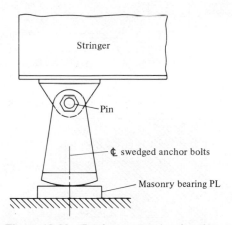

Figure 18-20 Rocker expansion bearing.

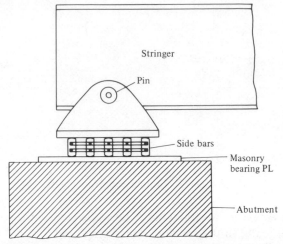

Figure 18-21 Expansion bearing with segmental rollers.

account the bridge deflection. Among the types of bearings that may be used are hinges, curved bearing plates, or some type of pin arrangement.

A hinge type of connection should allow end rotation of the members such as is provided by the pin of Fig. 18-22. When heavy loads are involved it may be necessary to provide the hinge with some type of lubrication system which will permit it to rotate freely and not wear quickly.

Actually, fixed bearings need to be designed for vertical and longitudinal forces, but practically the vertical forces are so much larger than the

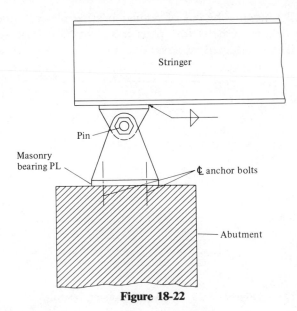

Figure 18-22

longitudinal ones that if they are designed for the vertical forces they surely will be strong enough to take the others. In some cases the specifications being used require the bearings to be designed for uplift, whether there is any computed uplift or not.

Problems

18-1. A three-lane, noncomposite, W section bridge has the cross section shown in the accompanying illustration. Assuming a simple span of 54 ft, design an interior beam using the 1977 AASHTO Specifications, A36 steel, and the HS20-44 loading.

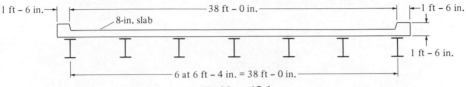

1 ft – 6 in. | 38 ft – 0 in. | 1 ft – 6 in.

8-in. slab

1 ft – 6 in.

6 at 6 ft – 4 in. = 38 ft – 0 in.

Problem 18-1

18-2. The through highway bridge shown in the accompanying illustration supports a three-lane roadway with the same cross section shown for Prob. 18-1 except that floor beams are also used. Design the stringers and floor beams for this bridge assuming the curb and railing outside the outside stringer weigh 180 lb/ft. Use the 1977 AASHTO Specifications, A36 steel, and the HS20-44 loading.

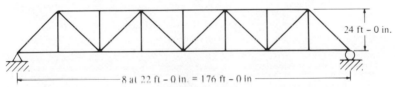

24 ft – 0 in.

8 at 22 ft – 0 in. = 176 ft – 0 in.

Problem 18-2

18-3. Compute the design forces in each of the members of the truss of Prob. 18-2.
18-4. Design the truss of Prob. 18-2 assuming the members are to be welded using the design forces obtained in Prob. 18-3.
18-5. Repeat Prob. 18-4 assuming the members are to be connected with 1-in. A325 high-strength friction bolts.
18-6. Establish the layout of a deck truss bridge of the type shown in the accompanying illustration using the 1977 AASHTO Specifications. Also,

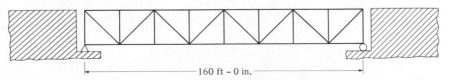

160 ft – 0 in.

Problem 18-6

establish the layout of the stringers. The bridge is to have four lanes of traffic and a 3 ft–0 in. sidewalk plus a handrail on each side. Assume the pavement is 8 in. thick and the curbing and sidewalk are 1 ft–6 in. thick.

18-7. Design the stringers and floor beams for the layout selected in Prob. 18-6 using A36 steel and the HS20-44 loading.

18-8. Design the members of the truss selected in Prob. 18-6 using A36 steel and the HS20-44 loading.

18-9. A light highway bridge is to be designed for a private industry to support the H10-44 loading. The roadway is to be 20 ft–0 in. between curbs and is to have 1 ft–6 in. wide curbs on each side of the same total thickness. Assuming a 6-in. thick floor slab, establish layout for stringers and floor beams and design them. Also design a pony truss (Warren) for a 100 ft–0 in. span using A36 steel and the 1977 AASHTO Specifications.

Chapter 19
Design
of Rigid Frames

19-1. INTRODUCTION

The rigid frame is a structure that has moment-resisting joints. The members are rigidly connected to each other at the joints to prevent relative rotation when the loads are applied. The advantages of these frames are economy, appearance, and saving in headroom. They can accomplish the same jobs which steel columns and trusses can accomplish and can do this without taking up nearly so much space. Rigid frames have proved to be very satisfactory for churches, auditoriums, field houses, armories, and other structures requiring large areas with no obstructions. Several of the more elementary types of rigid frames are shown in Fig. 19-1.

A rigid frame has on occasion been defined as an arch wrapped around a clearance diagram. Figure 19-2(a) shows a parabolic arch. The area to be enclosed is assumed to be rectangular as represented by the dotted lines in the figure. In part (b) the arch is bent into a shape which just includes the clearance diagram. A similar situation is shown in parts (c) and (d) of the figure.

Rigid frames can be economical for spans of from 25 or 30 ft to 200 or more feet. They can be either single span or multispan and single story or multistory. This chapter is confined to single span single-story rigid frames. The frames are generally spaced from about 15 to 35 ft on centers, depending on loads, type of building, and spans.

19-2. SUPPORTS FOR RIGID FRAMES

Hinged Supports

The supports at the bases of rigid-frame columns can theoretically be either hinged or fixed. Practically speaking, the hinge is almost always used. This is the type of support represented by anchor bolts passing

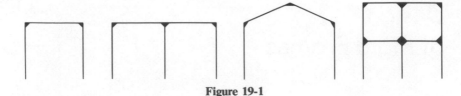

Figure 19-1

through a steel base plate into a concrete footing. To make the support act as close to a hinge as possible it is considered desirable to place the anchor bolts along a line corresponding to the neutral axis of the columns. Located on this line, which is perpendicular to the plane of the rigid frame, they will keep rotation resistance to a minimum. Should the bolts be placed near the corners of the bearing plate they would greatly increase the rotation resistance of the column base.

Although this type of hinge can supply a large vertical reaction component it can supply only a small horizontal one. When rigid frames of normal proportions have spans of approximately 60 to 80 ft or larger or when frames have small column height-to-span ratios, the horizontal reactions to be resisted become so large as to rule out the individual concrete spread footings. Either the footings (or the column bases) have to be tied together or the footings have to be supported by rock or some other rigid type of support. One solution which may be feasible for the smaller spans is to make use of the reinforcing bars in the floor slab by connecting them to the column bases.

A more common method is to use tie rods between the column bases to provide the necessary horizontal reaction components. These rods, which can be as large as 2 or 3 in. in diameter and which are generally connected to the column base plates, are probably equipped with turn-

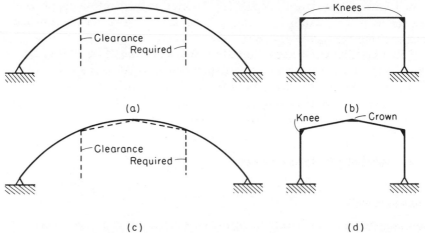

Figure 19-2 (a) Two-hinged arch. (b) Two-hinged rigid-frame arch. (c) Two-hinged arch. (d) Two-hinged rigid-frame arch.

Coliseum of Oregon State University, Corvallis, Ore. (Courtesy of Bethlehem Steel Company.)

Penn Fruit Company Building, Bala-Cynwyd, Pa. (Courtesy of The Lincoln Electric Company.)

buckles for adjusting them to the desired lengths. After the turnbuckles are adjusted to their proper positions the rods may be encased in concrete.

Tie rods are often initially tightened or prestressed to take (or counter-act) the dead-load reactions. Sometimes they are tightened by eye until the supports begin to give while on other occasions they may be tightened until attached gauges show the stress has reached a certain value. In addition to taking the horizontal reaction components, the tie rods help resist overturning of the foundations due to lateral loads.

Fixed Supports

If a small concrete spread footing is used alone, it will probably rotate appreciably (thus approximating hinge behavior) no matter how rigidly the frame is connected to the footing. Unless a rigid frame is anchored in rock or in an extremely large and rigid concrete footing there is little chance of having a fixed support. For almost any other kind of support, including piling, there is probably settlement and rotation. If support settlement is feasible the support is not fixed and the designer will be wise to assume his supports are hinged.

For the rare case where a frame does have fixed supports there will be smaller deflections and moments. The design of rigid frames with fixed supports, however, is more difficult because there are more redundants, and the results may be more questionable because of the difficulty of really achieving fixed supports.

19-3. RIGID-FRAME KNEES

Probably the proper design of the knees of a rigid frame is the most critical part of the design because maximum moments occur at the knees. As a matter of fact they must be capable of carrying shear, thrust, and moment. The result is that they must be stronger than the columns and girders and they may be deepened or stiffened to provide the necessary strength.

Figure 19-3 shows several methods of making rigid joints or knees to give them greater moment resistance. Connections can be made by rivet-ing, welding, or bolting but probably the cleanest-looking and most eco-nomical structures are obtained with welding. In designing rigid frames it is usually desirable to make as much use of standard rolled shapes as possible and to keep the large percentage of the welding (or other connec-tions) in the region of the knees.

A deepened knee made with straight flanges is shown in part (a) of Fig. 19-3, while a knee with curved flanges is shown in part (b). Usually the straight-tapered knees are more economical than the other types because they are easier to fabricate and they may also be more rigid. Very careful and expensive cutting is required for the knees with curved flanges.

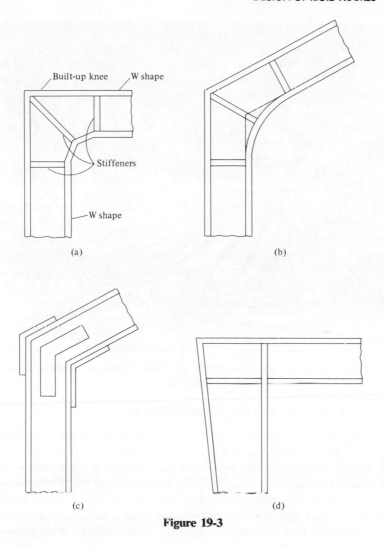

Built-up knee   W shape

Stiffeners

W shape

(a)   (b)

(c)   (d)

Figure 19-3

The curved knee, however, is probably more attractive and for long spans can be quite economical.

In part (c) of Fig. 19-3 is shown a knee that involves no change in the W sections. The necessary strengthening is obtained by adding plates to the flanges and web. This method of making up the knees is rarely economical and then only for short spans. Greater depth is usually required for economical knee design. A type of knee often used for steel rigid frame bridges is shown in part (d) of the figure. It has the disadvantage that continuous welds are needed for the full height of the column between the web and flanges.

Wollen Gymnasium at the University of North Carolina, Chapel Hill, N.C. (Courtesy of The Lincoln Electric Company.)

19-4.   APPROXIMATE ANALYSIS OF RIGID FRAMES

A two-hinged rigid frame or a rigid frame with fixed supports is statically indeterminate with the result that (as for any other statically indeterminate structure) the analysis is affected by the relative sizes of the members. Trial sizes or at least relative sizes need to be assumed for each of the members and the resulting structure analyzed to see if the assumed proportions are satisfactory. If the first size assumptions are poor, another set of sizes must be assumed and then checked, and so on. This trial-and-error process is referred to as design by successive approximations.

Unless fairly good initial member size assumptions are made, the problem can prove to be quite lengthy. There is, however, a good deal of published information on the analysis of rigid frames which should enable the designer initially to estimate very closely the moments in the frame he is designing. From these moments he can make some very good member-size assumptions which will greatly shorten the problem.

For this discussion the student is referred to the two-hinged rigid frame of Fig. 19-4. The vertical reactions for this frame can obviously be obtained by statics. Should the value of the horizontal reactions be available, the moment at any point in the frame can also be obtained by statics. Approximate expressions are available from which the values of these horizontal reactions can be estimated within a very few percent.

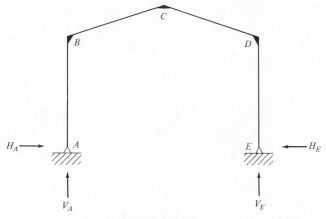

Figure 19-4

The published information is usually in the form of equations which give the values of H (the horizontal reaction components) for varying loading conditions. An example of this kind of information is shown in Fig. 19-5.[1]

From these equations the values of H can be estimated and the approximate moments at various points in the frame computed by statics. The member sizes can be estimated and will actually be the final sizes or very close to them if the equations are applied correctly. It should be noted that the values given for H by the formulas in Fig. 19-5 are actually good only for frames consisting of prismatic members. Practical rigid frames, however, will often be deepened at the knees. Deeper knees have greater moments with the result that the values of H are larger. Analysis also shows that these values vary with different column height-to-span ratios.

There is information available in several references which is helpful in estimating the increases in H for these factors. Martin P. Korn[2] suggests that H (as obtained by the formulas for constant section frames) be increased by 5% when the ratio of the frame height to span is 0.20 or more; by 10% when the ratio is 0.15; 15% for ratio of 0.12; etc. (The usual practice followed is not to increase the values of H when wind loads are involved.) In his book on rigid frames Korn presents tables that give values of the horizontal frame reactions as well as moments at the knees and crown for various loading conditions and different spans. Another valuable reference on the analysis and design of rigid frames is the book *Design of Welded Structures* published by the James F. Lincoln Arc Welding Foundation.[3]

---

[1] John D. Griffiths, *Single-Span Rigid Frames in Steel* (New York: AISC, 1948).
[2] Martin P. Korn, *Steel Rigid Frames Manual Design and Construction* (Ann Arbor, Mich.: J. W. Edwards, Inc., 1953), p. 15. A person doing rigid-frame design would be wise to obtain a copy of this well-written and interesting book.
[3] O. W. Blodgett, 1966 (Cleveland, Ohio) pp. 6.1-1 through 6.1-36.

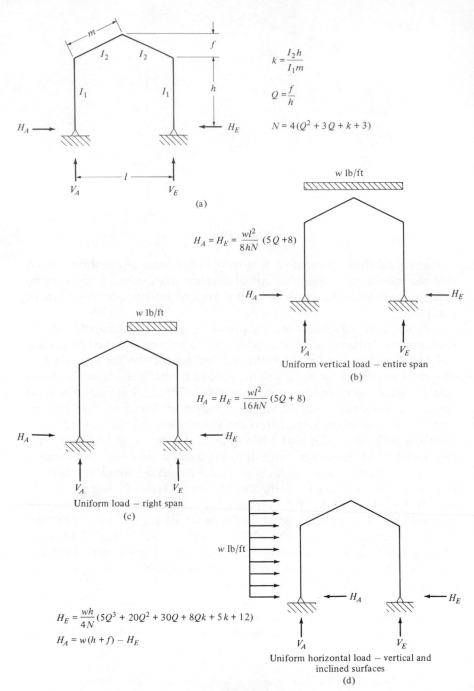

$$k = \frac{I_2 h}{I_1 m}$$

$$Q = \frac{f}{h}$$

$$N = 4(Q^2 + 3Q + k + 3)$$

$$H_A = H_E = \frac{wl^2}{8hN}(5Q + 8)$$

Uniform vertical load — entire span
(b)

$$H_A = H_E = \frac{wl^2}{16hN}(5Q + 8)$$

Uniform load — right span
(c)

$$H_E = \frac{wh}{4N}(5Q^3 + 20Q^2 + 30Q + 8Qk + 5k + 12)$$

$$H_A = w(h + f) - H_E$$

Uniform horizontal load — vertical and inclined surfaces
(d)

Figure 19-5

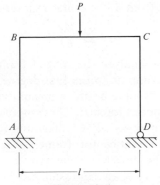

Figure 19-6

19-5. "EXACT" ANALYSIS OF RIGID FRAMES

After the member sizes are estimated it is necessary to analyze the frame by one of the so-called exact methods. The student has probably been exposed to slope deflection, moment distribution, column analogy, and other methods and she may want to know which one to use. The answer is probably the method in which she has the most confidence.

It is not the purpose of this section to explain several methods of analyzing statically indeterminate rigid frames, as such information can be found in several excellent textbooks on statically indeterminate structures. Only one general method of analysis is included and its application is limited to one-span hinged rigid frames. For this discussion the frame of Fig. 19-6 is considered.

Under load the distance from A to D represented by l in the figure changes because the roller at D moves slightly one way or the other. The amount of this deflection can be obtained by taking moments about D of the M/EI diagram from A to D (the second moment-area theorem) where

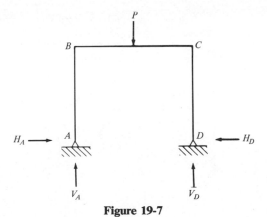

Figure 19-7

y is the vertical distance from a horizontal axis through the hinges.

$$\delta_L = \sum \frac{My}{EI} ds$$

It is now desired to analyze the hinged frame shown in Fig. 19-7, which is statically indeterminate to the first degree. The moment diagram for this frame equals the "simple beam" moment diagram plus the moment diagram due to the redundant force H. These moment diagrams are shown in Fig. 19-8 and are drawn as M/EI diagrams for deflection consider-ations. As is the usual practice for frames of this type, the moment diagrams are shown on the sides of the members on which tension occurs.

After examining Fig. 19-8, an expression can be written for the change in span length from A to D in the frame. In this expression, which equals zero, M' represents the "simple beam" moment at any point. It will be noted from Fig. 19-8 that if M' is considered positive the moment due to H will be of the opposite sign. The resulting expression is solved for H.

$$\sum \frac{My}{EI} ds = \sum \frac{M'y}{EI} ds + \sum \frac{(-Hy)(y)}{EI} ds = 0$$

$$H = \frac{\sum (M'y/EI) ds}{\sum (y^2/EI) ds}$$

The student familiar with column analogy will see that the preceding expression is the same as the column-analogy expression for hinged frames. The value of the numerator of this expression equals the moment of the area of the M'/EI diagram about the axis through the hinges. The denominator $\sum(y^2 ds/EI)$ is actually the "moment of inertia" of the frame about an axis through the hinges except that the width of each member is $1/EI$.

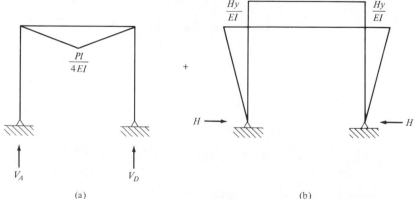

(a)   (b)

Figure 19-8 (a) "Simple beam" M/EI diagram. (b) M/EI diagram due to H.

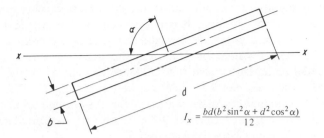

$$I_x = \frac{bd(b^2 \sin^2\alpha + d^2\cos^2\alpha)}{12}$$

Figure 19-9

Examples 19-1 and 19-2 briefly illustrate the analysis of two frames consisting of prismatic members. In these two problems the "moments of inertia" of the columns about the axis through the hinges are obtained with the usual expression for the moment of inertia about the base of a rectangle ($\frac{1}{3}bh^3$). For the horizontal girder in Example 19-1 the "moment of inertia" is obtained by merely using the transfer expression (Ad^2). Its moment of inertia about its own axis is considered negligible. For the inclined girders or rafters of Example 19-2 it is considered necessary to include the "moments of inertia" of the girders about their own axes. These values are computed by using the expression shown in Fig. 19-9 which was taken from the AISC Manual.

Welded rigid frame foundry, Bay City, Mich. (Courtesy of The Lincoln Electric Company.)

Should a member be nonprismatic and thus have varying moments of inertia along its cross section, it will be necessary to divide it up into sections of convenient ds lengths to make the necessary computations. The values of M' and I would be the average values for those ds lengths. Should small ds lengths be used the "moment of inertia" of the segments about their own axes can be neglected in computing the $\Sigma(y^2\,ds/EI)$ values. If E is constant it can be omitted from these calculations as it will be canceled. As a part of Example 19-3 a frame with deepened knees, and thus varying moments of inertia, is analyzed.

Example 19-1

Determine the value of H_D for the rigid frame shown in Fig. 19-10(a). The "simple beam" M'/EI diagram is drawn in part (b) of the figure.

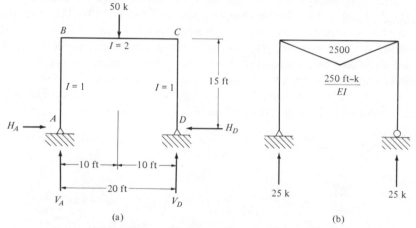

Figure 19-10   (a) Frame. (b) M'/EI diagram.

SOLUTION

$\Sigma(M'y/EI)\,ds =$ moment of area of "simple beam" M'/EI diagram about axis through hinges is:

$$\left(\frac{2500}{EI}\right)(15) = \frac{37,500}{E2} = \frac{18,750}{E}$$

$\Sigma(y^2/EI)\,ds =$ "moment of inertia" of the frame of $1/EI$ width about axis through hinges is:

$$\left(\frac{1}{3}\right)\left(\frac{1}{E1}\right)(15)^3(2) + \left(\frac{20}{E2}\right)(15)^2 = \frac{4500}{E}$$

$$H = \frac{\Sigma(M'y/EI)\,ds}{\Sigma(y^2/EI)\,ds} = \frac{18,750/E}{4500/E} = 4.16\text{ k}$$

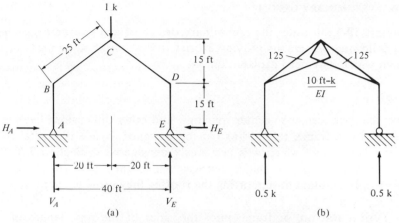

Figure 19-11 (a) Frame. (b) M'/EI diagram.

Example 19-2

Determine the value of H_E for the gabled frame shown in Fig. 19-11(a) for which all members are assumed to have the same moments of inertia. The simple beam M'/EI diagram is shown in part (b) of the figure.

SOLUTION

Table for $\sum \dfrac{M'y}{EI} ds$

Member	y	ds	$\dfrac{M'}{EI} ds$	$\dfrac{M'y}{EI} ds$
AB	—	15	—	—
BC	25	25	$125/E$	$3125/E$
CD	25	25	$125/E$	$3125/E$
DE	—	15	—	—
				$\sum = 6250/E$

Table for $\sum \dfrac{y^2}{EI} ds$

Member	I_0	Ad^2	$\dfrac{y^2}{EI} ds$
AB	—	—	$1,125/E$
BC	833	12,650	$13,483/E$
CD	833	12,650	$13,483/E$
DE	—	—	$1,125/E$
			$\sum = 29,216/E$

$$H = \frac{\sum (M'y/EI)\, ds}{\sum (y^2/EI)\, ds} = \frac{6250/E}{29,216/E} = 0.214 \text{ k}$$

19-6. PRELIMINARY DESIGN

Example 19-3 illustrates the preliminary design of a rigid-frame structure. The following paragraphs provide a brief discussion of some parts of the design which may need explanation.

PURLINS

After the span, center-to-center spacings, and other dimensions have been selected for a frame, the purlins can be designed. Space is not taken to review their design which was previously considered in Section 7-6. Perhaps, however, the two major services which they perform should be repeated. In addition to supporting the roofing they provide lateral support for the girders.

Purlins rest on, or frame into, the sides of the top flanges of the girders. They are probably much shallower than the girders and do not provide lateral support for the bottom girder flanges. Near the center of the frames, the top flanges of the girders are in compression due to positive moments and lateral support is provided by the purlins at their connection points. Near the knees of the frame, however, there will usually be negative

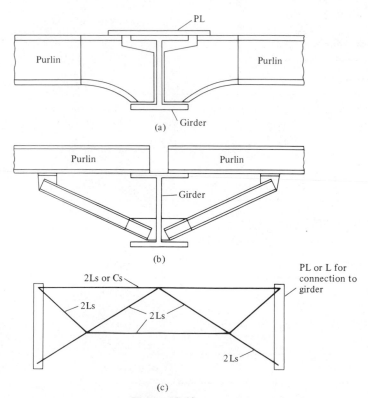

Figure 19-12

moments, and the ordinary depth purlins will not extend down a sufficient distance to provide lateral support for the bottom (now compression) flanges. For this reason some type of deepened purlin is needed.

Several methods of providing the necessary lateral support in the negative moment regions are shown in Fig. 19-12. In part (a) of the figure the very end of the purlin is deepened, while in part (b) braces are run from the purlins to the bottom of the girder. Another possibility is shown in (c) where a small truss with depths at its ends equal to the girder depths is used as the purlin. Outside purlins may very well be located just to the inside of the columns so they can be used to provide bracing to the inside flange.

Approximate Analysis of Frame

An approximate analysis of the frame can be made using formulas of the type given in Fig. 19-5. Any feasible load combinations which might be critical should be considered in the analysis. In Example 19-3 five different loading conditions are considered. In some frames there are other load combinations that should be considered in addition to the ones mentioned here such as crane loads or equipment loads. The AISC one-third allowable stress increase when wind loads are involved is again used by multiplying the combinations which include wind forces by $\frac{3}{4}$. The loading conditions considered here are as follows.

1. $DL + SL$
2. $DL + $ drift (or snow on one-half of the roof)
3. $\frac{3}{4}(DL + \frac{1}{2} SL + WL)$
4. $\frac{3}{4}(DL + \text{drift} + WL)$
5. $\frac{3}{4}(DL + SL + \frac{1}{2} WL)$

The $DL + SL$ combination is very often the controlling condition and some designers may not even make the calculations for the other combinations. This combination is not always the critical one and before the student starts omitting some of the feasible possibilities he should have a good deal of experience with frames. After such experience he may be able to recognize some load situations that do not have to be considered.

Drawing up the Knee

After the approximate analysis is made tentative moments can be calculated for the sizing of the members. For this design example a circular deepened knee is to be used. The greatest moments in the frame occur at the knees whereas the greatest moments in the columns and girders occur at their points of tangency with the knees. These points of tangency can be roughly estimated at this time and column and girder sizes estimated for the moments at those points. The exact sizes of the members are not

desired at this time, only their approximate depths, so that the knees can be drawn.

After the member depths are estimated the proportions of the knee can be established. Knee proportions are often established more on the basis of what is attractive than on the basis of structural needs. Korn (previous reference) recommends that a minimum radius of 2 $\frac{1}{2}$ times the depth of the column or girder (whichever is larger) be used for circular knees. He also says that from the standpoint of appearance there is too much knee if its depth exceeds 2 $\frac{1}{2}$ times the depth of the column or girder (whichever is larger).

Preliminary Design of Members

After the knee dimensions are established, the critical moments at the points of tangency in the columns and girders can be calculated. The columns and girders must be able to resist moments, shears, and axial forces. Although the shears and axial forces are usually quite small in relation to the moments they are of such magnitude that they need to be considered in design. It will be remembered from Chapter 8 that when the AISC Specification is used for such cases the designs are to be made with their interaction formulas, Formulas 1.6-1a, 1.6-1b, and 1.6-2, as applicable.

Using the AISC Specification it is also necessary to estimate effective lengths to calculate allowable stresses. For the design of the columns and girders it is assumed here that the effective lengths are equal to the unbraced lengths (thus $K = 1.0$). This assumption, which seems quite reasonable, is made for the following two reasons.

1. Axial loads are relatively small in these frames.
2. There is no doubt but that the effective lengths of the members in the plane of the frames are greater than their lengths ($K > 1.0$) because of sidesway. The actual values depend on the overall stiffness of the frame as discussed in Chapter 8; however the use of $K = 1.0$ yields very reasonable results because the least r is for the y axis (which is laterally braced).

For a detailed discussion of this subject the student is referred to Beedle et al., *Structural Steel Design* (New York: The Ronald Press Co., 1964), pp. 668–675.

"Exact" Analysis of Frame

After the tentative member sizes are selected the stresses are reviewed by an exact procedure such as the one described in Section 19-5. If the approximate formulas have been carefully applied it will probably be unnecessary to revise the design. The approximate formulas are so good

that many designers just omit this step. The only objection the author has to this omission is caused by the fact that inadvertently foolish math mistakes are often made and this procedure provides an independent check.

Example 19-3

Make a preliminary design of the frame shown in Fig. 19-13 using A36 steel and the AISC Specification. The frames are 18 ft–0 in. on centers and there are assumed to be seven equally spaced purlins on each side which provide lateral support for both flanges of the girders. The loads for which design is to be made are as follows.

DL (not including frame weight) = 16 lb/horizontal ft^2
assume girder weight = 80 lb/horizontal ft
snow load = 30 lb/horizontal ft^2
wind load = 20 lb/vertical ft^2

Analysis is to be made for each of the five load combinations mentioned earlier in this section.

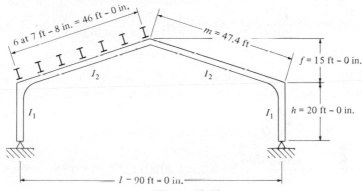

Figure 19-13

SOLUTION
Determining Approximate Values of H Using Equations of Fig. 19-5: Assuming $I_1 = I_2$ in Fig. 19-13:

$h = 20.0$ ft
$f = 15.0$ ft

$$m = \sqrt{(15)^2 + (45)^2} = 47.4 \text{ ft}$$

$$k = \frac{I_2 h}{I_1 m} = \frac{20.0}{47.4} = 0.422$$

$$Q = \frac{f}{h} = \frac{15.0}{20.0} = 0.75$$

$$N = 4(Q^2 + 3Q + k + 3) = 4(0.75^2 + 3 \times 0.75 + 0.422 + 3) = 24.94$$

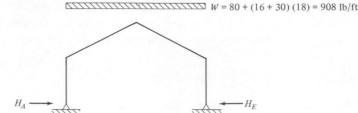

Figure 19-14

(a) $DL+SL$ (see Fig. 19-14): Value of H increased by 5% in accordance with Korn's recommendations:

$$H_A = H_E = 1.05 \frac{wl^2}{8hN}(5Q+8) = (1.05)\left(\frac{908\times90^2}{8\times20\times24.94}\right)(5\times0.75+8)$$

$$= 22{,}750 \text{ lb}$$

(b) Drift load (see Fig. 19-15): Value of H increased by 5% in accordance with Korn's recommendations:

$$H_A = H_E = 1.05 \frac{wl^2}{16hN}(5Q+8) = (1.05)\left(\frac{540\times90^2}{16\times20\times24.94}\right)(5\times0.75+8)$$

$$= 6740 \text{ lb}$$

(c) Wind load (see Fig. 19-16): Usual practice is not to increase the value of H for wind load to take into account effect of deepened knee:

$$H_E = \frac{wh}{4N}(5Q^3+20Q^2+30Q+8Qk+5k+12)$$

$$= \frac{(360)(20)}{(4)(24.94)}(5\times0.75^3+20\times0.75^2+30\times0.75$$

$$+8\times0.75\times0.422+5\times0.422+12)$$

$$= 3790 \text{ lb}$$

$$H_A = w(h+f) - H_E = (360)(20+15) - 3790 = 8810 \text{ lb}$$

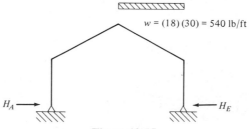

$w = (18)(30) = 540 \text{ lb/ft}$

Figure 19-15

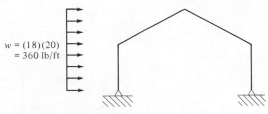

Figure 19-16

Estimating Frame Moments

(a) $DL + SL$ (see Fig. 19-17):

$$M_{\text{knee}} = (22.75)(20) = 455 \text{ ft-k}$$

$$M_{\text{crown}} = (40.9)(45) - (22.75)(35) - (45)(0.908)(22.5) = 120 \text{ ft-k}$$

(b) $DL + \text{drift}$:

$$M_{\text{knee}} = 320 \text{ ft-k}$$

$$M_{\text{crown}} = 87 \text{ ft-k}$$

(c) $\frac{3}{4} (DL + \frac{1}{2} SL + WL)$:

$$M_{\text{knee}} = 294 \text{ ft-k}$$

$$M_{\text{crown}} = 48 \text{ ft-k}$$

(d) $\frac{3}{4} (DL + \text{drift} + WL)$:

$$M_{\text{knee}} = 297 \text{ ft-k}$$

$$M_{\text{crown}} = 46 \text{ ft-k}$$

(e) $\frac{3}{4} (DL + SL + \frac{1}{2} WL)$:

$$M_{\text{knee}} = 399 \text{ ft-k}$$

$$M_{\text{crown}} = 78 \text{ ft-k}$$

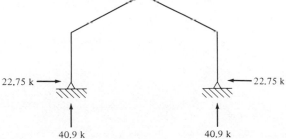

Figure 19-17

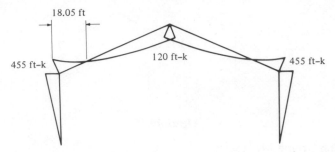

Figure 19-18   Moment diagram for critical moments $(DL+SL)$.

Preliminary Design of Frame:

(a) Laying out knee: Assuming $R=7.00$ ft and computing the angles and distances shown in Fig. 19-19. Preliminary calculations show that W24 sections can be reasonably used for column and girders.

(b) Trial column design:

$$\text{max moment (from Fig. 19-17)}=(22.75)(14.24)=324\text{ ft-k}$$

$$\text{max thrust}=V_A=40.9\text{ k}$$

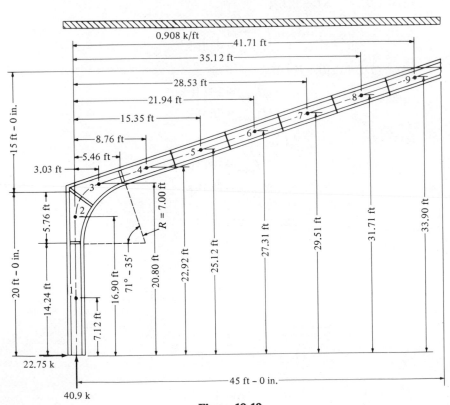

Figure 19-19

Try W24×94 ($A=27.7$, $L_c=9.6$, $L_u=15.1$, $S_x=222$, $r=1.98$)

$$f_a = \frac{40.9}{27.7} = 1.48 \text{ ksi}$$

$$\frac{Kl}{r} = \frac{(1)(12)(14.24)}{1.98} = 86.3$$

$$F_a = 14.65 \text{ ksi}$$

$$\frac{f_a}{F_a} = \frac{1.48}{14.65} = 0.101 < 0.150 \qquad \text{(use Formula 1.6-2 AISC)}$$

$$f_b = \frac{(12)(324)}{22} = 17.51 \text{ ksi}$$

$$F_b = 22 \text{ ksi}$$

$$\frac{f_a}{F_a} + \frac{f_b}{F_b} = \frac{1.48}{14.65} + \frac{17.51}{22.0} = 0.897 < 1.00 \qquad\qquad \text{OK}$$

(c) Trial girder design: Maximum moment from Fig. 19-19 is

$$-(22.75)(21.62) + (40.9)(5.46) - (5.46)(0.908)(2.73) = -282 \text{ ft-k}$$

$$\max T = (22.75)(\sin 71°35') + (40.9 - 5.46 \times 0.908)(\cos 71°35')$$
$$= 32.9 \text{ k}$$

Try W24×68 ($A=20.1$, $b_f=8.965$, $S_x=154$, $r=1.87$)

$$f_a = \frac{32.9}{20.1} = 1.637 \text{ ksi}$$

$$\frac{Kl}{r} = \frac{(1)(12)(7.67)}{1.87} = 49.3$$

$$F_a = 18.41 \text{ ksi}$$

$$\frac{f_a}{F_a} = \frac{1.637}{18.41} = 0.0889 < 0.15 \qquad \text{(use Formula 1.6-2 AISC)}$$

$$f_b = \frac{(12)(282)}{154} = 21.97 \text{ ksi}$$

$$L_c = 9.5 > 7.67 \text{ ft} \qquad \text{(use 24.0 ksi)}$$

$$\frac{f_a}{F_a} + \frac{f_b}{F_b} = \frac{1.637}{18.41} + \frac{21.97}{24.0} = 1.004 > 1.00 \qquad \text{(slightly overstressed)}$$

(d) Trial knee section: To reasonably fit in with the W24×94 column and the W24×68 girder flange and web dimensions, try $\frac{3}{4}$×9-in. flange plates and $\frac{1}{2}$-in. web plates. From table on next page

$$H = \frac{\sum (M' \, ds/EI)}{\sum (y^2 \, ds/EI)} = \frac{2(9,569,790)/E}{(2)(408,500)/E} = 23.4 \text{ k}$$

Segment	Section	x	y	M' Due to $DL+SL$	ds	I of section (ft^4)	$\dfrac{ds}{I}$	$\dfrac{M'y\,ds}{I}$	$\dfrac{y^2\,ds}{I}$
1	W24×76	0	7.12	0	14.24	0.129	110.5	0	5,600
2	Flange PLs $\frac{3}{4}\times 9$ and $\frac{1}{2}$ web	0.16	16.90	6.5	5.25	0.193	27.2	2,990	7,760
3	"	3.03	20.80	120	"	"	"	67,800	11,780
4	W24×68	8.76	22.92	323	6.94	0.088	78.9	508,000	36,000
5	"	15.35	25.12	521	"	"	"	1,032,000	49,790
6	"	21.94	27.31	676	"	"	"	1,456,000	58,850
7	"	28.53	29.51	796	"	"	"	1,853,000	68,710
8	"	35.12	31.71	878	"	"	"	2,196,000	79,340
9	"	41.71	33.90	918	"	"	"	2,454,000	90,670

$$\frac{1}{2}\Sigma = \frac{9,569,790}{E} \qquad \frac{1}{2}\Sigma = \frac{408,500}{E}$$

A check on column and girder sizes for the moments caused by an *H* of 23.4 k shows that they are satisfactory though girder is slightly over-stressed.

19-7.   FINAL DESIGN AND DETAILS

Knee Design

An exact theoretical determination of the stress distribution in the knees of a rigid frame is quite a complex problem because the stress variation is decidedly different from that in a straight beam subject to bending. The neutral axis does not fall at the middepth of the section because the compressive stresses accumulate going around the inside of the knee and the large percentage of the knee cross section will be placed in tension. The neutral axis for a rounded knee is roughly 25% of the distance from the bottom. This value varies a few percent with different roof pitches.

The usual designer will probably not go to a great deal of trouble in attempting to determine precise stresses in the knee. He will probably just use some approximate method for checking stresses. As the compressive stresses are rather high all around the inside curved surface and actually a little beyond the points of tangency, the usual practice is to place stiffeners at the points of tangency and at the knee.

Allen County War Memorial, Fort Wayne, Ind. (Courtesy of The Lincoln Electric Company.)

It is probably desirable to additionally stiffen the compression flanges of the knee to prevent local buckling. These stiffeners, which can be triangular in shape, need only to extend for about one-third of the web depth. They are usually placed at distances on centers equal to about $2\frac{1}{2}$ or 3 times the flange width.

Lateral Bracing

The student may have seen rigid frames where there was no bracing between the knees of the frame. She may think that the roof and the purlins provide sufficient lateral bracing and that the addition of special bracing is unnecessary. She must, however, be very careful because if the frame is not laterally stable, it may fail at very low computed stresses.

A plastic analysis of rigid frames will show that plastic hinges will form at the knees, after which there will be a redistribution of moments to other parts of the frame. For this redistribution to take place, however, it is essential for satisfactory lateral bracing to be provided in the vicinity of the plastic hinges. Bracing is also necessary for resisting wind loads perpendicular to the plane of rigid frames.

Cross bracing is usually not too satisfactory because it can interfere with the windows in the vertical walls. One type of bracing shown in Fig. 19-20 is frequently very satisfactory. This bracing, which might be connected to the diagonal stiffeners at the corners, is similar to the end portals in a through bridge and is often called *knee bracing*.

In the plane of the roof cross bracing is very satisfactory. Looking at the building from the side, the bracing might very well be arranged as shown in Fig. 19-21. Probably each of these members will be designed as a tension member capable of carrying the entire wind shear in its panel.

Design of Tie Rods

The gross cross-sectional area of the tie rod can be determined by dividing the maximum horizontal reaction H by the allowable tie rod stress (0.33 F_u). Should the rods not be encased in concrete it may be well to provide a little extra diameter to estimate the effect of corrosion. Incidentally the base plates can be made with the connections for the rods built into them.

Temperature Stresses

For the usual span, rigid-frame, temperature stresses are not normally considered as they are probably negligible unless spans in excess of about 200 ft are involved.

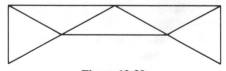

Figure 19-20

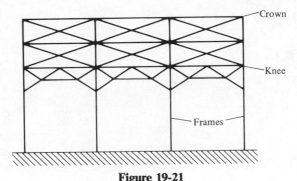

Figure 19-21

19-8. PLASTIC DESIGN OF RIGID FRAMES

The plastic design of rigid frames is felt to be a little beyond the scope of this book. The student, however, should realize that the plastic procedure will not only simplify analysis but will also provide considerable opportunities for economy. The AISC has published a booklet[4] which includes the design of single-span rigid frames by this method. The use of this excellent publication produces very realistic designs and abbreviates greatly the rather laborious elastic design procedure described in this chapter.

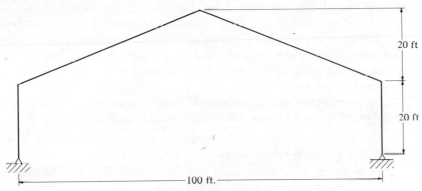

Problem 19-1

Problem

19-1. Design the rigid frame shown in the accompanying illustration using the AISC Specification and A36 steel. The frames are assumed to be spaced 20 ft–0 in. on centers and are subjected to a wind load of 20 psf of vertical projection and a snow load of 20 psf of horizontal projection.

[4]*Steel Gables and Arches* (New York: AISC, 1963).

Chapter 20
Multistory
Buildings

20-1. INTRODUCTION

Space is not taken in this introductory text to discuss the design of multistory buildings in detail. The material of this chapter is presented merely to give the student a general idea of the problems involved in the design of such buildings and not to present elaborate design examples.

Office buildings, hotels, apartment houses, and other buildings of many stories are quite common in the United States and the trend is toward an even larger number of tall buildings in the future. Available land for building in our heavily populated cities is becoming scarcer and scarcer and costs are becoming higher and higher. Tall buildings require a smaller amount of this expensive land to provide required floor space. Other factors contributing to the increased number of multistory buildings are new and better materials and construction techniques.

There are, on the other hand, several factors that may limit the heights to which buildings will be erected. These include the following.

1. Certain city building codes prescribe maximum heights to which buildings can be constructed.
2. Foundation conditions may not be satisfactory for supporting buildings of many stories.
3. Floor space may not be rentable above a certain height. Someone will always be available to rent the top floor or two of a 250-story building, but floor numbers 100 through 248 may not be so easy to rent.
4. There are several cost factors that tend to increase with taller buildings. Among these factors are elevators, plumbing, heating and air conditioning, glazing, exterior walls, wiring, etc.

Whether a multistory building is used for an office building, a hospital, a school, an apartment house, or whatever, the problems of design are generally the same. The construction is probably of the skeleton type in which the loads are transmitted to the foundation by a framework of steel beams and columns. The floor slabs, partitions, and exterior walls are all supported by the frame. This type of framing which can be erected to tremendous heights may also be referred to as beam-and-column construction. Bearing-wall construction is probably not used for buildings of more than a few stories although it has on occasion been used for buildings up to 10 or 12 stories. The columns of a skeleton frame are probably spaced 20, 25, or 30 ft on centers with beams and girders framing into them from both directions (see Fig. 13-1). On some of the floors, however, it may be necessary to have much larger open areas between columns for dining rooms, ballrooms, etc. For such cases very large beams (perhaps plate girders) may be needed to support column loads for many floors above.

It is usually necessary in multistory buildings to fireproof the members of the frame with concrete, gypsum, or some other material. The exterior walls are probably constructed with concrete or masonry units although an increasing number of modern buildings are being erected with large areas of glass in the exterior walls.

For these tall and heavy buildings the usual spread footings may not be sufficient to support the loads. If the bearing strength of the soil is high, steel grillage footings may be sufficient, but for poor soil conditions pile or pier foundations may be necessary.

For multistory buildings the beam-to-column systems are said to be superimposed on top of each other story by story or tier by tier. The column sections can be fabricated for one, two, or more floors with the

Broadview Apartments, Baltimore, Md. (Courtesy of The Lincoln Electric Company.)

two-story lengths probably being the most common. Theoretically, column sizes can be changed at each floor level but the costs of the splices involved usually more than cancel any savings in column weights. Columns of three or more stories in height are difficult to erect. The result is that the two-story heights work out very well most of the time.

20-2. COLUMN SPLICES

Column splices are usually placed 2 or 3 ft above floor levels to keep from interfering with the beam and column connections. Typical column splices are shown in Fig. 20-1. As shown in the figure, the column ends are usually

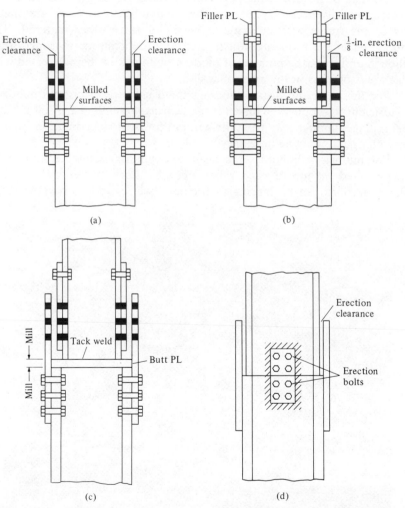

Figure 20-1 (a) Columns with same nominal depths. (b) Columns with small depth difference. (c) For columns of equal or different depths. (d) Splice made for all four sides of column.

milled so they can be placed firmly in contact with each other for purposes of load transfer. When the contact surfaces are milled a large part of the axial compression (if not all) can be transferred through the contacting areas. It is obvious that splice plates are necessary even though full contact is made between the columns and only axial loads are involved. They are even more necessary when consideration is given to the shears and moments existing in practical columns subjected to off-center loads, lateral forces, moments, etc.

There is obviously a great deal of difference between tension splices and compression splices. In tension splices all load has to be transferred through the splice, whereas in splices for compression members a large part of the load can be transferred directly in bearing between the columns. The splice material is then needed to transfer only the remaining part of the load.

The amount of load to be carried by the splice plates is difficult to estimate. Should the column ends not be milled, the plates should be designed to carry 100% of the load. When the surfaces are milled and axial loads only are involved, the amount of load to be carried by the plates might be estimated to be from 25 to 50% of the total load. If bending is involved perhaps 50 to 75% of the total load may have to be carried by the splice material.

The bridge specifications spell out very carefully splice requirements for compression members but the AISC does not. For example, the AREA says that compression member splices shall be designed to carry at least one-half of the total load. They further specify that compression members be spliced on all four sides.

Part (a) of Fig. 20-1 shows a splice that may be used for columns with the same nominal depths. The student will notice in the AISC Manual that wide-flange shapes of the same series (as W14) generally have the same inside distances between flanges although their total actual depths may vary as much as several inches (8.68 in. from a W14×730 to a W14×22). The type of splice shown in part (b) of the figure which has filler plates is used when the nominal depth of the top column is more than 2 in. less than that of the lower column.

Part (c) shows a type of splice which can be used for columns of equal or different nominal depths. For this type of splice the butt plate is shop-welded to the lower column and the clip angles used for field erection are shop-welded to the upper column. In the field the erection bolts shown are installed and the upper column is field-welded to the butt plate. The horizontal welds on this plate resist shears and moments in the columns.

Part (d) shows welded splices made on all four sides of the columns. The web splices are bolted in place in the field and field-welded to the column webs. The flange splices are shop-welded to the lower column and field-welded to the top column. The web plates may be referred to as *shear plates* and the flange plates as *moment plates*.

Welded column splice for Colorado State Service Building, Denver, Colo. (Courtesy of The Lincoln Electric Company.)

20-3. DISCUSSION OF LATERAL FORCES

For tall buildings, lateral forces must be considered as well as gravity forces. The usual practice is to consider wind forces in a building-frame design when the height is two or more times the least lateral dimension. The usual framing of buildings with a smaller height-width ratio has sufficient stiffness to resist forces of large magnitude without any special design provisions. Perhaps the common 2-to-1 ratio is a little high, however, for some modern buildings with their large areas of glass in the exterior walls.

High wind pressures on the sides of tall buildings produce overturning moments. These moments are probably resisted without difficulty by the axial strengths of the columns but the horizontal shears produced on each level may be of such magnitude as to require the use of special bracing or moment-resisting connections.

Unless they are fractured, the floors and walls provide a large part of the lateral stiffness of tall buildings. Although the amount of such resistance may be several times that provided by the lateral bracing, it is difficult to estimate and may not be reliable. Today so many modern buildings have light movable interior partitions, glass exterior walls, and lightweight floors that the steel frame should be assumed to provide all of the required lateral stiffness.

A building must not only be braced sufficiently laterally to prevent failure but also prevented from deflecting so much as to injure its various parts. Another item of importance is the provision of sufficient bracing to give the occupants a feeling of safety. They might not have this feeling in tall buildings which have a great deal of lateral movement in times of high winds. There have actually been tales of occupants of the upper floors of tall buildings complaining of seasickness on very windy days.

The horizontal deflection of a multistory building due to wind or seismic loading is called *drift*. It is represented by Δ in Fig. 20-2. Drift is measured by the *drift index*, Δ/h, where h is the height or distance to the ground.

The usual practice in the design of multistory steel buildings is to provide a structure with sufficient lateral stiffness to keep the drift index between approximately 0.0015 and 0.0030 radians for the worst storms that occur in a period of approximately 10 years. These so-called "10-year storms" might have wind in the range of about 90 mph depending on location and weather records. It is felt that when the index does not exceed these values, the users of the building will be reasonably comfortable.

In addition multistory buildings should be designed to withstand safely "50-year storms" and "100-year storms." For such cases however the drift index will be larger than the range mentioned with the result that the occupants will suffer some discomfort. The 1450-ft World Trade Center in New York City deflects or sways up to about 3 ft in "10-year storms" (drift index = 0.0021) while in hurricane winds it will sway up to about 7 ft (drift index = 0.0048).

Many areas of the world, including the western part of the United States, fall in earthquake territory, and in those areas it is necessary to consider seismic forces in design for tall or low buildings. During an earthquake there is an acceleration of the ground surface. This acceleration can be broken down into vertical and horizontal components. Usually the vertical component of the acceleration is assumed to be negligible but the horizontal component can be severe.

Most buildings can be designed with little extra expense to withstand the forces caused during an earthquake of fairly severe intensity. On the other hand, earthquakes during recent years have clearly shown that the

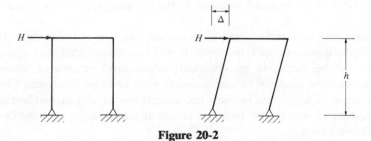

Figure 20-2

Chase Manhattan Bank Building, New York City. (Courtesy of Bethlehem Steel Company.)

average building which is not designed for earthquake forces can be destroyed by earthquakes that are not particularly severe. The usual practice is to design buildings for additional lateral loads (representing the estimate of the earthquake forces) which are equal to some percentage (5–10) of the weight of the building and its contents.

Many persons look upon the seismic loads to be used in design as being merely percentage increases of the wind loads. This is not altogether correct, however, as seismic loads are different in their action and are not proportional to the exposed area but to the building weight above the level in question.

The effect of lateral forces is to require the use of more steel. If the AISC Specification is used, however, it will be remembered that allowable stresses can be increased by one-third when wind or seismic loads are being considered alone or in combination with dead or live loads. Despite this increase in allowable stresses the lateral forces will probably require the addition of bracing or moment-resistant connections as described in the following section.

20-4. TYPES OF LATERAL BRACING

A steel building frame with no lateral bracing is shown in Fig. 20-3(a). Should the beams and columns shown be connected together with the standard ("simple beam") connections, the frame would have little resistance to the lateral forces shown. Assuming the joints to act as frictionless pins, the frame would be laterally deflected as shown in part (b) of the figure.

To resist these lateral deflections, the best, simplest, and most economical method from a theoretical standpoint is the insertion of full diagonal bracing as shown in part (c) of Fig. 20-3. From a practical standpoint, however, the student can easily see that in the average building full diagonal bracing would often be in the way of doors, windows, and other wall openings. Furthermore, many buildings have movable interior partitions and the presence of interior cross bracing would greatly reduce this flexibility. As diagonal bracing is the most direct, efficient, and economical it should be used whenever conditions permit. Usually it will only be convenient in solid walls in and around elevator shafts, stairwells, and other walls in which few or no openings are planned.

The most common method of providing resistance to lateral forces in tall buildings is the use of moment-resistant connections as illustrated in Fig. 20-4(a). This type of bracing is referred to as *bracket type bracing*. Various types of moment-resistant connections are discussed in Section 20-6 and shown in Fig. 20-7.

Other types of lateral bracing are shown in parts (b), (c), and (d) of Fig. 20-4. Each of these bracing types has the disadvantage that it is not very satisfactory for exterior walls which are largely glass. The result is an increased use of the bracket type bracing. The term *braced bent* is usually given to bents that are braced regardless of the method.

Knee braces shown in part (b) of the figure can be used in exterior walls unless the glass area is extremely large. They are also quite satisfactory for use in interior walls because they do not interfere with normal openings. The K bracing shown in part (d) can be used to advantage only

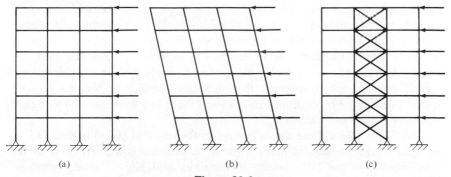

(a)   (b)   (c)

Figure 20-3

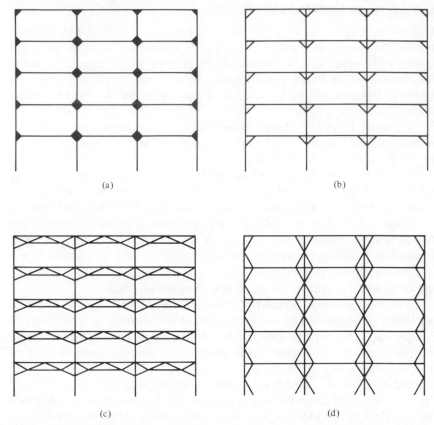

Figure 20-4 (a) Bracket-type bracing. (b) Knee braces. (c) Portal bracing. (d) K bracing.

where small openings are present in the walls. It might be noted that in the upper floors of a building the shearing forces are smaller and it may be possible to dispense with some or all of the lateral bracing on these levels.

Since the lateral forces are usually assumed to be equal in intensity against all sides of the building, the student may ask, "Should the same bracing be used in both directions of a building?" If the building is square or nearly square the answer is probably yes. When the length of a building is several times its least lateral dimension it is highly possible that the bracing will be left out in the long direction because the wind forces will be spread over so many columns that sufficient resistance will be present without bracing. For buildings in between these two extremes the designer will have to use her judgment.

In the usual building the floor system (beams and slabs) is assumed to be rigid in the horizontal plane and the lateral loads are assumed to be concentrated at the floor levels. Floor slabs and girders acting together provide considerable resistance to lateral forces. Investigation of steel buildings that have withstood high wind forces have shown that the floor slabs distribute the lateral forces so that all of the columns on a particular floor have equal deflections. When rigid floors are present they spread the lateral shears to the columns or walls in the building. When lateral forces are particularly large, as in very tall buildings or where seismic forces are being considered, certain specially designed walls may be used to resist large parts of the lateral forces. These walls are called *shear walls*.

It is not necessary to brace every panel in a building. Usually the bracing can be placed in the outside walls with less interference than in the inside walls where movable partitions may be desired. Probably the bracing of the outside panels alone is not enough and some interior panels may need to be braced. It is assumed that the floors and beams are sufficiently rigid to transfer the lateral forces to the braced panels. Three possible arrangements of braced panels are shown in Fig. 20-5. A symmetrical arrangement is probably desirable to prevent uneven lateral deflection in the building and thus torsion.

As previously indicated bracing around elevator shafts is usually permissible while for other locations it will often interfere with windows, doors, movable partitions, glass exterior walls, open spaces, etc. Should bracing be used around an elevator shaft as shown in Fig. 20-6(a) and should calculations show that the drift index is too large it may be possible to use a "hat truss" on the top floor as shown in part (b) of the figure. Such

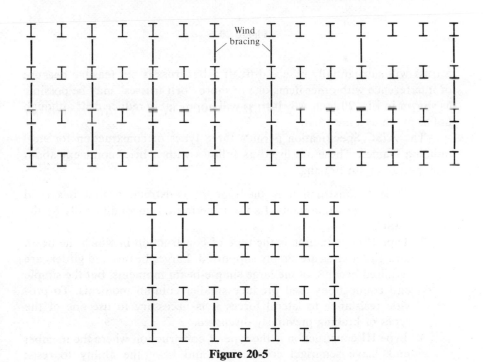

Figure 20-5

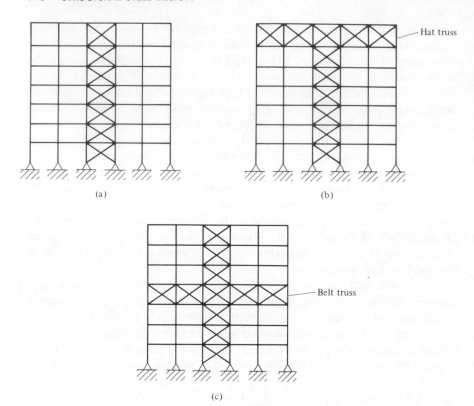

Figure 20-6

a truss will substantially reduce drift. If a hat truss is not feasible because of interference with other items one or more "belt trusses" may be possible as shown in Fig. 20-6(c). A belt truss will appreciably reduce drift although not as much as a hat truss.

The AISC Specification permits three types of construction for steel building frames. These are listed as follows with a brief comment about each pertaining to bracing.

1. Type I construction is the type of construction that has rigid moment-resisting joints. Its use is permitted unconditionally by the AISC.
2. Type II construction is the type of construction in which the beam and girder ends are simply supported. Larger beams and girders are required because of the large simple-beam moments, but the simple end connections used produce smaller column moments. To provide resistance to lateral forces it is necessary to use one of the types of bracing previously discussed.
3. Type III construction is the type of construction where the member ends have semirigid connections and have the ability to resist

certain moments in between the values of types I and II. The AISC says this type of construction will be permitted only if evidence is presented supporting the minimum values of restraint used in the design.

20-5. ANALYSIS OF BUILDINGS WITH DIAGONAL
WIND BRACING FOR LATERAL FORCES

Full diagonal cross bracing has been described as being the best and most economical type of wind bracing for tall buildings. Where this type of bracing cannot practically be used due to interference with windows, doors, or movable partitions, moment-resisting brackets are often used. Should, however, very tall narrow buildings (with height-to-least-width ratios of 5.0 or greater) be constructed, lateral deflections may become a problem with moment-resisting joints. The joints can satisfactorily resist the moments but the deflections may be excessive.

Should maximum wind deflections be kept under 0.002 times the building height there is little chance of injury to the building, says ASCE Subcommittee 31.[1] The deflection is to be computed neglecting any resistance supplied by floors and walls. To keep lateral deflections within this range it is necessary to use deep knee bracing or full diagonal cross bracing when the height-to-least-width ratio is about 5.0; and for greater values of the ratio, full diagonal cross bracing is essential.[2]

The student may often see bracing used in buildings in which he would think wind stresses were negligible. Such bracing stiffens up a building appreciably and serves the useful purpose of plumbing the steel frame during erection. Before bracing is installed the members of a steel building frame may be twisted in all sorts of directions. Connecting the diagonal bracing should pull them into their proper positions.

When diagonal cross bracing is used it is desirable to introduce initial tension into the diagonals. This prestressing will make the building frame tight and reduce its lateral deflection. Furthermore, the members can support compressive stress due to their pretensioning. Since the members can resist compression, the horizontal shear to be resisted will be assumed to split equally between the two diagonals. For buildings with several bay widths, equal shear distribution is usually assumed for each bay.

Should the diagonals not be initially tightened rather stiff sections should be used so they will be able to resist appreciable compressive forces. The direct axial forces in the girders and columns can be found by joints from the shear forces assumed in the diagonals. Usually the girder axial forces so computed are too small to consider but the values for

[1] "Wind Bracing in Steel Buildings," *Trans. ASCE* 105 (1940), pp. 1713–1739.
[2] L. E. Grinter, *Theory of Modern Steel Structures* (New York: Macmillan, 1962), p. 326.

columns can be quite important. The taller the building becomes the more critical are the column axial forces caused by lateral forces.

For types of bracing other than cross bracing similar assumptions can be made for analysis. The student is referred to pages 333–339 of *Theory of Modern Steel Structures* by L. E. Grinter (New York: Macmillan, 1962) for discussion of this subject.

20-6. MOMENT-RESISTING JOINTS

For a large percentage of buildings under 8 to 10 stories in height the beams and girders are connected to each other and to the columns with simple end framed connections of the types previously described in Chapter 12. As buildings become taller it is absolutely necessary to use a definite wind-bracing system or moment-resisting joints. Moment-resisting joints may also be used in lower buildings where it is desired to take advantage of continuity and the consequent smaller beam sizes and depths and shallower floor construction. Moment-resisting brackets may also be necessary in some locations for loads that are applied eccentrically to columns.

Several types of moment-resisting connections which may be used as wind bracing are shown in Fig. 20-7. The design of connections of these types was also presented in Chapter 12. The average design company through the years will probably develop a file of moment-resisting connections from their previous designs. When they have a wind moment of such and such a value they merely refer to their file and select one of their former designs which will provide the required moment resistance.

In the following paragraphs a few comments are made about each of the types of connections shown in Fig. 20-7. The letter preceding each of these paragraphs corresponds to the letter in the figure for that connection.

(a) The top-angle and seat-angle connection shown, sometimes called an *angle bracket*, provides the minimum acceptable type of wind connection. Even if web angles, shown dashed, are added, the moment resistance is not appreciably increased. This type of wind connection is not satisfactory for very tall buildings.

(b) The split-beam bracket is one of the most common types of wind connections used in multistory buildings. When large girder reactions have to be transferred by this excellent type of connection the seat may have to be stiffened as showed by the dashed line.

(c) This heavy bulky connection may be used to resist very large moments. Although it can reduce girder moments decidedly, its high fabrication cost probably cancels the saving in girder weight.

(d) In this figure a very good welded type of connection for resisting wind moments is shown. The simple clean lines of this connection and much smaller amounts of connection steel used should be noted. Although the connections of parts (a) and (b) of Fig. 20-7 are shown to be riveted or bolted, they can also be satisfactorily welded.

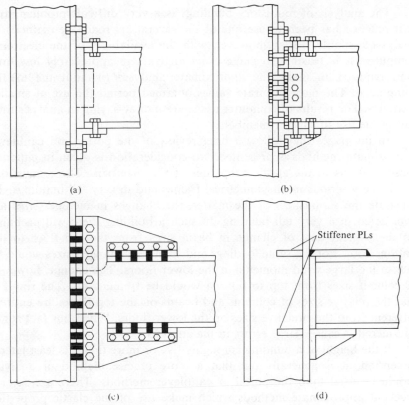

Figure 20-7 (a) Angle bracket. (b) Split-beam connection. (c) Heavy bracket. (d) Welded moment-resisting joint.

20-7. ANALYSIS OF BUILDINGS WITH MOMENT-RESISTING JOINTS FOR LATERAL LOADS

Approximate methods have been very popular for many years for analyzing multistory buildings for lateral forces. Among the several reasons for this popularity have been the following.

1. Large building frames are statically indeterminate to a very high degree and their analysis by an "exact" method is a lengthy and difficult problem.
2. The resistance to lateral forces supplied by the walls and floors in a tall building is difficult to estimate accurately and the results of analysis by an "exact" method are therefore not very precise.
3. Before the members of a statically indeterminate structure can be designed their forces have to be determined, but these forces are dependent upon their sizes. Application of an approximate analysis should yield forces from which very good estimates of member sizes can be made for preliminary design.

The analysis of multistory buildings is a very difficult problem, and past practice has been to use one of the several approximate methods of analysis available. Today, however, with the availability of the electronic computers it is feasible to make exact analyses in appreciably less time than required to make the approximate analyses (without the use of computers). The more accurate values obtained permit the use of smaller members. The results of computer usage are money-saving in analysis time and in the use of smaller members.

In the pages that follow a brief review of the portal and cantilever approximate methods is presented. No consideration is given in either of these methods to the elastic properties of the members. These omissions can be very serious in unsymmetrical frames and in very tall buildings. To illustrate the seriousness of the matter, the changes in member sizes are considered in a very tall building. In such a building there will probably not be a great deal of change in beam sizes from the top floor to the bottom floor. For the same loadings and spans the changed sizes would be due to the large wind moments in the lower floors. The change, however, in column sizes from top to bottom would be tremendous. The result is that the relative sizes of columns and beams on the top floors are entirely different from their relative sizes on the lower floors. When this fact is not considered it causes large errors in the analysis.

If the height of a building is roughly five or more times its least lateral dimension, it is generally felt that a more precise method of analysis should be used than the portal or cantilever methods. There are several excellent approximate methods which make use of the elastic properties and which give values closely approaching the results of the "exact" methods. These include the Factor method, the Witmer method of K percentages, and the Spurr method. Should an exact method be desired, the slope deflection and moment distribution methods are available. If the slope deflection procedure is used, the designer will have the problem of solving a large number of simultaneous equations. The problem is not so serious, however, if he uses a digital computer.

The Portal Method

The most common approximate method of analyzing tall building frames for lateral loads is the portal method. The chief advantage of this method, which is said to be satisfactory for most buildings up to 25 stories,[3] is its simplicity. A detailed description of this method, including the basis for the assumptions made, can be found in most structural analysis textbooks. Only a brief listing of the steps involved in its application is given in this section. These steps are as follows.

1. The horizontal shears on each level are arbitrarily distributed between the columns. One commonly used procedure is to assume

[3] "Wind Bracing in Steel Buildings," *Trans. ASCE* **105** (1940), p. 1723.

the shear divides between the columns in the ratio of one part to exterior columns and two parts to interior columns. Another common distribution (and the one used in the illustrative problem to follow) is to assume the shear V taken by each column is in proportion to the floor area it supports.

2. The moment M in each column is assumed to equal the column shear times half the column height (thus assuming a point of contraflexure at middepth).

3. The girder moments M are determined by joints by noting the sum of the girder moments at any joint equals the sum of the column moments at that joint. These calculations are easily made by starting at the upper left joint and working joint by joint across to the right; after which the next level is handled left to right, etc.

4. The shear V in each girder is assumed to equal its moment divided by half the girder length.

5. Finally, the column axial forces S are determined by summing up the beam shears and other column axial forces at each joint. These calculations are again handled conveniently by working from left to right and from the top floor down.

In Fig. 20-8 a building frame is shown which is to be analyzed by the portal method. The frames are assumed to be placed 20 ft–0 in. on centers and the exterior walls are assumed to be subjected to a wind pressure of 20 lb per vertical square foot. From this data the horizontal loads shown at each floor level are calculated.

This frame is analyzed in Fig. 20-9 by the portal method. The arrows shown on the figure give the direction of the girder shears and the column axial forces. The student can visualize the stress condition of the frame if he assumes the frame is tending to be pushed over from left to right by the wind, stretching the left exterior columns and compressing the right exterior columns.

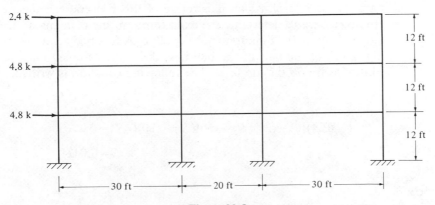

Figure 20-8

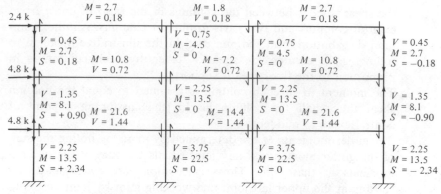

Figure 20-9 Frame analysis by the portal method.

The Cantilever Method

Another simple method of analyzing building frames for lateral forces is the cantilever method. This method is said to be a little more desirable for high narrow buildings than the portal method and may be used satisfactorily for buildings with heights not in excess of 25 to 35 stories.[4] It is not as popular as the portal method.

Instead of initially assuming the horizontal shears to be divided between the columns on each level in some proportion, the first assumption in the cantilever method pertains to the column axial forces. The axial force in a particular column is assumed to vary in direct proportion to its distance from the center of gravity of the group of columns on that level. With the wind blowing from left to right, the columns to the left of the center of gravity will be in tension while those to the right will be in compression. The steps involved in analyzing a building frame for lateral forces by the cantilever method are as follows.

1. To obtain the column axial forces moments are taken above an assumed plane of contraflexure through the middepth of the columns on each level. As an illustration of this procedure, moments are taken here to determine the axial forces in the columns on the top level of the building frame of Fig. 20-8. A free body is drawn in Fig. 20-10 of the part of the building above the assumed plane of contraflexure on the top level. The following equation is written.

$$\sum M_x = 0$$

$$(2.4)(6) + (30)(S) - (50)(S) - (80)(4S) = 0$$

$$S = 0.0424$$

[4] Op. cit.

The 20-story United Founders Life Tower in Oklahoma City, Okla. (Courtesy of The Lincoln Electric Company.)

"Topping out" of Blue Cross–Blue Shield Building in Jacksonville, Florida. (Courtesy of Owen Steel Company, Inc.) The Christmas tree is an old North European custom used to ward off evil spirits. It is also used today to show that the steel frame was erected with no lost time accidents to personnel.

2. The girder shears are determined by joints from the column axial forces.
3. The girder moments are determined by multiplying the girder shears by the half-girder lengths.
4. The column moments are found by joints from the girder moments.
5. The column shears are obtained by dividing the column moments by the half-column heights.

The analysis of the frame of Fig. 20-8 by the cantilever method is illustrated in Fig. 20-11. Arrows representing the directions of the column axial forces and the girder shears are again shown for the benefit of the student.

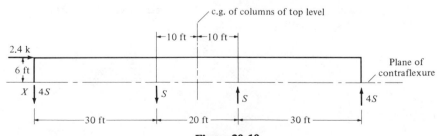

Figure 20-10

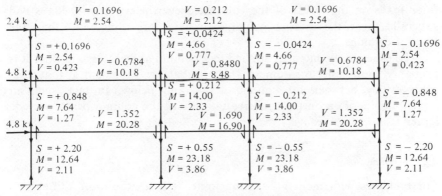

Figure 20-11 Frame analysis by the cantilever method.

20-8. ANALYSIS OF BUILDINGS FOR GRAVITY LOADS

Simple Framing

If simple framing is used, the design of the girders is quite simple because the shears and moments in each girder can be determined by statics. The gravity loads applied to the columns are relatively easy to estimate but the column moments may be a little more difficult. If the girder reactions on each side of the interior column of Fig. 20-12 are equal, no moment will theoretically be produced in the column at that level. This situation is probably not realistic because it is highly possible for the live load to be applied on one side of the column and not on the other (or at least be unequal in magnitude). The results will be column moments. If the reactions are unequal the moment produced in the column will equal the difference between the reactions times the distances to the c.g. of the column.

When computing the moments in exterior columns there may often be moments due to spandrel beams opposing the moments caused by the

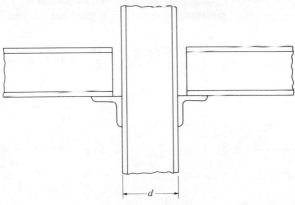

Figure 20-12

floor loads on the inside of the column. Nevertheless, gravity loads will generally cause the exterior columns to have larger moments than the interior columns.

To estimate the moment applied to a column above a certain floor it is probably reasonable to assume that the unbalanced moment at that level splits evenly between the column above and below. In fact, such an assumption is often on the conservative side as the column below may be larger than the one above.

Rigid Framing

For buildings with moment-resisting joints it is a little more difficult to estimate the girder moments and make preliminary designs. If the ends of each girder are assumed to be completely fixed, the moments for uniform loads are as shown in Fig. 20-13(a). Conditions of complete fixity are probably not realized, with the result that the end moments are smaller than shown in part (a) of the figure. There is a corresponding increase in the positive centerline moments approaching the simple moment $(wl^2/8)$ shown in part (b) of the figure. Probably a moment diagram somewhere in between the two extremes is more realistic. Such a moment diagram is represented by the dotted line of Fig. 20-13(a). A reasonable procedure is to assume a moment in the range of $1/10\,wl^2$, where l is the clear span.

When the beams and girders of a building frame are rigidly connected to each other a continuous frame is the result. From a theoretical standpoint an accurate analysis of such a structure cannot be made unless the entire frame is handled as a unit.

Before this subject is pursued further it should be realized that in the upper floors of tall buildings the wind moments will be small and the framed and seated connections described in Chapter 12 will provide sufficient moment resistance. For this reason the beams and girders of the upper floors may very well be designed on the basis of simple beam moments, while those of the lower floors may be designed as continuous members with moment-resistant connections because of the larger wind moments.

For the design of beams and girders it is necessary to consider two loading conditions: $(DL+LL)$ and $(DL+LL+WL)\frac{3}{4}$. In this latter ex-

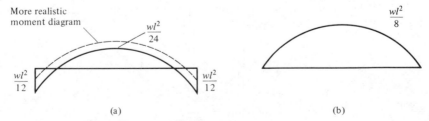

(a)   (b)

Figure 20-13   (a) Fixed-end beam. (b) Simple beam.

pression *WL* represents lateral forces whether due to wind or wind and earthquake and the $\frac{3}{4}$ factor represents the AISC one-third increase in allowable stresses when lateral forces are involved. For the upper floors the wind moments will not increase the girder sizes because of the allowable stress increase but on lower floors they will begin to cause increased sizes.

From a strictly theoretical viewpoint there are several live loading conditions which need to be considered to obtain maximum shears and moments at various points in a continuous structure. For the building frame shown in Fig. 20-14 it is desired to place live loads to cause maximum positive moment in span *AB*. A qualitative influence line for positive moment at the centerline of this span is shown in part (a) of the figure. This influence line shows that to obtain maximum positive moment at the centerline of span *AB* the live loads should be placed as shown in part (b) of the figure.

To obtain maximum negative moment at point *B* or maximum positive moment in span *BC*, other loading situations need to be considered. With the increased availability of electronic computers more of this detailed type of analysis is being done every day. The student, however, can see that unless the designer has access to a computer, she probably will not have sufficient time to go through all of these theoretical situations. Furthermore, it is doubtful if the accuracy of our analysis methods would justify all of the work anyway. It is, however, often feasible to take out two stories of the building at a time as a free body and analyze that part by one of the "exact" methods such as the successive correction method of moment distribution.

Before such an "exact" analysis can be made it is necessary to make an estimate of the member sizes probably based on the results of analysis by one of the approximate methods.

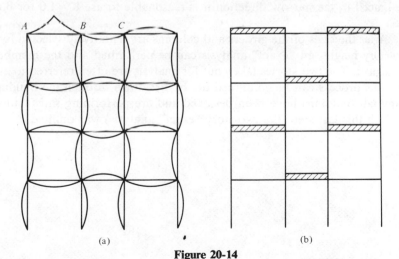

(a)   (b)

Figure 20-14

20-9.   DESIGN OF MEMBERS

Girders

The girders can be designed for the two load combinations mentioned in Section 20-8, $(DL+LL)$ and $(DL+LL+WL)\frac{3}{4}$. For the top several floors the second condition will not control design, but further down in the building it will control the sizes. As previously described in Chapter 13, the specifications may permit some considerable reduction in the live loads assumed to be supported by the girders.

Should simple framing be used the girders will be proportioned for simple beam moments plus the moments caused by the lateral loads. For continuous framing the girders will be proportioned for $wl^2/10$ (for uniform loads) plus the moments caused by the lateral loads. An interesting comparison of designs by the two methods is shown in pages 717–719 of Beedle et al., *Structural Steel Design* (New York: Ronald, 1964). It may be necessary to draw the moment diagram for the two cases (gravity loads and lateral loads) and add them together to obtain the maximum positive moment out in the span. The value so computed may very well control the girder size in some of the members.

A major part of the design of a multistory building is involved in setting up a column schedule which shows the loads to be supported by the various columns story by story. The gravity forces can be estimated very well while the shears, axial forces, and moments caused by lateral forces can be roughly approximated for the preliminary design from some approximate method (portal, cantilever, factor, or other).

If both axes of the columns are free to sway, the column effective lengths theoretically must be calculated for both axes as described in Chapter 8. If diagonal bracing is used in one direction, thus preventing sidesway, the K value will be less than 1.0 in that direction. When frames are braced in the narrow direction it is reasonable to use $K = 1.0$ for both axes.

After the sizes of the girders and columns are tentatively selected for a two-story height, an "exact" analysis can be performed and the members redesigned. The two stories taken out for analysis are often referred to as a *tier*. This process can be continued tier by tier down through the building. Many tall buildings have been designed and are performing satisfactorily in which this last step (the two-story "exact" analysis) was omitted.

Chapter 21
Plastic Analysis

21-1. INTRODUCTION

All designs performed in the previous chapters of this book were based on the elastic theory. The maximum load that a structure could support was assumed to equal the load that first caused a stress somewhere in the structure to equal the yield point of the material. Engineering structures have been designed for many decades by this method with satisfactory results. The design profession, however, has long been aware that ductile materials do not fail until a great deal of yielding occurs after the yield stress is reached. Chapters 21 and 22 are devoted to a consideration of the behavior of ductile steel structures in the plastic range.

When the stress at one point in a ductile steel structure reaches the yield point, that part of the structure will yield locally, permitting some readjustment of the stresses. Should the load be increased, the stress at the point in question will remain approximately constant, thereby requiring the less stressed parts of the structure to support the load increase. It is true that statically determinate structures can resist little load in excess of the amount that causes the yield stress to first develop at some point. For statically indeterminate structures, however, the load increase can be quite large; and these structures are said to have the happy facility of spreading out overloads due to the steel's ductility.

As early as 1914 a Hungarian, Dr. Gabor Kazinczy, recognized that the ductility of steel permitted a redistribution of stresses in an overloaded statically indeterminate structure.[1] In the United States Prof. J. A. Van den Broek introduced his plastic theory which he called limit design. This theory was published in a paper entitled "Theory of Limit Design" in February 1939 in the *Proceedings of the ASCE*.

In the plastic theory, rather than basing designs on the allowable-stress method the problem is handled by considering the greatest load that can be carried by the structure acting as a unit. The resulting designs are quite

[1] Lynn S. Beedle, *Plastic Design of Steel Frames* (New York: Wiley, 1958), p. 3.

interesting to the structural engineer as they offer several advantages. These include the following.

1. When the plastic-design procedure is used there can be considerable savings in steel (perhaps as high as 10 or 15% for some structures) as compared to a similar design made by the elastic procedure. It is true, however, that the necessary continuity connections required in plastic design may reduce the actual money savings somewhat.

2. Plastic design permits the designer to make a more accurate estimate of the maximum load that a structure can support, enabling him to have a better idea of the actual safety factor of the structure. The elastic method works very well for computing the stresses and strains for loads in the elastic range, but gives a very poor estimate of the actual collapse strength of a structure.

3. For many complicated structures plastic analysis is easier to apply than is elastic analysis.

4. Structures are often subjected to large stresses which are difficult to predict such as those caused by settlement, erection, etc. Plastic design provides for such situations by permitting plastic deformation.

Despite these several important advantages the acceptance of plastic design has been rather slow. Until recent years there has not been a great deal of information available concerning the ductility of steel. If a particular steel is brittle the plastic theory does not apply. A few decades ago when the design profession was just becoming interested in this theory there were several disastrous brittle failures of welded tanks and ships. (That also was the time when welded structures were beginning to gain popularity.) The acceptance of plastic design was slowed down to a walk for a long time, but in recent years much research has been performed in these areas, and plastic design is today gaining some acceptance.

Despite the great progress made in the field of plastic design the method still has some disadvantages with which the designer should be completely familiar. These include the following.

1. Plastic design is of little value for the high-strength brittle steels. (The method is just as applicable to high-strength steels as it is to those of lower strength structural grade steels as long as the steels have the required ductility.)

2. Plastic design today is not satisfactory for situations where fatigue stresses are a problem.

3. Columns designed by the plastic theory provide little savings.

4. Although for many statically indeterminate structures plastic analysis is simpler than elastic analysis, it should be realized that unstable plastic structures are more difficult to detect than are unstable elastic structures.

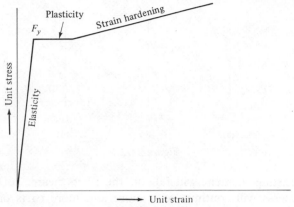

Figure 21-1

21-2. THEORY OF PLASTIC ANALYSIS

The basic theory of plastic design has been shown to be a major change in the distribution of stresses after the stresses at certain points in a structure reach the yield point. The theory is that those parts of the structure which have been stressed to the yield point cannot resist additional stresses. They instead will yield the amount required to permit the extra load or stresses to be transferred to other parts of the structure where the stresses are below the yield stress and thus in the elastic range and able to resist increased stress. Plasticity can be said to serve the purpose of equalizing stresses in cases of overload.

For this discussion the stress-strain diagram is assumed to have the idealized shape shown in Fig. 21-1. The yield point and the proportional limit are assumed to occur at the same point for this steel, and the stress-strain diagram is assumed to be a perfectly straight line in the plastic range. Beyond the plastic range there is a range of strain hardening. This latter range could theoretically permit steel members to withstand additional stress, but from a practical standpoint the strains occurring are so large that they cannot be considered.

21-3. THE PLASTIC HINGE

As the bending moment is increased at a particular section of a beam there will be a linear variation of stress until the yield stress is reached in the outermost fibers. An illustration of stress variation for a rectangular beam in the elastic range is shown in part (b) of Fig. 21-2. The *yield moment* of a cross section is defined as the moment that will just produce the yield stress in the outermost fiber of the section.

If the moment is increased beyond the yield moment the outermost fibers that had previously been stressed to their yield point will continue to have the same stress but will yield, and the duty of providing the necessary

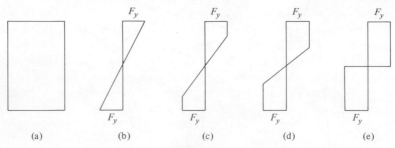

Figure 21-2

additional resisting moment will fall on the fibers nearer to the neutral axis. This process will continue with more and more parts of the beam cross section stressed to the yield point as shown by the stress diagrams of parts (c) and (d) of the figure, until finally a fully plastic distribution is approached as shown in part (e). When the stress distribution has reached this stage a *plastic hinge* is said to have formed because no additional moment can be resisted at the section. Any additional moment applied at the section will cause the beam to rotate with little increase in stress.

The *plastic moment* is the moment that will produce full plasticity in a member cross section and create a plastic hinge. The ratio of the plastic moment M_p to the yield moment M_y is called the *shape factor*. The shape factor equals 1.50 for rectangular sections and varies from about 1.10 to 1.20 for standard rolled-beam sections.

This paragraph is devoted to a description of the development of a plastic hinge in the simple beam shown in Fig. 21-3. The load shown is applied to the beam and increased in magnitude until the yield moment is reached and the outermost fiber is stressed to the yield point. The magnitude of the load is further increased with the result that the outer fibers begin to yield. The yielding spreads out to the other fibers away from the section of maximum moment as indicated in the figure. The length in which this yielding occurs away from the section in question is dependent on the loading conditions and the member cross section. For a concentrated load applied at the center line of a simple beam with a rectangular cross section, yielding in the extreme fibers at the time the plastic hinge is formed will extend for one-third of the span. For a W shape in similar circumstances yielding will extend for approximately one-eighth of the

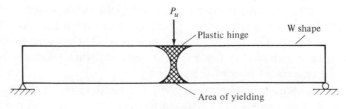

Figure 21-3

span. During this same period the interior fibers at the section of maximum moment yield gradually until nearly all of them have yielded and a plastic hinge is formed as shown in Fig. 21-3.

Although the effect of a plastic hinge may extend for some distance along the beam it is assumed to be concentrated at one section for analysis purposes. For the calculation of deflections and for the design of bracing, the length over which yielding extends is quite important.

When steel frames are loaded to failure, the points where rotation is concentrated (plastic hinges) become quite visible to the observer before collapse occurs.

21-4. THE PLASTIC MODULUS

The yield moment M_y equals the yield stress times the elastic modulus. The elastic modulus equals I/c or $bd^2/6$ for a rectangular section and the yield moment equals $F_y bd^2/6$. This same value can be obtained by considering the resisting internal couple shown in Fig. 21-4.

The resisting moment equals T or C times the lever arm between them, as follows.

$$M_y = \left(\frac{F_y bd}{4}\right)\left(\frac{2}{3}d\right) = \frac{F_y bd^2}{6}$$

The elastic section modulus can again be seen to equal $bd^2/6$ for a rectangular beam. The resisting moment of a rectangular section at full plasticity can be determined in a similar manner (see Fig. 21-5).

$$M_p = \left(F_y \frac{d}{2}b\right)\left(\frac{d}{2}\right) = \frac{F_y bd^2}{4}$$

The plastic moment is said to equal the yield stress times the plastic modulus. From the foregoing expression for a rectangular section, the plastic modulus Z can be seen to equal $bd^2/4$. The shape factor, which equals M_p/M_y, $F_y Z/F_y S$, or Z/S, is $(bd^2/4)/(bd^2/6) = 1.50$ for a rectangular section.

A study of the plastic modulus determined here shows that it equals the statical moment of the tension and compression areas about the neutral

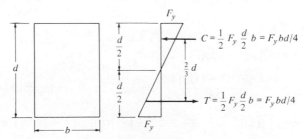

Figure 21-4

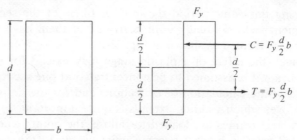

Figure 21-5

axis. Unless the section is symmetrical, the neutral axis for the plastic condition will not be in the same location as for the elastic condition. The total internal compression must equal the total internal tension. As all fibers are considered to have the same stress (F_y) in the plastic condition, the areas above and below the neutral axis must be equal. This situation does not hold for unsymmetrical sections in the elastic condition. Example 21-1 illustrates the calculations necessary to determine the shape factor for a tee beam and the ultimate uniform load w_u that the beam can support.

Example 21-1

Determine M_y, M_p, and Z for the steel tee beam shown in Fig. 21-6. Also calculate the shape factor and the ultimate uniform load (w_u) which can be placed on the beam for a 12-ft simple span. $F_y = 36$ ksi.

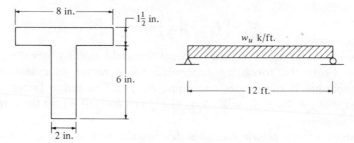

Figure 21-6

SOLUTION

Elastic calculations:

$$A = (8)\left(1\tfrac{1}{2}\right) + (6)(2) = 24 \text{ in.}^2$$

$$\bar{y} = \frac{(12)(0.75) + (12)(4.5)}{24} = 2.625 \text{ in. from top flange}$$

$$I = \left(\tfrac{1}{3}\right)(2)(1.125^3 + 4.875^3) + \left(\tfrac{1}{12}\right)(8)\left(1\tfrac{1}{2}\right)^3 + (12)(1.875)^2$$

$$= 122.4 \text{ in.}^4$$

$$S = \frac{I}{C} = \frac{122.4}{4.875} = 25.1 \text{ in.}^3$$

$$M_y = F_y S = \frac{(36)(25.1)}{12} = 75.3 \text{ ft-k}$$

Plastic calculations: Neutral axis is at base of flange.

$$Z = (12)(0.75) + (12)(3) = 45 \text{ in.}^3$$

$$M_p = F_y Z = \frac{(36)(45)}{12} = 135 \text{ ft-k}$$

$$\text{shape factor} = \frac{M_p}{M_y} \quad \text{or} \quad \frac{Z}{S} = \frac{45}{25.1} = 1.79$$

$$M_p = \frac{w_u l^2}{8}$$

$$w_u = \frac{(8)(135)}{(12)^2} = 7.5 \text{ k/ft}$$

The values of the plastic moduli for the standard steel beam sections are tabulated in the AISC Manual in the "Plastic Design Selection Table." These values will be frequently used in the pages to follow.

21-5. FACTORS OF SAFETY AND LOAD FACTORS

The factor of safety used in designing a particular structure should probably be selected only after a study is made of the uncertainties present in the design (a subject previously discussed in Chapter 1). Perhaps the usual practice, however, is to use a value that seems reasonable from the standpoint of past experience as expressed in the specifications being used. The usual safety factor considered to be present in elastic design is the one obtained by dividing the yield point of the steel by its working stress. For compact laterally supported beams of A36 steel this safety factor equals $\frac{36}{24} = 1.50$ using the AISC Specification.

If the safety factor is multiplied times the shape factor, the result is referred to as the *load factor*. For a typical W section the load factor equals the safety factor 1.50 times an average shape factor of W sections of about 1.12 equals 1.68 (say 1.70). In plastic design the working loads are multiplied by the load factor to give the estimated ultimate loads and members are proportioned with these loads on the basis of plastic or collapse strengths. This procedure is illustrated in the design examples of Chapter 22. In actual practice, load factors varying from roughly 1.70 to 2.00 are used depending on the individual designer's judgment of the particular conditions. The minimum values for load factors permitted by the AISC for design purposes are presented in Section 22-2.

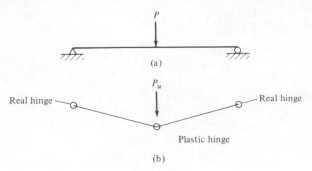

Figure 21-7

21-6.   THE COLLAPSE MECHANISM

A statically determinate beam will fail if one plastic hinge develops. To illustrate this fact, the simple beam of constant cross section loaded with a concentrated load at midspan shown in Fig. 21-7(a) is considered. Should the load be increased until a plastic hinge is developed at the point of maximum moment (underneath the load in this case) an unstable structure will have been created as shown in part (b) of the figure. Any further increase in load will cause collapse.

The plastic theory is of little advantage for statically determinate beams and frames but it may be of decided advantage for statically indeterminate beams and frames. For a statically indeterminate structure to fail it is necessary for more than one plastic hinge to form. The number of plastic hinges required for failure of statically indeterminate structures will be shown to vary from structure to structure, but may never be less than two. The fixed-end beam of Fig. 21-8 cannot fail unless the three plastic hinges shown in the figure are developed.

Although a plastic hinge may have formed in a statically indeterminate structure, the load can still be increased without causing failure if the geometry of the structure permits. The plastic hinge will act like a real hinge insofar as increased loading is concerned. As the load is increased there is a redistribution of moment because the plastic hinge can

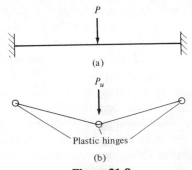

Figure 21-8

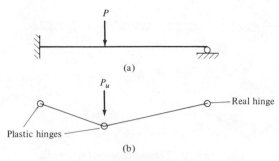

Figure 21-9

resist no more moment. As more plastic hinges are formed in the structure there will eventually be a sufficient number of them to cause collapse. Actually some additional load can be carried after this time before collapse occurs as the stresses go into the strain hardening range, but the deflections that would occur are too large to be permissible.

The propped beam of Fig. 21-9 is an example of a structure that will fail after two plastic hinges develop. Three hinges are required for collapse but there is a real hinge on the right end. In this beam the largest elastic moment caused by the design concentrated load is at the fixed end. As the magnitude of the load is increased a plastic hinge will form at that point.

The load may be further increased until the moment at some other point (here it will be at the concentrated load) reaches the plastic moment. Additional load will cause the beam to collapse. The arrangement of plastic hinges and perhaps real hinges which permit collapse in a structure is called the *mechanism*. Parts (b) of Figs. 21-7, 21-8, and 21-9 show mechanisms for various beams.

After observing the large number of fixed-end and propped beams used for illustration in this text, the student may form the mistaken idea that he will frequently encounter such beams in engineering practice. These types of beams, truthfully, are difficult to find in actual structures but are very convenient to use in illustrative examples. They are particularly convenient for introducing plastic analysis before continuous beams and frames are considered.

21-7. PLASTIC ANALYSIS BY THE EQUILIBRIUM METHOD

The method of plastic analysis known as the *equilibrium method* will be illustrated in this section for several beams. The analysis includes the computations of the plastic moments, the consideration of load redistribution after plastic hinges have formed, and the calculation of the ultimate loads that exist when the collapse mechanism is created.

As the first illustration the fixed-end beam of Fig. 21-10 is considered. This beam is assumed to support a load of 6.5 k/ft, including its own

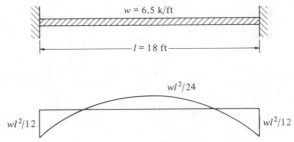

Figure 21-10   Elastic moment diagram.

estimated weight. A beam is selected by the elastic procedure using A36 steel and assuming full lateral support.

$$M = \frac{wl^2}{12} = \frac{(6.5)(18)^2}{12} = 175.5 \text{ ft-k}$$

$$S_{\text{reqd.}} = \frac{(12)(175.5)}{24} = 87.8 \text{ in.}^3$$

Use W18×50 ($S = 88.9$ in.^{3})

It is desired to determine the value of w_u, the ultimate uniform load that this W section can support before collapse.

The maximum moments in a uniformly loaded fixed-end beam in the elastic range occur at the fixed ends as shown in Fig. 21-10. If the magnitude of the uniform load is increased, plastic moments will eventually be developed at the beam ends as shown in Fig. 21-11(b). Although the plastic moment has been reached at the ends and plastic hinges formed, the beam cannot fail as it has, in effect, become a simple end-supported beam for further load increase as shown in part (c) of the figure.

The load can now be increased on this "simple" beam; and the moments at the ends will remain constant; but the moment out in the span

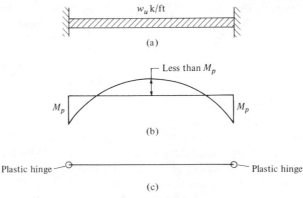

Figure 21-11

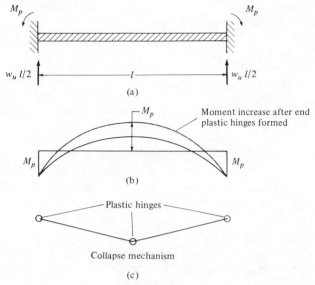

(a)

(b)

Moment increase after end
plastic hinges formed

Plastic hinges

Collapse mechanism

(c)

Figure 21-12

will increase as it would in a uniformly loaded simple beam. This increase is shown by the dotted line in Fig. 21-12(b). The load may be increased until the moment at some other point (here the beam center line) reaches the plastic moment. When this happens a third plastic hinge will have developed and a mechanism will have been created permitting collapse.

One method of determining the value of w_u is to take moments at the centerline of the beam (knowing the moment there is M_p at collapse). Reference is made here to Fig. 21-12(a) for the beam reactions. The value of Z was obtained from the AISC Manual.

$$M_p = -M_p + \left(w_u\frac{l}{2}\right)\left(\frac{l}{2} - \frac{l}{4}\right)$$

$$= \frac{w_u l^2}{16}$$

$$w_u = \frac{16 M_p}{l^2}$$

$$M_p = F_y Z = \frac{(36)(101)}{12} = 303 \text{ ft-k}$$

$$w_u = \frac{(16)(303)}{(18)^2} = 14.96 \text{ k/ft}$$

The plastic safety factor equals $14.96/6.50 = 2.30$. This value appears to be a more realistic value for a ductile steel structure than the one computed on the basis of the yield stress (about 1.5).

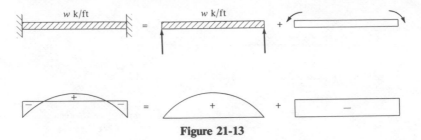

Figure 21-13

The same values could be obtained by considering the diagrams shown in Fig. 21-13. From structural analysis the student will remember that a fixed-end beam can be replaced with a simply supported beam plus a beam with end moments. Thus the final moment diagram for the fixed-end beam equals the moment diagram if the beam had been simply supported plus the end-moment diagram.

For the beam under consideration the value of M_p can be calculated as follows after studying Fig. 21-14.

$$2M_p = \frac{w_u l^2}{8}$$

$$M_p = \frac{w_u l^2}{16}$$

The propped beam of Fig. 21-15(a) has been designed by the elastic method to support a 50-k concentrated load at midspan. The W21×50 ($S=94.5$ in.^{3}, $Z=110$ in.^{3}) beam of A36 steel selected is now to be considered as a second illustration of plastic analysis. The elastic moment diagram for a propped beam loaded with a concentrated load P at midspan is shown in part (b) of the figure. From this diagram the maximum moment can be seen to occur at the fixed end. The concentrated load is assumed to be increased until the plastic moment is reached at the fixed end and a plastic hinge formed.

After this plastic hinge is formed the beam will act as though it is simply supported insofar as load increases are concerned, because it will have a plastic hinge at the left end and a real hinge at the right end. An increase in the magnitude of the load P will not increase the moment at the left end but will increase the moment out in the beam as it would in a simple beam. The increasing simple-beam moment is indicated by the dotted line in part (c) of Fig. 21-15. Eventually the moment at the concentrated load will reach M_p and a mechanism will form, consisting of two plastic hinges and one real hinge as shown in part (d).

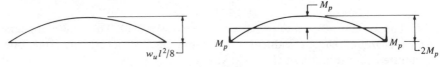

Figure 21-14

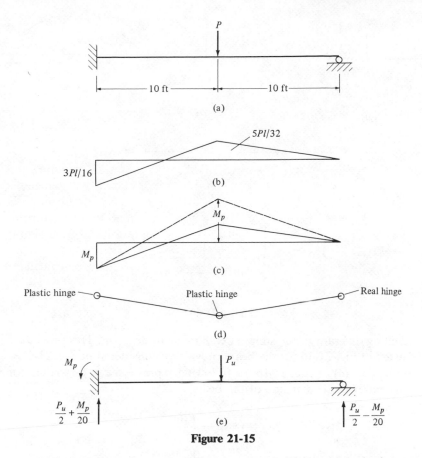

Figure 21-15

The value of the maximum concentrated load P_u which the beam can support can be determined by taking moments to the right or left of the load. Part (e) of Fig. 21-15 shows the beam reactions for the conditions existing just before collapse. Moments are taken to the right of the load as follows.

$$M_p = \left(\frac{P_u}{2} - \frac{M_p}{20} \right)(10)$$

$$= 3.33 P_u$$

$$P_u = 0.3 M_p$$

$$M_p = F_y Z = \frac{(36)(110)}{12} = 330 \text{ ft-k}$$

$$P_u = 99 \text{ k}$$

The other method described for handling the plastic analysis of a structure involved the drawing of the plastic-moment diagram on the structure after the structure had been made statically determinate by the creation of a sufficient number of plastic hinges and the equating of

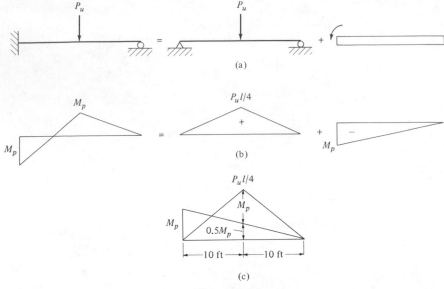

Figure 21-16

the resulting diagram to the simple-beam moment diagram. This procedure is repeated in Fig. 21-16 for the beam previously considered in Fig. 21-15.

From part (c) of Fig. 21-16 the following expression can be written for the moment at the center line of the beam.

$$M_p + 0.5M_p = \frac{P_u l}{4}$$

$$P_u = 0.3M_p$$

Another example is handled by a similar procedure in Fig. 21-17.

A propped beam with two concentrated loads is shown in Fig. 21-18(a). Design was performed by the elastic procedure for the 30-k and 50-k loads shown and a W24×84 ($S=196$ in.^{3}, $Z=224$ in.^{3}) was selected. It is now desired to determine the ultimate values of the two concentrated loads if they are increased at such a rate that they remain in the same proportion to each other. In part (b) of the figure the larger load at collapse is said to equal P_u and the smaller one $0.6P_u$.

As the loads are increased, the plastic moment will first be reached at the left end. After this plastic hinge forms the structure will be statically determinate since the right end is a real hinge. The loads may be increased until eventually another plastic hinge forms out in the span at one of the two concentrated loads. The point where the second plastic hinge will form is not obvious and it will be necessary to consider both possibilities.

Should the second plastic hinge be assumed to form at the point of application of the $0.6P_u$ concentrated load [see Fig. 21-18(d)] the values of

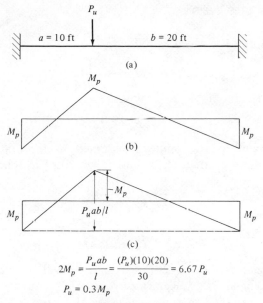

$$2M_p = \frac{P_u\,ab}{l} = \frac{(P_u)(10)(20)}{30} = 6.67\,P_u$$

$$P_u = 0.3\,M_p$$

Figure 21-17

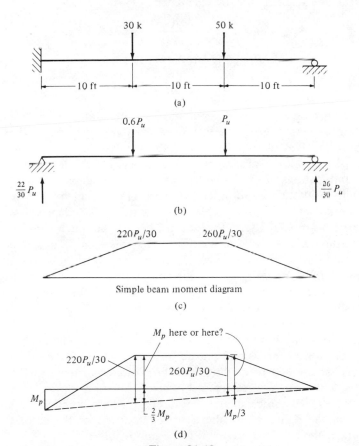

Figure 21-18

M_p and P_u can be determined as follows.

$$M_p + \frac{2}{3} M_p = \frac{220 P_u}{30}$$

$$M_p = 4.4 P_u$$

$$P_u = 0.227 M_p$$

If the second plastic hinge is assumed to form at the point of application of P_u, the values of M_p and P_u would be as follows.

$$M_p + \frac{M_p}{3} = \frac{260 P_u}{30}$$

$$M_p = 6.5 P_u$$

$$P_u = 0.154 M_p$$

The moment at the concentrated load P_u will obviously be greater than the moment at the concentrated load $0.6 P_u$ and the second plastic hinge will occur at P_u. The numerical value of P_u and the factor of safety for plastic analysis can now be calculated.

$$M_p = F_y Z = \frac{(36)(224)}{12} = 672 \text{ ft-k}$$

$$P_u = (0.154)(672) = 103.5 \text{ k}$$

$$\text{F.S.} = \frac{103.5}{50} = 2.07$$

The purpose of this problem is to show the student what to do when there is more than one feasible way in which a structure may collapse. The procedure involves the comparison of the various possibilities by making the necessary mathematical calculations for each. Actually there are some other combinations of hinges which might cause collapse for the beam of Fig. 21-18. Although these other cases are not critical for this particular beam the student will need to learn to consider all collapse possibilities. This same beam is analyzed in Section 21-8 by the much simpler virtual-work method and all of the possible mechanisms are considered. (The author felt that a consideration of all of the possibilities for the first example of this type using the equilibrium method would be confusing to the student and he only presented the two most obvious collapse possibilities to show how they would be compared to each other to determine which was the more critical.)

If there is any doubt in the student's mind as to whether or not the correct location of the second plastic hinge has been selected the moment diagrams for the two possibilities can be plotted as shown in Fig. 21-19. Should the student have selected the wrong location, the moment somewhere will exceed the calculated value of M_p. This situation is shown in part (c) of the figure where the assumed location of the second plastic hinge under the left load is shown to be impossible.

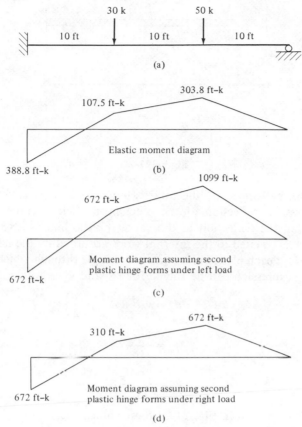

Figure 21-19

21-8. THE VIRTUAL-WORK METHOD

As the beams or frames become more complex the equilibrium method becomes more and more tedious to apply. A method is introduced in this section, called the *virtual-work method*, which will usually prove to be simpler and quicker than the equilibrium method. This procedure will be used to rework the problems previously considered in Section 21-7 and will be further illustrated in Chapter 22 for continuous beams and frames.

The structure is assumed to deflect through a small additional displacement after the ultimate load is reached. The work performed by the external loads during this displacement is equated to the internal work absorbed by the hinges. As a first illustration the uniformly loaded fixed-ended beam of Fig. 21-10 is considered. This beam and its collapse mechanism are redrawn in Fig. 21-20. Owing to symmetry, the rotations at the end plastic hinges are equal and they are represented by θ in the figure; thus the rotation at the middle plastic hinge will be 2θ.

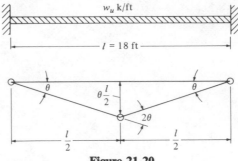

Figure 21-20

The work performed by the total external load ($w_u l$) is equal to $w_u l$ times the average deflection of the mechanism. The average deflection equals one-half the deflection at the center plastic hinge ($\frac{1}{2} \times \theta \times \frac{l}{2}$). The external work is equated to the internal work absorbed by the hinges or to the sum of M_p at each plastic hinge times the angle through which it works. The resulting expression can be solved for M_p and w_u as follows.

$$M_p(\theta + 2\theta + \theta) = w_u l \left(\theta \times \frac{l}{2} \times \frac{1}{2} \right)$$

$$M_p = \frac{w_u l^2}{16}$$

$$w_u = \frac{16 M_p}{l^2}$$

For the 18-ft span used in Fig. 21-10 these values become

$$M_p = \frac{(w_u)(18)^2}{16} = 20.25 w_u$$

$$w_u = \frac{M_p}{20.25}$$

Plastic analysis can be handled in a similar manner for the propped beam of Fig. 21-15. This beam is redrawn in Fig. 21-21 together with its

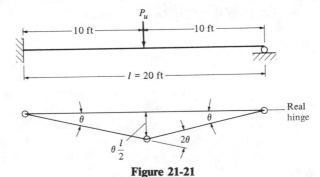

Figure 21-21

collapse mechanism. Again the end rotations are equal and are assumed to equal θ.

The work performed by the external load P_u as it moves through the distance $\theta l/2$ is equated to the internal work performed by the plastic moments at the hinges, noting that there is no moment at the real hinge on the right end of the beam.

$$M_p(\theta+2\theta)=P_u\left(\frac{l}{2}\right)$$

$$M_p=\frac{P_u l}{6} \qquad \text{(or 3.33 } P_u \text{ for the 20-ft beam shown)}$$

$$P_u=\frac{6M_p}{l} \qquad \text{(or 0.3 } M_p \text{ for the 20-ft beam shown)}$$

The fixed-end beam of Fig. 21-17 is redrawn in Fig. 21-22 together with its collapse mechanism and the assumed angle rotations. From this figure the values of M_p and P_u can be determined by virtual work as follows.

$$M_p(2\theta+3\theta+\theta)=P_u\left(2\theta\times\frac{l}{3}\right)$$

$$M_p=\frac{P_u l}{9} \qquad \text{(or 3.33 } P_u \text{ for this beam)}$$

$$P_u=\frac{9M_p}{l} \qquad \text{(or 0.3 } M_p \text{ for this beam)}$$

A person beginning the study of plastic analysis needs to learn to think of all the possible ways in which a particular structure might collapse. Such a habit is of the greatest importance when she begins to analyze more complex structures such as the frames of Chapter 22. In this light the plastic analysis of the propped beam of Fig. 21-18 is considered by the virtual-work method. The beam with its two concentrated loads is shown in Fig. 21-23 together with four possible collapse mechanisms and the necessary calculations. It is true that the mechanisms of parts (b), (d),

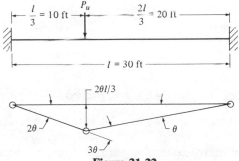

Figure 21-22

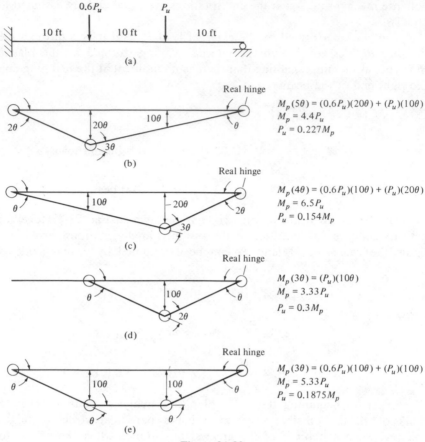

Figure 21-23

and (e) of the figure do not control, but such a fact is not obvious to the average student until she makes the virtual-work calculations for each case. Actually the mechanism of part (e) is based on the assumption that the plastic moment is reached at both of the concentrated loads simultaneously (a situation that might very well occur).

The value for which the collapse load is the smallest in terms of M_p is the correct value (or the value where M_p is the greatest in terms of P_u). For this beam the second plastic hinge forms at the P_u concentrated load and P_u equals $0.154\,M_p$.

21-9.   LOCATION OF PLASTIC HINGE FOR UNIFORM LOADINGS

There was no difficulty in locating the plastic hinge for the uniformly loaded fixed-end beam, but for other beams with uniform loads, such as propped or continuous beams, the problem may be rather difficult. For this discussion the uniformly loaded propped beam of Fig. 21-24(a) is considered.

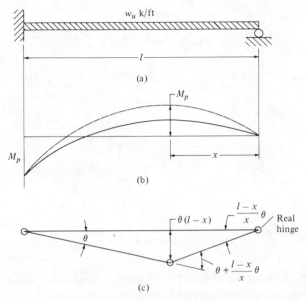

Figure 21-24

The elastic moment diagram for this beam is shown in part (b) of the figure. As the uniform load is increased in magnitude, a plastic hinge will first form at the fixed end. At this time the beam will, in effect, be a "simple" beam with a plastic hinge on one end and a real hinge on the other. Subsequent increases in the load will cause the moment to change as represented by the dotted line in part (b) of the figure. This process will continue until the moment at some other point (a distance x from the right support in the figure) reaches M_p and creates another plastic hinge.

The virtual work expression for the collapse mechanism of this beam shown in part (c) of Fig. 21-24 is written as follows.

$$M_p\left(\theta+\theta+\frac{l-x}{x}\theta\right)=(w_u l)(\theta)(l-x)\left(\frac{1}{2}\right)$$

Solving this equation for M_p, taking $dM_p/dx=0$, the value of x can be calculated to equal $0.414l$. This value is also applicable to uniformly loaded end spans of continuous beams with simple end supports as will be illustrated in Chapter 22.

The beam and its collapse mechanism are redrawn in Fig. 21-25 and the following expression for the plastic moment is written using the virtual-work procedure.

$$M_p(\theta+2.414\theta)=(w_u l)(0.586\theta l)\left(\frac{1}{2}\right)$$

$$M_p=0.0858w_u l^2$$

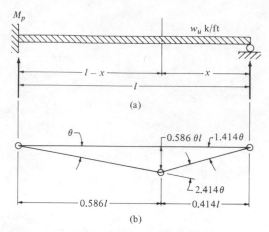

(a)

(b)

Figure 21-25

Problems

21-1 to 21-10. Find the values of S, Z, and the shape factor about the x axes for the sections shown in the accompanying illustrations.

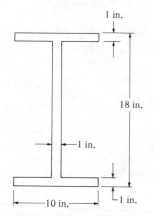

Problem 21-1 (*Ans.* 198.7, 234, 1.18)

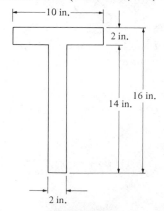

Problem 21-2

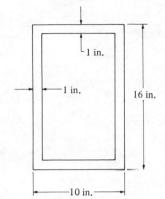

1 in.

1 in.

16 in.

10 in.

Problem 21-3 (*Ans*. 198, 248, 1.25)

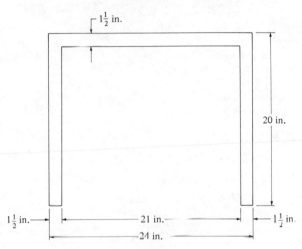

$1\frac{1}{2}$ in.

20 in.

$1\frac{1}{2}$ in. 21 in. $1\frac{1}{2}$ in.

24 in.

Problem 21-4

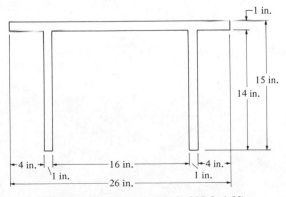

1 in.

15 in.

14 in.

4 in. 16 in. 4 in.

1 in. 1 in.

26 in.

Problem 21-5 (*Ans*. 114.8, 208.5, 1.82)

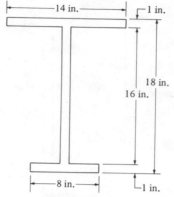

Problem 21-6

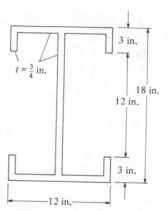

Problem 21-7   (*Ans*. 218.4, 254.4, 1.16)

Problem 21-8

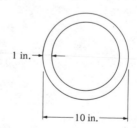

Problem 21-9   (*Ans*. 57.9, 81.3, 1.40)

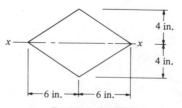

Problem 21-10

21-11   to 21-17. Determine the values of S, Z, and the shape factor about the x axes for the situations described.

21-11. A W30×173. (*Ans.* 533.9, 600.1, 1.124)

21-12. A W24×104 with one $\frac{3}{4}$ × 16 in. PL on each flange.

21-13. Two 8×6×$\frac{5}{8}$-in. Ls long legs vertical and back-to-back. (*Ans.* 19.7, 35.8, 1.82)

21-14. Two C10×30s back-to-back.

21-15. Four 8×8×$\frac{3}{4}$ in. Ls arranged as shown in the accompanying illustration. (*Ans.* 64.5, 104.2, 1.62)

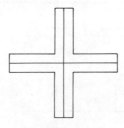

Problem 21-15

21-16. The section of Prob. 5-5(b).

21-17. The section of Prob. 5-12(b). (*Ans.* 205.5, 244.8, 1.19)

21-18. Rework Prob. 21-2 considering the y axis.

21-19. Rework Prob. 21-12 considering the y axis. (*Ans.* 96.4, 158.4, 1.64)

21-20. Rework Prob. 21-14 considering the y axis.

21-21   to 21-31. Select beam sections for the situations shown using the elastic theory, A36 steel, the AISC Specification, and assuming full lateral support. Calculate the factors of safety for the resulting beam sections according to the elastic and plastic theories. The loads shown in each figure are assumed to include the estimated beam weight.

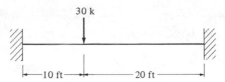

Problem 21-21 (*Ans.* Without 0.9 rule: W18×40, 1.54, 2.35. With 0.9 rule: W16×40, 1.46, 2.19)

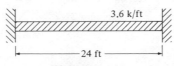

Problem 21-22

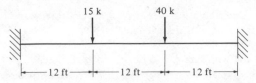

Problem 21-23 (*Ans.* Without 0.9 rule: W21×62, 1.50, 2.27. With 0.9 rule: W24×55, 1.35, 2.12)

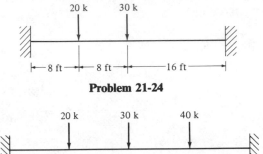

Problem 21-24

Problem 21-25 (*Ans.* Without 0.9 rule: W27×84, 1.55, 2.44. With 0.9 rule: W24×84, 1.43, 2.24)

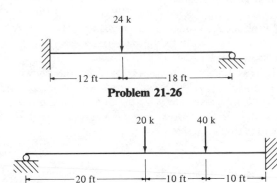

Problem 21-26

Problem 21-27 (*Ans.* Without 0.9 rule: W27×84, 1.55, 2.74. With 0.9 rule: W24×84, 1.43, 2.52)

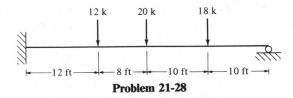

Problem 21-28

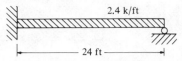

Problem 21-29 (*Ans*. Without 0.9 rule: W18×50, 1.54, 2.55. With 0.9 rule: W21×44, 1.42, 2.41)

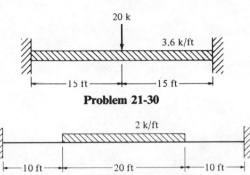

Problem 21-30

Problem 21-31 (*Ans*. Without 0.9 rule: W21×50, 1.55, 2.20. With 0.9 rule: W18×50, 1.45, 2.02)

Chapter 22
Plastic Analysis and Design

22-1. INTRODUCTION TO PLASTIC DESIGN

The shape factor multiplied by the safety factor was defined as the *load factor* in Chapter 21 and was shown to have an average value of approximately 1.70 for beams. To design a beam by the plastic method, the estimated working load is multiplied by the load factor; the plastic moment is calculated and a member is selected which furnishes the required plastic modulus $(Z = M_p / F_y)$. Example 22-1 presents a comparison of elastic and plastic designs for a uniformly loaded fixed-end beam.

Example 22-1

Design the beam shown in Fig. 22-1 by the plastic and elastic methods. The uniform load is 3 k/ft and includes the estimated beam weight. Use A36 steel, the AISC Specification, and assume full lateral support.

SOLUTION
Plastic analysis:

$$M_p(\theta + 2\theta + \theta) = (w_u l)\left(\frac{1}{2}\right)\left(\theta \frac{l}{2}\right)$$

$$= \frac{w_u l^2}{16}$$

Plastic design:

$$w_u = (1.70)(3) = 5.1 \text{ k/ft}$$

$$M_p = \frac{(5.1)(24)^2}{16} = 183.6 \text{ ft-k}$$

$$Z_{\text{reqd.}} = \frac{(12)(183.6)}{36} = 61.2 \text{ in.}^3$$

Use W18×35

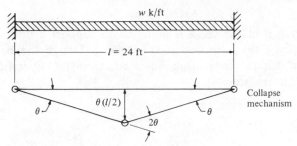

Figure 22-1

Elastic design:

$$-M = \frac{wl^2}{12} = \frac{(3)(24)^2}{12} = -144 \text{ ft-k}$$

$$+M = \frac{wl^2}{24} = \frac{(3)(24)^2}{24} = +72 \text{ ft-k}$$

$-M$ for design $= (0.9)(-144) = -129.6$ ft-k $\left.\begin{array}{l} \\ \\ \end{array}\right\}$ (AISC Section
$+M$ for design $= 72 + (0.10)\left(\dfrac{144+144}{2}\right) = +86.4$ ft-k $\left.\begin{array}{l} \\ \end{array}\right\}$ 1.5.1.4.1)

$$S_{\text{reqd.}} = \frac{(12)(129.6)}{24} = 64.8 \text{ in.}^3$$

Use W18×40. Percent weight saving by plastic design $= \frac{5}{40} = 12.5\%$

Plastic design is permissible for statically determinate structures but has little if any economic advantage. To illustrate this fact, the design of a simply supported beam by the two methods is presented in Example 22-2. The shape factor advantage (about 12%) has already been substantially used in the 10% higher allowable bending stress permitted by the AISC for compact laterally supported sections. As only one plastic hinge is needed to cause a statically determinate structure to be unstable, there is no redistribution of load after the hinge forms.

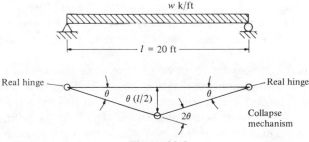

Figure 22-2

Example 22-2

The beam shown in Fig. 22-2 is assumed to support a uniform load of 4 k/ft (including the estimated beam weight). Design the beam by the plastic and elastic methods using A36 steel, the AISC Specification, and assuming full lateral support.

SOLUTION
Plastic analysis:

$$M_p(2\theta) = (w_u l)\left(\frac{1}{2}\theta\frac{l}{2}\right)$$

$$= \frac{w_u l^2}{8}$$

Plastic design:

$$w_u = (1.70)(4) = 6.8 \text{ k/ft}$$

$$M_p = \frac{(6.8)(20)^2}{8} = 340 \text{ ft-k}$$

$$Z_{\text{reqd.}} = \frac{(12)(340)}{36} = 113.3 \text{ in.}^3$$

Use W24×55

Elastic design:

$$M = \frac{(4)(20)^2}{8} = 200 \text{ ft-k}$$

$$S_{\text{reqd.}} = \frac{(12)(200)}{24} = 100 \text{ in.}^3$$

Use W24×55

If all of the plastic hinges needed to produce the collapse mechanism for a statically indeterminate structure form at the same time, plastic design will yield the same section as will elastic design. For such situations there is no load redistribution after the first hinge forms. This same situation was shown to occur in statically determinate structures where the formation of one plastic hinge was sufficient to cause collapse (see Example 22-2).

Three plastic hinges are required to form a mechanism for a fixed-end beam, but the elastic moment diagram shown in Fig. 22-3 shows that if such a beam is loaded with a concentrated load at its centerline the hinges will all theoretically form at the same time. Example 22-3 illustrates the design of this beam. The student does not have to worry about overlooking a situation where there is no redistribution or where all of the plastic hinges form simultaneously. The virtual-work procedure will yield the correct expression for M_p regardless of the order in which the hinges are formed.

Example 22-4 presents the design of a fixed-end beam loaded with a concentrated load at a point other than midspan. For this beam there is appreciable load redistribution after the first and second hinges form.

Example 22-3

Design the beam shown in Fig. 22-3 by the plastic and elastic methods using A36 steel, the AISC Specification, and full lateral support. Assume the load given takes into account the estimated beam weight.

SOLUTION
Plastic analysis:

$$M_p(\theta + 2\theta + \theta) = P_u(15\theta)$$

$$M_p = 3.75 P_u$$

Plastic design:

$$P_u = (1.70)(40) = 68 \text{ k}$$

$$M_p = (3.75)(68) = 255 \text{ ft-k}$$

$$Z_{\text{reqd.}} = \frac{(12)(255)}{36} = 85 \text{ in.}^3$$

Use W21×44

Elastic design (no advantage in 0.9 rule in AISC Section 1.5.1.4.1):

$$M = \frac{(40)(30)}{8} = 150 \text{ ft-k}$$

$$S_{\text{reqd.}} = \frac{(12)(150)}{24} = 75 \text{ in.}^3$$

Use W21×44

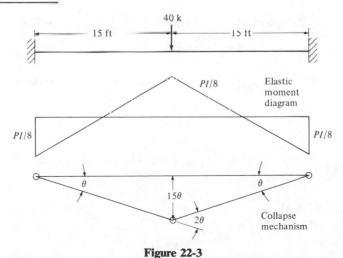

Figure 22-3

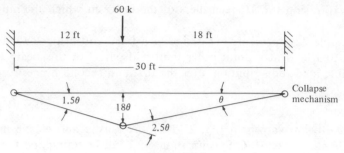

Figure 22-4

Example 22-4

Design the beam shown in Fig. 22-4 by the plastic and elastic methods using A36 steel, the AISC Specification, and assuming full lateral support. Assume the load given takes into account the estimated beam weight.

SOLUTION
Plastic analysis:
$$M_p(1.5\theta + 2.5\theta + \theta) = (P_u)(18\theta)$$
$$= 3.6P_u$$

Plastic design:
$$P_u = (1.70)(60) = 102 \text{ k}$$
$$M_p = (3.6)(102) = 367.2 \text{ ft-k}$$
$$Z_{\text{reqd.}} = \frac{(12)(367.2)}{36} = 122.4 \text{ in.}^3$$

<u>Use W24×55</u>

Elastic design: The elastic moment diagram is shown in Fig. 22-5.
$$-M \text{ for design} = (0.9)(-259) = -233.1 \text{ ft-k}$$
$$+M \text{ for design} = +207.4 + (0.10)\left(\frac{259 + 173}{2}\right) = +229 \text{ ft-k}$$
$$S_{\text{reqd.}} = \frac{(12)(233.1)}{24} = 116.6 \text{ in.}^3$$

Use W21 × 62 or W24×62. Percent weight saving by plastic design $= \frac{7}{62}$
$$= 11.3\%$$

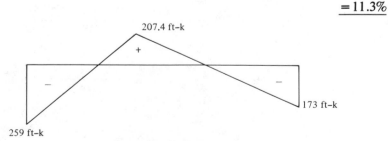

Figure 22-5

22-2. AISC REQUIREMENTS FOR PLASTIC DESIGN

Several of the more important requirements of the AISC Specification pertaining to plastic design are presented as follows.

1. Plastic design is permissible for simple and continuous beams and for braced and unbraced planar rigid frames. In addition, portions of structures rigidly constructed so as to be continuous over at least one interior support can be designed plastically.

2. The plastic design procedure can be used for steels having specified yield points up to 65 ksi or 448 MPa (A572, Grade 65).

3. The load factors used in design may not be less than 1.7 for the given dead and live loads nor less than 1.3 for those loads acting in conjunction with a load factor of 1.3 for any wind or seismic forces. (The 1.3 value is obtained by making use of the one-third increase in allowable stresses for wind and seismic loadings provided in AISC Section 1.5.6. This is the same thing as multiplying the load by $\frac{3}{4}$ or in this case the load factor).

4. The webs of beams and girders or columns that are not stiffened in some manner (as with diagonal stiffeners or doubler plates) shall be designed so that the ultimate shear, V_u, in kips does not exceed $0.55F_y td$ where t is the web thickness and d is the member depth.

5. The members of a plastically designed structure must be prevented from local and lateral buckling until the plastic hinges develop. To prevent such buckling the AISC provides minimum width-thickness ratios for the flanges of S, W, and similar built-up shapes as well as maximum permissible depth-thickness ratios for the webs of such members.

The minimum width-thickness ratios $(b_f/2t_f)$ are given in Section 2.7 of the AISC Specification and equal 8.5 for $F_y = 36$ ksi, 8.0 for $F_y = 42$ ksi, etc. In addition the width-thickness ratio of the flange plates in box sections and cover plates are not permitted to be greater than $190/\sqrt{F_y}$.

The depth-thickness ratio of the webs of members subject to plastic bending may not exceed the values given by the equations to follow in which P_y equals the member area times F_y.

$$\frac{d}{t} = \frac{412}{\sqrt{F_y}}\left(1 - 1.4\frac{P}{P_y}\right) \qquad \text{(AISC Formula 2.7-1a)}$$

when

$$\frac{P}{P_y} \leqslant 0.27$$

$$\frac{d}{t} = \frac{257}{\sqrt{F_y}} \qquad \text{(AISC Formula 2.7-1b)}$$

when

$$\frac{P}{P_y} > 0.27$$

6. To prevent lateral and torsional displacements at the plastic hinge locations associated with the collapse mechanism the members must be braced in those locations and at distances from the hinges not exceeding the value l_{cr} given by the following equation.

$$\frac{l_{cr}}{r_y} = \frac{1375}{F_y} + 25 \qquad \text{(AISC Formula 2.9-1a)}$$

when

$$+1.0 > \frac{M}{M_p} > -0.5$$

$$\frac{l_{cr}}{r_y} = \frac{1375}{F_y} \qquad \text{(AISC Formula 2.9-1b)}$$

when

$$-0.5 \geqslant \frac{M}{M_p} > -1.0$$

where

r_y = the r of the member about its weak axis, inches.
M = smaller moment at the ends of the unbraced length, ft-k.
M/M_p = end moment ratio which is positive when the member is bent in reverse curvature and negative when bent in single curvature

These requirements do not have to be followed in the vicinity of the last hinge to form but the lateral bracing requirements for elastic design must be met.

7. Web stiffeners are frequently required at plastic hinges to prevent hinges forming in a member at its point of connection to another member. For such situations it may be necessary to use web stiffeners as required by Section 1.15 of the AISC Specification.

8. If the edges of structural steel members in the vicinity of the plastic hinges are sheared, the edges should be ground, chipped, or planed smooth. Furthermore, holes in similar locations must be drilled or subpunched and reamed to full size.

9. There are several other AISC requirements for plastic design which refer in detail to columns, haunched members, connections, etc.

Example 22-5 illustrates the design of a uniformly loaded propped beam using the AISC Specification. The W section selected in this problem is a compact shape, but the ratios required for such shapes are checked to illustrate their application.

Example 22-5

Design a section for the span and loading shown in Fig. 22-6 using A36 steel and the AISC Specification. Lateral support is assumed to be provided at 4-ft intervals.

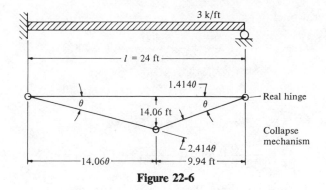

Figure 22-6

SOLUTION

Plastic Analysis:

$$M_p(\theta + 2.414\theta) = (w_u)(24)(14.06\theta)(\tfrac{1}{2})$$
$$= 49.4w_u$$

Plastic Design:

$$w_u = (1.70)(3) = 5.1 \text{ k/ft}$$
$$M_p = (49.4)(5.1) = 252 \text{ ft-k}$$
$$Z_{\text{reqd.}} = \frac{(12)(252)}{36} = 84 \text{ in.}^3$$

Try W21×44 ($A = 13.0$, $d = 20.66$, $b_f = 6.500$, $t_f = 0.450$, $t_w = 0.350$, $d/A_f = 7.06$ $r_y = 1.26$)

Checking AISC Requirements: Shear and moment diagrams just before collapse shown in Fig. 22-7.

1. Shear:
$$0.55F_y\, td = (0.55)(36.0)(0.350)(20.66) = 143.2 \text{ k} > 73.1 \text{ k} \qquad \text{OK}$$

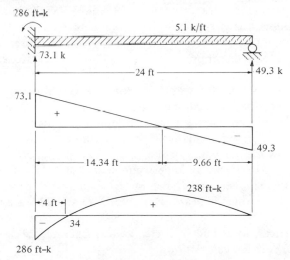

Figure 22-7

2. Local buckling:

$$\frac{b_f}{2t_f} = \frac{6.500}{(2)(0.450)} = 7.2 < 8.5 \qquad \text{OK}$$

$$\frac{P}{P_y} = \frac{0}{P_y} \text{ since no axial load}$$

Therefore, use AISC Formula 2.7-1a as follows

$$\frac{d}{t} = \frac{20.66}{0.350} = 59.0 < \frac{412}{\sqrt{36}} \left(1 - 1.4\frac{0}{P_y}\right) = 68.7 \qquad \text{OK}$$

3. Lateral buckling: Plastic hinge at support

$$1.0 > \frac{M}{M_p} = -\frac{34}{286} = -0.119 > -0.5$$

Therefore, use AISC Formula 2.9-1a as follows

$$\frac{l_{cr}}{r_y} = \frac{4 \times 12}{1.26} = 38.1 < \frac{1375}{36} + 25 = 63.2 \qquad \text{OK}$$

Interior plastic hinge is the last one to form (therefore, see AISC Section 1.5.1.4.1)

$$\left.\begin{array}{l} L_c = \dfrac{76.0 \times 6.50}{\sqrt{36}} = 82.3 \\[4mm] L_c = \dfrac{20,000}{7.06 \times 36} = 78.7 \end{array}\right\} \quad \text{Both} > 4 \times 12 = 48 \text{ in.} \qquad \text{OK}$$

22-3. CONTINUOUS BEAMS

Continuous beams are very common in engineering structures. Their continuity causes analysis to be rather complicated in the elastic theory, and even though one of the complex "exact" methods is used for analysis, the resulting stress distribution is not nearly so accurate as is usually assumed.

Plastic analysis is applicable to continuous structures as it is to one-span structures. The resulting values definitely give a more realistic picture of the limiting strength of a structure than can be obtained by elastic analysis. Continuous statically indeterminate beams can be handled by the virtual-work procedure or the equilibrium procedures as they were for the single-span statically indeterminate beams. All of the example problems of this chapter are handled by the virtual-work procedure. As an introduction to continuous beams Examples 22-6 and 22-7 are presented to illustrate two of the more elementary cases. These designs should be checked for shear, local buckling, etc., as was Example 22-5, but space is not taken to show the calculations.

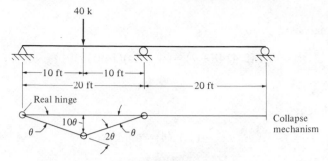

Figure 22-8

Example 22-6

Select a steel shape for the two-span beam shown in Fig. 22-8 using A36 steel, the AISC Specification, and assuming full lateral support.

SOLUTION
Plastic analysis:

$$M_p(2\theta+\theta)=(P_u)(10\theta)$$
$$=3.33P_u$$

Plastic design:

$$P_u=(1.70)(40)=68 \text{ k}$$
$$M_p=(3.33)(68)=226.7 \text{ ft-k}$$
$$Z_{reqd.}=\frac{(12)(226.7)}{36}=75.6 \text{ in.}^3$$

Use W18×40

Example 22-7

Select a section for the beam shown in Fig. 22-9. Use A36 steel, the AISC Specification, and assume full lateral support.

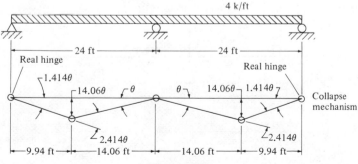

Figure 22-9

SOLUTION
Plastic analysis:

$$M_p(2.414\theta+\theta)=(w_u)(24)(14.06\theta)\left(\tfrac{1}{2}\right)$$
$$=49.4w_u$$

Plastic design:

$$w_u=(1.7)(4)=6.8 \text{ k/ft}$$
$$M_p=(49.4)(6.8)=336 \text{ ft-k}$$

$$Z_{\text{reqd.}}=\frac{(12)(336)}{36}=112 \text{ in.}^3$$

Use W18×55 or W24×55

Additional spans have little effect on the amount of work involved in the plastic design procedure. The same cannot be said for elastic design. Example 22-8 illustrates the design of a three-span beam which is loaded with a concentrated load on each span. The student from his knowledge of elastic analysis can see that plastic hinges will initially form at the first interior supports and then at the centerlines of the end spans, at which time each end span will have a collapse mechanism.

Example 22-8

Select a steel section for the beam and loading shown in Fig. 22-10 using A36 steel, the AISC Specification, and assuming full lateral support.

SOLUTION
Plastic analysis:

$$M_p(2\theta+\theta)=P_u(15\theta)$$
$$=5P_u$$

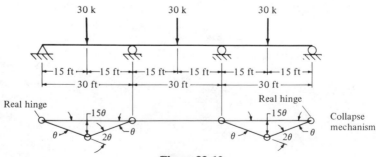

Figure 22-10

Plastic design:

$$P_u = (1.70)(30) = 51 \text{ k}$$

$$M_p = (5)(51) = 255 \text{ ft-k}$$

$$Z_{reqd.} = \frac{(12)(255)}{36} = 85 \text{ in.}^3$$

Use W21×44

A uniformly loaded two-span beam with equal spans was designed in Example 22-7. A two-span uniformly loaded beam with unequal spans is designed in Example 22-9. If the same size section is used for both spans, the longer span will obviously collapse first, and the shorter span will be decidedly overdesigned. Nevertheless this procedure is used in this example.

Example 22-9

Select a section for the beam shown in Fig. 22-11 using A36 steel, the AISC Specification, and assuming satisfactory lateral support.

SOLUTION
Plastic analysis:

$$M_p(\theta + 2.414\theta) = (24w_u)(14.06\theta)(\tfrac{1}{2})$$

$$= 49.4w_u$$

Plastic design:

$$w_u = (1.70)(6) = 10.2 \text{ k/ft}$$

$$M_p = (49.4)(10.2) = 504 \text{ ft k}$$

$$Z_{reqd.} = \frac{(12)(504)}{36} = 168 \text{ in.}^3$$

Use W24×68

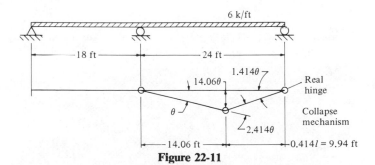

Figure 22-11

It may be more economical to select a section for the shorter span with its smaller moment and run it through both spans. Cover plates can be added in the longer span where the moment is greater than the plastic resisting moment of the section selected. The procedure is illustrated in Example 22-10 for the beam designed in Example 22-9. In the example a W21×44 with a plastic moment resistance of 286.2 ft-k is selected for the shorter span. The moment diagram for this beam (Fig. 22-13) shows that cover plates are needed for 15.64 ft in the right-hand span where the moment exceeds 286.2 ft-k.

This paragraph is devoted to the derivation of an expression for determining cover-plate sizes and is almost identical with the derivation presented in Section 16-1. For this discussion Z is the required plastic modulus for the entire moment, Z_s the plastic modulus for the section selected for the shorter span, d the depth of the section selected for the shorter span, t_p the thickness of one cover plate, and A_p is the area of one cover plate. An expression for the required area A_p can be written as follows.

$$Z = Z_s + 2A_p\left(\frac{d}{2} + \frac{t_p}{2}\right)$$

$$A_p = \frac{Z}{d+t_p} - \frac{Z_s}{d+t_p}$$

The cover plates should probably be extended a little distance beyond their theoretical points of cutoff. Some designers extend the plates 6 to 12 in. on each end, while others extend them until half of the strength of the plates is developed by the welds connecting them to the beam.

From a standpoint of economy some consideration should be given to using the W24×68 (selected for the 24-ft span in Example 22-9) for both spans. Its weight for the two spans totals 2856 lb while the weight of the W21×44 with the cover plates in the longer span (selected in Example 22-10) weighs 2426 lb. The use of cover plates saved 430 lb of steel but their connection costs may very well have exceeded the cost of 430 lb of structural steel.

Example 22-10

Select a section for the 18-ft span of the beam of Example 22-9 (see Fig. 22-12) and design the necessary cover plates for the 24-ft span.

SOLUTION
Plastic analysis for 18-ft span:

$$M_p(2.414\theta + \theta) = (18w_u)(10.55\theta)\left(\tfrac{1}{2}\right)$$

$$= 27.8w_u$$

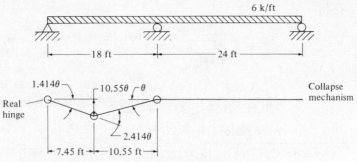

Figure 22-12

Plastic design for 18-ft span:

$$w_u = (1.70)(6) = 10.2 \text{ k/ft}$$
$$M_p = (27.8)(10.2) = 283.6 \text{ ft-k}$$
$$Z_{\text{reqd.}} = \frac{(12)(283.6)}{36} = 94.5 \text{ in.}^3$$

Use W21×44 $(Z = 95.4 \text{ in.}^3, d = 20.66)$

$$\text{actual } M_p = \frac{(36)(95.4)}{12} = 286.2 \text{ ft-k}$$

Drawing of moment diagram is shown in Fig. 22-13. Design of cover plates

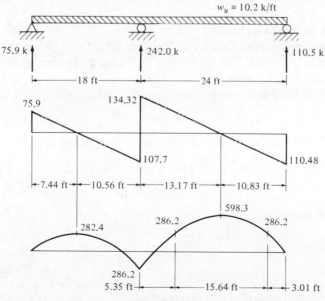

Figure 22-13

for long span:

$$Z_{\text{reqd.}} = \frac{(12)(598.3)}{36} = 199.4 \text{ in.}^3$$

Assuming $t_p = \frac{1}{2}$ in.:

$$A_p = \frac{199.4}{20.66 + 0.50} - \frac{95.4}{20.66 + 0.50} = 4.91 \text{ in.}^2$$

Try $10 \times \frac{1}{2}$-in. cover plates $\left(A_p = 5.0 \text{ in.}^2 \right)$

$$Z = 95.4 + (2)(5.0)(10.58) = 201.2 \text{ in.}^3 > 199.4 \text{ in.}^3 \qquad \text{OK}$$

Use $\frac{1}{2} \times 10$-in. cover plates 17.0 ft long

22-4. PLASTIC ANALYSIS OF FRAMES

The pin-supported frame of Fig. 22-14(a) is statically indeterminate to the first degree. The development of one plastic hinge will cause it to become statically determinate while the forming of a second hinge can create a mechanism. There are, however, several possible mechanisms that might feasibly occur in the frame. A possible beam mechanism is shown in part (c) of the figure, a sidesway mechanism is shown in part (d), and a possible combination beam and sidesway mechanism in part (e). The critical condition is the one that will cause the smallest value of P_u.

Example 22-11 presents the plastic analysis of the frame of Fig. 22-14. The distances through which the loads move in the virtual-work procedure should be carefully studied. *The solution of this problem shows that superposition does not apply to plastic analysis, a point of major significance.* The situation shown in part (c) of Fig. 22-14 added to the situation of part (d) does not equal that of part (e). In other words the effects of different loads cannot be determined separately and added together as they can in elastic analysis. The three-fourths AISC rule, representing a one-third increase in allowable stresses when lateral loads are involved, is used in this problem.[1]

Example 22-11

Determine an expression for the plastic moment in the frame of Fig. 22-14.

SOLUTION
Case (c) of figure:

$$M_p(\theta + 2\theta + \theta) = P_u(20\theta)$$

$$M_p = 5P_u$$

---

[1] The AISC Specification reflects the $\frac{3}{4}$ reduction by permitting the load factor of 1.7 for frames to be reduced to 1.3 when lateral loads are involved. Using these load factors in design it is unnecessary to use the $\frac{3}{4}$ value as done in Examples 22-11 through 22-13.

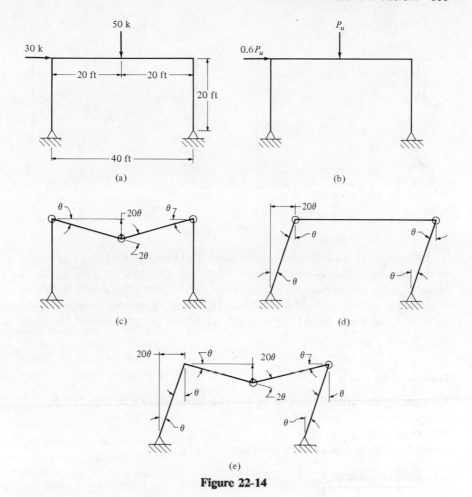

Figure 22-14

Case (d) of figure:

$$M_p(\theta+\theta)=\left(\tfrac{3}{4}\right)(0.6P_u)(20\theta)$$
$$M_p=4.5P_u$$

Case (e) of figure:

$$M_p(2\theta+2\theta)=\left(\tfrac{3}{4}\right)(0.6P_u)(20\theta)+\left(\tfrac{3}{4}\right)(P_u)(20\theta)$$
$$M_p=6P_u \quad \text{(controls)}$$

Critical value is the smallest value of P_u (or the largest value of M_p in terms of P_u).

$$M_p=6P_u$$
$$P_u=0.167M_p$$

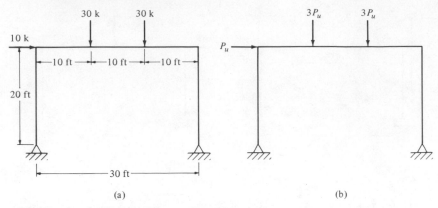

Figure 22-15

Example 22-12 is similar to Example 22-11 except that there are even more possible collapse mechanisms. Parts (a), (b), and (c) of Fig. 22-16 show possible beam collapse mechanisms, while part (d) shows a sidesway mechanism. Parts (e) and (f) of the figure show possible combination sidesway and beam mechanisms. Study carefully the angles used for parts (e) and (f).

Example 22-12

Determine the critical value of P_u for the frame shown in Fig. 22-15.

SOLUTION
Possible collapse mechanisms are shown in Fig. 22-16.

$$\text{Critical value of } P_u = (1/16.25)M_p$$

Example 22-13 illustrates the analysis of a frame with fixed supports. The fact that the frame has two more redundants than the hinged frames of the last two examples in no way makes its plastic analysis more difficult.

Example 22-13

Determine the critical value of P_u for the frame shown in Fig. 22-17.

SOLUTION
Possible collapse mechanisms are shown in Fig. 22-18.

$$\text{Critical value of } P_u = (1/15)M_p$$

For many frames the locations of the plastic hinges are not obvious. The usual procedure for such cases is to assume a mechanism and compute the

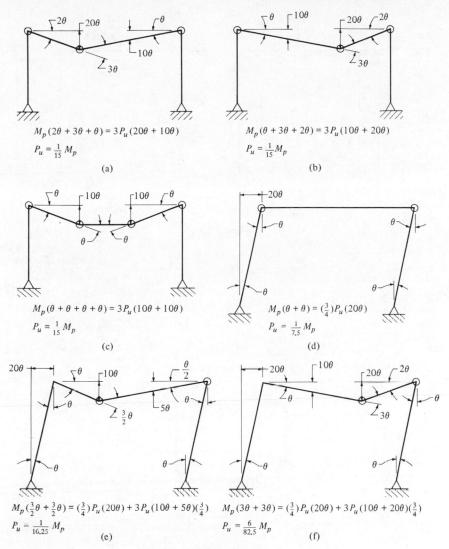

$$M_p(2\theta + 3\theta + \theta) = 3P_u(20\theta + 10\theta)$$
$$P_u = \frac{1}{15}M_p$$

(a)

$$M_p(\theta + 3\theta + 2\theta) = 3P_u(10\theta + 20\theta)$$
$$P_u = \frac{1}{15}M_p$$

(b)

$$M_p(\theta + \theta + \theta + \theta) = 3P_u(10\theta + 10\theta)$$
$$P_u = \frac{1}{15}M_p$$

(c)

$$M_p(\theta + \theta) = (\tfrac{3}{4})P_u(20\theta)$$
$$P_u = \frac{1}{7.5}M_p$$

(d)

$$M_p(\tfrac{3}{2}\theta + \tfrac{3}{2}\theta) = (\tfrac{3}{4})P_u(20\theta) + 3P_u(10\theta + 5\theta)(\tfrac{3}{4})$$
$$P_u = \frac{1}{16.25}M_p$$

(e)

$$M_p(3\theta + 3\theta) = (\tfrac{3}{4})P_u(20\theta) + 3P_u(10\theta + 20\theta)(\tfrac{3}{4})$$
$$P_u = \frac{6}{82.5}M_p$$

(f)

Figure 22-16

value of M_p. Should the correct hinge location have been assumed, the value of M_p will be the correct value. If the wrong hinge location was assumed the value of M_p will be too small.

Several other mechanisms can be assumed and M_p quickly computed for each. The largest value of M_p computed will be the one used. Although the mechanism finally used may not be exactly the right one, it will be satisfactory for all practical purposes. (It might be noted that the moment diagram can be drawn as a check for any one of the assumed conditions. If the moment at no point in the frame exceeds the calculated value of M_p, the correct mechanism has been used.)

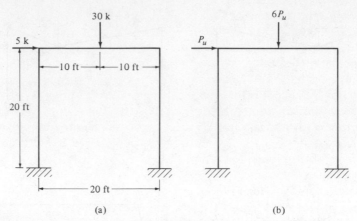

(a)   (b)

Figure 22-17

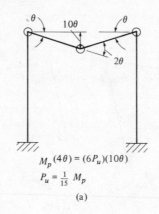

$$M_p(4\theta) = (6P_u)(10\theta)$$
$$P_u = \frac{1}{15}\,M_p$$

(a)

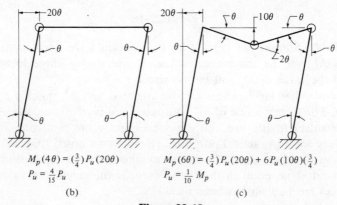

$$M_p(4\theta) = (\tfrac{3}{4})P_u(20\theta)$$
$$P_u = \frac{4}{15}P_u$$

(b)

$$M_p(6\theta) = (\tfrac{3}{4})P_u(20\theta) + 6P_u(10\theta)(\tfrac{3}{4})$$
$$P_u = \frac{1}{10}\,M_p$$

(c)

Figure 22-18

For a frame supporting concentrated and uniform loads, the expected positions of the plastic hinges will be at the concentrated loads. When a structure supports only uniform loads there will be a large (actually unlimited) number of possible hinge locations. After a few mechanism locations have been assumed and M_p determined for each, it is possible to narrow down the "critical hinge" location within a foot or two.

Shear stresses are often high within the boundaries of the rigid connection of two or more members whose webs lie in a common plane. Such a situation occurs at the rigid-frame knee in part (a) of Fig. 22-19. Since most of the moment in a beam is resisted by the flanges, an internal resisting couple is assumed to be located at the centers of the flanges with a lever arm of $0.95d_b$ as shown in part (b) of the figure. The magnitude of each of these forces V can be determined as $V = M_p/0.95d_b$.

This force is applied perpendicularly to the column as a shearing force. The column must have a shearing strength equal to V. If not, it will be necessary to use diagonal stiffeners as shown in part (c) of the figure, or reinforcing or doubler plates in contact with the web over the connection area.

The ultimate shearing strength of the column web $(0.55F_y td_c)$ must be equal to V or stiffening is required. From this information the required web thickness can be determined. In the expression developed, A_{bc} equals the planar area of the connection web $(d_b d_c)$. If M_p is assumed to be in foot-kips, the following expression, from Section 2.5 of the "Commentary on AISC Specification" can be developed.

$$0.55F_y td_c = \frac{M_p(12)}{0.95d_b}$$

$$t = \frac{23M_p}{A_{bc}F_y}$$

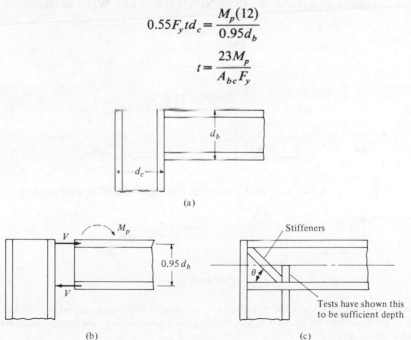

(a)

(b)   (c)

Figure 22-19

Should the shearing strength of the column web be insufficient, the diagonal stiffeners shown in part (c) of Fig. 22-19 can be used. These stiffeners act like the diagonals of a truss to carry the excess shear. The flange force V must be resisted by the web and the stiffener. An expression for the required area of the stiffener A_{st} is developed as follows.

$$\frac{M_p}{0.95d_b} = 0.55F_y td_c + F_y A_{st}\cos\theta$$

$$A_{st} = \frac{1}{\cos\theta}\left(\frac{M_p}{0.95d_b F_y} - 0.55td_c\right)$$

A similar discussion can be made for the elastic design of rigid knee connections and the AISC formula recommended in Section 1.5.1.2 of the "Commentary on AISC Specification."

Problems*

22-1 to 22-8. Select sections by the elastic and plastic methods for each of the beams shown.

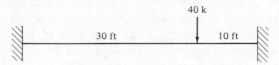

Problem 22-1 (*Ans.* W24×55, with 0.9, W24×55, W21×44 plastic)

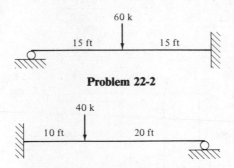

Problem 22-2

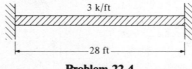

Problem 22-3 (*Ans.* W24×55, with 0.9, W24×55, W21×44 plastic)

Problem 22-4

Note: Use A36 steel, the AISC Specification, and assume full lateral support is provided for all of the beams and frames shown. The working loads given in each case include estimates of the structure weight.

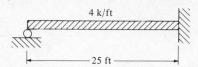

Problem 22-5 (*Ans.* W24×76, with 0.9, W24×68, W24×55 plastic)

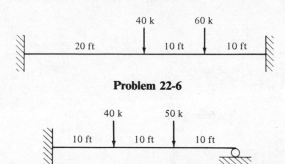

Problem 22-6

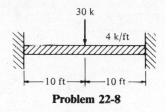

Problem 22-7 (*Ans.* W274×94, with 0.9, W27×84, W24×76 plastic)

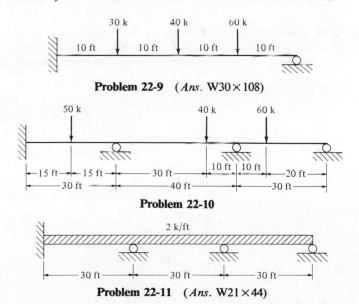

Problem 22-8

22-9 to 22-15. Select by the plastic method a single rolled section that will be satisfactory for the entire structure for each of the beams shown.

Problem 22-9 (*Ans.* W30×108)

Problem 22-10

Problem 22-11 (*Ans.* W21×44)

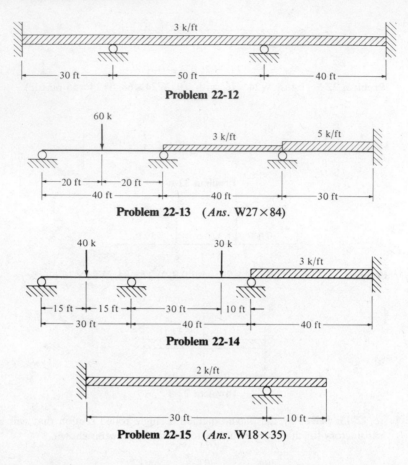

Problem 22-12

Problem 22-13 (*Ans.* W27×84)

Problem 22-14

Problem 22-15 (*Ans.* W18×35)

22-16 to 22-18. Select a single rolled section for the beams shown in the accompanying illustrations and draw shear and moment diagrams for the sections selected using the collapse loads used in design. Use plastic method only.

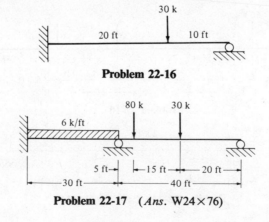

Problem 22-16

Problem 22-17 (*Ans.* W24×76)

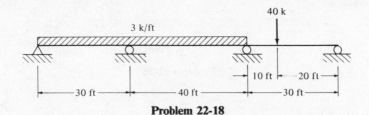

Problem 22-18

22-19 to 22-27. Select a single rolled section for each of the beams shown which will be satisfactory for the entire structure. Then select a section for the span that has the smallest moment and design cover plates as necessary for the other parts of the beam. Compare the weight differences between the two procedures. Use plastic method only.

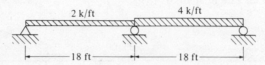

Problem 22-19 (*Ans*. W18×35, W14×22 with $\frac{1}{2}$×6×14 ft–0 in. PLs)

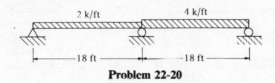

Problem 22-20

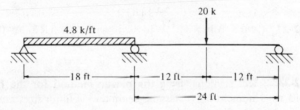

Problem 22-21 (*Ans*. W18×40, W16×31 with $\frac{1}{2}$×4×11 ft–0 in. PLs)

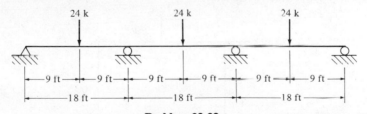

Problem 22-22

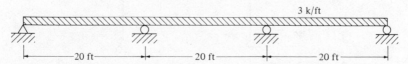

Problem 22-23 (*Ans*. W18×35, W16×26 with $\frac{3}{8}$×4×11 ft–0 in. PLs)

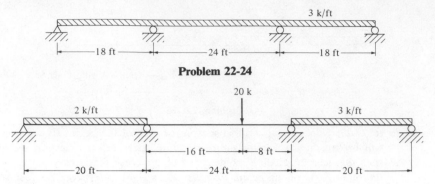

Problem 22-24

Problem 22-25 (*Ans.* W18×35, W14×22 with $\frac{1}{4}$×3×9 ft-0 in. PLs in span 1 and $\frac{5}{8}$×4×14 ft-0 in. PLs in span 3)

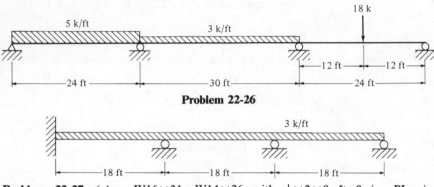

Problem 22-26

Problem 22-27 (*Ans.* W16×31, W14×26 with $\frac{1}{4}$×3×8 ft-0 in. PLs in span 3)

22-28 to 22-33. Select sections using the plastic method for the frames shown considering moment only. Assume beam and column sizes are the same for each frame.

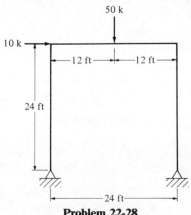

Problem 22-28

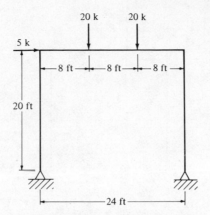

Problem 22-29 (*Ans.* W16×31)

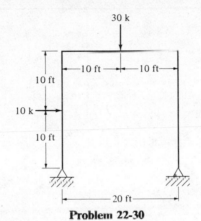

Problem 22-30

Problem 22-31 (*Ans.* W12×22 or W14×22)

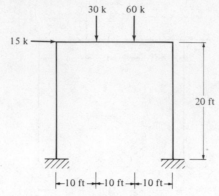

Problem 22-32

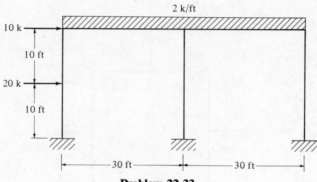

Problem 22-33

Chapter 23
Miscellaneous Topics

23-1. INTRODUCTION

In this chapter the author has placed a few topics that were only briefly mentioned in earlier chapters. It was felt that they either did not fit into these chapters or would have made them too long. The subjects presented here are as follows: the design of moment resisting column bases, ponding on flat roofs, and load factor design.

23-2. MOMENT RESISTING COLUMN BASES

Frequently column bases are designed to resist bending moments as well as axial loads. An axial load causes compression between a base plate and the supporting footing while a moment increases the compression on one side and decreases it on the other side. When the moments are relatively small the forces may be transferred to the footing through flexure of the base plate, but when the moments are large, stiffened or booted connections will be needed as described in this section.

Should the eccentricity of the load ($e = M/P$) be relatively small so that the load falls between the column flanges, the entire contact area between the plate and the supporting footing will remain in compression. For this situation the moment is usually sufficiently small to permit its transfer to the footing by bending of the base plate. The anchor bolts will not have calculable stresses but they are nevertheless considered necessary for good construction practice. They are definitely needed to hold the columns firmly in place and upright during the initial steel erection process. Temporary guy cables are of course necessary during erection. The anchor bolts cannot be logically designed but they should be substantial and capable of resisting unforeseen erection forces.

Should the eccentricity be sufficiently large so that the load falls outside of the column flange, there will be uplift on the other side of the column and one of the moment resisting connections of the types shown in Figure 23-1 will be needed. Anchor bolts capable of developing the

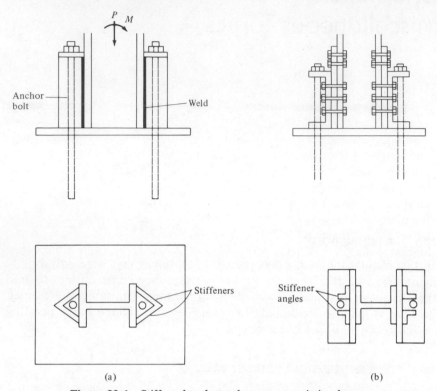

Figure 23-1 Stiffened or booted moment resisting bases.

moment will be connected to the column flanges and vertical stiffeners or boots will be used to provide sufficient room or length to transfer the anchor bolt forces to the column. The moment will be transferred from the column into the footing by means of the anchor bolts which will be embedded a sufficient distance into the footing to develop the anchor bolt forces. This distance should be calculated as required by reinforced concrete design methods. The booted connection shown in Fig. 23-1(a) is assumed to be welded to the column while the one of part (b) is assumed to be bolted to the column. The boots are generally made of angles or channels and are not as a rule connected directly to the base plate. Rather the moment is transmitted from the column to the foundation by means of the anchor bolts.

When a moment resisting or rigid connection between a column and its footing is used it is absolutely necessary for the supporting soil or rock beneath the footing to be appreciably noncompressible or the column base will rotate as shown in Figure 23-2. If this happens, the rigid connection between the column and the footing is useless. For the purpose of this section the subsoil is assumed to be capable of resisting the moment applied to it without appreciable rotation.

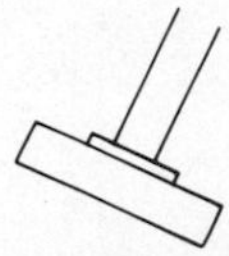

Figure 23-2

As a first numerical example a column base plate is designed for an axial load and a relatively small bending moment such that the resultant load falls between the column flanges. Assumptions are made for the width and length of the plate after which the pressures underneath the plate are calculated and compared to the permissible value. If the pressures are unsatisfactory the dimensions are changed and the pressures recalculated, until the values are satisfactory. The moment in the plate is calculated and the plate thickness determined. The critical section for bending is assumed to be at the center of the flange on the side where the compression is highest. Various designers will assume the point of maximum moment is located at some other point as at the face of the flange or the center of the anchor bolt.

The moment is calculated for a 1-in.-wide strip of the plate and is equated to its resisting moment. The resulting expression is solved for the required thickness of the plate as follows.

$$M = \frac{F_b I}{c} = \frac{(F_b)(\frac{1}{12})(1)(t)^3}{t/2}$$

$$t = \sqrt{\frac{6M}{F_b}}$$

Example 23-1

Design a moment resisting base plate to support a W14×120 column with an axial load of 300 k and a bending moment of 115 ft-k. A36 steel with $F_b = 27$ ksi and a concrete footing with $F_p = 1125$ psi is assumed.

SOLUTION

Using a W14×120 ($d = 14.48$ in., $b_f = 14.670$ in., $t_f = 0.940$ in.)

$$e = \frac{(12)(115)}{300} = 4.60 \text{ in.}$$

Therefore the resultant falls between the column flanges.

Try a 20×28 in. plate (after a few trials)

$$f = -\frac{P}{A} \pm \frac{Pec}{I} = -\frac{300}{(20)(28)} \pm \frac{(300)(4.60)(14)}{(\frac{1}{12})(20)(28)^3}$$

$$= -0.536 \pm 0.528 \begin{cases} -1.064 < 1.125 \text{ ksi} \\ -0.008 \text{ ksi} \end{cases} \qquad \text{OK}$$

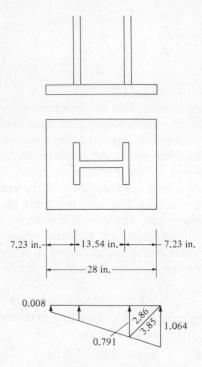

Taking moments to right at center of right flange

$$M = (2.86)\left(\frac{7.23}{3}\right) + (3.85)\left(\frac{2}{3} \times 7.23\right) = 25.45 \text{ in.-k}$$

$$t = \sqrt{\frac{6M}{F_b}} = \sqrt{\frac{(6)(25.45)}{27}} = 2.38 \text{ in.}$$

Use PL$2\frac{1}{2} \times 20$ in. $\times 2$ ft–4 in.

The moment considered in Example 23-2, which follows, is of such a magnitude that the resultant load falls outside the column flange. As a result there will be uplift on one side and the anchor bolt will have to furnish the needed tensile force to provide equilibrium.

In this design the anchor bolts are assumed to have no significant tension due to tightening. As a result they are assumed not to affect the force system. As the moment is applied to the column the pressure shifts toward the flange on the compression side. It is assumed that the resultant of this compression is located at the center of the flange.

Example 23-2

Repeat Example 23-1 with the same column and allowable stresses but with the moment increased from 115 ft-k to 225 ft-k.

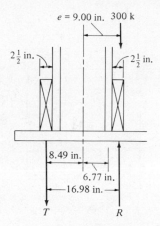

SOLUTION

Using a W14×120 ($d=14.48$ in., $b_f=14.670$ in., $t_f=0.940$ in.)

$$e=\frac{(12)(225)}{300}=9.00 \text{ in.}$$

Therefore, the resultant falls outside of the column flange.

Taking moments about the center of the right flange

$$(300)(9.00-6.77)-15.26T=0$$

$$T=43.84 \text{ k}$$

$$\text{anchor bolt area reqd.}=\frac{43.84}{22}=1.99 \text{ in.}^2$$

Use 2 in. diameter bolts (tensile stress area $=2.50$ in.^{2})

Checking pressure under the right flange

$$R=300+43.84=343.84 \text{ k}$$

$$f_a=\frac{343.84}{(14.67)(0.940)}=24.93 \text{ ksi}<27 \text{ ksi} \qquad \text{OK}$$

Approximate plate size assuming a triangular pressure distribution

$$\text{area reqd.}=\frac{343.84}{1.125/2}=611.3 \text{ in.}^2$$

Try a 30-in.-long plate

The load is $\frac{30}{2}-6.77=8.23$ in. from edge of plate and thus the pressure triangle will be 24.69 in. long and the required plate will be

$$\frac{343.84}{\frac{1}{2}\times1.125\times24.69}=24.76 \text{ in.} \qquad \underline{\text{say 26 in.}}$$

If the plate is made 26 in. wide the pressure zone will have an area of $26\times24.69=641.94$ in.^{2} and the maximum pressure will be twice the aver-

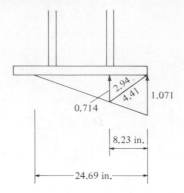

age pressure or

$$\frac{343.84}{641.94} \times 2 = 1.071 \text{ ksi}$$

Taking moments to right at center of flange

$$M = (2.94)\left(\frac{8.23}{3}\right) + (4.41)\left(\frac{2}{3} \times 8.23\right) = 32.26 \text{ in.-k}$$

$$t = \sqrt{\frac{(6)(32.26)}{27}} = 2.68 \text{ in.}$$

Use PL$2\frac{3}{4}$ in. $\times$ 26 in. $\times$ 2 ft–6 in.

In these examples the variation of stress in the concrete supporting the columns has been assumed to vary in a triangular or straight line fashion. It is possible to work with an assumed ultimate concrete theory where the concrete in compression under the plate is assumed to fail at a stress of $0.85 f_c'$. Examples of such designs are available in several texts.[1, 2]

23-3. PONDING

An introduction to the subject of ponding was presented in Section 7-4 of this text. The amount of water which accumulates on a flat roof is purely dependent on the flexibility of its framing. If the framing has little stiffness the ponding may very well cause the roof to collapse.

The AISC Specification presents a rational approach to the subject of ponding on roof surfaces that do not have sufficient slopes and/or drains to prevent the accumulation of water. This is the procedure that is described in this section.

---

[1]William McGuire, *Steel Structures* (Englewood Cliffs, N.J.: Prentice-Hall, 1968), pp. 987–1004.
[2]Carl L. Shermer, *Design in Structural Steel* (New York, N.Y.: Ronald Press, 1972), pp. 254–263.

In the AISC Commentary a more exact method is presented. This method should be applied when the check for stiffness required by the AISC Specification (described in this section) is not satisfied.

According to the AISC Specification (1.13.3) a roof is considered stable if it meets the following criteria.

$$C_p + 0.9C_s \leqslant 0.25 \quad \text{and} \quad I_d = \frac{25S^4}{10^6}$$

(for metal decks *not* for reinforced concrete)

where

$$C_p = \frac{32 L_s L_p{}^4}{10^7 I_p}$$

$$C_s = \frac{32 S L_s{}^4}{10^7 I_s}$$

L_p = column spacing in direction of girders or primary members, thus length of primary members, in feet

L_s = column spacing perpendicular to direction of girders, thus length, in feet, of secondary members

S = lateral spacing in feet of secondary members

I_p = moment of inertia (in.^{4}) of primary members

I_s = moment of inertia of secondary members (in.^{4})

I_d = moment of inertia of steel deck (in.^{4}) supported on secondary members

The AISC further states that the moment of inertia I_s must be decreased by 15% for trusses and steel joists. Furthermore they consider that a steel deck is to be considered a secondary member when it is directly supported by the primary members. Finally, the total bending stress due to dead loads, gravity live loads (if any), and ponded water may not exceed $0.80F_y$ for primary or secondary members. Example 23-3, which follows, illustrates the application of the AISC ponding provisions.

Example 23-3

The beams and girders shown in Fig. 23-3 support a flat 4-in. reinforced concrete roof slab. The slab is assumed to support a 6-psf built-up roof and a 30-psf live load. Using A36 steel and the AISC Specification check to see if the roof is adequate as regards ponding.

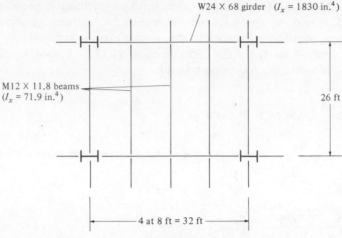

Figure 23-3

SOLUTION

$$C_p = \frac{32 L_s L_p{}^4}{10^7 I_p} = \frac{(32)(26)(32)^4}{(10)^7(1830)} = 0.0477$$

$$C_s = \frac{32 S L_s{}^4}{10^7 I_s} = \frac{(32)(8)(26)^4}{(10)^7(71.9)} = 0.1627$$

$$C_p + 0.9 C_s = 0.0477 + (0.9)(0.1627) = 0.1941$$

$$< 0.25 \qquad\qquad\qquad \text{OK}$$

Note: It is unnecessary to check I_d since the structural deck is reinforced concrete. The system is safe as regards ponding failure.

23-4 LOAD AND RESISTANCE FACTOR DESIGN

At the present time the AISC Specification consists of Part 1 dealing with allowable stress design and Part 2, which deals with plastic design. It is anticipated that within a very short time the AISC will add a Part 3 to their specification which will permit Load and Resistance Factor Design (or it may be that Part 2 will be replaced with this method). This method, abbreviated LRFD, is a design procedure that combines the calculation of ultimate or limit states of strength and serviceability with a probability based approach to safety.

In Canada a probability based LRFD method has already been adopted for both hot-rolled and cold-formed steel structures while much progress has been made in Europe toward the goal of formulating LRFD procedures for various national codes. Studies are being made in preparing similar procedures for timber and reinforced concrete structures in both

Europe and America.[3] As a result, great increases in the use of LRFD will seemingly be made during the next decade.

It is hoped that with LRFD (which is also referred to as a "limit states design") the designer will be able to make full use of his previous design experience and judgment as well as latest load test information. In LRFD the members and connections are proportioned so that strength and serviceability limit states exceed factored load combinations.

The basic criterion of LRFD is that the theoretical or nominal capacity of a member multiplied by an undercapacity or resistance factor of less than 1.0 must at least equal an analysis factor greater than 1.0 multiplied by the sum of each load effect times its load or overcapacity factor. This criterion can be expressed as follows.

(resistance factor)(nominal capacity of member)

⩾(analysis factor)(Σ of load effects times their respective load factors.)

For instance, in referring to the plastic moment strength the following type of expression may be written

$$0.86M_p \geqslant 1.1(1.1M_D + 1.4M_L)$$

The left-hand side of this expression refers to the resistance or capacity of the structure while the right-hand side refers to the load effects on the structure. This criterion can be expressed in the following form.

$$\phi R_n \geqslant \gamma_A \sum_{i=1}^{n} \gamma_i Q_i$$

The resistance side of the expression equals the theoretical or nominal capacity of the member (R_n) multiplied by the resistance or undercapacity factor (ϕ). R_n can represent moments, shears, axial forces, etc. The resistance factor ϕ is a number less than 1.0 which takes into account the uncertainties present in calculating the theoretical resistance or capacity of a member. Among these uncertainties are such items as variations in material properties (such as yield stress or ultimate tensile stress) and deviations in member thickness, depths, straightness, etc.

In the right-hand side of the equation the sum of the products of the load effects (called Q_i) and the overload factors (called γ_is) is multiplied by an analysis factor γ_A. The value of γ_A which is larger than 1.0 is selected to estimate the effect of the uncertainties of structural analysis. For instance, for analysis, the joints of a truss are usually assumed to consist of frictionless pins whereas they actually consist of welded or bolted connections that are definitely not frictionless. Almost all beam supports are assumed to be simple or fixed whereas they actually are somewhere in between.

[3]M. K. Ravindra and T. V. Galambos, "Load and Resistance Factor Design for Steel," *ASCE Journal of Structural Division* **104**, no. ST9 (September 1978), pp. 1337–1353.

The γ_i values are load factors usually larger than 1.0 which supposedly account for the uncertainties involved in estimating the magnitudes of dead and live loads. The value of γ_i used for dead loads is smaller than the γ_i used for live loads because the designer can estimate much more closely their magnitudes than he can the magnitudes of live loads. In this regard the student will notice that loads which remain in place for long time spans will be less variable in magnitude while those which are applied for brief periods such as wind loads will have larger variations. It is hoped that LRFD will make the designer more conscious of load variations than he is with the allowable stress and plastic design methods.

It is possible that γ_i for dead loads can occasionally be less than 1.0. One situation where this would be the case occurs where the dead load increases the resistance to some situation, as say, overturning.

At Washington University in St. Louis, Missouri, a research project was conducted on LRFD from 1969 through 1976 under the direction of T. V. Galambos and M. K. Ravindra. At the conclusion of the project, a document was published entitled *Proposed Criteria for Load and Resistance Factor Design of Steel Building Structures.*[4]

The proposed criteria in this document are not complete by themselves as they are meant to be subordinate to various parts of the AISC Specification. They deal with the design of the same kinds of structures covered by the present AISC Specification, that is, structural steel buildings fabricated with hot-rolled steel sections. It thus can be seen that this criteria does not cover bridge design, designs with cold-formed steel members, nor nuclear structure designs.

The criteria contains detailed design provisions including recommended ϕ and γ values for various types of designs as well as helpful charts, tables, and other design aids. Sufficient information is furnished to enable the design engineer to select reasonable ϕ and γ factors for unusual cases.

For the design of a beam by plastic moments the basic expression for LRFD may be written as follows.

$$\phi M_p \geqslant \gamma_A(\gamma_D M_D + \gamma_L M_L)$$

The values of ϕ and γ can be selected from the criteria previously mentioned or from other papers published by Galambos and Ravindra.[5, 6]

Using this information for beams subjected to dead and live loads, $\phi = 0.86$, $\gamma_A = 1.1$, $\gamma_D = 1.1$, and $\gamma_L = 1.4$. Thus the LRFD expression becomes

$$0.86 M_p \geqslant 1.1(1.1 M_D + 1.4 M_L)$$

[4] Research Report 45, C.E. Dept., Washington University, St. Louis, Mo., May 1976.
[5] T. V. Galambos and M. K. Ravindra, "Proposed Criteria for Load and Resistance Factor Design," *Engineering Journal*, AISC, **15**, no. 1 (First Quarter, 1978), pp. 8–17.
[6] T. V. Galambos and M. K. Ravindra, "Load and Resistance Factor Design for Steel," *ASCE Journal of the Structural Division* **104**, no. ST 9 (September 1978), pp. 1337–1353.

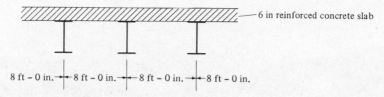

Figure 23-4

It is highly possible that the AISC, ACI, and other organizations may revise the values of these ϕ and γ values when LRFD is included in their specifications. There is at present some feeling among some designers that the present recommended values may be too liberal. Example 23-4, which follows, illustrates the design of a beam using LRFD.

Example 23-4

Using the plastic moment, select beams for the situation shown in Fig. 23-4 by LRFD. The beams have 40-ft simple spans. Use A36 steel, 150 lb/ft^3 concrete, $LL = 120$ lb/ft^2, and the following criterion with $\phi = 0.86$, $\gamma_A = \gamma_D = 1.1$, and $\gamma_L = 1.4$.

$$\phi M_p \geqslant \gamma_A(\gamma_D M_D + \gamma_L M_L)$$

SOLUTION

Loads and Moments: Dead Loads

$$\text{concrete slab} = (\tfrac{6}{12})(150)(8) = 600 \text{ lb/ft}$$
$$\text{est. beam wt.} = \ \ 90 \text{ lb/ft}$$
$$w_D = \overline{690 \text{ lb/ft}}$$
$$M_D = \frac{(0.69)(40)^2}{8} = 138 \text{ ft-k}$$

Live Loads

$$w_L = (120)(8) = 960 \text{ lb/ft}$$
$$M_L = \frac{(0.96)(40)^2}{8} = 192 \text{ ft-k}$$

Selection of beam

$$0.86 M_p = 1.1(1.1 M_L + 1.4 M_L)$$
$$0.86 M_p = (1.1)(1.1 \times 138 + 1.4 \times 192)$$
$$M_p = 538 \text{ ft-k}$$
$$Z \text{ reqd.} = \frac{(12)(538)}{36} = 179.3 \text{ in.}^3$$

Verrazano Narrows Bridge between Staten Island and Brooklyn, N.Y. (Courtesy of U. S. Steel Corporation.)

Use W24×76

It is to be realized that for a particular member there may be several load combinations that will have to be considered. The following are cases that might be checked for a particular situation.[7]

1. Dead load plus maximum live load

$$\phi R_n \geqslant 1.1(1.1DL + 1.4LL)$$

2. Dead load plus maximum instantaneous live load plus maximum wind load

$$\phi R_n \geqslant 1.1(1.1DL + 2.0LL_i + 1.6WL)$$

3. Dead load plus maximum live load plus maximum snow load

$$\phi R_n \geqslant 1.1(1.1DL + 2.0LL_i + 1.7SL)$$

[7]T. V. Galambos and M. K. Ravindra, "Proposed Criteria for Load and Resistance Factor Design," *Engineering Journal*, AISC, **15**, no. 1 (First Quarter, 1978), pp. 8–17.

A final advantage of LRFD is that it may help in some fashion to unify structural design for different materials such as structural steel, reinforced concrete, and timber. There may be a move in the direction of common load factors for these different materials, thus providing common reliability and safety values. Each discipline would of course have its own methods of computing limit states for their material.[8]

23-5. CONCLUSION OF TEXT

The author has attempted to include in this textbook only the elementary phases of structural steel design. His main purpose has been to try to interest the student in the subject because he feels that the longest step a person can take toward proficiency in any field is to become interested in the subject.

The student needs to realize that this book only begins to present the knowledge now available concerning structural steel design. A person going into this field should study a great deal to become familiar with this information and also to keep abreast of the latest developments. The amount of structural research under way at this time far exceeds the work done at any time in the past. Furthermore, steel research projects are being initiated at an ever-increasing rate.

Several current trends are underway in the steel design field and will apparently continue for some time. The designer must concentrate on these and other developments in order to progress in the field. These subjects include the following.

1. Vastly increased computer usage.
2. Continued development and applications of stronger and stronger steels.
3. An increasing trend toward single unit design such as the composite structures and orthotropic systems.
4. More plastic and or LRFD design applications.

[8] C. W. Pinkham and W. C. Hansell, "An Introduction to Load and Resistance Factor Design for Steel Buildings," *Engineering Journal*, AISC, 15, no. 1 (First Quarter, 1978), p. 7.

Index

Adachi, J., 413
Aesthetics, 32
Amplification factor, 198
Anderson, J.P., 218
ANSI, 383, 384

Basler, K., 139
Bateman, E. H., 226
Batho, C., 226
Beam-columns, 191–223
Beam connections, 332–361
  rigid, 335–337
  semirigid, 334–335
  simple, 332–334
Beams, 124–190
  bearing plates, 183–185,
    366–367
  camber, 161
  compact sections, 129–132
  continuous, 144–148
  cover-plated, 453–455
  deflections, 160–163
  floor, 124, 519
  girders, 124
  government anchors, 367
  holes, 132–135
  joists, 124
  lateral support, 135–144
  lintels, 124, 181–183
  partially compact, 131
  ponding, 162–163, 650–652
  purlins, 58–61, 169–171
  shear, 154–158
  shear center, 171–176
  spandrel, 124, 368
  stringers, 124, 519
  torsion, 166, 171–172,
    176–181
  unsymmetrical bending,
    163–168
  web crippling, 158–159

Bearing plates
  beams, 183–185
  columns, 114–118
  moment resisting, 645–650
Bearing-wall construction,
  365–367
Beedle, L. S., 25, 73, 77, 449,
  558, 590, 591
Bending and axial stress,
  191–223
Bessemer, Sir H., 4
Biggs, J. M., 413
Birkemoe, P. C., 231, 254,
  340
Block shear, 340–341
Blodgett, O. W., 549
Bolts, 224–270
  advantages, 227
  bearing-type connections,
    235, 239–241
  common (unfinished), 224–
    225, 274, 276–277
  direct tension indicator,
    230
  eccentric shear, 245–251
  failure, 234–235
  friction-type connections,
    235, 243–245
  high strength, 226–270
  installation, 227–231, 236
  interference, 225–226
  proofload, 227–228
  prying action, 253–257
  ribbed, 225–226
  shear and tension, 257–261
  shearing resistance, 231–244
  specifications, 235–237
  tension, 251–261
  turned, 225
  unfinished, 224–225, 274,
    276–277

Box sections, 176, 456
Bridges, 389–415, 517–542
  arch, 399–402
  beam, 393
  bearings, 537–541
  cantilever, 397–399
  deck, 390
  erection, 391–392
  floor systems, 390
  half-through, 390
  history, 389–390
  movable, 408
  orthotropic, 404
  plate girder, 393–394, 453–
    495
  prestressed steel, 405
  rigid frame, 403–404
  suspension, 402–403
  through, 390
Bridge trusses, 394–399, 517–
  542
  analysis, 525–531
  continuous, 396–397
  deflections, 536–537
  design, 517–542
  lateral bracing, 534–536
  Parker, 395
  Pratt, 394–395
  subdivided, 395–396
  Vierendeel, 404–405
  Warren, 394–395
  weights, 529–531
Buckling, 70–71, 77–81
  elastic, 84–85
  inelastic, 84–85
Building codes, 11
Building connections, 331–364
  rigid, 332, 335–337, 341–
    345
  semirigid, 332, 334–336,
    341–345

Building connections
(Continued)
  simple, 332–334
Burr, W. H., 89
Butler, L. J., 320
Butt joints, 233
Byran, C. W., 92

Camber, 161
Cantilever method, 584–587
Carbon, 4, 16–17, 22, 24
Chang, J. C. L., 404
Cochrane, V. H., 43
Coffin, C., 293
Columns, 70–123
  AASHTO formulas, 100
  AISC formulas, 97–100
  base plates, 114–118, 645–
    650
  Euler formula, 78–85
  Gordon-Rankine formula,
    94–96
  lacing and tie plates, 108–
    112
  lateral support, 107–108
  parabolic formulas, 92–94
  secant formula, 96–97
  sections used as, 73–77
  splices, 570–572
  straight line formulas, 89–92
  web stiffeners, 360–361
Compact sections, 129–132
Composite design, 416–452
  advantages, 417–418
  deflections, 449–450
  disadvantages, 418
  effective flange widths,
    419–420
  encased sections, 440–444
  history of, 416
  shear connectors, 426–434
  shoring, 418–419
  steel formed decks, 450
Compression members. *See*
  Columns
Cooper, P. B., 133
Cooper, T., 414
Crawford, S. F., 251
Critical buckling load, 79

Davy, H., 292
Deflections, 160–163, 449–
  450, 650–652
De Vries, K., 139
Double-modulus theory, 85
Driscoll, G. C., 431
Ductility, 2, 40

Earthquake forces, 386–387
Economy, 7, 28, 30–32, 126

Effective length, 80–81, 100–
  103
Elastic design, 17–18
Elastic limit, 14
Elastic strain, 15
Ellifritt, D. S., 208
Eney, W. J., 139
Engesser, F., 84–85
Equivalent axial load method,
  203–209
Euler formula, 78–85
Euler, L., 77
Eye-bars, 61

Factor of safety, 7–9
Failures of engineering structures, 9–10
Fairbanks, H. E., 538
Faraday, M., 293
Fatigue, 3, 62–64
Fireproofing, 3, 387–388
Fisher, J. W., 230, 238
Fleming, R., 503
Floor construction types for
  buildings, 369–381
  composite, 374–376
  concrete-pan, 376–377
  flat-slab, 378–379
  one- and two-way slabs,
    372–374
  open-web joists, 370–372
  precast concrete, 379–381
  steel decking, 378
  tile and block, 377
Fountain, R. S., 426, 435
Frames
  braced, 102
  unbraced, 103

Gage, 43
Galambos, T. V., 18, 653,
  654, 656
Gaylord, C. N., 165, 410
Gaylord, E. H., Jr., 165, 410
Gere, J. M., 139
Gilmor, M. I., 340
Girts, 58
Gordon, L., 95
Government anchors, 185
Griffiths, J. D., 549
Grinter, L. E., 503, 579, 580

Hall, J. R., 231
Hansell, W. C., 657
Herrschaft, D. C., 231
Higgins, T. R., 251
High-strength steels, 23–28

Historical data, 3–6
Hudson, C. W., 529
Hughes, J., 296
Hybrid structures, 28, 129

Impact, 413, 415
Interaction equations, 164–
  166, 194–210
Instantaneous center of rota-
  tion, 251, 320
Ironworkers, 30

Johnson, J. B., 92
Johnson, J. E., 48, 165, 251
Johnson, T. H., 81, 89
Johnston, B., 335
Johnston, B. G., 98, 139, 144

Kazinczy, G., 591
Kelly, W., 4
Korn, M. P., 549, 558
Kubo, G. G., 139
Kulak, G. L., 251, 320
Kussman, R. L., 133

Lacing, 76, 108–112
Lap joints, 232–233
Le-Wu-Lu, 218
Limit design. *See* Plastic
  analysis and design
Live loads
  buildings, 383–385
  highway bridges, 411–414
  railway bridges, 414–415
Load and resistance factor
  design (LRFD), 18, 652–
  657
Lothers, J. E., 335, 435, 499,
  529

Main members, 94
Maintenance, 2, 31–32
Marcus, S. H., 450
Marino, F. J., 163
McCormac, J. C., 505, 525,
  536
McGuire, W., 19, 251, 650
Mill buildings, 513–515
Modification factor, 198–203
Mount, E. H., 335
Multistory buildings, 568–590
  analysis, 579–589
  belt truss, 578
  design, 590
  drift index, 573
  hat truss, 577–578
  lateral bracing, 575–579
  lateral forces, 572–574

Munse, W. H., 254
Musschenbroek, P. V., 77

Nair, R. S., 254
Namyet, S., 413
Net areas, 40–49

Oliver, W. A., 282
Oyeledun, A. O., 230

Pal, S., 320
Partially compact sections, 131
Partitions, 382–383
Pinkham, C. W., 657
Pitch, 43
Plastic analysis and design, 17–18, 591–644
  advantages, 592
  AISC requirements, 623–626
  collapse mechanisms, 598–599
  continuous beams, 626–632
  disadvantages, 592
  equilibrium method, 599–607
  frames, 632–638
  load factor, 597, 623
  plastic hinge, 593–595
  plastic modulus, 595–597
  plastic moment, 594–595
  shape factor, 594
  virtual-work method, 607–612
  yield moment, 593
Plastic strain, 15
Plate girders, 453–495
  advantages, 458
  connectors, 468–471
  cover plates, 461–462, 465–468
  disadvantages, 458
  flanges, 461–463
  splices, 475–480
  stiffeners, 471–475, 484–485
  web, 459–461, 480–483
Ponding, 162–163, 650–652
Portal method, 582–584
Proportional elastic limit, 14
Proportional limit, 14
Prying action, 253–257
Purlins, 58–61, 169–171, 556–557

Rains, W. A., 388
Rankine, W. J. M., 95
Ravindra, M. K., 18, 653, 654, 656

Reduced modulus theory, 85
Reilly, C., 251
Residual stresses, 72–73
Rhude, M. J., 9
Rigid connections, 332, 335–337, 341–345
Rigid frames, 543–567
  analysis, 548–555
  defined, 543
  design, 556–558, 562–567
  knees, 546–547, 565–566
  lateral bracing, 566
  supports, 543–546
  tie rods, 566
Rivets, 271–289
  eccentric loads, 277
  prying action, 282
  shear and tension, 277–281
  specifications, 273–275
  tension, 281–282
  types, 273–274
Rods, 35–36, 57–61
Roof construction types, 381–382
Roof trusses, 496–516
  analysis, 503–509
  design, 509–511
  lateral bracing, 513–515
  pitch, 497, 500
  selection, 500–501
  slope, 497
  spacing, 501–502
  supports, 502
  types, 497–500
  weights, 502–503
Ryzin, G. V., 162

Safety, 7
Sag rods, 58–61
Salmon, C. G., 48, 165, 251
Sandhu, B. S., 218
Sawyer, M. H., 6
Scuppers, 163
Secondary member, 94
Section modulus, 125–126
Semirigid connections, 332, 334–336, 341–345
Shanley, F. R., 85
Shear, 154–158
Shear center, 171–176
Shear flow, 173
Shear lag, 47–49
Shedd, T. C., 529
Shermer, Carl L., 650
Simple connections, 332–334
Singleton, R. C., 426, 435
SI units, 32–34
Skeleton construction, 367–368

Slenderness ratios, 50, 71, 90–91
Slutter, R. G., 431
Smith, H. S., 389, 390, 399
Specifications, 10–12
Splices,
  columns, 570–572
  plate girders, 475–480
  tension members, 55–57
Steel, defined, 4, 16–17, 22
Steinman, D. B., 402–403, 415
Stetina, H. J., 332
Stiffened elements, 112–113
Strain-hardening, 15
Stress fluctuations, 40
Stress-strain diagrams, 14–17
Structural Stability Research Council, 98, 103
Structural steel sections, 19–22
Struik, J. H. A., 230, 238
Sweet's Catalog File, 370, 371, 380

Tall, L., 73
Tangent modulus theory, 84–85
Tension members, 35–69
Thürlimann, B., 139
Tie plates, 36–37, 76, 108–112
Timoshenko, S., 139
Torsion, 166, 171–172, 176–181
Transportation, 7, 30–31
Turneaure, F. E., 92

Uncertainties of design, 8
Unstiffened elements, 112–113
Upset rods, 57–58

Van den Broek, J. A., 591
Vierendeel, M., 404
Viest, I. M., 426, 435

Waddell, J. A. L., 529
Walls, 382
Web buckling, 160
Web crippling, 158–159
Web tear-out shear. See block shear
Welding, 290–330
  advantages, 291–292
  AISC requirements, 305–309
  classification, 297–299
  defined, 290
  electric arc, 292–295
  electrodes, 293–295
  end returns, 308

Welding (Continued)
  faying surface, 313
  fillers, 293
  fillet, 298, 303–304
  gas, 293
  groove, 298, 301–303
  inspection, 295–297
  joints, 299
  lands, 302–303

  partial-penetration, 303
  plug, 298
  position, 298–299
  shear and bending, 321–323
  shear and torsion, 318–321
  shielded arc (SAW), 294
  slot, 298
  submerged arc (SMAW),
    294–295, 304

  symbols, 299–301
  types, 292–295
Wilson, W. M., 282
Wind forces, 385–386, 413–415
Winter, G., 139
Woodward, J. H., 218

Yield point, 14
Yura, J. A., 214

81 82 83 84 9 8 7 6 5 4 3 2 1